Manned Spacecraft Design Principles

T0318388

Manned Spacecraft Design Principles

Pasquale M. Sforza

University of Florida

AMSTERDAM • BOSTON • HEIDELBERG • LONDON
NEW YORK • OXFORD • PARIS • SAN DIEGO
SAN FRANCISCO • SINGAPORE • SYDNEY • TOKYO

Butterworth-Heinemann is an imprint of Elsevier

Butterworth-Heinemann is an imprint of Elsevier
The Boulevard, Langford Lane, Kidlington, Oxford OX5 1GB, UK
225 Wyman Street, Waltham, MA 02451, USA

Notices

Knowledge and best practice in this field are constantly changing. As new research and experience
broaden our understanding, changes in research methods, professional practices, or medical treatment
may become necessary.

Practitioners and researchers must always rely on their own experience and knowledge in evaluating
and using any information, methods, compounds, or experiments described herein. In using such
information or methods they should be mindful of their own safety and the safety of others, including
parties for whom they have a professional responsibility.

To the fullest extent of the law, neither the Publisher nor the authors, contributors, or editors, assume
any liability for any injury and/or damage to persons or property as a matter of products liability,
negligence or otherwise, or from any use or operation of any methods, products, instructions, or ideas
contained in the material herein.

ISBN: 978-0-12-804425-4

British Library Cataloguing-in-Publication Data
A catalogue record for this book is available from the British Library.

Library of Congress Cataloging-in-Publication Data
A catalog record for this book is available from the Library of Congress.

For Information on all Butterworth-Heinemann publications
visit our website at http://store.elsevier.com/

Typeset by MPS Limited, Chennai, India
www.adi-mps.com

Printed and bound in the US

Working together
to grow libraries in
developing countries

www.elsevier.com • www.bookaid.org

Contents

Preface

This book builds upon a handbook used to support a one-semester senior undergraduate or entry-level graduate course intended to involve students in the preliminary design of a manned spacecraft and associated launch vehicle. The course followed a one-semester design course for aerospace engineering students devoted to design of a commercial jet transport. Commercial aircraft design is a relatively mature field and sufficient reference material is available to provide a secure mooring for student research and study. The manned spacecraft design course demands more from the student because the subject lacks a broad well-documented database and covers topics that rarely receive substantial attention in current curricula. An industrial approach is taken in order to help instill the spirit of the design process, which is that of making informed choices from an array of competing options and developing the confidence to do so. In the classroom setting, this design effort culminates in the preparation of a professional quality design report and an oral presentation describing the design process leading to the proposed space access vehicle. This report and presentation requires students to develop technical, time-management, and communication skills for a successful career.

The material is arranged in a manner that facilitates a team effort, the usual course of action, and also provides sufficient guidance to permit individual students to carry out a creditable design as part of independent study. Emphasis is placed on the use of standard, empirical, and classical methods in support of the design process in order to enhance understanding of basic concepts and to gain some familiarity with employing such approaches which are often encountered in practice. No particular computational approaches are specifically used, although student teams may choose to use available codes, and have done so, with varying degrees of success. CAD courses are generally required in engineering programs and their use is encouraged. My experience in teaching design over the years has led me to embrace the use of simple basic analyses and empiricisms so that students have the opportunity to learn some of those applied aerospace engineering skills that have been edged out of modern curricula by reductions in total credit hours and the perceived need for broadening of skills in other areas. Spreadsheet skills are quite sufficient to support the preliminary design process and such skills are quite valuable to those setting out on industrial careers. Class meetings in a university setting rarely provide more than 40 contact hours for explaining the design process and for conferring with the instructor. Thus there must be a substantial amount of time spent outside class in preparing the design.

This design handbook represents the cumulative efforts of the author over a number of years of offering this course at the University of Florida, and before that, at the Polytechnic Institute of Brooklyn, now New York University's Polytechnic School of Engineering. Because of the wide diversity of subject matter and techniques employed, errors are bound to appear. The responsibility for such errors is mine and I would appreciate learning of any so they may be corrected.

The author was just graduating from the Polytechnic Institute of Brooklyn when the first astronauts were launched into space. The science of hypersonic flight and access to space has since occupied a large portion of my professional life. I acknowledge a debt to the inspired research of many aerospace pioneers, especially Professor Antonio Ferri, champion of the scramjet, who lit the path to the stars. Finally, I am delighted to once again thank my wife, Anne, for her loving encouragement and support for my efforts in writing this book.

Pasquale M. Sforza
Highland Beach, FL, USA
sforzapm@ufl.edu

Introduction and Outline of a Spacecraft Design Report

This book has its origin in a handbook developed to support a one-semester course in which student teams were formed to carry out a preliminary design for a space transportation system capable of carrying astronauts to rendezvous with and dock at the International Space Station (ISS) for a given period of time, and then to return the crew safely to Earth. This mission, though limited in scope, forms a sound basis for considering the broader one of returning people to the moon and visiting other planets and asteroids.

I.1 SUBJECTS COVERED

Chapter 1 begins with the space race precipitated by the convergence of reduced weight nuclear weapons and increased payload capability of ballistic missiles. The geopolitical implications of putting satellites into orbit were great and the next step of safely sending humans back and forth to space bespoke a technical mastery that fired national pride and power. A brief description of the development of manned spaceflight starting from the Vostok launch by the USSR in 1961 and the Mercury capsule launched in reply by the USA in 1962 is presented. Over time, national competition became planetary cooperation culminating in the operation of the ISS. The outlook for the future of human spaceflight now ranges from interplanetary missions of discovery to space as a theme park for tourism.

Earth's atmosphere and the theoretical foundation for a model of the earth's atmosphere, but extendable to other planetary atmospheres, is the subject of Chapter 2. The 1976 US Standard Atmosphere is used as the model of choice in the book and the equations necessary to develop the appropriate data and procedures to define the atmospheric properties over the manned spacecraft flight envelope are derived. For convenience, tables are generated from these equations and are presented in SI units and also in English engineering units because their use is still quite pervasive, especially in the background literature. Standard atmospheric models are used to provide a common basis for comparing investigations by different teams and several such models currently in use are reviewed. Flight operations often encounter deviations in atmospheric properties from those predicted by standard models and must be considered in the design process.

The space environment in which a manned spacecraft must carry out part of its mission as it leaves the atmosphere is described in Chapter 3. The major influences on human space travel are shown to be the solar wind and the earth's gravitational and magnetic fields. The nature of solar radiation and the interaction of the solar wind of energetic protons and electrons with the earth's magnetic field are discussed. Gravitational constraints on achieving orbital motion and rarefied atmospheric drag effects on maintaining an orbit are illustrated. The energetic particles in the Van Allen radiation belts, the electron density in the ionosphere, and the increasing population of space debris are shown to pose dangers to astronaut health, space to ground communications, and spacecraft integrity.

A discussion of the nature of high-speed flight through the atmosphere in order to gain access to space is presented in Chapter 4. Various spaceplane projects aimed at human space-flight that have been pursued in recent years are described. The nature of flight trajectories in the sensible atmosphere is described and vehicles that have successfully operated in the human flight design space are identified. The goal of reusable spaceplanes that operate like conventional airplanes is placed in perspective by addressing the design issues associated with such craft. The transatmospheric manned missions which are likely to develop in the near future are discussed.

Chapter 5 presents the equations of orbital motion with particular reference to characteristics of earth orbits and the manner of altering those orbits. The basic ideas of conservation of energy and angular momentum for closed and open orbits are used to illustrate the maintenance of orbits and the achievement of escape from orbit for interplanetary missions. The ground track of orbits, effects of earth's rotation and precession, determination of longitude, the spacecraft horizon and effects on communication are analyzed. Interplanetary trajectories are discussed and the orbital transfer process for atmospheric entry is presented.

The most stressful part of human spaceflight is atmospheric entry from space and this is the subject of Chapter 6. The equations of motion for spacecraft entry into the atmosphere are derived and the general characteristics of gliding trajectories discussed. The spacecraft ballistic coefficient and earth's atmospheric density profile are important considerations in limiting the effects of deceleration on human physiology. These factors are combined in defining an entry trajectory corridor that lies within a restricted range of acceptable dynamic pressure levels. A similar corridor for assuring safe heating levels for entering spacecraft is also defined. Detailed studies of entry dynamics are divided into three categories: ballistic (zero-lift), low lift to drag ratio (L/D), and moderate L/D. This division may be illustrated by meteors and early space capsules (e.g., Mercury), next generation space capsules (e.g., Apollo and now Orion), and spaceplanes (e.g., the Space Shuttle Orbiter), respectively. The low-speed return and recovery analysis is divided into parachute-type systems like those used for capsule recovery and airplane-type systems like the Space Shuttle Orbiter.

Only when one is convinced of the ability to return humans safely from space is it appropriate to entertain the means for launching them into space. The general equations for launching space-craft into orbit are developed and the influence of thrust, lift, and drag are assessed in Chapter 7. Equations for the practical preliminary design case of constant thrust and negligible lift and drag are solved and permit determination of boost trajectories, burn-out velocities, and postburn-out booster trajectories. The selection of a launch system for a spacecraft of a given mass involves evaluating the use of one, two, or three stages and each is subjected to detailed analysis. The design choices are shown to be dependent upon the assumptions made concerning structure and engine weights, even in a preliminary design process. Because launch systems are large structures it is important to consider them at least as rigid bodies; the effects of elastic deformations are generally deferred to advanced stages of design. Moments of inertia, force and moment estimation procedures, and static longitudinal stability requirements are incorporated in developing the configuration and size of launch systems and thrust vector control requirements are assessed. Liquid and solid propellant rockets are discussed and the determination of tank size and weight is evaluated.

The flight mechanics of manned spacecraft during hypersonic entry are developed in Chapter 8. The application of Newtonian theory for normal stresses and the flat plate reference enthalpy (FPRE) method for shear stresses forms a simple but sound basis for spacecraft design. Because Newtonian theory depends on the local flow deflection, a unit problem for the local surface pressure may be defined involving the elemental surface area and the impinging velocity vector. Thus a relatively complex spacecraft design may be modeled as a collection of surface panels and simple summations used to generate the pressure forces and the associated moments. The FPRE method is also formulated as a unit problem so that frictional forces can be evaluated for both laminar and turbulent flow. A simple criterion for boundary layer transition is also provided. Blunt body space capsules and slender body spaceplanes are both treated in some detail. Because of the high temperatures of hypersonic flight, a development of the thermodynamic and transport properties of air relevant to atmospheric entry is given and useful approximations are illustrated. Both capsules and spaceplanes are treated as rigid bodies and their longitudinal and lateral static stability characteristics are evaluated. Some attention is also paid to the assessment of their dynamic stability characteristics. Aerodynamic and reaction control systems appropriate to entering spacecraft are described.

Thermal protection systems are crucial to safe and reliable human spaceflight and this topic is the subject of Chapter 9. The primary determinant of thermal protection is stagnation point heat transfer and attention is paid to the most useful correlations for design applications. The high temperatures ensure dissociation and ionization of air molecules so approximations are developed for the treatment of air chemistry. Heat transfer on a blunt hemispherical body and a spherically capped cone is addressed to highlight the different mechanisms at work. The analyses developed are extended to the design of heat shields for entry vehicles. Heat sink, melting ablator, and charring ablator heat shields are analyzed, as are active systems involving surface injection of coolants. The similarity parameters crucial to heat transfer effects are reviewed.

Chapter 10 covers spacecraft configuration design. The spacecraft environment and its effect on design are discussed and crew volume allowance in spacecraft is contrasted with passenger volume allowance on commercial, business, and military aircraft. The influence of mission duration on cabin configuration is considered. Vehicle mass characteristics as evidenced by successful manned spacecraft and the resultant ballistic coefficient are explored. Human factors in spacecraft design are assessed as the major areas of thermal control and management of the habitable volume. Environmental control and life support systems including heating, ventilating and air conditioning, air and water purification, waste management, fire and emergency control systems, and communications are discussed. Aspects of basic structural design issues along with space propulsion and power systems are described.

The definitions of safety and reliability and the means of apportioning mission reliability are discussed and the reliability function is introduced in Chapter 11. Failure rate models are used in the development of reliability estimates and apportionment goals are evaluated. Propulsion system reliability is used as an example of estimating mission success. An overview of probabilistic risk assessment is presented and the question of functional failures is addressed using the Space Shuttle as an example. The Weibull distribution and its role in risk and reliability studies are described.

Economic aspects of manned spaceflight including the costs of previous manned spaceflight programs like Apollo and the Space Shuttle are discussed in Chapter 12. An assessment of the general cost of spaceflight measured in terms of payload mass in orbit is carried out. The cost of spaceflight is evaluated on a component basis including development, production, flight operations, refurbishment, recovery, and insurance. General characteristics of the cost of various launch vehicle components are described.

The technical foundation for hypersonic flight back and forth across the atmosphere as used and discussed in the main text is presented in Appendix A including normal and oblique shocks, small disturbance theory, Prandtl—Meyer flow, Newtonian theory, and conical flow, among others.

Detailed configuration data on seven spaceplanes that have been flown or were the subject of wind tunnel tests is provided in tables and drawings in Appendix B. Flight characteristics of several of these vehicles are described based on the methods of calculation given in Chapter 8. This material is not readily available in one collection elsewhere.

I.2 AN APPROACH FOR A DESIGN COURSE

A basic mission profile that has been used successfully in a one-semester course involved the following tasks:

1. Lift-off from Earth's surface with a specified number of crew members
2. Climb/accelerate through the atmosphere to orbit at 400 km
3. Remain on orbit for two full orbits
4. Dock with ISS for a specified time
5. Undock and deorbit at 400 km altitude
6. Atmospheric entry at 100 km to descend and decelerate
7. Approach and landing on Earth's surface.

A design team comprised of six people was found to be most practical. This would include a program manager and five specialists, one in each of the following areas: aerodynamics, propulsion, trajectories and orbits, thermal protection, and configuration design. Each student submitted a request for one of these positions, in order of preference. Based on these requests, the instructor selected a specific role for each student and assigned them to a design team. Thus students about to embark on an engineering career would experience being part of a team of people previously not well known to them as would likely be the case in their first job. Exposure to the need for cooperating as part of a team was considered part of the educational process in design. Students in the course were in their last semester so that all had or were concurrently taking analysis courses in the areas listed. Therefore all had some exposure to the breadth of material to be covered and the design course was intended to bring much of it into better focus. The main product of the design course was the design report although oral presentation of the design report in a symposium setting with invited faculty and guests was an important adjunct.

I.2.1 PREPARATION OF THE DESIGN REPORT

One of the most important tasks facing every engineer is the preparation of a technical report. This may be a document like a proposal, which seeks to engage the interest of a sponsor to financially support the technical task proposed, or one describing the work that has been carried out in completing a technical task. Typically, engineers enjoy performing the technical work required to solve the particular problem at hand but often dread the planning, writing, and preparation of the technical document that describes the work.

A report is intended to present information clearly and in a manner that is both self-contained and interesting. Conceptually, report preparation is rather simple, being in essence an edited log of the work that has been, or is proposed to be, performed. Thus, it is convenient to keep a good journal of the work done along with the relevant background and illustrative material used. Though the technical work done may be well understood and appreciated by the engineer who carried it out day by day, this is not necessarily the case for other people who also need to know about that work. If the reader finds the report difficult to understand because the presentation is poor, then the engineer has wasted all the technical work done because the information cannot get beyond the person who actually did the work. Thus it is important to be sure that some basic requirements are met by the design report, such as the following: the reader should not have to search for important facts, the technical content should not be obscured by poor writing, and ambiguity should be avoided.

There is always some concern about the perspective of the report, that is, who is the reader? For design reports, there are generally three classes of readers: business and sales executives, technical managers, and technical staff engineers. To satisfy this broad group with one report, it is common to include an executive summary, a main text, and detailed appendices. Executives generally read the brief executive summary to clearly understand the general approach and results of the study. Technical management personnel read the executive summary and the main text so as to be able to guide the executive group as necessary. The technical staff needs all three sections since they may be called upon to review detailed questions from the other two groups who are involved in making major business decisions.

I.2.2 OUTLINE OF THE DESIGN REPORT

Aircraft and spacecraft companies generally have standard formats for reports used both within and outside the company. Though the details may vary from company to company, there is a general outline that tends to be followed. A layout of the chapters of a typical final report for a one-semester course along with a brief description of the material that should be addressed in each chapter follows. This tabular format provides a rubric for the instructor to use in grading the report.

No.	Headings	Comments
i	Cover/Title Page	
ii	Executive Summary	
iii	Table of Contents	
1	Mission and Market Survey	
	a. Mission specification	
	b. Market for mission	
	c. Competitive systems	
2	Flight Through the Atmosphere	
	a. Atmospheric characteristics	
	b. Ascent trajectories	
	c. Transfer orbits	
	d. Reentry trajectories	
	e. Guidance, navigation, and control	
3	Vehicle Configuration Design	
	a. Selection of vehicle type	
	b. Components and systems	
	c. Layout of vehicle	
	d. ECLSS	
	e. Power and thermal management	
4	Vehicle Weight Estimate	
	a. Payload	
	b. Weight breakdown by subsystem	
	c. Structural analysis	
5	Launch Stack Design	
	a. Staging considerations	
	b. Performance to LEO	
	c. Weight estimation	
	d. Engine selection	
	e. Drawings of the launch stack	
6	Vehicle Characteristics at High Speed	
	a. Newtonian theory for hypersonic flight	
	b. Drag polar and pitching moment	
	c. Stability	
	d. Heat transfer	
	e. Thermal protection system	
7	Vehicle Characteristics at Low Speed	
	a. Supersonic characteristics	
	b. Subsonic characteristics	
	c. Terminal operations	
8	Risk, Reliability, and Safety	
	a. System risk evaluation	
	b. System reliability	
9	Cost and Economic Analysis	
	a. Cost basis for evaluation	
	b. Cost and comparisons	
10	Conclusions and Recommendations	

Continued		
No.	**Headings**	**Comments**
11	Appendices	

I.3 SUGGESTIONS FOR REPORT PREPARATION

Report preparation should proceed in a timely manner; in a university setting this should be carried out regularly during the course presentation and not be delayed until the end of the semester. The reports should be prepared on a word processor and attention should be paid to regularly backing up files. Drawings are to be made on any CAD system available. Again it is important to remember to back up all files. The report should be submitted as a spiral bound or stapled document printed on standard white 8.5×11 inch paper. The pages must be numbered consecutively using a standard system throughout the report and should be consistent with the table of contents.

It may be useful to review some of the common errors in report preparation which are listed as follows:

1. Title page omitted or incomplete
2. Table of contents omitted or paginated incorrectly
3. An important section of the report, like the executive summary or a main text chapter, is omitted
4. References enumerated properly at the end of chapters but not cited in the text of the chapter, or vice versa
5. Figures captioned properly as they appear but not cited in the text, or vice versa
6. Pages are not numbered sequentially or are omitted entirely
7. Three-view drawings of the spacecraft and similar supporting drawings are omitted or are of poor quality
8. Printing is of poor quality or nonstandard fonts are employed
9. Spelling, grammar, and punctuation are poor
10. Material presented indicates that the author poorly understands it
11. Extensive quoting or plagiarizing of previously published material
12. Improper use of appendices; repetitive material, tables, etc. are to be incorporated into appendices so as to keep the flow of ideas smooth in the main text
13. Improper inclusion of calculations in the main text; there should be sufficient information, such as equations, to permit another engineer trained in the art to reproduce the results shown without actually incorporating numbers
14. Uneven emphasis among chapters, usually indicating varying degrees of effort among contributors
15. Too much repetition or discussion of irrelevant material
16. The conclusions reached by carrying out the work are not clearly stated.

Suggestions for preparing graphs and other figures are as follows:

1. Figures should be introduced as part of the text with appropriate numbering and captioning

2. The figures should be placed near to where they are described, so that they can be easily found
3. Figures should be self-contained so that, with the descriptive caption, their content can be easily grasped
4. The ordinate and abscissa of a graph must be clearly labeled, including units and scale divisions; a somewhat heavier line width may be used for the axes
5. Scales must be chosen appropriately so that the behavior being described in the text is apparent; autoscale features of graphing packages must be scrutinized in this regard
6. Analytic results should be indicated by lines that are distinguishable without using color since reports are often reproduced without benefit of color; different line styles, such as dotted and dot-dash, may be employed
7. Experimental data should be represented by discrete symbols; again avoid using symbols differing only in color
8. Maintain some degree of uniformity among the symbols and lines used on graphs dealing with essentially the same subject for ease of interpretation
9. Extrapolations and interpolations should be indicated by a change in the line style, such as changing from solid to dotted lines
10. Major grid lines in both coordinate directions should be shown.

MANNED SPACEFLIGHT

1.1 WHERE SPACE BEGINS

In the early days of space travel the renowned aerodynamicist Theodore von Karman suggested a useful definition for the edge of space: the altitude at which an airplane flying in a straight path at the orbital speed can no longer sustain its weight using only aerodynamic lift. This equilibrium condition on lift may be written in terms of the airplane's maximum lift coefficient $C_{L,max}$ as follows:

$$L = \frac{1}{2} C_{L,max} \rho(z) S V^2 = mg \tag{1.1}$$

From Eqn (1.1) the density must then be

$$\rho(z) = \frac{2}{C_{L,max} V^2} \left(\frac{mg}{S} \right)$$

Using the orbital velocity as $V = 7900$ m/s, as discussed subsequently, and a nominal wing loading of $mg/S = 3000$ N/m^2 (62.7 lb/ft^2), similar to that of the Space Shuttle Orbiter or the X-15 hypersonic research aircraft, the density (in kg/m^3) becomes

$$\rho(z) = \frac{9.61 \times 10^{-5}}{C_{L,max}}$$

Earth's atmosphere is discussed in Chapter 3 and standard atmospheric data shows that in the altitude range 50 km $< z <$ 70 km the density, in kg/m^3, is in the range $3 \times 10^{-4} > \rho > 8 \times 10^{-5}$. Therefore for a maximum lift coefficient in a reasonable range of about $0.3 < C_{L,max} < 1$ the corresponding altitude is between about 50 km and 70 km. Because the definition is somewhat arbitrary, the altitude of the edge of space is usually rounded off to $z_e = 100$ km and is often called the "Karman line." The Federation Aeronautique Internationale (FAI) uses the Karman line to define the official boundary between aeronautics and astronautics activities. The US Air Force (USAF) definition of an astronaut is a person who has flown more than 50 miles (approximately 80 km) high while NASA uses the FAI's 100 km figure. The mean radius of the earth is $R_E = 6371$ km so the edge of space is quite near ($z_e/R_E = 0.015$) and the gravitational acceleration in "near space" is essentially equivalent to that on the surface itself, as presumed in carrying out this analysis.

Manned Spacecraft Design Principles. DOI: http://dx.doi.org/10.1016/B978-0-12-804425-4.00001-5

1.2 STAYING IN SPACE

To stay at this edge of space we can no longer fly in a straight path like an airplane but instead we must follow a curved path. For simplicity we consider a circular path of radius $r = R_E + 100$ km, such that the centrifugal force mV^2/r balances the vehicle weight mg. The balance struck in this circular path we call "weightlessness": the net radial acceleration is zero so although the mass is fixed and the earth's gravity is still essentially the same as on its surface, the net force on the payload is zero. This scenario gave rise to the term "zero-g" which is not strictly correct; rather it is zero-net-acceleration. A circular orbit is used for clarity here but a more general treatment of orbits and trajectories is presented in Chapter 5.

We see then that this means the net radial acceleration $V^2/r = g$. For this equilibrium condition to exist, the velocity of the vehicle $V = (gr)^{1/2}$ where $r = R_E + z$ and $g = g(z)$. The earth's mean radius $R_E = 6371$ km and at the earth's surface $r = R_E$ and similarly $g = g_E$ so that $V = V_E \sim 7,900$ m/s and this is called the circular velocity. Strictly speaking, it is the velocity required to maintain a circular orbit at the surface of the earth. For the edge of space we chose the altitude $z_e = 100$ km arbitrarily and see that the exact altitude is not important in the calculation when $z/R_E \ll 1$. As will be shown in subsequent chapters, there is only a few percent variation in V between the surface and an altitude of 100 km.

Although the density at the edge of space is low, it isn't zero. The aerodynamic drag on a body can slow it down sufficiently to cause its orbit to decay sending the body on an entry trajectory down to the surface. The departure of a spacecraft from orbit is initiated by a retarding thrust force which forces the spacecraft into a high speed descent toward the earth. The details of the atmospheric entry process and its effects on astronauts are treated in detail in Chapter 6. Experience shows that at altitudes below about 150 km the drag on an unpowered orbiting body is great enough to cause the orbit to decay. Typical orbits for manned spacecraft, like that of the International Space Station (ISS), are situated at an altitude of about 400 km and are called low earth orbit (LEO). Even at that altitude a large spacecraft like the ISS experiences orbital decay and from time to time the onboard thrusters must be applied to boost the ISS back to its appropriate altitude.

1.3 GETTING INTO SPACE

To send a mass into space one must change its energy state by doing work on it. The details of the launch process for manned spacecraft are treated in Chapter 7. The total energy of a mass m is the sum of its potential and kinetic energies. If we define the potential energy to be zero at the surface of the earth, then the total energy is given by

$$me_t = me_p + me_k = mg_E R_E \left(1 - \frac{R_E}{r}\right) + \frac{1}{2} mV^2 \tag{1.2}$$

The velocity in a circular orbit is $V = (g_E R_E^2/r)^{1/2}$. Circular orbits are used here because they are simple to deal with and are not uncommon in practice. Note that this equation reduces to $V = V_E$ when $r = R_E$. In Chapter 5 we will generalize the analysis to include orbits with other shapes. Introducing the general form for the circular orbital velocity into Eqn (1.2) we find that the total energy per unit mass is

$$e_t = g_E R_E \left(1 - \frac{R_E}{r}\right) + \frac{1}{2} g_E \left(\frac{R_E^2}{r}\right) = g_E R_E \left(1 - \frac{1}{2}\frac{R_E}{r}\right) \tag{1.3}$$

The characteristic energy per unit mass in Eqn (1.3) is $g_E R_E = (7904 \text{ m/s})^2 = 6.247 \times 10^7$ J so that the total specific energy in Eqn (1.3) may be written in normalized form as follows:

$$\frac{e_t}{g_E R_E} = 1 - \frac{1}{2}\frac{R_E}{r} \tag{1.4}$$

Similarly, the normalized specific potential energy may be written as

$$\frac{e_p}{g_E R_E} = 1 - \frac{R_E}{r} \tag{1.5}$$

These forms for the specific energy are valid for other planetary bodies providing the proper values for the radius of the planet and the characteristic energy are substituted. A plot of the normalized energy variation as a function of the radius of the orbit is shown in Figure 1.1. The sum of the potential and kinetic specific energies yields the total specific energy for a circular orbit at the indicated radius from the center of the earth. Adding the escape energy to the orbital kinetic energy releases the body from the gravitational field of the earth. Using Figure 1.1 and Eqn (1.5) the normalized escape energy is found from the following relation:

$$1 - \frac{e_p}{g_E R_E} = \frac{R_E}{r} = \frac{1}{2}\frac{V_e^2}{g_E R_E} \tag{1.6}$$

Therefore the escape velocity is a function of distance from the planet and is given by

$$V_e = V_E \sqrt{2\frac{R_E}{r}} \tag{1.7}$$

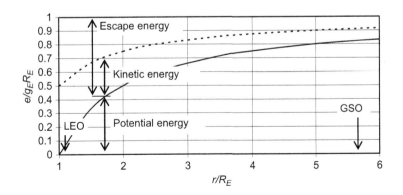

FIGURE 1.1

Energy diagram for earth showing the potential, kinetic, and orbital escape energy per unit mass as a function of radial distance. Low earth and geosynchronous orbit locations are indicated by LEO and GSO.

Table 1.1 Characteristics of Planets and Some Asteroids

Planetary Body	Radius (km)	Characteristic Energy (kJ/kg)	Escape Velocity (km/s)
Sun	6.96×10^5	1.9×10^8	618
Mercury	2,440	9,680	4.4
Venus	6,052	53,660	10.36
Earth	6,371	62,500	11.18
Moon	1,738	2,332	2.38
Mars	3,397	12,550	5.03
Jupiter	71,490	1,772,000	59.54
Saturn	60,330	630,100	35.5
Uranus	22,680	226,300	21.3
Neptune	27,610	276,100	23.5
Pluto	1,195	605	1.1
Ceres	469	130	0.51
Pallas	272	51.2	0.32
Vesta	262	31.3	0.25
Eros	8.4	0.018	0.006

From Eqn (1.7) we see that the escape velocity at the surface of the earth is $1.414V_E$ or about 11,200 m/s. The radius, characteristic energy, and escape velocity for planetary bodies in the solar system is presented in Table 1.1.

Recall that the characteristic energy for earth is 6.247×10^7 J/kg and this corresponds to the full vertical scale of Figure 1.1. The latent chemical energy content of a rocket fuel like RP-1 kerosene is about 43 MJ/kg so the characteristic energy is approximately equivalent to that which can be obtained from complete combustion of 1.45 kg of RP-1 which would require about another 3.8 kg of oxygen to carry out the combustion. Of course, the propellants must be carried aloft as well as a payload so a great deal of the energy available in the propellants is wasted in doing the work involved in carrying the propellants aloft. The details of the launch process are treated in Chapter 7. Also shown on Figure 1.1 are the relative locations of LEO, which, for $z = 400$ km, corresponds to $r/R_E = 1.062$ and geosynchronous orbit (GSO) at $r/R_E = 5.65$, while the moon, on this scale, is at $r/R_E = 60.33$.

1.4 THE FIRST FIFTY YEARS OF HUMAN SPACEFLIGHT

The space race was precipitated by the convergence of reduced weight nuclear weapons and increased payload capability of ballistic missiles. The geopolitical implications of putting satellites into orbit were great and the next step of safely sending humans back and forth to space bespoke a technical mastery that fired national pride and power. A brief description of the development of manned spaceflight follows, starting from the Vostok spacecraft launched by the USSR in 1961 and the Mercury spacecraft launched in reply by the USA in 1962.

The USSR launched Vostok, a 4700 kg spherical capsule carrying the first man into orbit in space. Because of the shape of the capsule the drag was low and lift was zero causing the Vostok to follow a ballistic trajectory during atmospheric entry, similar to that of the warhead of a ballistic missile, thereby subjecting the cosmonaut to high deceleration levels. Mercury, the first US space capsule, also carried a single astronaut and rode to orbit on an Atlas, the first US ICBM. The first stage of Atlas was essentially a propellant tank supported by internal pressure, just like a soft-drink can, containing RP-1, a kerosene-based fuel, and liquid oxygen (LOX). The Atlas's MA-5 propulsion system was comprised of a central core, or sustainer, engine, running continuously in union with two booster rockets that were disposed on either side of the sustainer. This was the first use of a one and one-half stage rocket, one in which the core engine fires throughout while the booster engines are expended at some earlier point in the flight. The Mercury capsule was shaped like a base-forward cone and therefore developed higher drag than the Vostok but still generated no lift and thus also entered on a ballistic trajectory subjecting the astronaut on board to the attendant high deceleration forces.

The USSR's Voskhod and the USA's Gemini were contemporaneous capsules that followed the Vostok and Mercury capsules and brought improvements in capsule design. Both had offset centers of gravity permitting the capsules to generate some degree of lift. The lift helped mitigate the deceleration loads while also affording greater spatial latitude in landing location. The Gemini capsule carried two crew members and was double the mass of its Mercury predecessor. It too flew into space on an ICBM launcher, the Titan II, which used non-cryogenic storable propellants, Aerozine 50, a hydrazine (N_2H_4) fuel derivative, and nitrogen tetroxide (N_2O_4) oxidizer. The Voskhod could carry two or three astronauts depending on whether or not spacesuits were worn. This spacecraft was followed by the Soyuz capsule which could carry up to three astronauts and started making trips to LEO in 1967. Variants of that basic design are in use to this day. Its mass, at about 6300 kg, represented a 40% growth over the original Vostok capsule.

The flights of the Apollo program burst upon the human spaceflight scene in 1968 in response to President Kennedy's vow to put a man on the moon before the decade (the 1960s) was out. The massive Saturn V rocket developed for this purpose was able to put almost 136,000 kg into LEO in anticipation of escape from earth's gravity for a trip to the moon. The first stage of the Saturn V alone incorporated 5 F-1 engines, the most powerful engines ever built, each producing 6.67 million Newtons (1.5 million lb) of thrust. The Apollo program succeeded in launching nine missions beyond the gravitational bonds of earth, with six landings on the moon permitting 12 astronauts to actually walk (or drive a lunar rover) on the moon's surface. The achievement of lunar landings came about during a period of budgetary constraints that ended the high-profile program in 1973. In the waning days of that program the powerful Saturn launch system was used to place a 77 metric ton space station, Skylab, in LEO. The large station had a three-man crew that occupied the station for a total of about 6 months out of its 6 year lifetime. The USSR launched their 18.2 metric ton Salyut 4 space station in 1974. It remained in orbit for about 2 years and was occupied by astronauts for a total of about 3 months. In 1975 a Soviet Soyuz capsule docked with an Apollo spacecraft marking the first joint USA−USSR mission. This was also the last flight of an Apollo spacecraft and the last manned mission flown by the USA until the start of test flights of the Space Shuttle in 1981.

The captivating idea of regular visits to space aboard a reusable spaceplane evolved into the US Space Shuttle program. Originally conceived as a completely reusable Space Transportation System (STS), the program was redefined to include only one truly reusable element, the Orbiter spaceplane because of budget limitations in the USA space program during the 1970s. However, this large flyable and reusable Orbiter vehicle was capable of carrying a 25,000 kg payload to LEO and the STS became a mainstay of US space travel. The first operational flight was in 1986 and the Space Shuttle ultimately made 135 trips with all but two being successful until its retirement in 2011. Several important new technologies appeared in the Space Shuttle Orbiter design. One improvement was the development of lightweight high-power (1650 kW/kg) turbopumps used to feed the cryogenic LH2/LOX propellants at extremely high pressures (22 MPa) to the Space Shuttle Main Engine (SSME or RS-25). The other was an innovative thermal protection system (TPS) for shielding the internal structure and crew of the Orbiter from excessive heating during atmospheric entry. The TPS was comprised of a mosaic of lightweight silicon-based insulating tiles covering most of the exposed surface area of the Orbiter and the application of high-temperature reinforced carbon composite materials on the nose and leading edge surfaces.

The Soviet Mir space station was the first modular space station and was constructed over the decade spanning 1986 to 1996. The 130 metric ton space station was operational until 2001 when it was de-orbited. The three-man crew registered an occupancy rate of about 80% during its lifetime. In 1998 the first component of the ISS was launched and the structure has been continuously inhabited by crews from many nations since 2000. The Space Shuttle serviced the ISS with 27 visits involving new modules and equipment and Soyuz has become the transporter of crew members. As of mid-2015, 547 astronauts have made it to space and of these, eight were civilian space tourists. National space competition has evolved into planetary cooperation with the operation the ISS. The outlook for the future of human spaceflight now ranges from interplanetary missions of discovery to vacation cruises for space tourists.

The progress of manned spaceflight is illustrated in Figure 1.2 by the mass placed in LEO since its beginning in 1961. The Soviet Buran, a large flyable spaceplane resembling the US Space

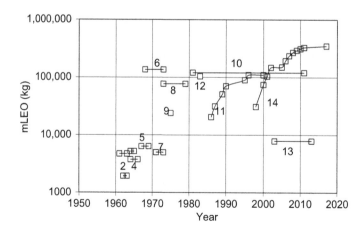

FIGURE 1.2

Manned spaceflight activities in terms of mass put into low earth orbit and years of operation. The spacecraft considered are: (1) Vostok, (2) Mercury, (3) Voskhod, (4) Gemini, (5) Soyuz, (6) Apollo, (7) Salyut 1, (8) Apollo-Skylab, (9) Apollo-Soyuz, (10) Space Shuttle, (11) Mir, (12) Buran, (13) Shenzhou, and (14) International Space Station.

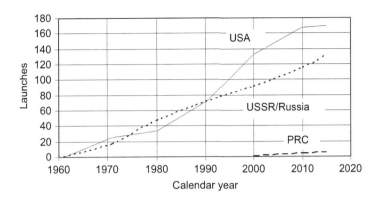

FIGURE 1.3

Cumulative launches of manned spacecraft as a function of the year of the launch. Two unsuccessful launches, STS-51-L and Soyuz T-10-1, are included in the totals.

Shuttle Orbiter in size and shape, is indicated as item 12 in Figure 1.2 because it was designed to carry a crew and its sole flight was successful, even though it was unmanned. The growing development of manned spaceflight is apparent in the cumulative number of launches as shown in Figure 1.3. For the first 50 years of manned spaceflight the USA averaged about 3.4 flights per year while the USSR/Russia averaged about 2.4 flights per year. China started flights in the last decade of that 50-year period and accomplished three. Between 2010 and 2015 the USA launched three flights but since the end of the Space Shuttle program in 2011 has launched no others while Russia launched 16 and China launched two. As will be discussed in Chapter 12, the economics of space flight places a severe financial burden on reusable launch systems if the frequency of flights is much less than 20 per year. The past history of manned space flight launches suggests that this required level of frequency is unlikely to be met unless there is a paradigm change for space travel. Only the 8 Apollo flights to the moon or its vicinity involved manned spaceflight escape from earth's gravity. If interplanetary trips with human crews are considered, the travel time involved is so great that trip frequency must necessarily be small. As far as LEO flights and suborbital flights are concerned, the only business case for high-frequency manned spaceflight is space tourism, which, by definition, is a round-trip travel service.

1.5 THE NEAR FUTURE OF HUMAN SPACEFLIGHT

It appears that the nations that have had manned spaceflight experience, the United States, Russia, and China are gravitating toward moving human spaceflight beyond earth's gravitational pull. The general target in the near future seems to be a return to the moon or flight to an asteroid, although the ultimate goal is Mars. All parties recognize that round trip flights to the moon are achievable but that the larger problem is long-term stays in space whether on, or in orbit about, the surface of a celestial body. Some of the advances required involve the development of regenerative life support systems, the use of indigenous materials for propellant replenishment, and the capability for manufacturing and assembly in microgravity.

Because essentially the same near-term outlook for human spaceflight is shared by the current participants, the approach being considered by NASA will be used to illustrate the path ahead. NASA describes its plan for human spaceflight under three broad topics headings:

- Earth Reliant: this is the near-earth region of space where missions last 6−12 months and travel between orbit and earth is measured in hours. This region has been fairly well conquered and because the earth is nearby it serves as a less risky arena to practice the basics of interplanetary space exploration.
- Proving Ground: this is the region beyond earth's gravitational grasp with missions lasting 1−12 months and round-trip travel times of around 10 days. This region of space exposes astronauts to larger doses of solar and cosmic radiation more typical of that which would be encountered on a trip to Mars. This is a riskier training ground but a more realistic one for dealing with problems that may arise on interplanetary missions. A specific mission to be pursued here to address some of these issues is NASA's Asteroid Redirect Mission (ARM).
- Mars Ready: this represents the state of preparedness for an interplanetary mission that lasts 2−3 years with travel time to return to earth being measured in months. The long travel times require that the vehicle and crew be "earth-independent."

The specific objective of ARM is to launch a robotic spacecraft to a near-earth asteroid to capture a small asteroid 2−6 m in diameter or, failing that, to collect a large boulder several metric tons in mass from a larger asteroid and to transport it to an orbit around the moon. Astronauts aboard an Orion crew exploration vehicle (CEV) can then visit the orbiting chunk of asteroid and remove samples for return to earth where they can be studied in detail. This mission, planned for the 2020s, is part of the larger plan to develop experience with long duration missions beyond LEO and to investigate advanced technologies necessary for a human mission to Mars in the 2030s. An array of possible asteroid missions is given in Table 1.1. The heading "scientific" in Table 1.2 illustrates the fundamental objective of the ARM while the heading "mitigation" refers to the development of those techniques which will safely deflect asteroids that pose a collision threat to earth (NRC, 2010). The column headed "commercial" is aimed at the possible discovery of resources, like valuable minerals or water, on asteroids whose recovery would promote commercial space activity.

Table 1.2 Classification of Missions to Near-Earth Asteroids

Mission Type	Scientific	Mitigation	Commercial
Flyby	Observation with limited data return	Direct interaction	Superficial assessment
Rendezvous	Surface exploration using transponders and tomography	Orbit tracking and adjustment	Detailed assessment, in situ sampling, and limited resource recovery
Return	Probe and sample return with detailed characterization	Multiply interactive orbit monitoring, tracking, and adjustment	Large-scale mining

Solar electric propulsion (SEP) will be an important component of the robotic spacecraft sent to the near-earth asteroid for sample collection. In the region between Earth and Mars collecting solar radiation with advanced photovoltaic cells will provide sufficient power for an ion engine. Using an electric field to accelerate ions to produce thrust, rather than a nozzle to accelerate hot gas, is a much more fuel-efficient process, as described by Sforza (2012). Therefore the mass of propellant required with SEP is much less than would be the case for chemical propulsion and this is a major attraction for a long trip out to a near-earth asteroid. The trade-off for the increased fuel efficiency of an ion engine is reduced thrust in comparison to a chemical rocket engine which translates into increased travel time. In March 2015, NASA's Dawn spacecraft, powered by an ion engine, entered an orbit around Ceres, the largest of the dwarf planets in the solar system. The unmanned Dawn spacecraft was launched in 2007 and visited Vesta, the second largest asteroid, in 2011 and 2012. Although ion propulsion has a history in space exploration, the lower thrust levels currently achieved (<1 N) would require trip times well beyond that acceptable for manned missions. The development of much larger thrust ion engines is one of the advances expected to arise from pursuit of the ARM.

Successful rendezvous with the asteroid, recovery of a suitable sample from its surface, transportation of the sample to the vicinity of the moon, and inserting it into the proper lunar orbit demands perfection in trajectory and navigation techniques beyond those presently available. The development of these skills in pursuit of the ARM is necessary to deal with the transport of cargo ships to Mars as well as the establishment of precise orbits around that planet. There are similar concerns involved in the rendezvous and docking of the Orion CEV with the robotic spacecraft which must be addressed.

Human factors play strongly into NASA's long-term goals of human spaceflight. The habitability of the spacecraft cabins for missions lasting months or more is a primary concern as is the general physiological requirements for maintaining the health and well-being of the astronauts. A discussion of spacecraft cabin design is presented in Chapter 10. One of the advances in life support system technology needed in the build-up to interplanetary exploration is the primary life support system (PLSS), that is, the spacesuit. Currently, the spacesuit must be maintained and refurbished on earth and that is not acceptable for deep space missions. In addition to improved ease of maintainability the spacesuits must deal more effectively with CO_2 removal, humidity control, and oxygen regulation. NASA expects that the proving ground provided by the ARM will permit realistic testing of various spacesuit designs prior to any interplanetary travel.

For the planned asteroid missions, NASA is developing the heavy-lift Space Launch System (SLS) to replace the capability once provided by the Space Shuttle. The Block 1 version SLS will use a cluster of three SSME and two five-segment Shuttle-type solid rocket boosters to launch 70 metric tons (70,000 kg or 154,500 lb) into orbit, perhaps as early as 2020. Also in planning is the larger Block 2 version of the SLS which will use a cluster of five SSME and two strap-on solid rocket boosters to launch 130 metric tons, although it may be premature to speculate on this version of the launch vehicle. In the interim only Russia and China are capable of placing human crews in orbit. Russia continues to refine its Soyuz spacecraft and is increasing efforts to develop technology aimed at supporting human exploration beyond LEO. Similarly, China is investing in heavy-lift launchers, like the Long March 5 and Long March 9, which are comparable in lift capability to NASA's SLS Block-1 and Block-2 vehicles.

Commercial interest in human spaceflight is currently centered on the development of new orbital spacecraft to fill NASA's requirements for space exploration as well as new suborbital spacecraft to fill the perceived demand for space tourism. Boeing and SpaceX are developing new orbital space capsules, the CST-100 and the Dragon, respectively, that can satisfy the operational requirements currently filled by Lockheed Martin's Orion capsule. All the capsules rely on conventional parachute-assisted landings. Sierra Nevada is building a reusable orbital spaceplane, the Dream Chaser, based on early USAF and NASA research on hypersonic lifting bodies so as to have the capability of making runway landings. XCor Aerospace's two-person Lynx suborbital reusable spaceplane aims to provide the sole passenger a minute of microgravity during a flight to about 60 km in altitude. Later versions promise the passenger astronaut status by climbing to 100 km and providing around 3 min of microgravity. Blue Origin's New Shepard reusable capsule spacecraft is designed to carry a crew of three to an altitude of 100 km and to provide up to 4 min of microgravity. The cryogenic propellant engine of the New Shepard can be throttled down by a factor of five so as to permit a vertical, tail-down, landing of the propulsion module after the crew capsule is ejected for a parachute-assisted landing. Perhaps the most familiar of the new reusable suborbital spacecraft is Virgin Galactic's SpaceShipTwo (SS2), a spaceplane which has flown supersonically and at altitudes over 20 km. The commercial interest in space tourism is based on an apparently significant consumer demand for suborbital flights even though tickets are reputed to cost (in 2015) around $200,000. The planned altitude capability for vehicles providing commercial space tourist flights is 100 km, the minimum for which passengers can claim astronaut status.

1.6 NOMENCLATURE

ARM	asteroid redirect mission
C_L	lift coefficient
e	energy per unit mass
FAI	Federation Aeronautique Internationale
GSO	geosynchronous orbit
g	gravitational acceleration
g_E	gravitational acceleration at the earth's surface
ICBM	intercontinental ballistic missile
ISS	international space station
L	lift
LEO	low earth orbit
m	mass
PLSS	primary life support system
PRC	People's Republic of China
R_E	radius of the earth
r	radial distance
S	reference area for airplane lift
SEP	solar electric propulsion
SSME	Space Shuttle main engine
STS	space transportation system (popularly known as the US Space Shuttle)
USA	United States of America

USSR	Union of Soviet Socialist Republics
V	velocity
V_E	circular orbit velocity at the surface of the earth
V_e	escape velocity
z	altitude
z_e	altitude of the edge of space
ρ	density

1.6.1 SUBSCRIPTS

k	kinetic
max	maximum
p	potential
t	total

REFERENCES

NRC (2010). *Defending planet earth − near earth objects surveys and hazard mitigation strategies* Washington, DC: National Research Council, National Academies Press.

Sforza, P. M. (2012). *Theory of aerospace propulsion* Oxford, UK: Elsevier.

EARTH'S ATMOSPHERE

2.1 THE ATMOSPHERIC ENVIRONMENT

Designing vehicles to remain in orbit or to traverse the atmosphere requires sound knowledge of the environment through which they will pass. The pressure, density, and temperature of the environment influence the magnitude of aerodynamic forces and thermal stresses on the vehicle while the composition of the environment affects chemical and electronic interactions with vehicle surface materials.

The Earth's atmosphere may be considered to be comprised of a number of concentric spherical shells characterized by noticeably different temperature gradients, each separated from the next by relatively thin transition zones in which the temperature gradient changes sign. A useful representation of the Earth's atmosphere appears in Figure 2.1 in which four distinct layers are identified, each in terms of its temperature gradient.

2.1.1 VERTICAL STRUCTURE ACCORDING TO TEMPERATURE

The first layer starts at the Earth's surface and extends up to about 8–14.5 km (5–9 mi). This layer is called the troposphere and within it the temperature continually drops from the sea level value down to around −50°C. The tropopause is a relatively thin layer in which the temperature gradient increases to zero and becomes positive, separating the troposphere from the next layer. Together the tropopause and the troposphere constitute the lower atmosphere.

The second layer is the stratosphere. It starts just above the troposphere and within it the temperature gradient is positive and the temperature gradually increases to about 0°C at an altitude of about 50 km (31 mi). The temperature in this region increases due to the absorption of ultraviolet radiation from the sun. The ozone layer, which absorbs and scatters the solar ultraviolet radiation, lies within the stratosphere. The stratopause is the thin transition zone separating the stratosphere from the next layer. Within it the temperature gradient drops to zero and then becomes negative again.

The third layer is the mesosphere. It starts just above the stratopause and extends to an altitude of about 90 km (56 mi). In this layer, the temperature gradient is negative and as the altitude increases the temperature drops to levels as low as −100°C. The mesopause is the thin transition region which separates the mesosphere from the next layer. The region, including

Manned Spacecraft Design Principles. DOI: http://dx.doi.org/10.1016/B978-0-12-804425-4.00002-7

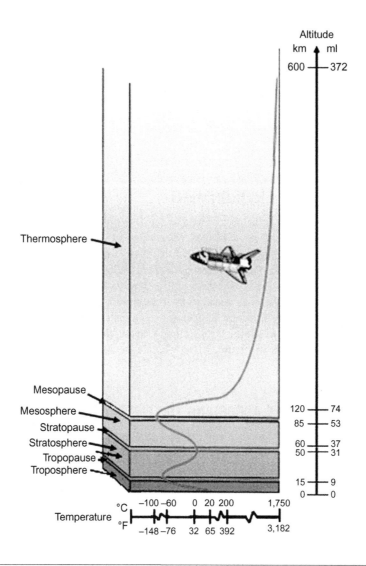

FIGURE 2.1

Description of the Earth's atmosphere in terms of temperature (NASA).

the stratosphere, the stratopause, the mesosphere, and the mesopause, is generally called the middle atmosphere. Note that aerodynamic entry from space is usually considered to begin at the mesopause, around 100 km, the nominal edge of the middle atmosphere.

The fourth layer of the atmosphere is the thermosphere which starts just above the mesosphere and extends to an altitude of about 600 km (372 mi). The temperature increases with altitude because of solar excitation and therefore the temperature level reached depends upon the degree of solar activity. Temperatures range from a low of about 600 K to as high as 2000 K. The mean temperature reached is usually taken as 1000 K. This layer is known as the upper atmosphere. As indicated by the presence of the Space Shuttle Orbiter in Figure 2.1, the thermosphere is the region of low earth orbit (LEO).

The region known as the exosphere starts at the top to the thermosphere and therefore is off the scale of Figure 2.1. The exosphere continues on ultimately merging with the interplanetary gases of space. In this region, hydrogen and helium are virtually the only species present and then only at extremely low densities.

2.1.2 VERTICAL STRUCTURE ACCORDING TO COMPOSITION

The composition of the first 100 km of the atmosphere is essentially constant and by mole fraction it is comprised of 78% nitrogen, 21% oxygen, and 1% other gases (Argon 0.93%, CO_2 0.03%, and neon, helium, krypton, hydrogen, xenon, and ozone in increasingly smaller amounts). For most thermochemical purposes, the atmosphere is considered a binary mixture of 79% nitrogen and 21% oxygen. This fixed composition approximation makes the temperature distribution a reliable means for dividing up the various important regions in the atmosphere. The pressure and density in the regions of specified temperature behavior may be readily determined from the equation of state and the conditions of hydrostatic equilibrium.

Beyond 100 km, diffusive separation of species dominates and atmospheric composition depends on altitude. This region is therefore called the heterosphere. The concentration of heavier molecular species like N_2 and O_2 decreases rapidly through the thermosphere, but dissociation of molecular oxygen leaves a substantial concentration of atomic oxygen present. This active species can have deleterious effects on spacecraft surface material. In the exosphere, above about 600 km, the only species present are helium and atomic hydrogen. Thus the heterosphere is marked by a continual decrease in molecular weight, going from 28.96 at the mesopause down to less than 4 in the exosphere.

2.2 EQUATION OF STATE AND HYDROSTATIC EQUILIBRIUM

If one considers the atmosphere to behave as a perfect gas, then

$$p = \rho R T = \rho \frac{R_u}{W} T \tag{2.1}$$

The molecular weight of the mixture of atmospheric gases is essentially constant up to 100 km and is given by $W_m = 28.96$, and the atmospheric gas constant $R = 0.287$ kJ/kg-K (or 1716 ft^2/s^2-R). Beyond 100 km, the molecular weight drops substantially with altitude and must be taken into account. The hydrostatic equation for the atmosphere is

$$dp = -\rho g dz \tag{2.2}$$

Because the gravitational acceleration depends on altitude, a new altitude function may be defined. This is the *geopotential* altitude, h, and it is related to the geometric altitude by the equation

$$g_E dh = g dz$$

Here $g_E = 9.807$ m/s^2 (or 32.15 ft/s^2) is the gravitational acceleration at the surface of the Earth, $z = 0$. The gravitational acceleration varies with altitude according to Newton's law of gravitation and may be written as

$$g = \frac{g_E R_E^2}{(R_E + z)^2} = g_E \frac{1}{\left(1 + \dfrac{z}{R_E}\right)^2}$$

The mean radius of the Earth is taken to be $R_E = 6371$ km so that for LEOs, where the altitude is around 400 km, the ratio $z/R_E \sim 0.06$, and therefore the difference between h and z is relatively small. Integrating the relation between h and z yields the geopotential altitude

$$h = z \frac{1}{1 + \dfrac{z}{R_E}}$$

Thus the difference between the geometric and geopotential altitudes for a LEO of up to 400 km is less than 6.3%. Using the equation of state in the hydrostatic equilibrium equation yields

$$\frac{dp}{p} = -g \frac{dz}{RT} = -g_E \frac{dh}{RT} \qquad (2.3)$$

2.3 THE 1976 U.S. STANDARD ATMOSPHERE MODEL

The state properties of a point in the atmosphere are functions of space and time and can vary considerably. In order to carry out scientific and engineering operations in a manner that contributes to repeatability and consistency, it is advantageous to have a standard atmosphere to use as a basis. The World Meteorological Organization defines such a standard as "A hypothetical vertical distribution of atmospheric temperature, pressure, and density which by international agreement and for historical reasons, is roughly representative of year-round mid-latitude conditions." A discussion of atmospheric structure and standard models may be found, for example, in the Handbook of Geophysics (1985).

The standard atmosphere used in this book is the 1976 U.S. Standard Atmosphere, unless otherwise indicated. The U.S. Committee on Extension to the Standard Atmosphere (COESA) developed the 1976 version of the U.S. Standard Atmosphere (COESA, 1976). Rocket and satellite data along with conventional meteorological instruments and perfect gas theory were used to develop a standard profile for the thermodynamic and state properties of the atmosphere from sea level to a geometric altitude of 1000 km. This model is an idealized, steady-state representation of the atmosphere in the mid-latitudes during a period of moderate solar activity. Properties tabulated include temperature, pressure, density, gravitational acceleration, scale height, number density, mean particle speed, mean collision frequency, mean free path, mean molecular weight, sound speed, dynamic viscosity, kinematic viscosity, thermal conductivity, and geopotential altitude.

Note that the 1976 U.S. Standard Atmosphere model chosen for use in this book considers dry air and does not account for diurnal or day-to-day variations in the characteristics of the atmosphere, not to mention latitude, season, or degree of solar activity. The U.S. Standard Atmosphere Supplements (COESA, 1966) includes tables of temperature, pressure, density, sound speed, viscosity, and thermal conductivity for five northern latitudes (15°, 30°, 45°, 60°, and 75°), for summer and winter conditions.

2.3.1 **THE LOWER AND MIDDLE ATMOSPHERE: 0–100 KM**

The fundamental premise of a standard atmosphere is the postulation of an appropriate relation for temperature in terms of geopotential altitude, $T=T(h)$. Using this profile, one may integrate the hydrostatic equation (2.3) to find the pressure $p=p(h)$ and then use the equation of state to determine the density $\rho=\rho(h)$. The chemical composition of the atmosphere is essentially constant up to around 100 km altitude because global-scale, or turbulent, mixing is dominant up to around that altitude. Thus this spherical shell of the atmosphere is called the homosphere, a region in which the composition of the air is homogeneous. Above 100 km, molecular diffusion dominates and the possibility of chemical segregation based upon molecular mass becomes increasingly important. The molecular weight of atmospheric air is constant at $W = 28.96$ up to 86 km and drops slightly, about 2%, to 28.40 at $z=100$ km. Note that water vapor is neglected in this model. Because water's molecular weight is 18, its presence would reduce the mixture molecular weight leading to a lower air density for a given pressure and temperature. However, the typical level of humidity is such that its effect is not only slight in magnitude ($<1\%$) but also essentially confined to the troposphere. With this assumption of constant composition, the gas constant is invariant and the hydrostatic and state equations may be easily integrated.

The 1976 U.S. Standard Atmosphere defines atmospheric layers, each with $T=T_i + \lambda_i(h - h_i)$, where T_i is the temperature of the start of layer i, h_i is the altitude at the start of layer i, and λ_i is the lapse rate, i.e., dT/dh, in that layer. Integration of the hydrostatic equation for non-zero λ yields

$$p = p_i \left[\frac{T_i}{T_i + \lambda_i(h - h_i)} \right]^{\frac{g_E}{R\lambda_i}} \tag{2.4}$$

In isothermal layers where $\lambda=0$, the temperature $T=T_i=$constant, and the pressure is instead given by

$$p = p_i \exp\left[-\frac{g_E}{RT_i}(h - h_i) \right] \tag{2.5}$$

The temperature at the Earth's surface, $h=z=0$, is taken as $T= 15°C = 288.15$ K and $g_E/R = 34.17$ K/km. The properties of this atmospheric model in the various layers are given in Table 2.1. The first 98.45 km in geopotential altitude, or 100 km in geometric altitude, may be divided into nine layers defined by different temperature variations. The convention for defining the thermal layers is to use geopotential altitude up to a geometric altitude $z = 86$ km, where the geopotential altitude $h = 84.85$ km. The hydrostatic equation (2.3) is more readily integrated in terms of h because g_E is constant. At this height above the surface, the difference between the geopotential and geometric altitudes remains small. Furthermore, in the region below 86 km the composition of the atmosphere is constant, and therefore the gas constant R is also constant and the state equation (2.1) yields the density directly. In practical operations of aircraft and spacecraft, the geometric altitude is of major importance and as the distance above the surface of the earth increases, the difference between geopotential and geometric altitude continually increases, therefore tables of thermodynamic and transport properties are usually listed as a function of geometric altitude.

The atmospheric layers in Table 2.1 are listed in a manner which conforms to this convention. The temperature is assumed constant from $z = 86$ km to $z = 92$ km and thereafter the temperature

Table 2.1 Definition of the Layers in the 1976 Model Atmosphere[a]

Layer	Geopotential Altitude, h (km)	Geometric Altitude, z (km)	Lapse Rate, λ_i (K/km)	Thermal Type
1	0	0	−6.5	Neutral
2	11	11.019	0	Isothermal
3	20	20.063	+1.0	Inversion
4	32	32.162	+2.8	Inversion
5	47	47.351	0	Isothermal
6	51	51.413	−2.8	Neutral
7	71	71.802	−2.0	Neutral
8	84.85	86	0	Isothermal
9	90.69	92	+1.03	Inversion
10 on	98.45	100	Increasing	Inversion

[a]The 8th and 9th layer data are approximations based on other atmospheric data.

continues to rise linearly to $z = 100$ km. In this region, the molecular weight of the air drops about 2%, from 28.96 at 86 km to 28.40 at 100 km. Beyond $z = 100$ km, the temperature of the model atmosphere monotonically increases to an asymptotic value of 1000 K at 500 km. The corresponding state conditions in this outer region of the model atmosphere will be discussed subsequently. It should be noted that the asymptotic temperature in this outer region can vary widely depending on solar activity. For example, the value of about 1500 K illustrated in Figure 2.1 corresponds to a period marked by a high level of solar activity.

The distribution of pressure, in kPa, in the various layers is then given by the following:

Layer 1 (0−11 km):

$$p = 101.33 \left(\frac{288.15}{288.15 - 6.5\, h} \right)^{\frac{34.17}{-6.5}}$$

Layer 2 (11−20 km):

$$p = 22.64 \exp \left(\frac{-34.17[h - 11]}{216.65} \right)$$

Layer 3 (20−32 km):

$$p = 5.474 \left(\frac{216.65}{216.65 + [h - 20]} \right)^{\frac{34.17}{1}}$$

Layer 4 (32−47 km):

$$p = 0.8680 \left(\frac{228.65}{228 + 2.8[h - 32]} \right)^{\frac{34.17}{2.8}}$$

Layer 5 (47−51 km):

$$p = 0.1109 \exp\left(\frac{-34.17[h-47]}{270.65}\right)$$

Layer 6 (51−71 km):

$$p = 0.06694\left(\frac{270.65}{270.65-2.8[h-51]}\right)^{\frac{34.17}{-2.8}}$$

Layer 7 (71−84.85 km):

$$p = 0.003957\left(\frac{214.65}{214.65-2[h-71]}\right)^{\frac{34.17}{-2}}$$

Layer 8 (84.85−90.69 km)

$$p = 0.0003733 \exp\left(-\frac{34.17}{186.95}(h-84.85)\right)$$

Layer 9 (90.69−98.45 km)

$$p = 0.0001288\left(\frac{186.95}{186.95+1.03(h-90.69)}\right)^{\frac{34.17}{1.03}}$$

2.3.2 PROPERTIES OF THE LOWER AND MIDDLE ATMOSPHERE

The profile of the temperature of the homosphere, as described in Table 2.1, is shown in Figure 2.2. It is clear that the temperature is constant at about 216 K in the lower reaches of the stratosphere, in the altitude range of 10−20 km. This is the region of high speed manned flight, from commercial jet airliners to military aircraft up to the Mach 3 SR-71 Blackbird. The sound speed may be found from its definition for a perfect gas

$$a^2 = \left(\frac{\partial p}{\partial \rho}\right)_s = \gamma RT = \gamma \frac{R_u}{W} T \tag{2.6}$$

Because the speed of sound is proportional to the square root of the temperature, and the temperature through the standard atmosphere has little variation, the speed of sound is relatively constant up to 100 km altitude, as can be seen in Figure 2.3. It is common to assume a constant value for preliminary design purposes and this can be used with an error on the order of $\pm 10\%$ over the whole range.

The dynamic viscosity μ, in the units of Pa-s, and the thermal conductivity k, in the units of W/m-s, may be calculated from the equations given, for example, by Handbook of Geophysics (1985), and shown below:

$$\mu = 1.458 \times 10^{-6}\left[\frac{T^{3/2}}{T+110.4}\right] \tag{2.7}$$

$$k = 2.6502 \times 10^{-3}\left[\frac{T^{3/2}}{T+245.4 \times 10^{-12/T}}\right] \tag{2.8}$$

Graphs of the variation of these important gas properties are presented in Figures 2.4 and 2.5.

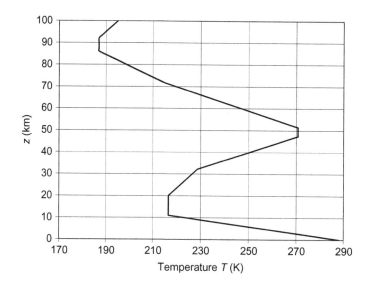

FIGURE 2.2

The 1976 U.S. Standard Atmosphere temperature distribution for the first 100 km of geometric altitude.

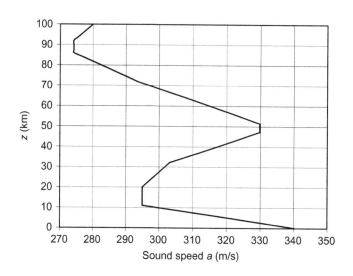

FIGURE 2.3

The 1976 U.S. Standard Atmosphere sound speed distribution for the first 100 km of geometric altitude.

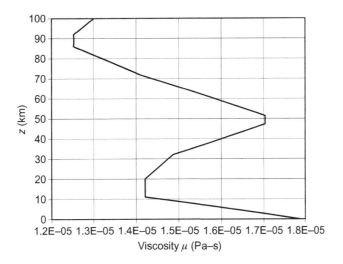

FIGURE 2.4

The 1976 U.S. Standard Atmosphere viscosity distribution for the altitude range between sea level and 100 km.

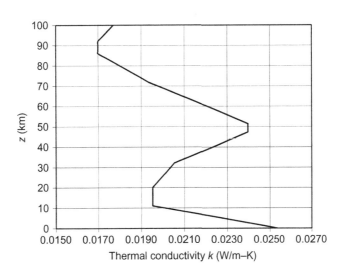

FIGURE 2.5

The 1976 U.S. Standard Atmosphere thermal conductivity distribution for the altitude range between sea level and 100 km.

The temperature range in the first 100 km of the atmosphere is small, 288−195 K, with an average of 230 K as shown in Figure 2.2. Thus the specific heat may be taken to be constant at $c_p = 1.005$ KJ/kg-K. The Prandtl number, which is the ratio of kinematic viscosity $\nu = \mu/\rho$ to the thermal diffusivity $\alpha = \rho c_p/k$, may be written as

$$\text{Pr} = \frac{\mu c_p}{k} \approx 0.732 \tag{2.9}$$

Because μ/k varies little with temperature, as does c_p, the Prandtl number is essentially constant with a range of $0.715 < \text{Pr} < 0.748$ up to $z = 100$ km. Choosing the average temperature of this band of the atmosphere to be $T_{av} = 230$ K results in the number given above in Eqn (2.9). The ratio of specific heats, like the gas constant, depends upon the composition of the air, but for altitudes below 100 km $\gamma = 1.4$ and $R = 0.287$ kJ/kg-K.

Knowing the temperature and pressure at each altitude in the model atmosphere permits the calculation of the density from the state equation (2.1). The variation of the pressure and density is shown in Figures 2.6 and 2.7, respectively. The altitude range for these graphs is divided into two bands: one from 0 to 50 km and one from 50 to 100 km in order to more clearly underline the point that the distribution of both pressure and density is approximately exponential in nature. The great difference in the magnitude of the pressure and density in the two altitude ranges is also emphasized in this presentation of the data.

2.3.3 ATMOSPHERIC SCALE HEIGHT

An interesting feature of the atmosphere may be obtained by normalizing Eqn (2.3) in terms of the atmospheric pressure ratio $\delta = p/p_{sl}$ and a characteristic length H as follows:

$$\frac{d\delta}{\delta} = \left(\frac{g_E H}{RT}\right) d\left(\frac{h}{H}\right)$$

We may set the characteristic length, or scale height, of the atmosphere to be

$$H = \frac{RT}{g_E}$$

Then the hydrostatic equation becomes

$$\frac{d\delta}{\delta} = \frac{dh}{H}$$

Obviously, if both R and T were constant with altitude the pressure profile would be exponential and given by

$$\delta = \exp\left(-\frac{h}{H}\right) \tag{2.10}$$

In the first 100 km of the atmosphere, the difference between h and z is less than 2% so that in Eqn (2.10) h may be replaced with z with little loss in accuracy. Thus H could be considered a scale height over which the pressure drops by a factor e^{-1}, or, in other words, H is the e-folding distance for the pressure. The state equation (2.1) indicates that, for constant T and R, the atmospheric pressure ratio δ is equal to the atmospheric density ratio $\sigma = \rho/\rho_{sl}$. As mentioned

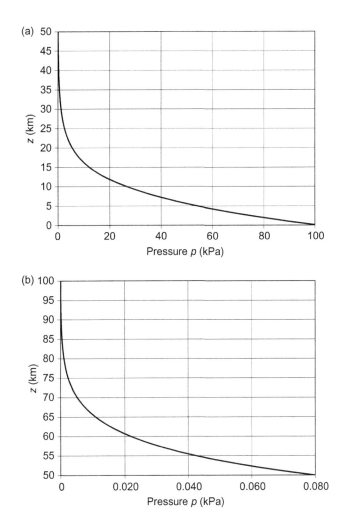

FIGURE 2.6

The 1976 U.S. Standard Atmosphere pressure distribution for the altitude range between (a) sea level and 50 km and (b) 50 and 100 km.

previously, the first 100 km is characterized by constant composition so that we may set $H = 0.2926T$ where H is measured in kilometers. The temperature, though not constant, varies fairly little up to $z = 100$ km so that $\sigma \sim \delta$. Thus, the behavior of the density should be close to exponential in character and this is illustrated by Figure 2.8 in which the standard atmosphere density ratio is compared to the following exponential approximation:

$$\sigma = \exp\left(-\frac{z}{H}\right) \tag{2.11}$$

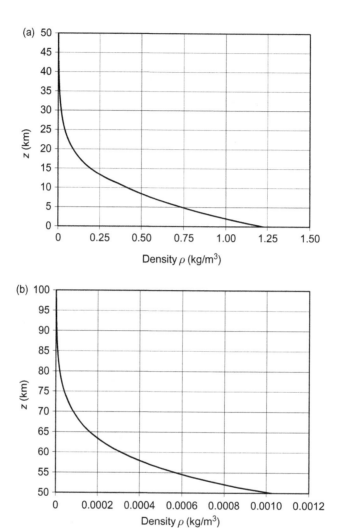

FIGURE 2.7

The 1976 U.S. Standard Atmosphere density distribution for the altitude range between (a) sea level and 50 km and (b) 50 and 100 km.

The scale height chosen here, $H = 7.16$ km, which corresponds to an average temperature $T_{av} = 245$ K, is a value commonly used in spacecraft reentry preliminary design analyses.

2.3.4 THE UPPER ATMOSPHERE: 100–500 KM

The kinetic theory of gases defines the mean free path λ as the average distance between collisions of gas molecules. The variation of λ with altitude in the standard atmosphere is shown in

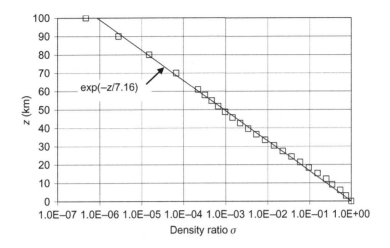

FIGURE 2.8

Comparison of the 1976 U.S. Standard Atmosphere density ratio to the exponential approximation of Eqn (2.11) with $H = 7.16$ km.

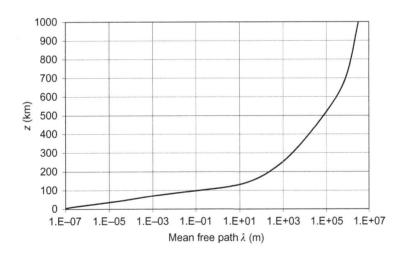

FIGURE 2.9

The 1976 U.S. Standard Atmosphere mean free path distribution for the altitude range between sea level and 1000 km.

Figure 2.9. It increases exponentially in the inner and middle atmosphere from about 10^{-7} to 10^{-1} m at $z = 100$ km. Thereafter it grows more slowly, reaching to several tens of kilometers by the end of the thermosphere, $z = 500$ km.

At the high altitudes of the thermosphere, the large mean free path assures that the frequency of collisions between molecules becomes small. Under these increasingly rarefied conditions, the

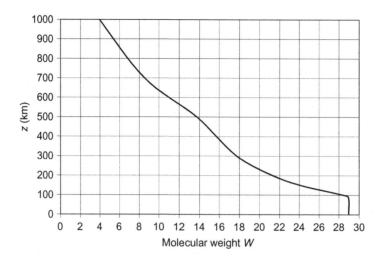

FIGURE 2.10

The 1976 U.S. Standard Atmosphere molecular weight distribution for the altitude range between sea level and 1000 km.

rapidly falling number of collisions means that each chemical species acts independently under the force of gravity. Thus the different species become segregated in altitude according to their mass. The molecular weight of the atmospheric gas therefore drops with altitude as shown in Figure 2.10. The dissociation of molecular oxygen in the thermosphere is a large factor in reducing the mean molecular weight in that region. Note that between $z = 300$ km and $z = 500$ km, the molecular weight is that of atomic oxygen plus or minus 12.5%. The presence of atomic oxygen in the region of LEO constitutes a threat to the integrity of spacecraft surfaces because of its chemical activity. Once in the exosphere, $z > 500$ km, only helium and atomic hydrogen are sensibly present. However, during periods of high solar activity the thermosphere can have high levels of molecular nitrogen in addition to atomic oxygen. Though molecular nitrogen is highly energized by the heightened solar radiation, it is difficult to dissociate. The presence of these species may also be present in amounts comparable to, or greater than, that of helium.

The molecules present may be considered to have a Maxwell−Boltzmann distribution of velocities with the mean molecular velocity being the measure of temperature according to the relation:

$$\bar{a} = \sqrt{\frac{8R_u T}{\pi W}} \tag{2.12}$$

The temperature profile for the 1976 U.S. Standard Atmosphere up to 500 km is shown in Figure 2.11. The temperature for this range of altitude in the model approaches 1000 K asymptotically.

The pressure ratio and density ratio through the entire range of the model atmosphere, from $z = 0$ km to $z = 1000$ km is shown in Figure 2.12. It is clear that in the thermosphere the pressure and density profiles no longer follow an exponential variation with altitude. Figure 2.9 shows that at the edge of the thermosphere, $z = 500$ km the mean free path $\lambda \sim 100$ km, or on the order of the scale height there. Indeed, the edge of the thermosphere is often designated as the altitude at which the mean free path is equal to the local scale height.

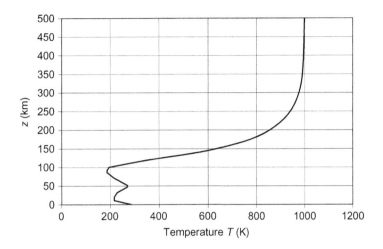

FIGURE 2.11

The 1976 U.S. Standard Atmosphere temperature distribution for the altitude range between sea level and 500 km. The temperature remains constant at 1000 K from 500 to 1000 km.

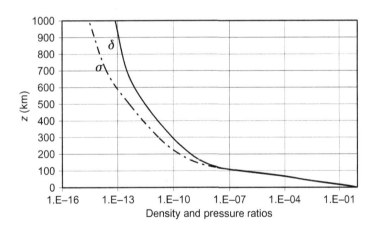

FIGURE 2.12

The 1976 U.S. Standard Atmosphere pressure ratio δ and density ratio σ distribution for the altitude range between sea level and 1000 km.

The temperature ratio above 100 km may be approximated to $\pm 7\%$ by the curve fit below:

$$\theta \approx 3.477 - 2.85 \exp\left[-\left(\frac{z-100}{70}\right)\right] \tag{2.13}$$

This relation is compared to the standard atmosphere profile in Figure 2.13. As mentioned previously, the temperature in the thermosphere is sensitive to solar activity. Thus, in practice, the standard value of 1000 K in the thermosphere may be bracketed by extremes of 600 K on the low side to 1900 K on the high side.

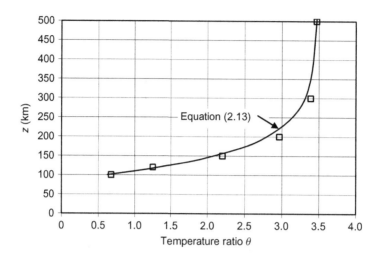

FIGURE 2.13

Comparison of the 1976 U.S. Standard Atmosphere temperature ratio (open symbols) to the exponential approximation of Eqn (2.13).

2.4 FLOW PROPERTIES USING THE ATMOSPHERIC MODELS

Design and analysis of spacecraft that traverse the atmosphere requires information on free stream flow parameters like unit Reynolds number (Reynolds number per unit length), Mach number, and dynamic pressure and these involve only the atmospheric properties discussed in the previous sections. It is often convenient to have representations for these properties in an analytic form, and a number of such approximations are presented subsequently.

2.4.1 REYNOLDS NUMBER AND MACH NUMBER

The Reynolds number is a dimensionless similarity parameter that provides a measure of the relative importance of pressure force to viscous force in a flow. The definition and several alternative forms are shown as follows:

$$\text{Re} = \frac{\rho V l}{\mu} = \frac{\rho V^2}{\mu \left(\frac{V}{l}\right)} = \frac{aMl}{\nu} \tag{2.14}$$

In this equation, l is a characteristic length of the body in the flow. The Mach number is the ratio of the local speed V of a gas to the sound speed a in that gas at the local conditions and is defined as

$$M = \frac{V}{a} \tag{2.15}$$

The sound speed a was defined previously in Eqn (2.6). The ratio of the Mach number to the unit Reynolds number Re/l is a function of atmospheric properties alone:

$$\frac{M}{(Re/l)} = \frac{\nu}{a} \tag{2.16}$$

As mentioned previously, the kinetic theory of gases defines the mean free path λ as the average distance between collisions of gas molecules. Liepmann and Roshko (2002) show that this characteristic length for collisions may be expressed in terms of the mean speed of the molecules $\bar{a}$ and the kinematic viscosity or the thermal conductivity as follows:

$$\lambda \sim \frac{\nu}{\bar{a}} \sim \frac{k}{\bar{a}} \tag{2.17}$$

The mean molecular speed is proportional to and on the same order of magnitude as the sound speed as shown below:

$$a = \sqrt{\gamma RT} = \bar{a}\sqrt{\frac{\pi\gamma}{8}} \tag{2.18}$$

The general variation of both Ml/Re and λ with altitude up to $z = 100$ km is shown in Figure 2.14. The relation between the two lengths may be expressed, within $\pm 7\%$ be the following expression:

$$\frac{M}{(Re/l)} \approx \frac{3}{5}\lambda \tag{2.19}$$

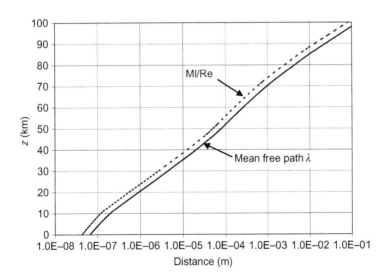

FIGURE 2.14

The variation of the parameter Ml/Re and the mean free path λ is shown as a function of geometric altitude.

We may rewrite the expression in Eqn (2.19) as follows:

$$\frac{M}{\text{Re}} \approx \frac{3}{5}\frac{\lambda}{l} = \frac{3}{5}Kn \tag{2.20}$$

The parameter $Kn = \lambda/l$ introduced in Eqn (2.20) is the Knudsen number, the ratio of the mean free path in the gas to the characteristic dimension of the body immersed in the flow of that gas. As pointed out by Liepmann and Roshko (2002), the Knudsen number is an indicator of the nature of the molecular collisions occurring in the gas. If $Kn \ll 1$ collisions between molecules and other molecules are dominant and the flow may be considered a continuum. However, if $Kn \gg 1$ collisions between molecules and body surfaces are dominant and the flow is considered a molecular flow.

A simple analytic approximation for Ml/Re in the units of meters is given below where z is measured in kilometers:

$$\frac{Ml}{\text{Re}} = 3.906 \times 10^{-8} \exp\left(\frac{z}{7.16}\right) \tag{2.21}$$

This expression is only accurate to about $\pm 25\%$ up to about 90 km altitude, with higher errors in the 90−100 km range. A modification of this model, with the same units is given by

$$\frac{Ml}{\text{Re}} = 3.906 \times 10^{-8} \exp\left(\frac{z}{7.16}\right)\left[1 + 0.27 \sin\left(\frac{\pi z}{29}\right)\right]^{-1} \tag{2.22}$$

The parameter Ml/Re given by Eqn (2.22) is compared to that computed using the actual model atmosphere in Figure 2.15 and is accurate to within $\pm 10\%$ up to 90 km altitude. For a characteristic length $\lambda = 1$ m the quantity $M/\text{Re} \leq 0.1$ all the way up to $z = 100$ km. Therefore it is reasonable to expect the flow around a spacecraft of such size or larger to be in a continuum flow.

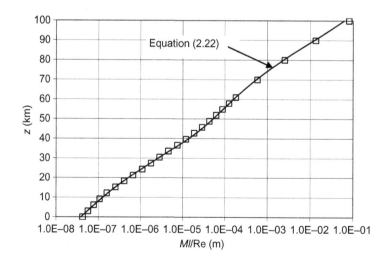

FIGURE 2.15

Variation with altitude of the ratio of Mach number to unit Reynolds number using the 1976 Standard Atmosphere (open symbols) and the exponential approximation in Eqn (2.22).

The question of how to denote a continuum flow may be investigated from the point of view of the ratio of relaxation time for molecular collisions to the residence time of the flow over the body. This may be expressed in terms of the variables already discussed as follows:

$$\frac{M^2}{\mathrm{Re}} = \frac{V^2}{a^2} \frac{\nu}{Vl} = \frac{\nu/a^2}{l/V} \approx \frac{3}{5} \frac{\lambda/a}{l/V} = \frac{3}{5} \sqrt{\frac{8}{\pi\gamma}} \frac{\lambda/\bar{a}}{l/V} \approx \frac{\tau_{coll}}{\tau_{res}} \tag{2.23}$$

If the time between collisions of molecules in a fluid particle τ_{coll} is much smaller than the residence time of the fluid particle in the flow field over the body τ_{res} we may consider the fluid field to be continuous. Therefore a more stringent requirement for continuum flow may be written as

$$\frac{M^2}{\mathrm{Re}} \approx \frac{3}{5} M \cdot Kn \ll 1 \tag{2.24}$$

2.4.2 DYNAMIC PRESSURE

Inertia forces in a flow are proportional to the dynamic pressure

$$q = \frac{1}{2}\rho V^2 = \frac{1}{2}\gamma p M^2 \tag{2.25}$$

The ratio of dynamic pressure to the square of the Mach number depends only upon the gas composition and state as given by

$$\frac{q}{M^2} = \frac{1}{2}\gamma p \tag{2.26}$$

The general variation of this ratio with altitude is shown in Figure 2.16. Also shown in the figure is an approximate analytic correlation given as follows:

$$\frac{q}{M^2} = 0.88\left[1 + 0.175 \sin\left(\frac{\pi z}{24.4}\right)\right]\exp\left(-\frac{z}{7.16}\right) \tag{2.27}$$

Here q is measured in kilopascals and z in kilometers. On the scale shown in Figure 2.16, the difference between the curves using the 1976 Standard Atmosphere and the approximate model of Eqn (2.27) is not apparent, but the accuracy of Eqn (2.27) is less than $\pm 8\%$ except at the two ends of the range.

The dynamic pressure is related to the lift of a vehicle of weight W and reference (usually planform) area S so that during steady cruising flight the dynamic pressure may be expressed as

$$q = \frac{L}{C_L S} = \frac{1}{C_L}\left(\frac{W}{S}\right)_{cruise} \tag{2.28}$$

Therefore for a given wing (or, more generally, planform) loading the dynamic pressure encountered fixes the required lift coefficient C_L for equilibrium flight. It is useful to display contours of constant dynamic pressure q as a function of altitude and Mach number. This can be readily accomplished by means of the analytic approximations presented above, the results of which are displayed in Figures 2.17 and 2.18 for different flight Mach number ranges. Figure 2.17 covers the complete manned flight range including $M \sim 25$ for reentry from LEO all the way up to $M = 40$ which is the upper range for reentry from a lunar mission. Figure 2.18 illustrates the range of flight Mach numbers for manned missions at high supersonic and low hypersonic cruise conditions.

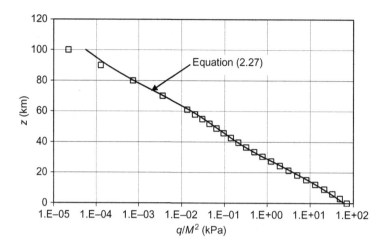

FIGURE 2.16

Variation with altitude of the ratio of dynamic pressure q (in kPa) to the square of the Mach number for the 1976 Standard Atmosphere and the approximation in Eqn (2.27).

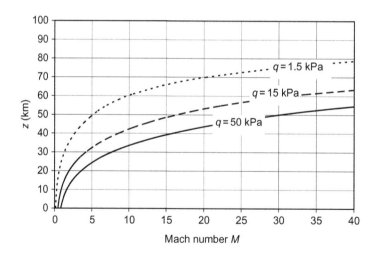

FIGURE 2.17

Contours of constant dynamic pressure for practical flight conditions at high Mach numbers.

2.4.3 ATMOSPHERIC PROPERTY CURVE FITS

It is often advantageous to have reasonably accurate analytic expressions for the atmospheric properties to use in design studies. In many flight applications, particularly spacecraft launch and reentry, the atmospheric density is approximated by a simple exponential function of altitude. However, this usually yields relatively high errors if used over too wide an altitude range. In this section, we present

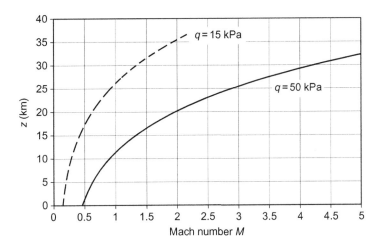

FIGURE 2.18

Contours of constant dynamic pressure for practical flight conditions at low Mach numbers.

several expressions for the thermodynamic properties in the constant composition portion of the atmosphere $0 < z < 100$ km. The temperature ratio may be approximated by the following expression:

$$\theta = 0.9 - 0.00165z + 0.12 \sin\left(\pi\frac{z - 35}{35}\right) \tag{2.29}$$

Results using this equation are compared to the actual temperature ratio profile in Figure 2.19. The error is less than about $\pm 4\%$ between 6 and 100 km (20 and 300 kft). Equation (2.29) may be used to calculate the sound speed as well according to

$$a = a_{sl}\sqrt{\theta} \tag{2.30}$$

The results for sound speed are compared to the actual profile in Figure 2.20 and the error is less than about $\pm 2\%$.

A reasonable approximation to the density profile of atmospheric density is given by

$$\sigma = \frac{\rho(h)}{\rho(0)} = 1.065\left[1 + 0.18\sin\left(\frac{\pi z}{28.48}\right)\right]\exp\left(-\frac{z}{7.12}\right) \tag{2.31}$$

The pressure ratio may be obtained using the state equation so that

$$\delta = \frac{p}{p_{sl}} = \sigma\theta \tag{2.32}$$

The accuracy of the pressure and density profiles using Eqns (2.29), (2.31), and (2.32) is illustrated in Figure 2.21. For the range $0 < z < 18$ km, the errors are within $\pm 8\%$ while for the range 18 km $< z < 82$ km, the errors are within $\pm 3\%$. Beyond about 80 km, the approximation equations give value which become considerably higher. It is clear from Figure 2.12 that above $z = 100$ km the decay in density is no longer exponential. However, as mentioned at the outset of this chapter, the aerodynamic effects of greatest importance occur within the first 100 km of the atmosphere.

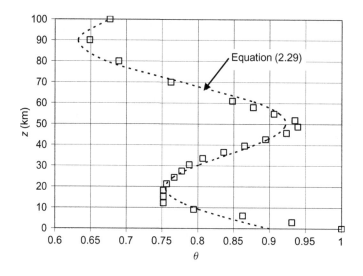

FIGURE 2.19

Variation with altitude of the temperature ratio θ (open symbols) for the 1976 Standard Atmosphere and the approximation (dotted line) in Eqn (2.29).

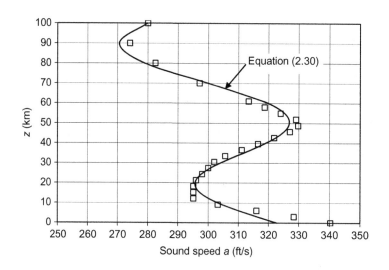

FIGURE 2.20

Variation with altitude of the sound speed a (open symbols) for the 1976 Standard Atmosphere and the approximation (solid line) in Eqn (2.30).

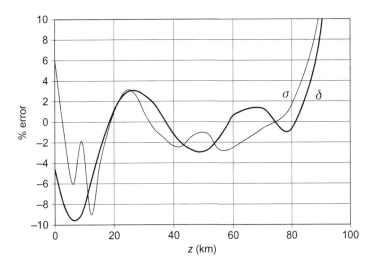

FIGURE 2.21

Comparison of the errors between the approximate and actual atmospheric pressure and density ratios.

2.5 TABLES OF ATMOSPHERIC PROPERTIES

2.5.1 TABLES IN SI UNITS

The properties of the 1976 U.S. Standard Atmosphere are presented in SI units in the following tables. Tables 2.2–2.9 present thermodynamic and transport properties for various altitude ranges and increments. Table 2.10 presents the basic parameters used in the U.S. Standard Atmosphere when using SI units.

2.5.2 TABLES IN ENGLISH UNITS

The properties of the 1976 U.S. Standard Atmosphere are presented in English units in the following tables. Tables 2.11 and 2.12 present thermodynamic and transport properties for the altitude range $0 < z < 200$ kft in 10 kft increments and $200 < z < 280$ kft in 20 kft increments. Table 2.13 presents the basic parameters used in the U.S. Standard Atmosphere when using English units.

2.6 OTHER MODEL ATMOSPHERES

AIAA (2010) is a guide to over 70 selected Earth and planetary atmospheric models covering a range of altitudes up to 4000 km. The guide includes descriptions of the content and technical basis of the models, their uncertainties and limitations, and sources of computer code. NASA has compiled a catalog and archive of atmospheric models in a coordinated modeling center, NASA Goddard (2013), which may be found on a website which includes options to run models and view

Table 2.2 Thermodynamic Properties of the 1976 U.S. Standard Atmosphere for Altitudes between $z=0$ km and $z=30$ km in Intervals of 1 km

z (km)	T (K)	p (kPa)	ρ (kg/m^3)	θ	δ	σ
0	288.15	1.013E+02	1.225E+00	1.000E+00	1.000E+00	1.000E+00
1	281.65	8.988E+01	1.112E+00	9.774E−01	8.873E−01	9.075E−01
2	275.15	7.951E+01	1.007E+00	9.549E−01	7.848E−01	8.217E−01
3	268.66	7.012E+01	9.093E−01	9.324E−01	6.922E−01	7.423E−01
4	262.17	6.166E+01	8.194E−01	9.098E−01	6.087E−01	6.689E−01
5	255.68	5.405E+01	7.364E−01	8.873E−01	5.336E−01	6.012E−01
6	249.19	4.722E+01	6.601E−01	8.648E−01	4.661E−01	5.389E−01
7	242.70	4.111E+01	5.900E−01	8.423E−01	4.058E−01	4.817E−01
8	236.22	3.565E+01	5.258E−01	8.198E−01	3.520E−01	4.292E−01
9	229.73	3.080E+01	4.671E−01	7.973E−01	3.041E−01	3.813E−01
10	223.25	2.650E+01	4.135E−01	7.748E−01	2.616E−01	3.376E−01
11	216.77	2.270E+01	3.648E−01	7.523E−01	2.241E−01	2.978E−01
12	216.65	1.936E+01	3.112E−01	7.519E−01	1.911E−01	2.541E−01
13	216.65	1.642E+01	2.640E−01	7.519E−01	1.621E−01	2.156E−01
14	216.65	1.386E+01	2.229E−01	7.519E−01	1.368E−01	1.819E−01
15	216.65	1.163E+01	1.870E−01	7.519E−01	1.148E−01	1.527E−01
16	216.65	9.704E+00	1.560E−01	7.519E−01	9.580E−02	1.274E−01
17	216.65	8.044E+00	1.293E−01	7.519E−01	7.941E−02	1.056E−01
18	216.65	7.565E+00	1.216E−01	7.519E−01	7.468E−02	9.930E−02
19	216.65	6.467E+00	1.040E−01	7.519E−01	6.384E−02	8.489E−02
20	216.65	5.529E+00	8.890E−02	7.519E−01	5.458E−02	7.258E−02
21	217.58	4.727E+00	7.569E−02	7.551E−01	4.667E−02	6.178E−02
22	218.57	4.047E+00	6.451E−02	7.585E−01	3.995E−02	5.266E−02
23	219.57	3.467E+00	5.500E−02	7.620E−01	3.422E−02	4.490E−02
24	220.56	2.972E+00	4.694E−02	7.654E−01	2.934E−02	3.831E−02
25	221.55	2.549E+00	4.008E−02	7.689E−01	2.516E−02	3.272E−02
26	222.54	2.188E+00	3.425E−02	7.723E−01	2.160E−02	2.796E−02
27	223.54	1.880E+00	2.930E−02	7.758E−01	1.856E−02	2.392E−02
28	224.53	1.616E+00	2.507E−02	7.792E−01	1.595E−02	2.047E−02
29	225.52	1.390E+00	2.148E−02	7.826E−01	1.373E−02	1.753E−02
30	226.51	1.197E+00	1.841E−02	7.861E−01	1.182E−02	1.503E−02

results using an online interface. Several other commonly used atmospheric models are briefly described as follows:

ISO (1982) presents information on atmospheric properties at levels between the surface and 80 km including day-to-day variability caused by season, latitude, and longitude. Systematic variation in the atmospheric properties is shown for altitudes up to 80 km by a family of models, comprising tropical, subtropical, mid-latitude, subarctic, and arctic regions. In addition,

Table 2.3 Transport Properties of the 1976 U.S. Standard Atmosphere for Altitudes between $z=0$ km and $z=30$ km in Intervals of 1 km

z (km)	a (m/s)	μ (N-s/m^2)	ν (m^2/s)	k (W/m-K)
0	340.29	1.789E − 05	1.461E − 05	2.533E − 02
1	336.43	1.758E − 05	1.581E − 05	2.481E − 02
2	332.53	1.726E − 05	1.715E − 05	2.430E − 02
3	328.58	1.694E − 05	1.863E − 05	2.378E − 02
4	324.59	1.661E − 05	2.027E − 05	2.326E − 02
5	320.54	1.628E − 05	2.211E − 05	2.273E − 02
6	316.45	1.595E − 05	2.416E − 05	2.220E − 02
7	312.30	1.561E − 05	2.646E − 05	2.167E − 02
8	308.10	1.527E − 05	2.904E − 05	2.114E − 02
9	303.85	1.493E − 05	3.196E − 05	2.060E − 02
10	299.53	1.458E − 05	3.525E − 05	2.006E − 02
11	295.15	1.422E − 05	3.899E − 05	1.952E − 02
12	295.07	1.422E − 05	4.568E − 05	1.950E − 02
13	295.07	1.422E − 05	5.384E − 05	1.950E − 02
14	295.07	1.422E − 05	6.379E − 05	1.950E − 02
15	295.07	1.422E − 05	7.601E − 05	1.950E − 02
16	295.07	1.422E − 05	9.111E − 05	1.950E − 02
17	295.07	1.422E − 05	1.099E − 04	1.950E − 02
18	295.07	1.422E − 05	1.169E − 04	1.950E − 02
19	295.07	1.422E − 05	1.367E − 04	1.950E − 02
20	295.07	1.422E − 05	1.599E − 04	1.950E − 02
21	295.70	1.427E − 05	1.885E − 04	1.958E − 02
22	296.38	1.432E − 05	2.220E − 04	1.967E − 02
23	297.05	1.438E − 05	2.614E − 04	1.975E − 02
24	297.72	1.443E − 05	3.074E − 04	1.983E − 02
25	298.39	1.448E − 05	3.614E − 04	1.992E − 02
26	299.05	1.454E − 05	4.244E − 04	2.000E − 02
27	299.72	1.459E − 05	4.981E − 04	2.008E − 02
28	300.38	1.465E − 05	5.841E − 04	2.017E − 02
29	301.05	1.470E − 05	6.844E − 04	2.025E − 02
30	301.71	1.475E − 05	8.014E − 04	2.033E − 02

temporal and spatial variations and frequency distributions of observed temperatures and densities are given. This atmospheric model is also that used by the International Civil Aviation Organization (ICAO).

Picone, Hedin, Drob, and Aikin (2002) discuss the Naval Research Laboratory's Mass Spectrometer, Incoherent Scatter Radar Extended (NRLMSISE-00) model which describes the neutral temperature and species densities in Earth's atmosphere. The two digit extension refers

Table 2.4 Thermodynamic Properties of the 1976 U.S. Standard Atmosphere for Altitudes between $z=0$ km and $z=50$ km in Intervals of 2 km

z (km)	T (K)	p (kPa)	ρ (kg/m³)	θ	δ	σ
0	288.15	1.013E + 02	1.225E + 00	1.000E + 00	1.000E + 00	1.000E + 00
2	275.15	7.951E + 01	1.007E + 00	9.549E − 01	7.848E − 01	8.217E − 01
4	262.17	6.166E + 01	8.194E − 01	9.098E − 01	6.087E − 01	6.689E − 01
6	249.19	4.722E + 01	6.601E − 01	8.648E − 01	4.661E − 01	5.389E − 01
8	236.22	3.565E + 01	5.258E − 01	8.198E − 01	3.520E − 01	4.292E − 01
10	223.25	2.650E + 01	4.135E − 01	7.748E − 01	2.616E − 01	3.376E − 01
12	216.65	1.940E + 01	3.119E − 01	7.519E − 01	1.915E − 01	2.546E − 01
14	216.65	1.417E + 01	2.278E − 01	7.519E − 01	1.399E − 01	1.860E − 01
16	216.65	1.035E + 01	1.664E − 01	7.519E − 01	1.022E − 01	1.359E − 01
18	216.65	7.564E + 00	1.216E − 01	7.519E − 01	7.467E − 02	9.929E − 02
20	216.65	5.529E + 00	8.890E − 02	7.519E − 01	5.458E − 02	7.257E − 02
22	218.57	4.047E + 00	6.450E − 02	7.585E − 01	3.995E − 02	5.265E − 02
24	220.56	2.971E + 00	4.693E − 02	7.654E − 01	2.933E − 02	3.831E − 02
26	222.54	2.188E + 00	3.425E − 02	7.723E − 01	2.160E − 02	2.796E − 02
28	224.53	1.616E + 00	2.507E − 02	7.792E − 01	1.595E − 02	2.047E − 02
30	226.51	1.197E + 00	1.841E − 02	7.861E − 01	1.181E − 02	1.503E − 02
32	228.49	8.889E − 01	1.355E − 02	7.930E − 01	8.775E − 03	1.106E − 02
34	233.74	6.632E − 01	9.885E − 03	8.112E − 01	6.547E − 03	8.069E − 03
36	239.28	4.984E − 01	7.256E − 03	8.304E − 01	4.920E − 03	5.923E − 03
38	244.82	3.770E − 01	5.365E − 03	8.496E − 01	3.722E − 03	4.380E − 03
40	250.35	2.871E − 01	3.994E − 03	8.688E − 01	2.834E − 03	3.261E − 03
42	255.88	2.199E − 01	2.994E − 03	8.880E − 01	2.171E − 03	2.444E − 03
44	261.40	1.694E − 01	2.258E − 03	9.072E − 01	1.673E − 03	1.843E − 03
46	266.92	1.313E − 01	1.714E − 03	9.263E − 01	1.296E − 03	1.399E − 03
48	270.65	1.023E − 01	1.317E − 03	9.393E − 01	1.010E − 03	1.075E − 03
50	270.65	7.977E − 02	1.027E − 03	9.393E − 01	7.875E − 04	8.382E − 04

to the year of release of the model. It is based on a large set of data from satellites, rockets, and radars having wide temporal and spatial distribution and has been extensively tested against experimental data. The model has a flexible mathematical formulation and is valid for use from ground level to the exosphere.

Bowman, Tobiska, Marcos, and Valladares (2008) present JB2006, an empirical atmospheric density model which describes the neutral temperature and the total density in the Earth's thermosphere and exosphere above 120 km. It includes a representation of the semiannual variation, and is claimed to provide a more accurate model of the total density than previous

Table 2.5 Thermodynamic Properties of the 1976 U.S. Standard Atmosphere for Altitudes between $z = 50$ km and $z = 100$ km in intervals of 2 km

z (km)	T (K)	p (kPa)	ρ (kg/m^3)	θ	δ	σ
50	270.65	7.977E − 02	1.027E − 03	9.393E − 01	7.875E − 04	8.382E − 04
52	269.03	6.221E − 02	8.055E − 04	9.337E − 01	6.141E − 04	6.575E − 04
54	263.52	4.833E − 02	6.389E − 04	9.145E − 01	4.771E − 04	5.216E − 04
56	258.02	3.736E − 02	5.044E − 04	8.954E − 01	3.688E − 04	4.117E − 04
58	252.52	2.872E − 02	3.962E − 04	8.763E − 01	2.835E − 04	3.234E − 04
60	247.02	2.196E − 02	3.096E − 04	8.573E − 01	2.167E − 04	2.528E − 04
62	241.53	1.669E − 02	2.407E − 04	8.382E − 01	1.647E − 04	1.965E − 04
64	236.04	1.260E − 02	1.860E − 04	8.191E − 01	1.244E − 04	1.519E − 04
66	230.55	9.460E − 03	1.429E − 04	8.001E − 01	9.338E − 05	1.167E − 04
68	225.07	7.052E − 03	1.092E − 04	7.811E − 01	6.962E − 05	8.910E − 05
70	219.58	5.220E − 03	8.282E − 05	7.621E − 01	5.153E − 05	6.761E − 05
72	214.26	3.836E − 03	6.236E − 05	7.436E − 01	3.787E − 05	5.091E − 05
74	210.35	2.801E − 03	4.638E − 05	7.300E − 01	2.765E − 05	3.786E − 05
76	206.45	2.033E − 03	3.431E − 05	7.165E − 01	2.007E − 05	2.801E − 05
78	202.54	1.467E − 03	2.524E − 05	7.029E − 01	1.448E − 05	2.060E − 05
80	198.64	1.052E − 03	1.846E − 05	6.894E − 01	1.039E − 05	1.507E − 05
82	194.74	7.500E − 04	1.342E − 05	6.758E − 01	7.404E − 06	1.095E − 05
84	190.84	5.310E − 04	9.693E − 06	6.623E − 01	5.242E − 06	7.912E − 06
86	186.95	3.733E − 04	6.957E − 06	6.488E − 01	3.686E − 06	5.679E − 06
88	186.95	2.615E − 04	4.872E − 06	6.488E − 01	2.581E − 06	3.977E − 06
90	186.95	1.833E − 04	3.415E − 06	6.488E − 01	1.809E − 06	2.788E − 06
92	186.95	1.285E − 04	2.394E − 06	6.488E − 01	1.268E − 06	1.954E − 06
94	188.89	9.027E − 05	1.665E − 06	6.555E − 01	8.911E − 07	1.359E − 06
96	190.83	6.365E − 05	1.162E − 06	6.623E − 01	6.284E − 07	9.486E − 07
98	192.77	4.505E − 05	8.142E − 07	6.690E − 01	4.448E − 07	6.646E − 07
100	195.08	3.201E − 05	5.726E − 07	6.757E − 01	3.160E − 07	4.674E − 07

models, including the NRLMSISE-00. Comparisons with accurate daily density drag data previously calculated for many satellites suggest it is valid for use from an altitude of 120 km to the exosphere. For 400 km altitude, the standard deviation of the JB2006 model is reduced to 10% for periods of low geomagnetic storm activity.

CIRA-86 (1986) is the COSPAR International Reference Atmosphere (CIRA). It provides empirical models of atmospheric temperature and densities from 0 to 2000 km as recommended by the Committee on Space Research (COSPAR). In the thermosphere (above 100 km), CIRA-86 is identical with the NRLMSIS-86 model. For the lower part (0−120 km), the model consists of tables of the monthly mean values of temperature and zonal wind for the latitude

Table 2.6 Transport Properties of the 1976 U.S. Standard Atmosphere for Altitudes between $z=0$ km and $z=50$ km in Intervals of 2 km

z (km)	a (m/s)	μ (N-s/m²)	ν (m²/s)	k (W/m-K)
0	340.29	1.789E − 05	1.461E − 05	2.533E − 02
2	332.53	1.726E − 05	1.715E − 05	2.430E − 02
4	324.59	1.661E − 05	2.027E − 05	2.326E − 02
6	316.45	1.595E − 05	2.416E − 05	2.220E − 02
8	308.10	1.527E − 05	2.904E − 05	2.114E − 02
10	299.53	1.458E − 05	3.525E − 05	2.006E − 02
12	295.07	1.422E − 05	4.557E − 05	1.950E − 02
14	295.07	1.422E − 05	6.239E − 05	1.950E − 02
16	295.07	1.422E − 05	8.540E − 05	1.950E − 02
18	295.07	1.422E − 05	1.169E − 04	1.950E − 02
19	295.07	1.422E − 05	1.367E − 04	1.950E − 02
20	295.07	1.422E − 05	1.599E − 04	1.950E − 02
22	296.38	1.432E − 05	2.220E − 04	1.967E − 02
24	297.72	1.443E − 05	3.074E − 04	1.983E − 02
26	299.05	1.454E − 05	4.244E − 04	2.000E − 02
28	300.38	1.465E − 05	5.841E − 04	2.017E − 02
30	301.71	1.475E − 05	8.014E − 04	2.033E − 02
32	303.02	1.486E − 05	1.096E − 03	2.050E − 02
34	306.49	1.514E − 05	1.531E − 03	2.093E − 02
36	310.10	1.543E − 05	2.126E − 03	2.139E − 02
38	313.66	1.572E − 05	2.930E − 03	2.185E − 02
40	317.19	1.601E − 05	4.007E − 03	2.230E − 02
42	320.67	1.629E − 05	5.441E − 03	2.275E − 02
44	324.11	1.657E − 05	7.338E − 03	2.320E − 02
46	327.52	1.685E − 05	9.831E − 03	2.364E − 02
48	329.80	1.704E − 05	1.294E − 02	2.394E − 02
50	329.80	1.704E − 05	1.659E − 02	2.394E − 02

range of $\pm 80°$. The lower portion of the model was merged with NRLMSIS-86 at 120 km altitude. The model accurately reproduces most of the characteristic features of the atmosphere such as the equatorial wind and the general structure of the tropopause, stratopause, and mesopause.

Justus and Leslie (2008) present the NASA Manned Space Flight Center (MSFC) global reference atmospheric model (GRAM). It provides global geographical variability and complete altitude coverage (surface to orbital altitudes) as well as seasonal and monthly variability of the

Table 2.7 Transport Properties of the 1976 U.S. Standard Atmosphere for Altitudes between $z=50$ km and $z=100$ km in Intervals of 2 km

z (km)	a (m/s)	μ (N-s/m^2)	ν (m^2/s)	k (W/m-K)
50	329.80	1.704E − 05	1.659E − 02	2.394E − 02
52	328.81	1.696E − 05	2.105E − 02	2.381E − 02
54	325.43	1.668E − 05	2.610E − 02	2.337E − 02
56	322.01	1.640E − 05	3.252E − 02	2.292E − 02
58	318.56	1.612E − 05	4.068E − 02	2.248E − 02
60	315.07	1.584E − 05	5.114E − 02	2.203E − 02
62	311.55	1.555E − 05	6.461E − 02	2.158E − 02
64	307.99	1.526E − 05	8.203E − 02	2.112E − 02
66	304.39	1.497E − 05	1.047E − 01	2.067E − 02
68	300.74	1.467E − 05	1.344E − 01	2.021E − 02
70	297.06	1.438E − 05	1.736E − 01	1.975E − 02
72	293.44	1.408E − 05	2.258E − 01	1.930E − 02
74	290.75	1.387E − 05	2.990E − 01	1.897E − 02
76	288.04	1.365E − 05	3.978E − 01	1.864E − 02
78	285.30	1.343E − 05	5.321E − 01	1.831E − 02
80	282.54	1.321E − 05	7.156E − 01	1.798E − 02
82	279.75	1.298E − 05	9.677E − 01	1.764E − 02
84	276.94	1.276E − 05	1.316E + 00	1.731E − 02
86	274.09	1.253E − 05	1.801E + 00	1.697E − 02
88	274.10	1.253E − 05	2.572E + 00	1.697E − 02
90	274.10	1.253E − 05	3.670E + 00	1.697E − 02
92	274.10	1.253E − 05	5.235E + 00	1.697E − 02
94	275.52	1.265E − 05	7.597E + 00	1.714E − 02
96	276.93	1.276E − 05	1.098E + 01	1.730E − 02
98	278.33	1.287E − 05	1.581E + 01	1.747E − 02
100	279.73	1.298E − 05	2.267E + 01	1.764E − 02

thermodynamic variables and wind components. The model includes the ability to simulate spatial and temporal perturbations in the thermodynamic variables due to turbulence and other atmospheric perturbation phenomena.

Drob et al. (2008) describes the Horizontal Wind Model (HWM) as an empirical global model of horizontal winds in the mesosphere and thermosphere (middle and upper atmosphere). The model is used to provide the necessary wind information, accounting for time, space, and geophysical variations, to complement atmospheric models for the thermodynamic properties.

Table 2.8 Thermodynamic Properties of the 1976 U.S. Standard Atmosphere for Altitudes between $z=86$ km and $z=1000$ km

Altitude, z (km)	Molecular Weight, W	Temperature, T (K)	Pressure, p (kPa)	Density, ρ (kg/m³)	θ	δ	σ
86	28.96	186.9	3.73E − 04	6.96E − 06	6.49E − 01	3.68E − 06	5.68E − 06
88	28.92	187.0	2.61E − 04	4.87E − 06	6.49E − 01	2.58E − 06	3.98E − 06
90	28.88	187.0	1.83E − 04	3.41E − 06	6.49E − 01	1.81E − 06	2.79E − 06
92	28.80	186.9	1.29E − 04	2.40E − 06	6.49E − 01	1.27E − 06	1.96E − 06
94	28.72	189.0	9.05E − 05	1.67E − 06	6.56E − 01	8.93E − 07	1.36E − 06
96	28.67	191.0	6.38E − 05	1.16E − 06	6.63E − 01	6.30E − 07	9.50E − 07
98	28.64	193.1	4.52E − 05	8.15E − 07	6.70E − 01	4.46E − 07	6.65E − 07
100	28.56	195.1	3.21E − 05	5.73E − 07	6.77E − 01	3.17E − 07	4.68E − 07
120	28.48	360.0	2.54E − 06	2.22E − 08	1.25E + 00	2.51E − 08	1.81E − 08
150	24.10	634.4	4.54E − 07	2.08E − 09	2.20E + 00	4.48E − 09	1.69E − 09
200	21.30	854.6	8.47E − 08	2.54E − 10	2.97E + 00	8.36E − 10	2.07E − 10
300	17.73	976.0	8.77E − 09	1.92E − 11	3.39E + 00	8.66E − 11	1.56E − 11
500	13.75	999.2	3.02E − 10	5.22E − 13	3.47E + 00	2.98E − 12	4.26E − 13
700	8.53	1000.0	3.19E − 11	3.07E − 14	3.47E + 00	3.15E − 13	2.51E − 14
1000	3.94	1000.0	7.71E − 12	3.56E − 15	3.47E + 00	7.61E − 14	2.91E − 15

Table 2.9 Gas Kinetic Properties of the 1976 U.S. Standard Atmosphere for Altitudes between $z=100$ km and $z=1000$ km

Altitude, z (m)	Mean Free Path, λ (m)	Molecular Weight, W (kg/kmol)	Number Density, n (1/m³)	Collision Frequency, f (1/s)	Mean Molecular Speed, $\bar{a}$ (m/s)	Scale Height, H (km)
100	1.42E − 01	28.40	1.19E + 19	2.7E + 03	381.4	6.01
150	3.30E + 01	24.10	5.19E + 16	2.3E + 01	746.5	23.38
200	2.42E + 02	21.30	7.18E + 15	3.9E + 00	921.6	36.18
300	2.60E + 03	17.73	6.51E + 14	4.2E − 01	1080	51.19
400	1.60E + 04	15.98	1.06E + 14	7.2E − 02	1149	59.68
600	2.80E + 05	11.51	5.95E + 12	4.8E − 03	1356	88.24
800	1.40E + 06	5.54	1.23E + 12	1.4E − 03	1954	193.9
1000	3.10E + 06	3.94	5.44E + 11	7.5E − 04	2318	288.2

Table 2.10 Parameters Used in the U.S. Standard Atmosphere for Altitudes Between $z=0$ km and $z=100$ km

Property	Symbol	Value
Universal gas constant	R_u	8.3145 J/mol-K
Molecular weight of air	W	28.964 kg/kg-mol
Gas constant for air	R	0.28706 kJ/kg-K
Gravitational acceleration	g_E	9.8067 m/s^2
Ratio of specific heats	γ	1.40
Radius of the Earth	R_E	6371 km
Sea level density	ρ_{sl}	1.225 kg/m^3
Sea level temperature	T_{sl}	288.16 K
Sea level pressure	p_{sl}	101.3 kPa

Table 2.11 Thermodynamic Properties of the 1976 U.S. Standard Atmosphere for Altitudes between $z = 0$ and $z = 280{,}000$ft Altitude

z kft	T R	p lb/ft^2	ρ lb-s^2/ft^4	θ	δ	σ	z km
0	519	2.12E + 03	2.38E − 03	1.000	1.000E + 00	1.000E + 00	0.0
10	483	1.46E + 03	1.76E − 03	0.931	6.878E − 01	7.386E − 01	3.0
20	447	9.73E + 02	1.27E − 03	0.863	4.599E − 01	5.332E − 01	6.1
30	412	6.30E + 02	8.19E − 04	0.794	2.975E − 01	3.447E − 01	9.1
40	390	3.93E + 02	5.87E − 04	0.752	1.858E − 01	2.471E − 01	12.2
50	390	2.44E + 02	3.64E − 04	0.752	1.151E − 01	1.531E − 01	15.2
60	390	1.51E + 02	2.26E − 04	0.752	7.137E − 02	9.492E − 02	18.3
70	392	9.37E + 01	1.39E − 04	0.756	4.429E − 02	5.857E − 02	21.3
80	398	5.85E + 01	8.57E − 05	0.767	2.765E − 02	3.606E − 02	24.4
90	403	3.68E + 01	5.31E − 05	0.777	1.738E − 02	2.236E − 02	27.4
100	409	2.33E + 01	3.32E − 05	0.788	1.100E − 02	1.396E − 02	30.5
110	418	1.48E + 01	2.07E − 05	0.807	7.011E − 03	8.692E − 03	33.5
120	434	9.60E + 00	1.29E − 05	0.836	4.537E − 03	5.428E − 03	36.6
130	449	6.31E + 00	8.19E − 06	0.865	2.982E − 03	3.446E − 03	39.6
140	464	4.21E + 00	5.28E − 06	0.894	1.988E − 03	2.222E − 03	42.7
150	479	2.84E + 00	3.46E − 06	0.924	1.343E − 03	1.454E − 03	45.7
160	487	1.94E + 00	2.32E − 06	0.939	9.176E − 04	9.770E − 04	48.8
170	485	1.33E + 00	1.60E − 06	0.935	6.283E − 04	6.717E − 04	51.8
180	470	9.04E − 01	1.12E − 06	0.906	4.271E − 04	4.713E − 04	54.9
190	455	6.07E − 01	7.77E − 07	0.877	2.868E − 04	3.270E − 04	57.9
200	440	4.02E − 01	5.33E − 07	0.848	1.901E − 04	2.242E − 04	61.0
220	410	1.69E − 01	2.41E − 07	0.790	8.003E − 05	1.013E − 04	67.1
240	382	6.69E − 02	1.02E − 07	0.736	3.161E − 05	4.296E − 05	73.2
260	360	2.49E − 02	4.03E − 08	0.694	1.178E − 05	1.696E − 05	79.2
280	339	8.76E − 03	1.51E − 08	0.653	4.140E − 06	6.337E − 06	85.3

Table 2.12 Transport Properties of the 1976 U.S. Standard Atmosphere for Altitudes between $z = 0$ and $z = 280$ kft

z kft	a ft/s	μ lb-s/ft²	ν ft²/s	k Btu/hr-ft-R	z km
0	1.116E+03	3.738E−07	1.573E−04	0.01464	0.00
10	1.077E+03	3.535E−07	2.014E−04	0.01373	3.05
20	1.037E+03	3.325E−07	2.624E−04	0.01280	6.10
30	9.947E+02	3.108E−07	3.793E−04	0.01186	9.14
40	9.679E+02	2.970E−07	5.057E−04	0.01127	12.2
50	9.679E+02	2.970E−07	8.161E−04	0.01127	15.2
60	9.679E+02	2.970E−07	1.316E−03	0.01127	18.3
70	9.708E+02	2.985E−07	2.144E−03	0.01133	21.3
80	9.775E+02	3.019E−07	3.523E−03	0.01148	24.4
90	9.841E+02	3.053E−07	5.745E−03	0.01163	27.4
100	9.907E+02	3.087E−07	9.304E−03	0.01177	30.5
110	1.003E+03	3.148E−07	1.524E−02	0.01203	33.6
120	1.021E+03	3.242E−07	2.513E−02	0.01244	36.6
130	1.038E+03	3.333E−07	4.070E−02	0.01284	39.6
140	1.056E+03	3.424E−07	6.481E−02	0.01323	42.7
150	1.073E+03	3.513E−07	1.016E−01	0.01363	45.7
160	1.082E+03	3.559E−07	1.533E−01	0.01384	48.8
170	1.080E+03	3.548E−07	2.222E−01	0.01378	51.8
180	1.063E+03	3.460E−07	3.089E−01	0.01339	54.9
190	1.045E+03	3.371E−07	4.337E−01	0.01300	57.9
200	1.028E+03	3.280E−07	6.155E−01	0.01261	61.0
220	9.922E+02	3.095E−07	1.285E+00	0.01181	67.1
240	9.575E+02	2.917E−07	2.856E+00	0.01105	73.2
260	9.302E+02	2.777E−07	6.888E+00	0.01046	79.2
280	9.022E+02	2.634E−07	1.749E+01	0.00987	85.3

Table 2.13 Parameters Used in the U.S. Standard Atmosphere for Altitudes between $z=0$ kft and $z=280$ kft

Property	Symbol	Value
Universal gas constant	R_u	1545.36 ft-lb/lb-mol-R
Molecular weight of air	W	28.96 lb/lb-mol
Gas constant for air	R	53.36 ft-lb/lb-R
Gravitational acceleration	g_E	32.174 ft/s²
Ratio of specific heats	γ	1.40
Radius of the Earth	R_E	20.903×10^6 ft
Sea level density	ρ_{sl}	0.002377 lb-s²/ft⁴
Sea level pressure	p_{sl}	2116 lb/ft²
Sea level temperature	T_{sl}	518.67R

2.7 NOMENCLATURE

a	sound speed
$\bar{a}$	mean molecular speed
C_L	lift coefficient
c_p	specific heat
g	acceleration of gravity
g_E	acceleration of gravity at $z = 0$
H	atmospheric scale height
h	geopotential altitude
Kn	Knudsen number
k	gas conductivity
L	lift
l	length
M	Mach number
Pr	Prandtl number
p	pressure
q	dynamic pressure
R	gas constant
Re	Reynolds number
R_E	mean radius of the earth
R_u	universal gas constant
S	reference area
T	temperature
V	velocity
W	weight of vehicle or molecular weight
z	geometric altitude
δ	pressure ratio p/p_{sl}
γ	ratio of specific heats
σ	density ratio ρ/ρ_{sl}
λ	temperature lapse rate, K/km
ρ	gas density
μ	gas viscosity
θ	temperature ratio T/T_{sl}
ν	kinematic viscosity μ/ρ

2.7.1 SUBSCRIPTS

i	index identifying an atmospheric layer
sl	sea level conditions

REFERENCES

AIAA (2010). *AIAA guide to reference and standard atmosphere models (G-003C-2010e)* American Institute of Aeronautics and Astronautics. eISBN: 978-1-60086-784-2, http://dx.doi.org/10.2514/4.867842.

Bowman, B. R., Tobiska, W. K., Marcos, F. A., & Valladares, C. (2008). The JB2006 empirical thermospheric density model. *Journal of Atmospheric and Solar-Terrestrial Physics, 70,* 774−793.

CIRA-86 (1986). The COSPAR International Reference Atmosphere (CIRA-86). Committee on Space Research (COSPAR), NCAS British Atmospheric Data Centre. 2006. <http://badc.nerc.ac.uk/view/badc.nerc.ac.uk__ATOM__dataent_CIRA>.

COESA (1966). *U.S. Standard Atmosphere Supplements.* Washington, DC: National Oceanic and Atmospheric Administration, NASA, and USAF, U.S. Government Printing Office.

COESA (1976). *U.S. Standard Atmosphere.* Washington, DC: National Oceanic and Atmospheric Administration, NASA, and USAF, U.S. Government Printing Office.

Drob, D. P., et al. (2008). An empirical model of the earth's horizontal wind fields: HWM07. *Journal of Geophysical Research, 113,* A12304. Available from http://dx.doi.org/10.1029/2008JA013668.

Handbook of Geophysics (1985). In J. A. Jursa (Ed.), *Handbook of geophysics and the space environment.* Bedford, MA: Air Force Geophysics Laboratory, USAF.

ISO (1982). *Reference atmospheres for aerospace use* International Standards Organization, ISO 5878:1982.

Justus, C. J., & Leslie, F. W. (2008). The NASA MSFC Earth Global Reference Atmospheric Model—2007 Version. NASA/TM—2008—215581.

Liepmann, H. W., & Roshko, A. (2002). *Elements of gasdynamics.* Dover, NY.

NASA Goddard (2013). *Model web catalog and archive.* Community Coordinated Modeling Center, NASA Goddard Space Flight Center. <http://ccmc.gsfc.nasa.gov/modelweb/models_home.html>.

Picone, J. M., Hedin, A. E., Drob, D. P., & Aikin, A. C. (2002). NRLMSISE-00 empirical model of the atmosphere: Statistical comparisons and scientific issues. *Journal of Geophysical Research, 107*(A12), 1468. Available from http://dx.doi.org/10.1029/2002JA009430.

Schlatter, T. W. (2010). *Atmospheric composition. Encyclopedia of aerospace engineering.* NY: John Wiley and Sons. Available from http://dx.doi.org/10.1002/9780470686652.eae319.

Von Karman, T. (1954). On the foundation of high speed aerodynamics. In W. R. Sears (Ed.), *General theory of high speed aerodynamics* (Vol. VI, pp. 3—30). Princeton, NJ: Princeton University Press, High Speed Aerodynamics and Jet Propulsion.

THE SPACE ENVIRONMENT

The space environment begins at the earth's surface, proceeds out through the atmosphere, and into the planetary regions of outer space. Starting from the surface of the earth the first effects felt by a manned spacecraft are those due to the atmosphere itself. Details of the earth's atmosphere were the subject of Chapter 2 and here we briefly examine the broader environment in which a manned spacecraft must carry out its mission. A useful discussion of the effects of the space environment on spacecraft is presented by James, Norton, and Alexander (1993). A more detailed exposition of the space environment may be found in Anderson and Mitchell (2005).

3.1 GRAVITATIONAL EFFECTS

The Newtonian theory of gravitation accurately describes the general behavior of bodies moving through space and will be discussed in detail in Chapter 5. One result of interest is that the gravitational acceleration toward the center of the earth is given by

$$g = \frac{m_E G}{r^2} = \frac{m_E G}{R_E^2 \left(1 + \dfrac{z}{R_E}\right)^2} = \frac{g_E}{\left(1 + \dfrac{z}{R_E}\right)^2} \tag{3.1}$$

Here the mass of the earth is $m_E = 5.9736 \times 10^{24}$ kg, the universal gravitational constant $G = 6.67259 \times 10^{-11}$ m^3/kg-s^2, and z is the altitude above the mean radius of the earth $R_E = 6371$ km. For altitudes of 400 km, the error in the standard gravitational acceleration at the surface of the earth (9.807 m/s^2), is only about -6%. Thus the gravitational acceleration in typical manned orbits is approximately that at the earth's surface and the astronauts onboard are not in "zero gravity." Instead they are in radial force equilibrium between the acceleration of gravity toward the center of the earth and the centripetal acceleration due to the curvature of the orbit. This weightlessness poses some human health problems in orbit particularly loss of bone and muscle mass, among others (Gunga, 2015). In addition, there are spacecraft issues, particularly the feeding of liquid propellants under weightless conditions because the location of the propellant in the tank is uncertain. Positive displacement expulsion mechanisms which use metallic or elastomeric diaphragms or movable pistons are generally required to assure proper propellant flow as described by Huzel and Huang (1992).

Manned Spacecraft Design Principles. DOI: http://dx.doi.org/10.1016/B978-0-12-804425-4.00003-9

Assuming, for simplicity, a circular orbit and equating the centripetal and gravitational accelerations yields the equilibrium velocity

$$\frac{V^2}{r} = g \tag{3.2}$$

Expanding r and g leads to the velocity, in m/s, as follows:

$$V = \sqrt{\frac{g_E}{\left(1 + \frac{z}{R_E}\right)^2} R_E \left(1 + \frac{z}{R_E}\right)} = \frac{7904}{\sqrt{1 + \frac{z}{R_E}}} \tag{3.3}$$

Thus the equilibrium velocity in a circular orbit is slightly less than that evaluated at the surface of the earth and decreases as the orbit radius grows. A rule of thumb for low earth orbits (LEOs) then is that the orbital velocity $V \sim V_E = 7.9$ km/s, the orbital velocity calculated at the surface of the earth.

In a circular orbit the period of one revolution is

$$\tau = 2\pi \frac{r}{V} = 2\pi \sqrt{\frac{r}{g}} = 2\pi \sqrt{\frac{R_E}{g_E}} \left(1 + \frac{z}{R_E}\right)^{3/2} = 5067 \left(1 + \frac{z}{R_E}\right)^{3/2} \tag{3.4}$$

In a typical 400 km orbit $\tau = 92.52$ min so another rule of thumb for LEO is that the typical orbital period is about 90 min.

3.2 GAS DENSITY AND DRAG EFFECTS

Drag on a launch vehicle in ascent and during the stressful period of entry into the atmosphere is a concept that is now, well into the space age, quite familiar. Later chapters will concentrate on such factors and their effects on design. Drag is usually defined as

$$D = \frac{1}{2} C_D \rho S V^2 \tag{3.5}$$

The drag coefficient C_D and the reference area S upon which the drag force is based are dependent upon the geometry of the vehicle. The velocity of the vehicle can be known with some accuracy by means of onboard or ground-based measurement techniques. Least well-defined is the density ρ at a point in the atmosphere and the lack of accuracy in that property implies an inability to accurately determine the major force acting on the spacecraft. The idea that the density of the atmosphere continually decreases with altitude is easily accepted because we imagine that if we go high enough we ultimately reach the so-called vacuum of space. However, although a vacuum suggests complete emptiness, space has an array of particles moving through it. The difference in space is that those particles are generally quite far apart so that they rarely collide with each other. The particles can and do collide with spacecraft with results that can affect the performance of a spacecraft as well as be harmful to humans and hardware.

Let us assume, to begin with, that the density drops off exponentially with altitude, a well-founded assumption, as shown in the previous chapter. Then the atmospheric density might be written as $\rho = \rho_{sl}\exp(-z/H)$ where H is a constant with the same units as the altitude z. The weight of a column of air acting on an area A at sea level may be estimated by integrating the weight of a small height of that column $\rho g_E A dz$ from $z = 0$ to an arbitrarily large distance $z = h$. The result is $W = \rho_{sl} g_E HA[1 - \exp(-h/H)]$ and the pressure at sea level is therefore $p_{sl} = W/A = \rho_{sl} g_E H[1 - \exp(-h/H)]$. If we let h become unbounded then the sea level pressure becomes $p_{sl} = \rho_{sl} g_E H$ and using standard values we find $H = 8.43$ km. In Chapter 2, we pointed out that a good fit to the atmospheric density data is $H = 7.16$ m, but the small difference is irrelevant in the current global analysis. Thus the weight of the entire atmosphere atop a 1 m^2 area is about equal to the weight of a column of sea level density air of height 8.43 km, about the height of Mount Everest.

The atmosphere is comprised mainly of nitrogen and oxygen gases which are considered to be collections of atoms and molecules of given mass. In any volume, the mass contained is equal to the sum of the masses of the particles within with the volume. Density may be defined as the ratio of the mass of the gas contained in a volume to the magnitude of that volume. Starting with a very large sample volume and working down to smaller volumes we would find that the calculated density varies with volume because of different spatial distributions of particles. However, we may consider an experimental limiting process where we define

$$\rho = \lim_{l \to 0} \frac{m}{l^3}$$

In this process of using smaller and smaller sample volumes the effect of spatial variations disappears and the density will be well-defined over a range of ever smaller sample volumes. In this regime of small sample volumes, we can ignore the individual particles and instead consider the mass of the gas in the volume to be uniformly distributed throughout the volume. This is the concept of a continuum. Of course, this experimental limiting process gives a well-defined density in the limit as l approaches zero but at some point the actual sample volume becomes so small that there may be only one particle in the sample volume at the time of the measurement. In this case, the average distance between two particles must be about equal to the characteristic dimension of the box l and we can call this average distance the mean free path of a particle λ. The ratio of this length λ to the length l of the side of the volume is called the Knudsen number

$$Kn = \frac{\lambda}{l}$$

When $Kn \ll 1$, there are so many particles so close together in the volume l^3 that a density may be defined and a great number of collisions between them will therefore occur in a given time period. The science of continuum gas dynamics is based on situations in which $Kn \ll 1$.

Obviously, if we've allowed the density to decrease with altitude the number of particles in a given volume must also decrease as the volume is reduced. If we ascend high enough there will soon be very few particles within the volume until we have $Kn \sim 1$ and the continuum assumption no longer applies. At this altitude and higher the nature of gas flow is called free molecular flow and is in the realm of kinetic theory of gases rather than the continuum theory of gases. In the earth's atmosphere, the mean free path increases roughly linearly from 10^{-7} m at sea level to about 10^{-1} m or 10 cm at 100 km and then increases less rapidly to about 10^4 m at the typical LEO of 400 km and then to almost 10^7 m as z increases beyond 1000 km. A spacecraft with a typical characteristic transverse dimension of about 5 m will have $Kn = 2 \times 10^{-8}$ at sea level and 2×10^{-2} at $z = 100$ km and 2×10^6 at $z = 400$ km and therefore would pass from continuum flow to rarefied flow to free molecular flow during the ascent to orbit.

As mentioned previously, the velocity and its change with time can be measured and the mass of a spacecraft is known so that in the formal drag equation given earlier the information lacking is $C_D \rho S$. The reference area S is a specified area and can be chosen arbitrarily. Often the projected frontal area of a spacecraft is selected for this role. That means that ρC_D is the unknown and therefore the measured variation of orbital speed may be used to measure this quantity using the equation $D = -m(dV/dt)$. In the case of continuum gas dynamics ($Kn \ll 1$), the drag coefficient depends upon physical details of the flow as represented by similarity parameters like the Mach and Reynolds numbers as well as the body geometry and can be quite accurately specified.

In high altitude atmospheric studies, it had been traditional for some time to use a value of $C_D = 2.2$ even though it was appreciated that under rarefied conditions the drag coefficient was dependent on the details of the collisions of individual particles with the surface of the spacecraft. Gaposchkin and Coster (1988) present a discussion of the various atmospheric and drag coefficient models that have been used in analyzing satellite drag. Moe and Moe (2005) discuss the important role played by atomic oxygen adsorbed on satellite surfaces and present calculations of satellite drag coefficients using parameters of gas–surface interactions measured in orbit. They show drag coefficients for some simple practical shapes which all tend to increase with altitude between 150 and 300 km. The sphere shows values of C_D varying between 2.1 at 150 km and 2.3 at 300 km with the other shapes showing similar trends with somewhat higher values. Experience indicates that orbits lower than around 100 km cannot be sustained because of drag effects while those above about 500 km encounter negligible free molecular drag but the effects of solar radiation pressure, earth-reflected radiation, and earth-emitted infrared radiation then assume greater importance. The atmosphere expands and contracts in response to solar heating so that the density distribution is dependent upon season, time of day, and latitude. Standard atmospheres, discussed previously in Chapter 2, are defined as general guides for design purposes and variations from the standard distribution must be accounted for in advanced design studies and mission planning.

Our equation for the deceleration along the orbital path is

$$\frac{dV}{dt} = -\frac{D}{m} = -\rho_{sl} \exp\left(-\frac{z}{H}\right) \frac{C_D V^2}{2m}$$

Using Eqn (3.3), the traditional value $C_D = 2.2$, and the value $H = 8.34$ km obtained previously, the drag to surface area ratio becomes

$$\frac{D}{S} = \frac{1}{2}\left(1.225 \text{ kg/m}^3\right)\exp\left(-\frac{z}{8.34}\right)2.2\frac{V_E^2}{1 + \frac{z}{R_E}}$$

For a LEO of 200 km altitude, the drag per unit area is found to be $D/S = 0.0031$ N/m^2. The drag of a 1000 kg vehicle with a reference area $S = 3$ m^2 would be 0.0093 N yielding a deceleration of $a = D/m = 0.0093$ N/1000 kg $= 9.3 \times 10^{-6}$ m/s^2 or about $9.5 \times 10^{-7} g_E$. If that's the only force acting on the vehicle then $dV/dt = -9.3 \times 10^{-6}$ and $V = V_E - 9.310^{-6}t$. Assuming 100 days (8.64×10^6 s) on orbit the velocity $V = 7904 - [9.3 \times 10^{-6}(8.64 \times 10^6 \text{ s})] = 7324$ m/s. The equation of motion normal to the flight path is discussed in Chapters 6 and 7 and may be written as

$$\frac{d\gamma}{dt} = \frac{\cos\gamma}{V}\left(\frac{V^2}{r} - g\right) \tag{3.6}$$

The auxiliary relation for the radial velocity is

$$\frac{dr}{dt} = V \sin \gamma \qquad (3.7)$$

The flight path angle γ is the angle between the velocity vector and the local horizon (measured positive in the positive r-direction). In a circular orbit $\gamma = 0$ and Eqn (3.6) reduces to Eqn (3.2) which expresses the equilibrium condition. However, if drag causes V to be reduced at a given radius r, the term in square brackets on the right-hand side of Eqn (3.6) becomes negative, causing γ to become increasingly negative as time proceeds. Equation (3.7) shows that because the flight path angle γ is negative, r is decreasing meaning that the orbit of 200 km cannot be maintained. Thus the orbit will decay in a spiraling path until the spacecraft enters the atmosphere and burns up somewhere along the entry trajectory.

Another phenomenon arising from the extremely low density and pressure at spacecraft altitudes is that of outgassing, whereby molecules of various materials escape from the parent material, much like the process of evaporation. The loss of this constituent can substantially alter the material properties of the item from which it escaped. In addition, the escaping molecules can impact other components of the spacecraft, coating them with a film that compromises their performance. Campbell and Scialdone (1993) present a compilation of outgassing data that is useful for selecting spacecraft materials.

3.3 **THE SUN**

The extreme heat of the corona of the sun causes some of the particles within it to accelerate to speeds in excess of the restraining effect of the gravitational attraction of the sun. This outflow of particles is composed primarily of ionized hydrogen (electrons and protons) along with a small amount of helium nuclei (alpha particles) and a trace amount of heavier ions. The gas is actually plasma because while it is essentially electrically neutral in charge the presence of large numbers of free electrons makes it an excellent conductor of electricity. The particle flow emitted by the sun represents corpuscular radiation called the solar wind. At 1 astronomical unit (AU, the distance of the earth from the sun, approximately 150 million km) the solar wind speed is in the range of 350–700 km/s, the proton density is on the order of $1/cm^3$ to $10/cm^3$, and the proton flux typically varies between $\sim 2 \times 10^8$ and $\sim 4 \times 10^8$ $cm^{-2} s^{-1}$. The sun also emits electromagnetic radiation extending from radio frequency (RF) to x-ray frequency. The mean value of the electromagnetic energy flux incident on a plane placed normal to the sun's rays is called the solar constant and has the value of 1.361 kW/m^2 at a distance of 1 AU.

The sun rotates with a period of around 27 days causing the solar wind to be emitted in a spiraling fashion much like the water droplets from a spinning garden sprinkler. The emission process is not completely uniform with localized violent eruptions of particles occurring on portions of the sun's surface at irregular intervals. These eruptions, which may last for minutes or several hours, are known as solar flares and carry much higher concentrations of particles and therefore are more energetic than the mean solar wind. They can be disruptive of communications on earth as well as posing a significant danger to crew members and hardware in space. English, Benson, Bailey, and Barnes (1973) discuss the mission planning protocols developed for dealing with solar flares during the Apollo missions. Implications of solar flares on spacecraft design are discussed in Chapter 11. Other high energy particles, mainly protons and atomic nuclei, originate outside the solar system, possibly from supernovae, and are called cosmic rays. They do not pose a great threat because of their low flux but collision of these particles with electronic equipment may cause damage. Shielding against this threat is generally counterproductive because of mass restrictions. Instead designers turn to redundancy, radiation-hardened critical components, etc.

Thermal radiation from the sun must also be considered because thermal management of spacecraft environments is complicated by the difficulty of rejecting heat to the space environment. Direct and reflected sunlight on spacecraft surfaces raises the temperature dramatically while shadowed surfaces radiate heat to the vacuum of space. One simple solution is to provide a slow and steady spin to the spacecraft so that surfaces are intermittently in or out of the sunlight. Hughes and Tennyson (1992) develop models for determining surface temperatures on spacecraft based on their orbital motion and attitude. In general, the surfaces may be maintained at an average temperature similar to that in the stratosphere. This does not account for heat sources within the spacecraft due to human presence and electrical equipment which must be treated by other means as discussed in Chapter 10.

3.4 THE MAGNETIC FIELD

The earth's magnetic field surrounds the planet with field lines that emanate from the south magnetic pole out into space then bend back returning to merge into the north magnetic poles forming what is called the magnetosphere. If there were nothing to interfere with these magnetic field lines of the magnetosphere they would form an axisymmetric array like that imagined for a simple bar magnet as shown in Figure 3.1. The bar magnet would be tilted from the earth's rotational axis by about 11°. The plane perpendicular to the dipole is called the magnetic equator and that is tilted by the same amount from the geographical equator.

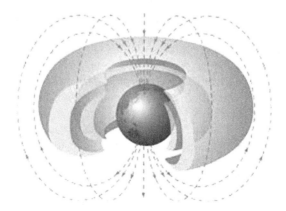

FIGURE 3.1

The shape of earth's magnetic field with no outside disturbances (NASA).

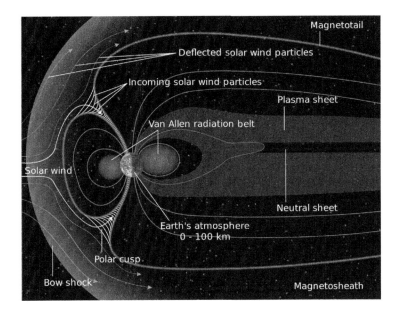

FIGURE 3.2

A schematic diagram showing the various effects occurring in near-earth space (NASA).

The strength of the magnetic field is weakest at the geomagnetic equator and on that plane at the surface of the earth it is about 3×10^4 nT or about 0.3 gauss. Household utility magnets have strength of about 100 G. The field strength drops off slowly with altitude to about half the surface value at an altitude of 1000 km. The magnetic field strength increases with latitude reaching about twice the magnetic equatorial level near the poles. The symmetry of the earth's magnetic field as shown in Figure 3.2 is in evidence close to the earth but further away from the surface the interaction of the magnetosphere with a solar wind of charged particles disturbs its symmetry.

A rather complicated diagram which incorporates many aspects of space physics in near-earth space is shown in Figure 3.2. In that diagram may be seen the deformation of the earth's magnetic field lines due to interaction with the continuing stream of particles ejected by the sun.

The magnetic field is important in shielding the surface of the earth from the solar wind and in certain respects acts like a solid blunt body in a supersonic flow. There is seen a bow shock wave and the compression of the magnetic field lines in the flow upstream of the earth. The magnetic pressure of the upstream field lines balances the solar pressure at about $10R_E$ and may be thought of as a stagnation point. This boundary may vary by $\pm 3R_E$ or more because of variations in the solar wind. Downstream of the earth the field lines are stretched out in a wake extending $200R_E$ or more. The magnetopause may be thought of as a boundary between the earth's magnetic field and the solar wind. The region between the magnetopause and the bow shock is considered the magnetosheath through which the magnetically deflected solar wind flows. The magnetotail is the downstream region containing the so-called plasma sheet, a region of high density, high energy plasma. It may extend out past 300 earth radii. Within the plasma sheet is the neutral sheet wherein the magnetic field line directions away from the South Pole are separated from those toward the North Pole.

3.5 VAN ALLEN RADIATION BELTS

One of the benefits of exploring space with satellites was demonstrated by the discovery of the Van Allen radiation belts, two toroidal regions surrounding the earth in the magnetosphere illustrated in Figure 3.2. These donut-shaped volumes are populated by charged particles, protons and electrons, trapped by the earth's magnetic field. The belts are named after Dr James Van Allen, who inferred their existence from data gathered by Explorer I, a US satellite launched in 1958. A more detailed illustration of the Van Allen belts is shown in Figure 3.3.

The inner Van Allen radiation belt is dominated by high energy protons with energies up to 100 MeV (200 MeV is the average energy released in the nuclear fission of one U-235 atom) and extends from an altitude of about 1000–6000 km and over a latitude range of about 40°N to about 40°S. The outer Van Allen belt which is dominated by high energy electrons (0.1–10 MeV) begins at about 18,000 km (depending on latitude) and extends to 60,000 km. The upper boundary is dependent upon the activity of the sun. When electrons, protons, and perhaps some other charged particles encounter Earth's magnetic field, many of them are trapped by the field. They bounce back and forth between the magnetic north and south poles, following the magnetic field lines.

The radiation in the belts can be harmful, but it is noted that LEOs are typically below the inner radiation belt and there is little health hazard. However, the radiation belts are variable in extent depending on solar activity and there are latitudes where the inner belt can at times come close to the orbit of a manned spacecraft like the International Space Station (ISS). Traveling through the Van Allen belts on the way to outer space is best done rapidly to reduce exposure times to human crew members. Once beyond the magnetosphere the crew is continuously exposed to cosmic rays. Careful consideration of crew protection from radiation hazards was a hallmark of the Apollo program. English et al. (1973) describe the problems of radiation protection of space flight and the solutions they found which kept the Apollo crew members safe.

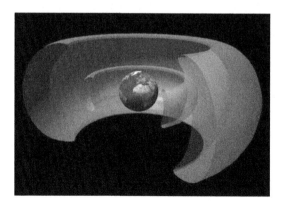

FIGURE 3.3

A schematic diagram showing general shape of the inner and outer Van Allen radiation belts (NASA).

3.6 **THE IONOSPHERE**

In the altitude band that lies between 60 km and about 500 km the atmosphere is characterized by a high density of ionized particles. Solar x-rays and ultraviolet radiation from the sun photoionize the gases in this region which means that the degree of ionization depends upon details like solar activity, latitude, time of day, etc. The ionosphere is divided into three basic layers denoted by the letters D, E, and F. The D layer lies in the region between the upper stratosphere and through the mesosphere ending at about 90 km and has the lowest electron density. The E layer lies between 90 km and 150 km, and the F layer between 150 km and about 500 km. The F layer covers the altitude band populated by LEO satellites. A schematic diagram of the ionosphere, as defined by electron density level, is overlaid with the atmosphere, as defined by temperature, in Figure 3.4.

The critical frequency (in Hz) at or below which a radio signal cannot propagate is given as $f_c = 9(n_e)^{1/2}$ where n_e is the electron number density, electrons per m^3. In the F layer, Figure 3.4 shows a nominal peak of about 10^6 electrons per cm^3 (10^{12} electrons per m^3) so that $f_c = 9$ MHz. This value may vary with solar activity latitude, etc., but there are other atmospheric factors that limit the propagation of signals in the range up to several hundred megahertz. Most radio signals in the 300 MHz (UHF) to 300 GHz (EHF) band will propagate through the ionosphere. These frequencies are used for communication between spacecraft and ground stations but effects of atmospheric absorption and scattering may sometimes compromise performance.

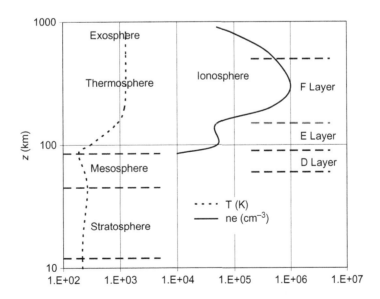

FIGURE 3.4

Schematic diagram of the ionosphere in terms of electron density overlaid with the atmosphere in terms of temperature.

3.7 METEOROIDS AND ORBITAL DEBRIS

Meteoroids are naturally occurring space debris, particles of various sizes which orbit the sun. Orbital debris is the remains of once-functioning man-made spacecraft that orbit the earth. NASA reports that there are half a million pieces of debris larger than a marble and 20,000 of those are larger than a softball. Debris smaller than this are most dangerous because they cannot be tracked. Space Shuttle Orbiter windows have been damaged by particles that were later determined to be paint flecks. As discussed in Chapter 11, the safety procedure is avoidance for larger pieces of space debris. The spacecraft and debris orbits are tracked and if necessary the spacecraft's orbit is changed slightly to avoid the collision.

Below altitudes of about 300 km, the effect of drag mentioned in Section 3.2 tends to cause the orbits of space debris to decay and burn up in the atmosphere. On the other hand high altitude orbits, like geosynchronous orbits at 35,740 km, are characterized by lower orbital velocity, around 40% of that at lower orbits. Therefore closing speeds for impact would also be smaller and because the energy is proportional to the cube of the closing velocity such impacts would be far less serious.

NASA's Long Duration Exposure Facility (LDEF) was designed to provide long-term data on the space environment and its effects on space systems and operations. This roughly cylindrical structure 9.1 m long and 4.3 m wide with a mass of almost 10,000 kg carried 57 experiments. It was launched on the Space Shuttle in 1984 and retrieved almost 6 years later after completing 32,422 orbits. By this time, the orbit of the LDEF had decayed to about 320 km and it was close to entering the Earth's atmosphere. One of the areas of interest was space debris impacts. Stuckey (1993) reported that examination of the surfaces of the LDEF showed over 34,000 impacts of 50 microns or greater the largest being 0.57 cm in diameter. The ratio of the number of impacts on the leading edge surfaces to the trailing edge surfaces of the LDEF were found to be in the range of 10−20 depending upon the size of the craters.

3.8 SPACECRAFT CHARGING

Electric charges may build up on the surface of a spacecraft and create an electrostatic potential with respect to the plasma through which it is traveling. Likewise the surface charge distribution may be nonuniform if the surface is nonuniformly conductive resulting in differential charging. Interaction of the plasma with the surface and solar photoelectric effects are the main causes of spacecraft charging. On the sun side of the spacecraft, the photoelectric effect dislodges electrons from the surface leaving it with a net positive charge. On the shadow side, the plasma interaction dominates and usually liberates protons leaving a net negative charge. The charge can continue to build up until the discharge potential is reached and a discharge, or electric arc, occurs which can damage electrical circuits or cause structural damage.

3.9 USEFUL CONSTANTS, ACRONYMS, AND CONVERSIONS

Mass of the earth $m_E = 5.9736 \times 10^{24}$ kg

Radius of the earth $R_E = 6371$ km

Standard gravitational acceleration at the surface of the earth $g_E = 9.807$ m/s^2

Universal gravitational constant $G = 6.67259 \times 10^{-11}$ m^3/kg-s^2

Earth−moon distance = 384,400 km

1 AU = 150 million km, the mean distance between the earth and the sun

1 gauss = 10^5 nanotesla = 10^5 nT

1 mi = 1609.3 m = 0.8684 nm (nautical miles)

3.10 NOMENCLATURE

C_D drag coefficient

D drag

f_c critical frequency

G universal gravitational constant

g gravitational acceleration

K_n Knudsen number λ/l

H scale height of the atmosphere

l length

m mass

m_E mass of the earth

n_e electron number density

R_E radius of the earth

r radius measured from the center of the earth

S reference area for drag force

t time

V velocity

V_E circular orbit velocity evaluated at the surface of the earth

z altitude measured from the surface of the earth

γ flight path angle measured between the velocity vector and the local horizon

λ mean free path

τ orbital period

3.10.1 SUBSCRIPT

E evaluated at the surface of the earth

REFERENCES

Anderson, B. J., & Mitchell, D. G. (2005). The space environment. In V. L. Pisacane (Ed.), *Fundamentals of space systems* (2nd ed.). New York, NY: Oxford.

Campbell, W. A., & Scialdone, J. J. (1993). Outgassing data for selecting spacecraft materials. NASA RP-1124, Revision 3.

English, R. A., Benson, R. E., Bailey, J. V., & Barnes, C. M. (1973). Apollo experience report—protection against radiation. NASA TN D-7080.

Gaposchkin, E. M., & Coster, A. J. (1988). Analysis of satellite drag. *The Lincoln Laboratory Journal, 1*(2), 203−224.

Gunga, H.-C. (2015). *Human physiology in extreme environments* New York, NY: Elsevier.

Hughes, P. C., & Tennyson, R. C. (1992). Long duration exposure facility surface temperatures. *Journal of Spacecraft and Rockets, 29*(1), 96−101. Available from http://dx.doi.org/10.2514/3.26319.

Huzel, D. K., & Huang, D. H. (1992). *Modern engineering for the design of liquid propellant engines* Reston, VA: American Institute of Aeronautics and Astronautics.

James, B. F., Norton, O. W., & Alexander, M. B. (1993). The natural space environment: Effects on spacecraft. NASA Reference Publication 1350.

Moe, & Moe, M. (2005). Gas−surface interactions and satellite drag coefficients. *Planetary and Space Science, 53*, 793−801.

Stuckey, W. K. (1993). Lessons learned from the long duration exposure facility. Aerospace Corporation Report No. TR 93(3935)-7.

MANNED HYPERSONIC MISSIONS IN THE ATMOSPHERE

4.1 TRANSATMOSPHERIC MANNED MISSIONS

In Chapter 1, we noted that the edge of space is defined as the altitude $z_e = 100$ km. The general aerodynamic and thermodynamic characteristics of manned hypersonic missions carried out in the atmosphere below this altitude are shown in Figure 4.1.

- The rising line labeled ascent to orbit includes two missions characterized by a duration of about 3 min $< t <$ 10 min, a Mach number range of $0 < M < 8$, a stagnation temperature range of 500 K $< T_t <$ 2000 K, and little or no atomic oxygen concentration. The missions depicted include actual ascent to orbit such as the Space Shuttle ($M = 5$) as well as suborbital operations like those of the X-15 ($M = 6$) and current commercial space vehicle designs.

- The first oval region includes supersonic cruising flight in the Mach number range of $1.5 < M < 4$. This is typical of military aircraft like the SR-71 ($M = 3.3$), the X-B70A ($M = 2.8$) and commercial aircraft like the Concorde SST ($M = 2.04$) and the Soviet Tu-144 SST ($M = 2.35$).

- The second oval region covers the generic mission of extended hypersonic cruise in the atmosphere. This mission is characterized by a duration of $10-100$ min, a Mach number range of $4 < M < 12$, a stagnation temperature range of 500 K $< T_t <$ 4000 K, and little to moderate oxygen atom concentration. This region was expected to be the domain of the hypersonic transport or "Orient Express" type of aircraft, onc which has yet to be realized. Rarefied gas dynamic effects are important at altitudes above that at which the rarefaction parameter $V' = \sqrt{C^* \, \mathrm{Re}} = 0.007$. Here $Re = Vl/\nu$ is the Reynolds number based on the characteristic vehicle dimension l and $C^* = \rho^* \mu^* / \rho_e \mu_e$ is the Chapman–Rubesin factor. All these parameters are discussed in detail in Section 8.4. Because typical vehicle dimensions are likely to be greater than 3 m, the effects of low density will be minimal in this flight regime.

- The third oval region describes the generic mission of lifting flight for either entry into orbit or return from orbit. The LEO return mission is characterized by a duration of around 10 min, Mach numbers in the range of $10 < M < 25$, stagnation temperatures in the range 1500 K $< T_t <$ 5500 K, and substantial oxygen dissociation and moderate nitrogen dissociation. The lunar return mission is characterized by a duration of about 15 min, Mach numbers in the range of $10 < M < 36$, stagnation temperatures in the range 1500 K $< T_t <$ 9500 K, and substantial oxygen and nitrogen dissociation as well as moderate ionization. As mentioned previously, rarefaction effects are important above

Manned Spacecraft Design Principles. DOI: http://dx.doi.org/10.1016/B978-0-12-804425-4.00004-0

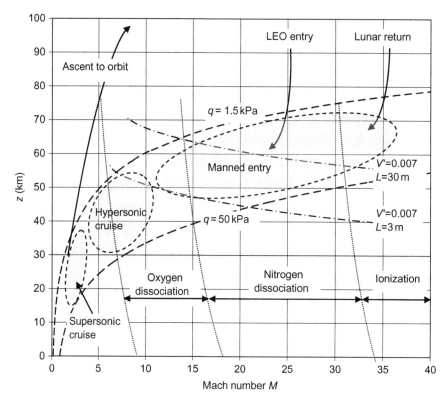

FIGURE 4.1

Flight envelopes for manned hypersonic missions. The region between the dashed lines $q=1.5$ kPa and $q = 50$ kPa define the safe manned flight corridor and those between the dash−dot curves of $V'=(C/Re)^{1/2} = 0.007$ define the maximum altitude for fully continuum flow for characteristic body length L. The regions of oxygen and nitrogen dissociation and ionization are delineated by the nearly vertical dotted lines.

the lines of constant $V' = 0.007$. Because vehicles of interest are likely to be on the order of 3 m (Apollo capsule) or 30 m (Space Shuttle Orbiter), or more in length, the effects of low density flight will occur only above about 50 km in altitude.

4.2 TRANSATMOSPHERIC VEHICLES

Although expendable launch vehicle have been transporting humans to space with some degree of frequency and safety for over 50 years, the lure of the reusable launch vehicle (RLV) has not waned. The Space Transportation System (STS), known familiarly as the Space Shuttle, was intended to be a reusable launch system which would be flown regularly in much the same way as commercial aircraft.

FIGURE 4.2

Artist's conception of the X-20 manned glider during atmospheric entry. The overall length was 10.8 m, the span 6.32 m, and the height of the fins 2.6 m (USAF).

FIGURE 4.3

Artist's concept of the X-30 NASP flying through Earth's atmosphere on its way to low-Earth orbit (NASA).

Budgetary problems in the early 1970s forced the design into the mold of the X-20 Dynasoar, illustrated in Figure 4.2, a 1957−1963 program aimed at building a manned glider that could be launched into space on a rocket and then return to earth like an airplane. The X-20 was canceled before a prototype was built and the plan of a reusable STS was instead transformed into the reusable Space Shuttle Orbiter. Thirty years later, the X-30 National Aerospace Plane (NASP) shown in Figure 4.3 was envisioned to be an airbreathing scramjet vehicle that would fly from an airport to orbit and back. It too is only an artist's rendering because, like the X-20, it never was built. Since the inception of the STS there have been more than a few attempts to design and build a reusable space launch system and these are reviewed in Table 4.1.

The McDonnell−Douglas DC-X Clipper was a test bed for a vertical lift-off and landing, single stage to orbit vehicle. An 18,900 kg version 12 m high and 4.1 m wide was built to demonstrate the feasibility of the vertical lift-off and tail-down landing concept that was the staple of science fiction

Table 4.1 U.S. Reusable Space Launcher Projects

Vehicle	Agency	Period	Design LEO Payload (kg)	Stages	Status
X-30 National Aerospace Plane	DARPA-USAF-NASA	1986–1993	4500	1	Never flew
DC-X Delta Clipper	SDIO-NASA	1991–1996	9000	1	Low-speed demo flights
X-33 Venture Star	NASA-Lockheed Martin	1996–2001	29,000	1	Never flew
Rascal	DARPA	2000–2005	136	2	Never flew
Alasa	DARPA-Boeing	2012–2016	45	2	Flight planned for 2016

FIGURE 4.4

The DC-X, or Delta Clipper, shown during powered tests to study the feasibility of tail-down landing of a vertical lift-off spacecraft (NASA).

spaceships. A total of 12 powered tests of taking off vertically and landing vertically were carried out and a photograph of the craft during testing is shown in Figure 4.4. As noted in Table 4.1 the program, part of the research done under the Space Defense Initiative Office (SDIO), popularly known as Star Wars, was canceled as the overall program faced budgetary pressures and was realigned.

The idea of an RLV was still a much sought after goal and the next attempt at pursuing that goal took some hard-learned lessons from the demise of the X-30 NASP project. In this case, a

FIGURE 4.5

An artist's conception of the X-33 Venture Star, a half-scale suborbital rocket-powered spaceplane intended to test technologies for a reusable spaceplane.

half-scale rocket-powered suborbital vehicle, the X-33 Venture Star, was aimed at being a test bed and demonstrator of those technologies deemed required for a successful full-scale RLV. The X-33 design was based on three advanced concepts: the lifting body (Kempel, Painter, & Thompson, 1994), the linear aerospike rocket engine (Sforza, 2012), and a metallic thermal protection system (Dorsey, Chen, Wurster, & Poteet, 2004). The vehicle was planned to be launched vertically and controlled autonomously to an altitude of over 90 km and a Mach number of about 13, and returned to base to land horizontally like an airplane. The intent was to have a subscale, therefore more affordable, test vehicle that although being limited to suborbital flight could nevertheless experience the more stressful aspects of orbital entry. As seen in Table 4.1, this concept was never brought to the stage of cutting metal (Figure 4.5).

There was a successful reusable unmanned spaceplane, the Boeing X-37B, which was launched to LEO four times aboard an Atlas V launch vehicle, the first being in April, 2010. This craft has the ability to stay on-orbit for long periods, with one flight lasting about 15 months. Autonomous de-orbit, entry, and landing on a runway have all been achieved successfully. A discussion and photograph of this craft appears in Section 12.6.3. As is obvious from the payload column in Table 4.1, only the first three projects were to carry humans into space. The technical difficulties of a reusable launch system and associated expenditures sapped the enthusiasm and confidence of the funding agencies and the manned projects were canceled. The last project in Table 4.1, Alasa (Airborne Launch Assist Space Access), is essentially a refinement of the 1985 antisatellite test in which an ASM-135 missile launched at an altitude of about 13 km from an F-15 climbing at $M = 0.92$ successfully intercepted and destroyed a satellite in orbit at 550 km altitude. The interest in a fully reusable space launch system remains and, as will be seen in later chapters, the arguments for such a system are rooted in the idea of achieving inexpensive access to space. However, in addition to ever-present technical obstacles, the stumbling block for serious development of such a system is the lack of sufficient demand to make the economics of reusability realizable.

4.3 **FLIGHT TRAJECTORIES IN THE ATMOSPHERE**

Trajectories for human flight within the sensible atmosphere are shown in Figure 4.6 for essentially the same flight envelope as discussed in Figure 4.1. Note that the flight corridor for a single stage to orbit vehicle like the NASP extends through a region of high-speed atmospheric flight too stressful to be considered useful for military cruise missions (NRC, 1998). The intent of Figure 4.6 is to emphasize that vertical rocket ascent takes minutes instead of the tens of minutes for an airbreathing fly-to-orbit vehicle. The Space Shuttles (135 flights with the last in 2011), and other rocket ascents, typically climb out of the sensible atmosphere at relatively low Mach number, then rotate to level flight attitude and accelerate in the very low drag near-space environment. Aerodynamic loads on vertical ascent vehicles are relatively light, as discussed in Chapter 7.

Of course, there are many possible paths within the acceptable NASP manned flight corridor since such a spaceplane hasn't yet been flown. The 119 X-15 flights in the early 1960s were measured in minutes and only the SR-71 Blackbird was a true cruise vehicle flying at supersonic speeds for times on the order of a couple of hours or so, and also capable of routine (subsonic) in-flight refueling to extend range. The XB-70 Valkyrie was an experimental supersonic bomber while the Soviet Burya was an unmanned ramjet-powered intercontinental cruise missile. The two successful flights of the air-launched and rocket-boosted X-43A at Mach numbers of nominally 6.8 and 9.7 were encouraging for space plane development. The scramjet engine in the X-43A used hydrogen fuel and can be seen during a wind tunnel test in Figure 4.7. Subsequent flights of the air-launched and rocket-boosted X-51, which used kerosene-based jet fuel in its scramjet engine,

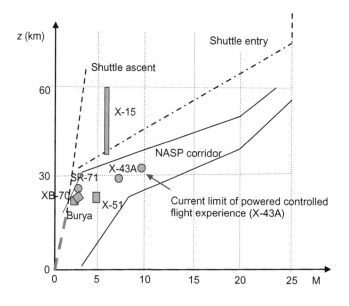

FIGURE 4.6

Trajectories for manned flight in the lower atmosphere. Variations on this chart appeared many times in the 1980s and 1990s as part of discussions of the NASP concept (Schweikart, 1998).

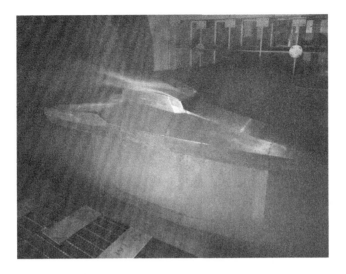

FIGURE 4.7

Full-scale model of the X-43A mounted upside down on a strut for testing in NASA Langley's 8 ft (2.4 m), high-temperature wind tunnel. The scramjet inlet is clearly visible at the midsection of the model.

FIGURE 4.8

The X-51A Wave Rider hypersonic scramjet vehicle and aft-mounted rocket booster carried on a wing pylon of a B-52 aircraft in preparation for an airborne launch (USAF).

reached $M = 5$ taking another step forward for hypersonic airbreathing flight. This vehicle is shown in Figure 4.8 mounted under the wing of a B-52 in preparation for an airborne launch.

A closer look at the flight corridor is provided in Figure 4.9. The dashed lines describe constant dynamic pressure, $q = 0.5\gamma pM^2$ (in kPa or kN/m^2) that are descriptive of the corridor acceptable for manned vehicles. In equilibrium cruising flight, constant q lines are equivalent to lines of constant $(W/S)/C_L$. An important design consideration for a transatmospheric vehicle is the ability to generate

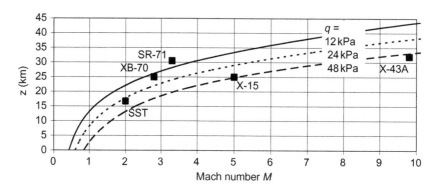

FIGURE 4.9

The low-speed end of the envelope showing the effect of wing loading on altitude performance.

reasonable values of C_L in trimmed hypersonic flight while keeping the wing loading (W/S) at a reasonable value. The wing loading affects the landing speed achievable by the vehicle. For example, in order to maintain landing speeds of around 185kts (95 m/s), the nominal landing speed of the X-15, we must have $[(W/S)/C_L]_{land} \sim 5.51$ kN/m² (115 lb/ft²). The Space Shuttle Orbiter, whose nominal landing speed is around 170kts (87 m/s), is not shown in Figure 4.9 because it does not cruise, it instead decelerates continuously along a trajectory generally within the indicated corridor.

4.4 REUSABLE SPACEPLANE DESIGN ISSUES

The take-off and landing requirements set stringent demands on the vehicle characteristics such that the desired cruise conditions may be met. Sforza (2014) illustrates this interplay and its effect on limiting the design space using an engine selection chart like that shown in Figure 4.10.

4.4.1 AERODYNAMIC DESIGN ISSUES

The landing lift coefficient determines the allowable landing wing loading (W/S)$_l$ for a given landing speed and deceleration (and therefore field length). The landing capabilities of high-speed spaceplanes are discussed in some detail in Section 6.8. Although the wing planform area S is fixed, the landing weight is considerably less than the take-off weight because of the large fuel fraction needed for hypersonic vehicles and therefore the wing loading in take-off is substantially larger than that in landing. For this reason, take-off is the major problem for airplane-like operation of reusable spaceplanes. Because high-speed wings are generally not well-suited for low-speed flight, research on high lift devices is necessary to augment performance. One solution is to use low wing loadings but this leads to large wing surfaces and, because skin friction actually drives the high-speed cruise performance of slender vehicles, this approach tends to be counter-productive.

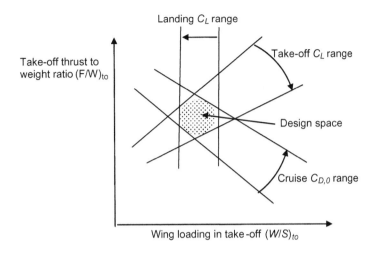

FIGURE 4.10

Engine selection chart defining design space for the vehicle. Arrows denote direction of increasing magnitude of the coefficients shown.

The take-off constraint is one of field length, and this may be satisfied by using a high take-off thrust to weight ratio $(F/W)_{to}$, a solution which often results in higher fuel usage and/or over-design of the propulsion system. Again, reduced wing loading helps, but poses the skin friction problem mentioned previously. High lift devices, the conventional solution, can help, but other techniques, like rocket assisted take-off, or some other expendable component to aid acceleration to appropriate lift-off speed may afford improvements at acceptable cost.

The cruise constraint arises from the need to achieve compatibility between engine thrust capability for essentially sea level take-off as well as for cruise at high altitude. The important parameter here is the vehicle's zero-lift or profile drag coefficient $C_{D,0}$ at the cruise condition. Powered slender vehicles don't generally have base drag issues; instead it is skin friction that dominates, as discussed in Chapter 8. Although there has been substantial effort in understanding transition and attempting to delay it, there hasn't been similar effort applied to arriving at new ideas, mechanisms, or materials helpful in reducing skin friction at high speed. Such capability would relax the wing loading constraint which, in turn, would alleviate the landing and take-off problem.

4.4.2 OPERATIONAL DESIGN ISSUES

Take-off and landing requires controllable low-speed operation and additional systems, like retractable landing gear capable of handling heavy take-off weights that then are over-designed for much lighter landing weights due to high fuel usage. Take-off at low fuel weight with subsequent in-flight refueling trades the landing gear problem for operational problem of tanker availability. How much time and fuel is sacrificed by getting the tanker and the hypersonic vehicle in place at the refueling altitude and loading all the mission fuel? The fuel fraction required for long range cruise is large and the compromises required by in-flight refueling will likely have a strong impact on mission time. Using available information it seems likely that hypersonic cruise vehicles will have fuel fractions around 60−70% of gross take-off weight (GTOW) because $M = 3$ aircraft like the XB-70 and the SR-71 are already in the 50−60% range. Similarly, payload fraction will likely be approaching that of rockets, i.e., on the order of about 10% GTOW, which is also the approximate payload fraction for the XB-70 and SR-71 vehicles.

Thermal management becomes an issue because for a stipulated range a higher flight Mach number has higher heating rates but for shorter time durations than lower Mach number flight. In addition to proper selection of materials, the high skin temperatures will affect ground handling and turnaround times. The turn radius for hypersonic vehicles is large (200−2000 km) and grows like the square of the Mach number. Various studies on global reach bombers suggest that a hypersonic bomber may have to just continue on around the globe to return to a base in the continental United States. This issue, and others, was assessed in NRC (1998). Acceleration to supersonic speeds after take-off must be done over water because of sonic boom issues and flying at subsonic speed just to get over water can use up valuable fuel and mission time. One may locate a base near a coast to avoid this difficulty but this is likely to actually require two bases, one on each coast to accommodate the round-the-world flights that might be necessary if a turnaround is not possible.

These design issues militate against the development of a spaceplane that takes off and lands with the ease and frequency of an airplane. Instead, spaceplane design has evolved into the concept of a spacecraft that is boosted by a rocket but lands on a runway like an airplane, available for refurbishment and reuse. In addition, efforts are aimed at inventing means to recover all or part of the initial rocket stage in a manner that is both economical and amenable to refurbishment and reuse, like the spaceplane. In this sense, the goal of developing an integrated vehicle that is fully reusable has been transformed into one of developing a system of reusable components that can be reintegrated for reuse. The simplest visualization of this approach is the idea of a fly-back rocket booster. The demands of flying a large flexible empty tank back to base like an airplane after launching a spaceplane has its own set of design problems so a simpler solution is sought. The idea of recovering the Space Shuttle's solid rocket booster motors (SRBM) from a parachute-assisted ocean landing was a step in that direction, but proved to be time-consuming and costly. Svitak (2015) describes some current efforts: recovering an entire core stage (SpaceX), returning only the liquid-fueled rocket motor and associated systems as part of a simple winged recovery system (Airbus), jettisoning the engines and flying them by parafoil for an in-air recovery by helicopter (United Launch Alliance).

4.4.3 PROPULSION DESIGN ISSUES

A relatively simple assessment of the propulsion issues crucial to the development of a reusable spaceplane that can take off like an airplane, fly to suborbital or orbital altitudes, and then return and land like an airplane may be made by considering a long range hypersonic ($M \gg 1$) cruise mission. The cruise range (in km) of an aircraft may be estimated from the following form of the Breguet equation (Sforza, 2012):

$$R \cong -0.3 I_{sp}\left(M\frac{L}{D}\right)\ell n\left(1 - \frac{W_f}{W_1}\right) \tag{4.1}$$

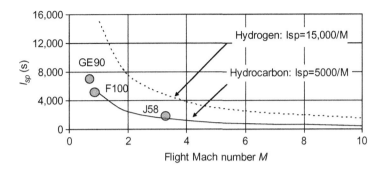

FIGURE 4.11

Typical behavior of the specific impulse for airbreathing jet engines operating with two different kinds of fuel.

The specific impulse I_{sp} (see Section 7.2.1) represents the contribution of the propulsion system to the range performance of the vehicle. For airbreathing jet engines, the specific impulse may be represented as follows:

$$I_{sp,j} = \frac{k_j}{M} \qquad (4.2)$$

Here the coefficient k_j depends upon the type of fuel used, as illustrated in Figure 4.11. As discussed in Chapter 7, the specific impulse for rockets is not a function of flight Mach number although it is a weak function of flight altitude.

The lift to drag ratio L/D represents the contribution of the aerodynamic configuration to the range performance of the vehicle. The aerodynamic characteristics of hypersonic flight vehicles are discussed in some detail in Chapter 8. The commonly used formulation for L/D suggested by Thorne and published by Kuchemann and Weber (1968) is given by

$$\frac{L}{D} = k_a \frac{M+3}{M} \qquad (4.3)$$

This relation, with $k_a = 4$, has been used widely but without any significant data to support it. An examination of available data led to the results shown in Figure 4.12 which suggest that k_a should be closer to 2.5 at high Mach numbers.

As is apparent from Figure 4.12, available flight test data, though sparse, do not support the use of $k_a = 4$ in the high supersonic and hypersonic regimes, but the general trend that shows L/D dropping off rapidly with Mach number until the hypersonic regime where it levels out at a relatively low value of about 2. The data and sources are presented in Table 4.2. As noted in Table 4.2, the SR-71 data is not a flight measurement but is inferred from the Breguet equation using the following data: $R = 3685$ mi, $V = 2181$ mph ($M = 3.2$), $W = 140,000$ lb at start of cruise with 80,000 lb of fuel burned during cruise and a thrust specific fuel consumption (TSFC) of 2.19 per hour (data for J-58 with afterburner). Another inference may be obtained using the data from the record speed run from New York to London which covered 3461 mi at an average speed of 1807 mph, with one slow-down for refueling over the Atlantic Ocean. This would give values of $M = 2.65$ and $L/D = 5$. The X-15 is characterized by a fairly constant value of $L/D = 2.4$ for trimmed flight over the range of Mach number $2 < M < 6$ according to Saltzman and Garringer (1966). This apparently occurs because of the offsetting effects of decreasing base drag and increasing induced drag as the Mach number increases from the transonic regime.

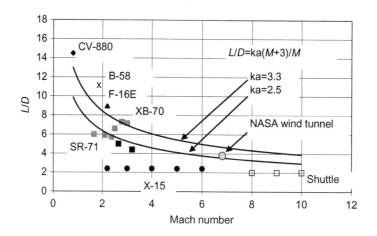

FIGURE 4.12

Flight test data for various aircraft as a function of Mach number and Thorne's empirical relation with values of k_a different from the usually quoted value of 4. The NASA wind tunnel data point is taken from Fetterman, Henderson, Bertram, and Johnston (1965).

Table 4.2 Lift to Drag Ratios of Various Aerospace Vehicles

Aircraft	Mach Number	L/D	Source
CV-880	0.8	14.5	Heffley and Jewell (1972)
B-58	2	11.3	Loftkin (1985)
F-16E	2.2	9	Whitford (1987)
XB-70	2.8	7.5	Heffley and Jewell (1972)
	$1.65 < M < 3$	$5.9 < L/D < 7.5$	Arnaiz (1977)
SR-71	3.2	4.4	Inferred from Breguet equation
	2.65	5	Inferred from record speed run
X-15	$2 < M < 6$	2.4	Saltzman and Garringer (1966)
Space Shuttle	$M > 4$	2	Kirsten, Richardson, and Wilson (1983)
NASA wind tunnel	6.8	4	Fetterman et al. (1965)

The single point labeled NASA Wind Tunnel is representative of data reported in Fetterman et al. (1965). They present wind tunnel data for a variety of slender bodies tested at $M=6.8$ and a Reynolds number based on length of about 1.5 million. The bodies were characterized by a nondimensional volume parameter $v^{2/3}/S$, where v is the volume of the body and S is the planform area. The point chosen is for a body for which the volume parameter is $v^{2/3}/S = 0.2$ which is approximately the value for the X-15. It appears that for bodies with reasonable volume a hypersonic $L/D = 4$ may be credible. Discussion of the volumetric parameter and its influence on L/D is presented in Chapter 8.

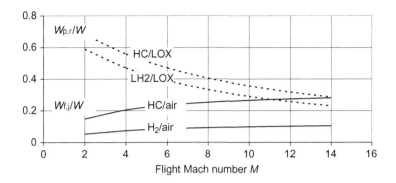

FIGURE 4.13

Comparison of cruise fuel fractions for rocket and jet engines powered vehicles for a range of 5000 km and equivalent *L/D*. Propellant combinations are indicated in the figure.

Assuming that the range, Mach number, and *L/D* are equal for a rocket-powered and a jet-powered aircraft, we may use Eqn (4.1) to determine the propellant fraction required by the rocket-powered vehicle in terms of that required by the jet-powered vehicle with the following result:

$$\frac{W_{p,r}}{W} = 1 - \left(1 - \frac{W_{f,j}}{W}\right)^{k_j/Mk_r} \tag{4.4}$$

Considering a relatively long range case where $R = 5000$ km and using $k_a = 2.5$ in Eqn (4.3) leads to the results shown in Figure 4.13 where for hydrogen fuel $I_r = 450$ s and $I_j = (15{,}000 \text{ s})/M$ and for hydrocarbon fuel $I_r = 350$ s and $I_j = (5000 \text{ s})/M$. The comparison shows that as Mach number increases the differences in required propellant fraction between the two types of propulsion systems decrease. Reduced propellant fraction translates into increased payload. Recall that the propellant for the airbreathing jets is fuel alone while that for rockets includes both fuel and oxidizer. However, it must also be noted that flight operation of airbreathing jets has not been demonstrated beyond $M = 6$ for hydrocarbon (HC) fueled jets or $M = 10$ for hydrogen fueled jets. The differences for $M < 12$ clearly illustrate the attractiveness of using airbreathing ramjets and scramjets for high-speed cruise. This improvement in fuel fraction carries over to boost acceleration and explains the reasons for the interest in and continued study of hypersonic airbreathing engines.

Though propulsion systems like the air-turbo-ramjet (ATR) and the variable-cycle turbo-ram-rocket (VCTRR), among others, combine rockets and ramjets, they rely heavily on turbomachinery. Turbomachines are complex and expensive and therefore do not lend themselves to expendable systems. In addition, since the turbomachine must provide substantial thrust at low speeds as well as at high speeds, it is necessarily large in cross-section, a characteristic that is a burden in supersonic and hypersonic flight. An additional difficulty with airbreathing engines like ramjets and scramjets is that although there are few moving parts, these generally do not include the inlet, which isn't necessarily a simple component and varying inlet geometry is often economically counter-productive. For these reasons, there is no current engine that can operate effectively over a broad Mach number spectrum and this is the major requirement for a spaceplane that can operate like a conventional airplane.

4.5 TRANSATMOSPHERIC FLIGHT MISSIONS IN THE NEAR FUTURE

The hypersonic rocket or airbreathing manned space access mission may serve military, commercial, and civilian government (e.g., NASA) customers. The space access mission necessarily requires acceleration through the atmosphere to approximately $M = 25$ to establish a low earth orbit (LEO) but high altitude or suborbital missions can be accomplished at low hypersonic Mach numbers, e.g., the X-15 at $M = 5$ or 6. Similarly, high Mach number cruise can provide long range high-speed transportation.

In the late 1950s, it became apparent that the ballistic missile community had seized the lead from manned bombers in global reach arguments. Ferri (1961), a pioneer of hypersonic flight and supersonic combustion ramjet (scramjet) propulsion, suggested that "The ideal vehicle for space investigations is... a vehicle that is able to take off as an airplane... and be used again for successive missions." This is the usual argument for a reusable transatmospheric vehicle. Though the US military has an interest in hypersonic flight, it doesn't appear to emphasize human crews. Indeed, the USAF continues to use the rocket-boosted unmanned X-37 reusable spaceplane for long duration space missions related to intelligence gathering which is one manifestation of their interest in space access. There is also interest in rapid response strike capability, but this too appears to be primarily an autonomous unmanned vehicle task involving hypersonic cruise within the atmosphere.

Although the launch market forecast for satellites during the period 2013−2022 appears reasonably strong according to Caceres (2015), launches of manned spacecraft are likely to be no more than 5 per year, as shown in Section 12.5.1. A study by Futron (2002) projected that, by 2021, suborbital space tourism could bring in over $700 million a year in revenue by flying 15,000 passengers on suborbital flights and $300 million a year in revenue by flying 60 passengers into LEO. The order of magnitude of ticket prices at the time for suborbital flights was set at $100,000 and would decline to about half that by 2021. For orbital flights the ticket price would be $20 million at the start, dropping to about $5 million by 2021. In a follow-up study, Futron (2006) modified the projections for suborbital flights downward somewhat while increasing the initial ticket price to the range of $200,000 but maintaining the drop to the originally forecast $50,000. Much of the basis for the perceived strength of the space tourism demand is the growth of the "high-income segment" of the population and the desire for exciting theme-related vacation adventures. Tauri (2012) presented a 10-year forecast of demand for reusable suborbital vehicles considering that in addition to space tourism such vehicles would serve the following additional markets: basic and applied research, education, small satellite deployment, media, and public relations. In recognition of the likelihood of growing numbers of space tourists, the FAA (2014) generated recommendations for proper practices to ensure human spaceflight occupant safety.

Two astronaut extravehicular activity (EVA) missions to the Hubble Space Telescope validated the feasibility of humans carrying out repairs in space (Waltz, 1993, 1998). Shooman and Sforza (2002) proposed that the International Space Station be used as a base for the repair, refurbishment, and refueling of satellites in LEO and showed that the cost of operating a satellite for one lifetime plus replenishment of propellant and replacement of electromechanical attitude control components, batteries, and solar panels for a second lifetime can be considerably less than launch of a satellite and its replacement. This on-orbit approach to extending satellite lifetime is currently gaining renewed interest as a spin-off of NASA's Commercial Resupply Services program for the ISS (Morring, 2015). It is suggested that this effort at commercializing activities in space will expand the demand for transatmospheric flight and lead to more affordable space access.

4.6 NOMENCLATURE

$C_{D,0}$	zero-lift drag coefficient
C_L	lift coefficient
D	drag
EVA	extravehicular activity or spacewalk
F	thrust
GTOW	gross take-off weight
HC	hydrocarbon
I_{sp}	specific impulse
ISS	International Space Station
kts	knots = 0.514 m/s
k_a	constant equation (4.3)
k_j	constant equation (4.2)
L	lift
LEO	low earth orbit
LH_2	liquid hydrogen
LOX	liquid oxygen
M	Mach number
NASP	National Aerospace Plane
p	pressure
q	dynamic pressure $\gamma p M^2/2$
R	range
Re	Reynolds number
RLV	reusable launch vehicle
S	reference surface area for aerodynamic forces
SSTO	single stage to orbit
STS	Space Transportation System, popularly known as the Space Shuttle
T	temperature
T_t	total, or stagnation, temperature
TSTO	two stages to orbit
t	time
V	velocity
V'	rarefaction parameter
W	weight
z	altitude
γ	ratio of specific heats
ρ	density
υ	volume
μ	viscosity

4.6.1 SUBSCRIPTS

e	edge of space
f	fuel
j	airbreathing jet engines
p	propellant
1	start of cruise
r	rocket engine
to	take-off

REFERENCES

Arnaiz, H. H. (1977). Flight-measured lift and drag characteristics of a large, flexible, high supersonic cruise airplane. NASA TM X-3532.

Caceres, M. (June, 2015). Expanding customer base for space payloads. *Aerospace America*, 22−24.

Dorsey, J. T., Chen, R. R., Wurster, K. E., & Poteet, C. C. (2004). Metallic thermal protection system requirements, environments, and integrated concepts. *Journal of Spacecraft and Rockets, 41*(2), 162−172.

FAA (2014). *Recommended practices for human space flight occupant safety, version 1.* Washington, DC: Federal Aviation Administration, Office of Commercial Space Transportation.

Ferri, A. (1961). Possible directions of future research in air breathing engines. In A. L. Jaumotte, A. H. Lefebvre, & A. M. Rothrock (Eds.), *Combustion and propulsion, fourth AGARD colloquium, Milan, Italy, 1960* (pp. 3−67). New York, NY: Pergamon Press.

Fetterman, D. E., Henderson, A., Bertram, M. H., & Johnston, P. J. (1965). Studies relating to the attainment of high lift-drag ratios at hypersonic speeds. NASA TN D-2956.

Futron (2002). *Space tourism market study.* Bethesda, MD: Futron Corporation.

Futron (2006). *Suborbital space tourism demand revisited.* Bethesda, MD: Futron Corporation.

Heffley, & Jewell (1972). NASA CR-2144, December (also in Appendix F of McCormick, *Aerodynamics, Aeronautics, and Flight Mechanics*).

Kempel, R. W., Painter, W. D., & Thompson, M. O. (1994). Developing and flight testing the HL-10 lifting body: A precursor to the Space Shuttle. NASA Reference Publication 1332.

Kirsten, P. W., Richardson, D. F., & Wilson, C. M. (1983). Predicted and flight test results of the performance, stability and control of the space shuttle from reentry to landing. In *Shuttle performance: Lessons learned, NASA conference publication 2283, Part I.*

Kuchemann, D., & Weber, J. (1968). An analysis of some performance aspects of various types of aircraft designed to fly over different ranges at different speeds. *Progress in Aerospace Sciences, 9*, 329−456.

Loftkin, K. (1985). Quest for performance—the evolution of modern aircraft. NASA SP-468.

Morring, F. (March 12, 2015). Jupiter space tug could deliver cargo to the moon. *Aviation Week & Space Technology.*

NRC (1998). *Review and evaluation of the air force hypersonic technology program.* Washington, DC: National Research Council, National Academy Press.

Saltzman, E. J. & Garringer, D. J. (1966). Summary of full-scale lift and drag characteristics of the X-15 airplane. NASA TN D-3343.

Schweikart, L. (1998). *The hypersonic revolution, case studies in the history of hypersonic technology, Volume III, the quest for the orbital jet, The National Aerospace Plane Program (1983−1995),* Air Force History and Museums Program, Bolling AFB.

Sforza, P. M. (2012). *Theory of aerospace propulsion.* Waltham, MA: Elsevier.

Sforza, P. M. (2014). *Commercial aircraft design principles.* Waltham, MA: Elsevier.

Shooman, M. L., & Sforza, P. M. (2002). A reliability driven mission for space station. In *2002 annual reliability and maintainability symposium*, Philadelphia, PA.

Svitak, A. (2015). In return. *Aviation Week & Space Technology*, June 22−July 5.

Tauri (2012). *Suborbital reusable vehicles: A 10 year forecast of market demand* Alexandria, VA: The Tauri Group.

Waltz, D. M. (1993). *On-orbit servicing of space systems* Malabar, FL: Krieger Publishing Co.

Waltz, D. M. (1998). *Supplement to on-orbit servicing of space systems* Malabar, FL: Krieger Publishing Co.

Whitford, R. (1987). *Design for air combat* UK: Jane's Information Group, Ltd., p. 59.

ORBITAL MECHANICS

5.1 SPACE MISSION GEOMETRY

Orbital mechanics has its foundation in three important developments which took place over the last half of the seventeenth century:

- Tyco Brahe's detailed observations of the motion of the planets.
- Kepler's laws describing the nature of the orbits of the planets in a manner consistent with Brahe's observational data. The first law is that the orbits are ellipses with the sun at the focus, the second is that the radius vector between the sun and any planet sweeps out equal areas in equal times, and the third, that the square of the orbital period of any planet is proportional to the cube of its mean distance from the sun.
- Newton's law of gravitation which laid the theoretical foundation for the prior observations and is discussed in the following section.

This chapter aims to develop those features of orbital mechanics which are useful in the design and operation of manned spacecraft.

5.1.1 ORBITS AND HOW THEY WORK

Newton's law of gravitation states that two bodies with masses m_1 and m_2 attract each other with a force proportional to the product of their masses and inversely proportional to the square of the distance r between them:

$$\vec{F} = G\frac{m_1 m_2}{r^2}\hat{r}$$

The quantity $\hat{r}$ is the unit vector in the direction connecting the centers of the masses and G is the universal gravitational constant, $G = 6.673 \times 10^{-11}$ m^3/kg $\cdot$ s^2.

If several bodies are present in space they will all have an effect on each other. However, for the case of a spacecraft orbiting the Earth, Pisacane (2005) shows that the effect of other bodies is negligible for most practical purposes. For example, the effect of the sun and the moon on the motion of a spacecraft in the vicinity of the Earth is 3–4 orders of magnitude smaller than the effect of the Earth alone. Thus for basic design purposes only two masses need to be considered,

Manned Spacecraft Design Principles. DOI: http://dx.doi.org/10.1016/B978-0-12-804425-4.00005-2

the spacecraft mass m and the mass of the planetary body being orbited m_P. This is the classical two-body problem in a central force field. The application of Newton's second law of motion leads to the following equation:

$$\ddot{\vec{r}} = -\frac{G(m_P + m)}{r^2}\hat{r}$$

(5.1)

The spacecraft mass is much smaller than the planetary mass $m \ll m_P$ and both bodies are considered to be spherical bodies of uniform density so that we may consider them point masses. Pisacane (2005) also notes that the departure from sphericity in most planetary bodies is very small. A typical departure from spherical shape in a planet is the flattening at the poles or oblateness. This feature of the Earth, for example, contributes a change of only 0.1% to its gravitational potential and may also be neglected for basic design studies. Then the gravitational acceleration g toward the center of the Earth is

$$g = \frac{m_E G}{r^2}$$

(5.2)

5.1.2 CLOSED ORBITS IN A CENTRAL FORCE FIELD

With the assumptions made previously for the case of a spacecraft orbiting a planetary body the trajectory of the spacecraft is a conic section, i.e., a hyperbola, a parabola, or an ellipse. The only closed trajectory for the spacecraft is an ellipse with its origin at the center of the planetary body, as shown in Figure 5.1. The general equation for the trajectory is

$$r = r_p \frac{1 + e}{1 + e \cos\theta} = \frac{a(1 - e^2)}{1 + e \cos\theta}$$

(5.3)

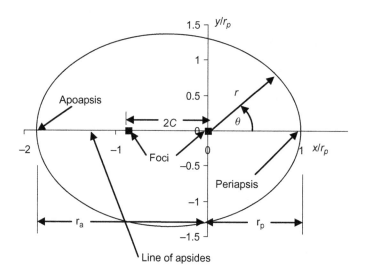

FIGURE 5.1

Elliptical orbit for eccentricity $e = 0.3$. The foci of the ellipse are shown as shaded symbols.

The quantity r_p denotes the periapsis, the shortest distance between the spacecraft and the center of the planetary mass, and e is the eccentricity, which is defined as follows:

$$e = \frac{C}{a} = \frac{r_a - r_p}{r_a + r_p} \tag{5.4}$$

Here the quantity C denotes half the distance between the foci of the ellipse and the semimajor axis is $a = (r_a + r_p)/2$ so that $a = C + r_p$. The quantity r_a is the apoapsis, which defines the largest distance between the spacecraft and the center of the planetary mass. The periapsis and the apoapsis lie along the line of apsides which coincides with the major axis of the ellipse. For orbits around the Earth, the periapsis is called the perigee and the apoapsis is called the apogee.

The velocity of the orbiting spacecraft and the period of its motion are of primary interest for design purposes. The velocity is given by what is classically called the *vis viva* equation

$$V^2 = \left[\frac{k}{m} \left(\frac{2}{r} - \frac{1}{a} \right) \right] \tag{5.5}$$

For the case where the planetary body is the Earth

$$\left(\frac{k}{m} \right)_E = m_E G = (398,600 \text{ km}^3/\text{s}^2) \tag{5.6}$$

We have assumed that the spacecraft mass is essentially constant, though in practice there may be small changes due to, for example, burning some fuel for control or orientation purposes. The period of the motion is the time required to complete one orbit and is given by

$$\tau = \frac{2\pi}{f} = 2\pi \sqrt{a^3 \left(\frac{k}{m} \right)^{-1}} = \frac{2\pi r_p^{3/2}}{\sqrt{\frac{k}{m}}} \left(\frac{a}{r_p} \right)^{3/2} \tag{5.7}$$

Here f is the average angular speed of the orbiting spacecraft. Note that the case of a circular orbit, $r = r_p = r_a = $ constant, the velocity and the period are also constant. In the case of a circular orbit at the surface of the Earth, $r = R_E$, the constant orbital speed is $V_E = 7.909$ km/s and the orbital period is $\tau_E = 84.35$ min. As the radius of the circular orbit increases the orbital velocity decreases and the orbital period increases. For a circular orbit at an altitude 200 km above the surface of the Earth the velocity drops to 7.788 km/s while the orbital period rises to 88.35 min.

5.1.3 EARTH ORBIT CHARACTERISTICS

Focusing attention on Earth orbits we may set the radial coordinate r equal to the sum of the Earth's radius R_E and the altitude above the Earth's mean surface z as follows:

$$r = R_E + z \tag{5.8}$$

Two representative orbits around the Earth are shown in Figure 5.2, a circular orbit where the eccentricity $e = 0$ and an elliptic orbit with an eccentricity $e = 0.3$. The velocity and period are dependent upon e, the eccentricity of the orbit, and the perigee of the orbit, r_p. Using Eqns (5.3) and (5.4) in Eqns (5.5) and (5.7) yields the following:

$$V^2 = \frac{1}{r_p} \left(\frac{k}{m} \right) \left[\frac{2(1 + e \cos \theta)}{1 + e} - (1 - e) \right] \tag{5.9}$$

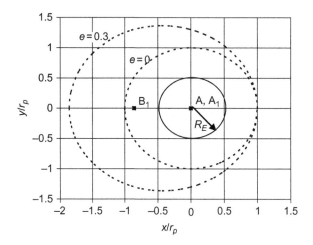

FIGURE 5.2

Schematic diagram of a circular orbit ($e = 0$) and an elliptical orbit ($e = 0.3$) around Earth. Here A denotes the center of the circular orbit while A_1 and B_1 denote the foci of the elliptical orbit.

$$\tau = \frac{2\pi}{\sqrt{\dfrac{k}{m}}} \left(\frac{r_p}{1-e}\right)^{3/2} \tag{5.10}$$

As an example, consider the International Space Station (ISS) which is in an elliptical orbit with an apogee altitude $z_a = 437$ km and a perigee altitude $z_p = 361$ km. The eccentricity of the orbit is found from Eqn (5.4) to be

$$e = \frac{r_a - r_p}{r_a + r_p} = \frac{z_a - z_p}{2R_E + z_a + z_p} = 0.0056$$

The orbital period of the ISS is then found from Eqn (5.10) to be $\tau = 5544$ s or 92.39 min.

The defining orbital parameters are r_p and e, the perigee and eccentricity of the orbit, respectively. Ideally, these parameters may be chosen arbitrarily for spacecraft, the only apparent limitation being the ability of the launching system to provide the necessary orbital velocity V. However, there are other limitations on the orbital parameters which must be considered in practice.

The first limitation, which applies to all satellites, is that arising from the aerodynamic drag due to the presence of the atmosphere. Although the atmospheric density decays rapidly with altitude, as discussed in Chapter 2, it does erode the kinetic energy of the satellite. The diminution of the velocity over time ultimately causes the orbit to decay and start the satellite on a fall to earth unless a thrust force is provided to accelerate it and raise the satellite back to the desired orbit. The magnitude of the drag force depends not only upon the local atmospheric density, i.e., the orbital altitude, but also on the scale and geometry of the satellite. For a spacecraft of the geometry and scale of a typical space capsule, like the Apollo command module, the relation between the minimum orbital altitude z, in kilometers, and the mission time t, in hours, is approximately given by

$$z_d = 100t^{0.15} \tag{5.11}$$

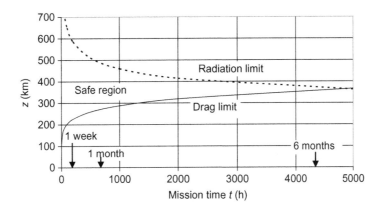

FIGURE 5.3

The safe altitude corridor for Apollo-sized manned spacecraft is shown as a function of mission duration. The region bracketed by radiation effects (dotted line) and drag effects (solid line) is safe for extended mission times.

This relation is illustrated in Figure 5.3 where it is clearly shown that even for missions of short duration the orbital altitude should be at least 200 km.

The second limitation is that arising from the need to ensure the safety of the crew from dangerous Van Allen radiation belts at high altitudes. Danger from exposure to radiation is a function of both orbital altitude and the extent to which the materials and construction of the spacecraft provide protective shielding. Obviously, the additional weight due to material used solely to provide shielding affects the performance of the spacecraft. For an Apollo-sized space-craft, the relation between the maximum altitude for maintaining radiation safety z_r, in kilo-meters, and the mission time t, in hours, when radiation shielding material of about 20 kg/m^2 is provided is approximately given by

$$z_r = 1300t^{-0.15} \tag{5.12}$$

This relation is also illustrated in Figure 5.3 where it is clearly shown that even for missions of short duration the orbital altitude should be less than about 600 km. It may be noted that the ISS orbits at an altitude in the range of about 350–450 km altitude, within the safe corridor. Of course, because of its large scale the ISS orbit decays due to drag and must be periodically boosted back to its nominal orbit.

If an eccentric orbit is considered, it too is subject to the same limitations in terms of its apogee and perigee altitudes. Using Eqn (5.4) with the outside limits of 600 km for the altitude at apogee and 200 km for the altitude at perigee yields an orbit of eccentricity $e = 400/(2R_E + 800) = 0.0296$ which is still an essentially circular orbit. Consider again the practical case of the ISS and assume it to be in a circular orbit, rather than the actual elliptical one with $e = 0.0056$. We would find the orbital period to be 5497 s for a circular orbit of altitude $z = z_p = 361$ km and 5590 s for one of altitude $z = z_a = 437$ km compared to the actual value of 5544 s. The difference in period is ± 46 s or about 0.83%. Thus we see that for practical purposes in preliminary design it is reasonable to exploit the simplicity of considering manned spacecraft orbits to be circular.

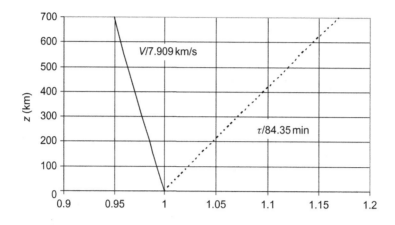

Velocity V and period τ for circular orbits around the earth at altitude z; V and τ are normalized with respect to their values at $z = 0$.

Under the assumption of a circular orbit of nominal radius $r = R_E + z$ the earth, Eqns (5.9) and (5.10) become

$$V = \frac{1}{R_E^{1/2}\left(1+\dfrac{z}{R_E}\right)^{1/2}}\left(\frac{k}{m}\right)^{1/2} = \frac{7.909}{\left(1+\dfrac{z}{R_E}\right)^{1/2}} \tag{5.13}$$

$$\tau = 2\pi\sqrt{\frac{R_E^3\left(1+\dfrac{z}{R_E}\right)^3}{(k/m)}} = 5061\left(1+\frac{z}{R_E}\right)^{3/2} \tag{5.14}$$

The variation of the orbital velocity and period is shown in nondimensional form in Figure 5.4. Note that the orbital velocity decreases only about 5% over the altitude range while the period increases by about 17%.

5.1.4 IN-PLANE ORBITAL TRANSFER: INTERSECTING ORBITS

Disturbing the equilibrium of an orbit by applying a thrust force will result in changing the space-craft's velocity. Once the force is removed the spacecraft will assume a new orbit. For simplicity, we first consider an in-plane transfer of a spacecraft of mass m from orbit 1 to an intersecting orbit 2 as shown schematically in Figure 5.5. We assume that a thrust force F is applied impulsively to the spacecraft at point A in orbit 1 so that

$$F\Delta t = m\Delta V \tag{5.15}$$

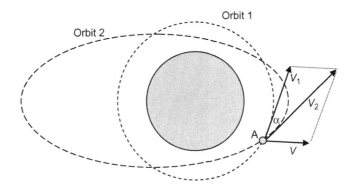

FIGURE 5.5

Velocity change for transfer between intersecting orbits 1 and 2 which both lie in the same plane.

The velocity change imparted to the spacecraft produces exactly the required velocity for orbit 2 at point A and the spacecraft commences to travel on orbit 2. The velocity vector diagram in Figure 5.5 shows that

$$\vec{V_1} + \vec{\Delta V} = \vec{V_2}$$

The increment of velocity magnitude needed is given by

$$(\Delta V)^2 = V_1^2 + V_2^2 - 2V_1 V_2 \cos \alpha \qquad (5.16)$$

Here V_1 is the velocity at point A on orbit 1 and V_2 is the velocity at point A on orbit 2 and α is the angle between V_1 and V_2 at point A. Note that the minimum value of ΔV occurs when $\alpha = 0$, that is, when the orbits are just tangent at point A, the transfer point.

The most common transfer between intersecting orbits for human spaceflight is the de-orbit process for initiating entry into the atmosphere and descent to the surface. It will be made clear in Chapter 6 that deceleration and heating limitations for human spaceflight require flight path angles at entry to be very small, usually in the range of $-0.5° < \gamma < -2.5°$. This necessity is of practical importance because ΔV needed will be close to the minimum possible, as shown in Eqn (5.16).

5.1.5 IN-PLANE ORBITAL TRANSFER: NONINTERSECTING ORBITS

In the case of in-plane transfer between nonintersecting orbits, the minimum energy transfer takes place along an intermediate partial orbit known as the Hohmann transfer ellipse. As might be expected, this transfer makes use of the minimum ΔV transfer process between intersecting orbits described above. Here, however, it is applied twice, once at a point on the initial orbit and then again at a point on the desired final orbit. The points of application are special: they lie on the line of apsides of the Hohmann ellipse with the initial impulse being applied at the periapsis and the final impulse at the apoapsis as illustrated in Figure 5.6. Thus the initial point of application of an impulsive thrust force to achieve ΔV_1 is at P where the periapsis radial coordinate is $r_p = r_1$.

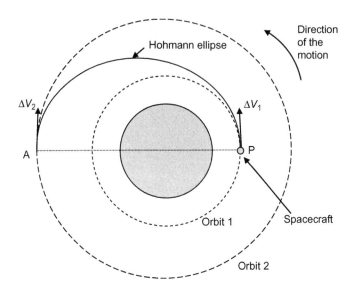

FIGURE 5.6

Hohmann transfer to raise altitude of spacecraft from orbit 1 to orbit 2. Points A and P are the apoapsis and periapsis of the Hohmann ellipse at which points the impulsive forces are applied, resulting in the velocity increments shown.

Likewise, the final point of application of an impulsive thrust force to achieve ΔV_1 is at A where the apoapsis radial coordinate is $r_a = r_2$. The velocity changes required at points P and A are as follows:

$$\Delta V_1 = V_P - V_1 \tag{5.17}$$

$$\Delta V_2 = V_A - V_2 \tag{5.18}$$

Using Eqns (5.5) and (5.6) yields

$$\Delta V_1 = \frac{631.34}{\sqrt{r_1}}\left[\sqrt{\frac{r_2}{a}} - 1\right] \tag{5.19}$$

$$\Delta V_2 = \frac{631.34}{\sqrt{r_2}}\left[\sqrt{\frac{r_1}{a}} - 1\right] \tag{5.20}$$

Recall that the quantity a is the semimajor axis of the orbit

$$a = \frac{1}{2}(r_P + r_A) = \frac{1}{2}(r_1 + r_2) \tag{5.21}$$

When raising the orbital altitude, which is the case illustrated in Figure 5.6, the final orbital radius $r_2 > a$ which means that, according to Eqn (5.19), the spacecraft at point P requires a positive impulsive increment in velocity ΔV_1 to start it climbing to the higher altitude. When the spacecraft arrives at point A it requires a negative impulsive increment in velocity because the initial orbital radius $r_1 < a$ and therefore, according to Eqn (5.20), $\Delta V_2 < 0$.

Because there are no dissipative forces being considered the entire process is reversible. Therefore if we wish to decrease the orbital altitude from that of orbit 2 to that of orbit 1 the same Hohmann transfer may be used if we simply reverse the directions of all the velocities. Using Figure 5.6 again, we may imagine the spacecraft to now be traveling in a clockwise direction and the sense of velocities ΔV_1 and ΔV_2 to be reversed. Thus, when the spacecraft is at point A an impulsive retrograde thrust is applied producing a negative velocity increment ΔV_2 slowing down the spacecraft and starting it along the Hohmann ellipse. Then when it arrives at point P an impulsive prograde thrust would be applied producing a positive velocity increment ΔV_2 accelerating the spacecraft to the velocity appropriate to the lower orbit 1.

The time required to carry out the Hohmann transfer τ_H is just the time needed to complete half of an elliptical orbit and may be obtained by calculating half the orbital period given by Eqn (5.7). This yields the following result for τ_H in minutes:

$$\tau_H = \frac{1}{2}\tau = \pi\sqrt{\frac{(r_1+r_2)^3}{8(k/m)}} = 42.17\left(1+\frac{z_1+z_2}{2R_E}\right)^{3/2} \tag{5.22}$$

Considering the transfer between the two extreme practical orbital altitudes of 200 and 600 km for example, the Hohmann transfer time is about 46.2 min or around half the time it takes for the spacecraft to complete a typical orbit.

5.2 ENERGY AND ANGULAR MOMENTUM IN ORBITS

Taking the scalar product of the velocity vector $\frac{d\vec{r}}{dt}$ and Eqn (5.1) yields

$$\frac{d\vec{r}}{dt}\cdot\frac{d}{dt}\left(\frac{d\vec{r}}{dt}\right) + \frac{m_E G}{r^2}\left(\frac{d\vec{r}}{dt}\cdot\hat{r}\right) = \frac{1}{2}\frac{d}{dt}\left(\frac{dr}{dt}\right)^2 + m_E G\frac{d}{dt}\left(\frac{-1}{r}\right) \tag{5.23}$$

This equation may be written as

$$\frac{d}{dt}\left(\frac{1}{2}V^2 - \frac{m_F G}{r}\right) = 0 \tag{5.24}$$

The integral of Eqn (5.24) is a constant of the motion

$$\frac{1}{2}V^2 - \frac{m_E G}{r} = \frac{U}{m} = \frac{1}{m}(U_{kin} + U_{pot}) = const. \tag{5.25}$$

The total energy per unit mass, or specific energy, of the orbiting body is a constant equal to the sum of the specific kinetic and potential energies of the body. In the same fashion, we may take the vector product of the radius vector $\vec{r}$ and Eqn (5.1) to obtain

$$\vec{r}\times\frac{d\vec{V}}{dt} + \frac{m_E G}{r^2}(\vec{r}\times\hat{r}) = \vec{r}\times\frac{d\vec{V}}{dt} = 0 \tag{5.26}$$

Because $\vec{V}\times\frac{d\vec{V}}{dt} = 0$, we may rewrite Eqn (5.26) as follows:

$$\frac{d}{dt}(\vec{r}\times\vec{V}) = \frac{d}{dt}(\vec{h}) = 0 \tag{5.27}$$

The integral of Eqn (5.27) is another constant of the motion given by

$$\vec{h} = \vec{r} \times \vec{V} = \text{const.} \tag{5.28}$$

The vector $\vec{h}$ is the angular momentum of the body around the central point.

Thus, as might be expected for a conservative force field like a gravitational field the total specific energy and the angular momentum of the body moving in it are constants when no other forces or torques are applied to it.

5.2.1 CONSERVATION OF ENERGY

Using Eqns (5.2) and (5.6), the potential energy of a spacecraft of mass m orbiting the earth on an orbit with a radial vector r may be written as

$$U_{pot} = \int_{R_E}^{r} mg dr = \int_{R_E}^{r} m \frac{k}{r^2} dr = mg_E R_E \left(1 - \frac{R_E}{r}\right) \tag{5.29}$$

Here the potential energy of the spacecraft is taken to be zero at the surface of the earth. Using Eqns (5.3) and (5.4), we may transform Eqn (5.29) to yield the potential energy per unit mass of the spacecraft in terms of the perigee r_p and eccentricity e of the orbit as follows:

$$\frac{U_{pot}}{m} = g_E R_E \left(1 - \frac{R_E}{r_p} \frac{1 + e \cos \theta}{1 + e}\right) \tag{5.30}$$

Using Eqns (5.5) and (5.6), the kinetic energy of that spacecraft may be written as

$$U_{kin} = \frac{1}{2} mV^2 = \frac{1}{2} m \left(\frac{k}{m}\right) \left(\frac{2}{r} - \frac{1}{a}\right) = \frac{1}{2} mg_E R_E \left(2 \frac{R_E}{r} - \frac{R_E}{a}\right) \tag{5.31}$$

Here we also used Eqns (5.2) and (5.6) which yield

$$g_E R_E = \frac{m_E G}{R_E} = \frac{1}{R_E} \left(\frac{k}{m}\right) = \frac{3.986 \times 10^{14}}{R_E} \tag{5.32}$$

Here we take $m_E G = 3.986 \times 10^{14}$ m³/s² and $m_E G/R_E = 62.56 \times 10^6$ m²/s² = 62.56 MJ/kg. In a similar fashion, the kinetic energy per unit mass of the spacecraft may also be written in terms of the perigee r_p and eccentricity e of the orbit as follows:

$$\frac{U_{kin}}{m} = \frac{1}{2} g_E R_E \left(\frac{R_E}{r_p}\right) \frac{1 + e \cos \theta}{1 + e} \left(2 - \frac{1 - e^2}{1 + e \cos \theta}\right) \tag{5.33}$$

Equations (5.30) and (5.33) illustrate that the potential energy and kinetic energy vary with position on the orbit. However, summing Eqns (5.30) and (5.33) yields the total energy per unit mass of the spacecraft as follows:

$$\frac{U}{m} = g_E R_E \left[1 - \frac{R_E}{r_p} \frac{1 - e^2}{2(1 + e)}\right] \tag{5.34}$$

The total energy of the spacecraft is a constant whose magnitude depends only on the perigee and the eccentricity. For a given perigee, the energy is a minimum for a circular orbit, $e = 0$.

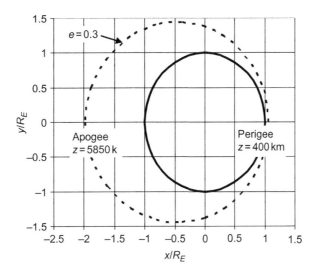

FIGURE 5.7

An elliptic orbit with eccentricity $e = 0.3$ and a perigee of $r_p = 400$ km.

As an example, consider an elliptic orbit with eccentricity $e = 0.3$ and a perigee of $r_p = 400$ km for which Eqns (5.3) and (5.4) yield the orbit shown in Figure 5.7. The variation of the specific potential and kinetic energies for this orbit is shown in Figure 5.8. The kinetic energy of the spacecraft is highest at perigee and lowest at apogee while the behavior of the potential energy is just the opposite. The total energy per unit mass of the spacecraft for this orbit is

$$\frac{U}{m} = g_E R_E \left[1 - 0.35 \frac{R_E}{r_p} \right] = 41.9 \text{ MJ/kg}$$

As mentioned previously, circular, or nearly circular, earth orbits are the more usual case. For a circular orbit $e = 0$ and $r_p = R_E + z_O$ and the specific potential and kinetic energies become

$$\frac{U_{pot}}{m} = g_E R_E \left[1 - \left(1 + \frac{z_O}{R_E} \right)^{-1} \right]$$

$$\frac{U_{kin}}{m} = \frac{1}{2} g_E R_E \left(1 + \frac{z_O}{R_E} \right)^{-1}$$

The energy distribution as a function of distance z_O from the earth's surface to a spacecraft in a circular orbit is shown in Figure 5.9. The approximate regions of the inner and outer Van Allen radiation belts are also shown in Figure 5.9. The former is characterized by high-energy protons and the latter by high-energy electrons. On the scale shown in Figure 5.9, human spaceflight is necessarily confined to altitudes very close to the origin for radiation safety reasons mentioned previously. The spacecraft total energy requirement increases rather rapidly beyond low earth orbit (LEO) providing another reason for favoring LEO for human spaceflight.

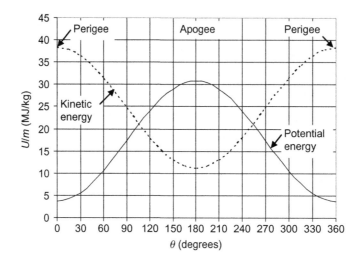

FIGURE 5.8

Variation of the specific potential and kinetic energies for the orbit of Figure 5.7 is shown as a function of the polar angle θ.

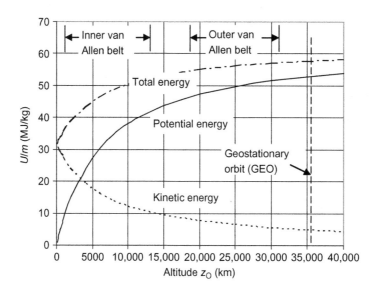

FIGURE 5.9

The specific energy distribution as a function of distance z_O from the earth's surface to a spacecraft in a circular orbit ($e = 0$).

An orbit that is useful for unmanned earth observation and communication satellites is the geostationary earth orbit (GEO) whose altitude is shown in Figure 5.9. The GEO is an equatorial ($i = 0$) circular orbit ($e = 0$) situated at a radial distance (r_{GEO}) such that the satellite's rotational rate equals that of the earth, thereby ensuring that the same area of the earth is always in view. The earth's period is 23.93 h or 86,164 s and Eqn (5.14) may be solved to yield a GEO altitude $z_{GEO} = 35,800$ km ($r_{GEO} = 42,170$ km).

5.2.2 CONSERVATION OF ANGULAR MOMENTUM

The angular momentum is a vector defined as follows:

$$\vec{h} = \vec{r} \times \vec{V} \tag{5.35}$$

The angular momentum vector is normal to the plane formed by the radius and velocity vectors and therefore normal to the plane of the orbit. If no external torque is applied, the angular momentum is a constant of the motion. The magnitude of the angular momentum may be expressed as

$$h = rV \cos \phi \tag{5.36}$$

The relationship between the variables is illustrated in Figure 5.10. When the normal component of velocity $V \sin \phi$ is in the direction opposite to that of the radius vector the flight path angle ϕ is considered negative. Negative flight path angles are encountered during atmospheric entry which is discussed in detail in Chapter 6. Note that for circular orbits the velocity is normal to the radius and therefore $\phi = 0$.

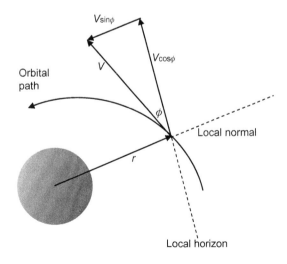

FIGURE 5.10

The orbital flight path is shown with the local normal direction defined by the radius vector with the local horizon perpendicular to that direction. The flight path angle is also defined in the diagram.

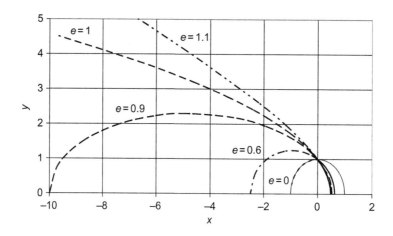

FIGURE 5.11

Orbits with semilatus rectum $p = 1$ for various eccentricities and origin at the focus. Note that the vertical scale is stretched.

The relationship between the magnitude of the angular momentum and the geometry of the orbit is shown by Bate, Mueller, and White (1971) to be given by

$$p = \frac{h^2}{m_E G} \tag{5.37}$$

Here p is the semilatus rectum of the orbit, the perpendicular distance between the major axis of the orbit and the orbit at the focus. The orbit may thereby also be defined by

$$r = \frac{p}{1 + e \cos \theta} \tag{5.38}$$

The relative shape and size of orbits with equal angular momentum given by $p = 1$ is shown in Figure 5.11 for different eccentricities. The orbit is an ellipse for $e < 1$, a parabola for $e = 1$, and a hyperbola for $e > 1$. From Eqn (5.37), we see that the semilatus rectum for elliptical orbits is proportional to the semimajor axis of the orbit as given by

$$p = a(1 - e^2) \tag{5.39}$$

The elliptical and circular orbits are closed curves while parabolic and hyperbolic orbits are open. In the case of circular orbits ($e = 0$), the semilatus rectum $p = a = r$, while for parabolic and hyperbolic orbits the semimajor axis is infinite underlining the notion that the orbits are open.

5.2.3 OPEN ORBITS: PARABOLIC ORBITS AND ESCAPE SPEED

The total energy per unit mass of a spacecraft is constant and if we use Eqn (5.29) to define the potential energy we find

$$\frac{U}{m} = \frac{1}{2}V^2 + g_E R_E \left(1 - \frac{R_E}{r}\right) = const. \tag{5.40}$$

Equation (5.38) shows that for parabolic orbits ($e = 1$) the radius goes to infinity as θ approaches π because the angular momentum h, and therefore p, is constant, according to Eqn (5.37). It follows that the velocity V of the spacecraft must approach zero as r approaches infinity to maintain the angular momentum h at a constant value. At the same time, the total energy of the spacecraft must also remain constant. Equating the energy at some initial point on the orbit, denoted by the subscript 1, to that at far away where r is approaching infinity yields

$$\frac{1}{2}V_1^2 + g_E R_E \left(1 - \frac{R_E}{r}\right) = \lim_{r \to \infty} \left[\frac{1}{2}V^2 + g_E R_E \left(1 - \frac{R_E}{r}\right)\right] = g_E R_E \qquad (5.41)$$

We may say that the spacecraft located at r_1 can escape the gravitational influence of the earth by achieving escape velocity, that is, the velocity satisfying Eqn (5.41) or

$$V_1 = V_{esc} = \sqrt{2\frac{g_E R_E^2}{r_1}} = \sqrt{2\frac{m_E G}{r_1}} \qquad (5.42)$$

The same analysis used here for earth orbits may be applied to other planets by replacing m_E with the mass of the planet considered. Data for the basic properties of the moon and the planets of the solar system are given in Table 5.1.

The concept of the escape speed is convenient for determining when an orbit changes its character from closed to open. The notion of escape refers to the inability of the central gravitational force field to keep the spacecraft under its influence and has it return, i.e., to "close" the orbit. As noted in the case of a translunar orbit increasing the insertion speeds provide diminishing returns in reducing travel time and will also be more costly in terms of propellant consumption.

Table 5.1 Planetary Data from NSSDC (2013)

Planet	Orbital Period (years)	Solar Distance (10^6 km)	m/m_E[d]	Radius (km)	g/g_E	Escape Velocity[e] (km/s)
Sun[a]	–	–	332,948	6.96×10^8	28.1	617.7
Mercury	0.241	57.9	0.0553	2440	0.378	4.3
Venus	0.615	108.2	0.815	6052	0.907	10.4
Earth	1.000	149.6	1.000	6378	1.000	11.2
Moon	0.0748[b]	0.384[c]	0.0123	1738	0.166	2.4
Mars	1.881	227.9	0.107	3396	0.377	5.0
Jupiter	11.86	778.6	317.8	71,492	2.36	59.5
Saturn	29.43	1434	95.2	60,268	0.916	35.5
Uranus	83.75	2873	14.5	25,559	0.889	21.3
Neptune	163.7	4495	17.1	24,764	1.12	23.5
Pluto	248	5870	0.0022	1195	0.059	1.1

[a]*Data for the sun taken from Pisacane (2005).*
[b]*Earth orbital period.*
[c]*Mean distance from the earth.*
[d]*$m_E = 5.97 \times 10^{24}$ kg.*
[e]*From planet surface or 1 bar pressure level of gas giants.*

5.2.4 OPEN ORBITS: HYPERBOLIC ORBITS AND EXCESS SPEED

In the special case of a parabolic orbit, the velocity of the spacecraft eventually decreases to zero far from the central body. However, at the location r_1 the spacecraft may be accelerated to a speed greater than the escape speed, $V_1 > V_{esc}$, and will then be left with excess speed as it proceeds far from the central body. The orbit of the spacecraft would then be described by a hyperbola. Applying the requirement of constant energy once more yields

$$\frac{1}{2}V_1^2 + g_E R_E \left(1 - \frac{R_E}{r_1}\right) = \lim_{r \to \infty} \left[\frac{1}{2}V_\infty^2 + g_E R_E \left(1 - \frac{R_E}{r}\right)\right] = \frac{1}{2}V_\infty^2 + g_E R_E \tag{5.43}$$

Now, using the definition of escape speed from Eqn (5.42) in Eqn (5.43), we see that far from the central body the spacecraft has a speed given by

$$V_\infty^2 = V_1^2 - V_{esc}^2 \tag{5.44}$$

Unmanned spacecraft launched on interplanetary missions must have a velocity in excess of the escape velocity in order to have sufficient excess velocity to permit it to reduce transit time. Without a need for return to earth the hyperbolic orbit is acceptable. Asteroids and comets typically travel at speeds in excess of the escape speeds of most of the planets in the solar system, as given in Table 5.1. The higher escape speeds of giant planets like Jupiter and Saturn are indicative of their large gravitational attraction, suggesting that they can influence the path of such bodies.

5.3 ORBITAL TRANSFER FOR ATMOSPHERIC ENTRY

As will be discussed in Chapter 6, the flight path angle γ, that is, the angle between the velocity of the spacecraft and the local horizon, at the entry interface z_e must be very small for safe human spaceflight. The entry interface marking the top of the aerodynamically sensible atmosphere is generally set at the arbitrary altitude $z_e = 100$ km. The entry flight path angle for manned spacecraft should lie in the range $-10° < \gamma_e < 0°$ for safe entry. In Figure 5.12, we show the spacecraft in orbit at r_1 firing a retrorocket at $\theta = 180°$ imparting a $\Delta V < 0$ to the spacecraft. The spacecraft is thereby put on a Hohmann transfer ellipse toward a new, but fictitious, circular orbit at r_3. This transfer path is designed to cross the entry interface $z_e = 100$ km at the desired flight path angle γ. The subsequent motion of the spacecraft is covered in detail in Chapter 6.

To determine the ΔV required we first calculate the Hohmann transfer orbit by choosing a value for the fictitious orbit at r_3 and calculating the eccentricity e using Eqn (5.4) to obtain

$$e = \frac{r_1 - r_3}{r_1 + r_3} \tag{5.45}$$

Because the center of the earth is one of the foci of the transfer ellipse and the entry interface $r_e < r_1$ the value of the fictitious final orbit must lie in the range $0 < r_3 < r_1$. The semimajor axis of the transfer ellipse is

$$a = \frac{1}{2}(r_1 + r_3) \tag{5.46}$$

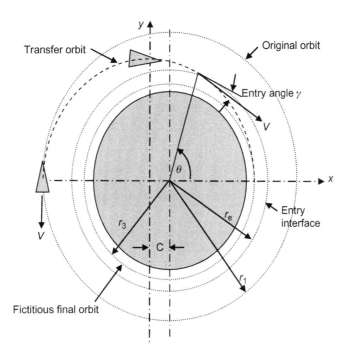

FIGURE 5.12

Transfer from circular orbit at r_1 to an ideal new orbit at r_3. A retrorocket impulse produces the required ΔV to permit the spacecraft to cross the entry interface r_2 at the flight path angle γ.

The distance between the earth's center and that of the transfer ellipse is

$$C = ae = \frac{1}{2}(r_1 - r_3) \tag{5.47}$$

Using Eqn (5.3) to calculate the value of $\theta = \theta_e$ at which $r(\theta_e) = r_e$ yields

$$\theta_e = \arccos\left[\frac{a}{r_e}\left(\frac{1-e^2}{e}\right) - \frac{1}{e}\right] \tag{5.48}$$

The flight path angle at the entry interface γ_e is given by

$$\gamma_e = \frac{\pi}{2} - \theta_e + \arctan\left[\left(\frac{ae}{r_e \sin \theta_e} - \frac{1}{\tan \theta_e}\right)(1 - e^2)\right] \tag{5.49}$$

This angle is the difference between the angles of the tangents to the circular and elliptic curves at the point of entry r_e, q_e. Thus, for every value of z_3 in the permissible range one may find the point at which the spacecraft enters the atmosphere and the flight path angle at entry. The variation of the entry flight path angle with $\pi - \theta$, the polar angle of entry measured from the point of application of ΔV is shown in Figure 5.13 for initial circular orbital values of $r_1 = 300$, 400, and 500 km. Obviously, for the small entry flight path angles desired for human spaceflight, the impulsive retrorocket burn must be applied almost halfway around the globe.

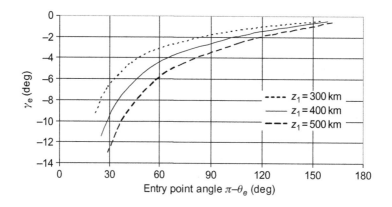

FIGURE 5.13

Variation of the entry flight path angle with $\pi - \theta_e$, the polar angle of entry measured from $\theta = \pi$, the point of application of ΔV for several initial orbital altitudes.

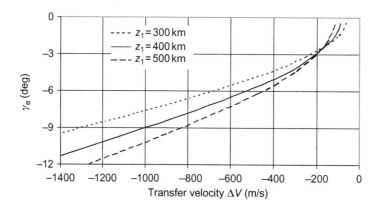

FIGURE 5.14

Variation of the entry flight path angle with the required impulsive change in velocity ΔV for several initial orbital altitudes.

The magnitude of the velocity change needed to drop out of orbit for the above cases is shown in Figure 5.14. Relatively small values of ΔV are needed to achieve shallow flight path angles but these values increase rapidly for steeper angles. The fuel mass needed to achieve the required ΔV may be determined for the impulsive thrust cases by noting that Eqn (5.15) states that the impulse equals the change in momentum of the spacecraft

$$Fdt = (\dot{m}_p g_E I_{sp})dt = dm_p g_E I_{sp} = mdV \tag{5.50}$$

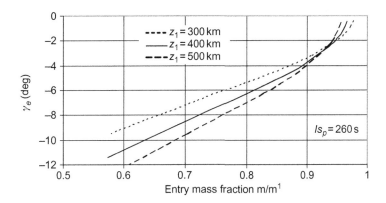

FIGURE 5.15

The mass fraction of the spacecraft as it crosses the entry interface m_e/m_1 is shown as a function of the entry angle γ_e for various initial altitudes z_1 and a retrorocket specific impulse of 260 s.

Noting that $dm_p = -dm$ because the mass of the spacecraft is decreasing only because of the consumption of propellant permits one to integrate Eqn (5.50) to obtain the so-called rocket equation

$$\frac{m_e}{m_1} = \exp\left(-\frac{\Delta V}{g_E I_{sp}}\right) \tag{5.51}$$

In Eqn (5.51), m_1 is the mass of the spacecraft in orbit at r_1 and m_e is the mass of the spacecraft as it crosses the entry interface at $r = r_e$. The quantity I_{sp} is the specific impulse of the retrorocket which represents its specific fuel consumption according to

$$I_{sp} = \frac{F}{\dot{m}_p g_E} \tag{5.52}$$

The specific impulse is measured in seconds and for operational solid rocket motors that would be used in retrorocket systems a typical value is $I_{sp} = 260$ s. The mass fraction of the spacecraft as it crosses the entry interface m_e/m_1 is shown as a function of the entry angle γ_e in Figure 5.15 for various initial altitudes z_1. For very small entry angles only 5% of the mass of the spacecraft in the initial orbit is needed to position it properly for entry. The propellant mass requirements become quite large as the entry angle steepens beyond about $-4°$.

5.4 THE GROUND TRACK OF AN ORBIT

Figure 5.16 shows a spacecraft orbiting the earth at a velocity V in the direction of the rotation of the earth. The plane of the orbit passes through the center of the earth and is inclined at an angle i to the equator. The ground track of the orbit is the trace of the intersection between the plane of the orbit and the surface of the earth. This trace is also known as the locus of the so-called subsatellite points (SSPs) of the orbit.

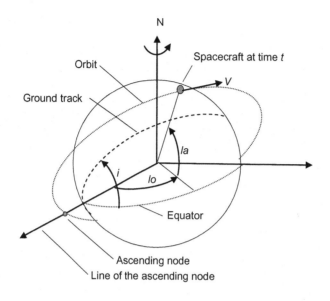

FIGURE 5.16

A spacecraft is shown orbiting the earth at a velocity V in the direction of the rotation of the earth.

5.4.1 DEFINING THE ORBIT

The orbit is defined by the polar coordinates r, θ where θ is the true anomaly of the orbit and measures the angle between the radius r at a point on the orbit and that at the periapsis (perigee for earth orbits) as shown in Figure 5.1. In Figure 5.16, the radius r to the spacecraft's orbit sweeps out a plane that passes through the center of the earth at which one of the foci is located. The plane of the orbit is shown making an angle l_a with the equatorial plane thereby defining the latitude of the spacecraft's location on the ground track over the earth. Similarly, the angle l_o measured from the axis of the ascending node to the projection of r on the equatorial plane defines the relative longitude of the spacecraft's location over the earth.

The ascending node of the orbit of the spacecraft is the point at which the SSP crosses the equator while moving in a northerly direction and defines the start of an orbit. The periapsis, or perigee for earth orbits, where $r = r_p$, is shown in Figure 5.17 along with ω, the argument of the periapsis, which defines the angle between the last ascending node and r_p. The true anomaly θ is measured from periapsis so the equation of the orbit may be written as

$$r = r_p \frac{1 + e}{1 + e \cos(\theta - \omega)} \tag{5.53}$$

The time at which the SSP crosses the ascending node while moving in a northerly direction may be taken as a reference, say $t = 0$, and all subsequent positions of the spacecraft may be referred to it.

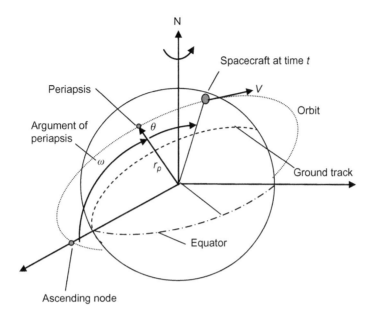

FIGURE 5.17

A spacecraft is shown orbiting the earth at a velocity V in the direction of the rotation of the earth. The periapsis (perigee for earth), the radius at periapsis r_p, and ω, the argument of the periapsis, are also shown.

The time measured from periapsis may be obtained with the aid of the eccentric anomaly, which is the angle E shown in Figure 5.18. The relationship between the eccentric anomaly and the true anomaly is given by

$$\cos E = \frac{e + \cos \theta}{1 + e \cos \theta} \tag{5.54}$$

This relation for the eccentric anomaly always places E in the same half plane as θ so that the selection of the proper value of E from $\cos E$ is well-defined.

Kepler's second law states that in an orbit equal areas are swept out by the radius vector in equal times. Thus the ratio of the area swept out in going to any location θ from $\theta = 0$, or periapsis, to the total area of the orbit πab is equal to the ratio of the time required for that movement to the period of the orbit. As shown by Bate et al. (1971), the eccentric anomaly is useful in carrying out the calculation of the swept-out area resulting in the following relation for the time interval between the time $t(\theta)$ and the time of periapsis $t(0) = t_p$:

$$t - t_p = \sqrt{\frac{a^3}{m_E G}}(E - e\sin E) \tag{5.55}$$

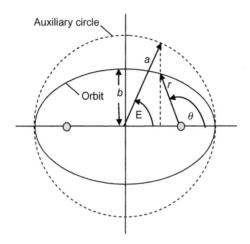

FIGURE 5.18

The auxiliary circle with radius equal to the semimajor axis of the orbit a is shown along with the definition of the eccentric anomaly E.

Note that for a circular orbit ($e = 0$), periapsis is undefined so Eqn (5.55) does not apply. However, using Kepler's second law we can readily write, defining $t_p = t(0) = 0$, that

$$t - t_p = \frac{\theta}{2\pi}\tau = \theta\sqrt{\frac{a^3}{m_E G}} \tag{5.56}$$

For example, consider the orbit in Figure 5.7 which has an eccentricity $e = 0.3$, a perigee $r_p = 6778$ km (perigee altitude $z_p = 400$ km) and an apogee $r_a = 12{,}590$ km. From Eqn (5.21), the semimajor axis is found to be $a = 9684$ km. According to Eqn (5.10), this orbit has an orbital period $\tau = 158.0$ min.

- A true anomaly $\theta = 1139°$ corresponds to $n = \theta/360 = 3.164$ orbits or $\theta_4 = 59.04°$ in the fourth orbit. The eccentric anomaly is given by Eqn (5.54) and for this point in the orbit

$$\cos E = \frac{0.3 + \cos(59.04°)}{1 + 0.3\cos(59.04°)} = 0.7056$$

Because θ_4 lies between $0°$ and $180°$, E must also fall in that range. This corresponds to $E = 45.13° = 0.7876$ rads so that t_4, the time elapsed measured from the time of periapsis at the start of the fourth orbit, $t_{p,4}$, is calculated from Eqn (5.55) as

$$t_4 - t_{p,4} = \sqrt{\frac{(9684 \text{ km})^3}{3.989 \times 10^5 \text{ km}^3/\text{s}^2}}\,[0.7876 - 0.3\sin(0.7876)] = 867.6 \text{ s}$$

The total time elapsed since the first periapsis is $t = 867.6 \text{ s} + 3\tau = 14.46$ min $+ 3$ (158 min) $= 488.5$ min.

• Later on in the same (fourth) orbit, say where the true anomaly is $\theta = 1367°$, which corresponds to $n = 3.797$ orbits or $\theta_4 = 287.0°$ in the fourth orbit, the eccentric anomaly is

$$\cos E = \frac{0.3 + \cos(287°)}{1 + 0.3 \cos(287°)} = 0.5445$$

Again, because θ_4 lies between $180°$ and $360°$, E must also fall in that range. This corresponds to $E = 360° - 57.00° = 5.287$ rads so that the time from periapsis, calculated from Eqn (5.55) is

$$t_4 - t_{p,4} = \sqrt{\frac{(9684 \text{ km})^3}{3.989 \times 10^5 \text{ km}^3/\text{s}^2}} [5.287 - 0.3 \sin(303°)] = 8356 \text{ s}$$

The total time elapsed since the first periapsis is $t = (t_4 - t_{p,4}) + 3\tau = 139.2 \text{ min} + 3$ (158 min) = 613.2 min.

The true anomaly is measured from the periapsis, but the periapsis of an orbit is not necessarily at an ascending node so that to find the time required to go from the ascending node, that is, the start of an orbit, to a prescribed true anomaly θ we must account for the time necessary to travel from the ascending node to periapsis, that is, the time required to traverse the arc of the orbit subtended by ω, the argument of periapsis shown in Figure 5.17. This is important because, as we shall subsequently show, the location of the ascending node on a rotating planet changes with each orbit and therefore so does ω.

Bate et al. (1971) show that the time elapsed in moving from one point on an orbit defined by the true anomaly θ_0 to another point on the orbit defined by the true anomaly θ may be written as follows:

$$t(\theta) - t(\theta_0) = \sqrt{\frac{a^3}{m_E G}} [2n_p \pi + (E - e \sin E) - (E_0 - e \sin E_0)] \tag{5.57}$$

In Eqn (5.57), the quantity n_p denotes the number of times the spacecraft passed through periapsis in traveling from θ_0 to θ. Applying Eqn (5.57) to the two example cases quoted previously will provide equivalent solutions. In the case of circular orbits the situation is simpler. For example, if we consider a circular orbit similar to the elliptic one considered previously, the perigee, apogee, and semimajor axis are all given by $r_p = 6771$ km (perigee altitude $z_p = 400$ km). According to Eqn (5.10), this orbit has an orbital period $\tau = 92.52$ min, or about 58.5% of the period of the $e = 0.3$ orbit.

• For a true anomaly $\theta = 1139°$, or 3.164 orbits, which corresponds to being $59.04°$ from the start of the fourth orbit the elapsed time is $t = \theta\tau/360 = 292.7$ min.
• Later on in the same (fourth) orbit, say where the true anomaly is $\theta = 1367°$, which corresponds to 3.797 orbits, or $287.0°$ from the start of the fourth orbit the elapsed time is $t = \theta\tau/360 = 351.3$ min.

5.4.2 THE SPACECRAFT'S LATITUDE

The latitude angle l_a at any point in the spacecraft's orbit is illustrated in Figure 5.16 and is given by spherical trigonometry as

$$l_a = \arcsin(\sin i \sin \theta) \tag{5.58}$$

The incidence angle of the plane of the orbit is always in the range $0 \leq i \leq 180°$ and therefore, according to Eqn (5.58), $\sin i$ is a constant ≤ 1 multiplying $\sin \theta$. Because θ is measured rising to the north from the line of the ascending node, for $0° < \theta < 180°$ the spacecraft is in the northern hemisphere and a computer or calculator will typically return a value for the latitudinal angle in the range $0° < l_a < 90°$. The latitude of the spacecraft is then defined as $L_a = l_a N$. However, for true anomalies in the range $180° < \theta < 360°$ the spacecraft is over the southern hemisphere and a computer or calculator will typically return the value $-90° < l_a < 0°$. The latitude of the spacecraft in this case is defined as $L_a = -l_a S$. Note that the rotation of the earth plays no role in the determination of the latitude because Eqn (5.58) always yields the same answer for the latitude angle l_a when θ increases by 2π every revolution.

Using the example orbit mentioned previously (Figure 5.7) with the further assumption that the orbital inclination $i = 30°$ leads to

- For $\theta = 1139°$ and $t = 488.5$ min we find $\sin l_a = \sin(30°)\sin(1139°) = 0.4236$ or $l_a = 25.38°$. The latitude of the spacecraft is then $L_a = 25.38°N$.
- At the later time of $t = 613.2$ min, where $\theta = 1367°$, we find $\sin l_a = -0.4782$ which yields the latitudinal angle $l_a = -28.56°$. The latitude is $L_a = -(-28.56°)S = 28.56°S$.

Note that the latitude depends solely on the angles θ and i so that the latitudes for the circular orbits have the same values of latitude as the $e = 0.3$ cases. However, the times at which the latitudes indicated are achieved are different, the circular orbit achieving latitude $25.38°N$ in 292.7 min and $28.56°S$ in 351.3 min.

5.4.3 THE SPACECRAFT'S LONGITUDINAL ANGLE

The determination of the longitude of the spacecraft ground track requires closer attention because of effects due to the rotation and oblateness of the earth, as well as the convention of using east and west longitudes. As shown in Figure 5.16, the longitudinal angle l_o (which is *not* the conventional earth-based longitude) lies on the equatorial plane and is always measured in an easterly direction. It is that angle between the line of the last ascending node and the projection of the spacecraft's radius vector on the equatorial plane is given by

$$\sin l_o = \frac{\tan l_a}{\tan i} \tag{5.59}$$

For the example $e = 0.3$ elliptical orbits at 488.5 and 613.2 min, we find that $\sin l_o = \tan(25.38°)/\tan(30) = 0.8216$ or $l_o = 55.26°$ and $\sin l_o = \tan(-28.56°)/\tan(30) = -0.9428$ or $l_o = -70.52°$, respectively. However, we wish to measure the longitudinal angle continuously from $0°$ to $360°$ so we have to refine the solution to Eqn (5.59).

We consider several points: (i) from Eqn (5.58) we see that the magnitude of the latitude angle l_a must be equal to or less than that of the orbital inclination angle i; (ii) the determination of l_o is confined to a single orbit so that within that orbit $0° \leq l_o \leq 360°$; (iii) to exploit the rotation of the earth in launching spacecraft we are generally interested in direct, or prograde, (easterly) orbits so that $0° \leq i \leq 90°$ ($i = 0$ denotes an equatorial orbit while $i = 90°$ denotes a polar orbit); (iv) because there are two possible values of l_o for a $\sin l_o > 0$ and two possible values of l_o for a $\sin l_o < 0$ we must be able to select that value of l_o appropriate to the position of the satellite.

This last point requires some careful consideration and can be illustrated by the following development. The true anomaly of the spacecraft at a point in any orbit may be defined by $\theta_n = \theta - 360°$ $(n - 1)$, where $n = 1, 2, 3\ldots$ represents the number of the orbit. The spacecraft rises in north latitude to a maximum, then descends in latitude, passing the equator, achieves a maximum in south latitude, and finally decreases in south latitude until it passes through the next ascending node, thereby initiating the next orbit. Thus while l_a varies from $0°$ to a maximum less than $90°$ in north latitude back down to $0°$, the longitudinal angle l_o varies from $0°$ to $180°$. Then, when l_a of the spacecraft continues moving from $0°$ to a maximum less than $90°$ in south latitude and finally back to $0°$, the longitudinal angle l_o varies from $180°$ to $360°$, thereby completing the nth orbit.

We may track this last detail by following the sum of the argument of periapsis and the true anomaly in an orbit. We find $\sin l_o$ using Eqn (5.59) noting that, based on the definitions of the angles l_a and i, the range of $\sin l_o$ will be $-1 \le \sin l_a \le 1$ yielding the basic result $-90° < l_o < 90°$. However, the longitudinal angle in any orbit lies in the range $0° \le l_o \le 360°$. Therefore we introduce a corrected longitudinal angle λ in the nth orbit as follows:

$$
\begin{aligned}
0° &< \theta_n + \omega_n < 90° \rightarrow \lambda_n = l_o \\
90° &< \theta_n + \omega_n < 180° \rightarrow \lambda_n = 180° - l_o \\
180° &< \theta_n + \omega_n < 270° \rightarrow \lambda_n = 180° + l_o \\
270° &< \theta_n + \omega_n < 360° \rightarrow \lambda_n = 360° - l_o
\end{aligned}
\tag{5.60}
$$

Continuing with our running example defined in Section 5.4.1, noting we assumed from the outset of this example that the argument of periapsis $\omega = 0$ for all orbits we find the following:

- At $t = 488.5$ min we are in the fourth orbit with $\theta_4 = 59.04°$ at latitude $25.38°$N resulting in $l_o = 55.26°$. Therefore from Eqn (5.60) with $\theta_4 = 59.04°$, we set $\lambda = 55.26°$.
- At the later time $t = 613.2$ min we are still in the fourth orbit but with $\theta_4 = 287.0°$ at latitude $28.56°$S and $l_o = -70.52°$. Using Eqn (5.60), we set $\lambda = 360 + (-70.52°) = 289.5°$.

We will deal with including the argument of periapsis ω and its change with time in the discussion which follows.

5.4.4 EFFECT OF THE EARTH'S ROTATION ON LONGITUDE

Although we have now defined a suitable longitudinal angle λ in the nth orbit, we haven't yet determined the actual longitude of the SSP of the spacecraft because we haven't considered the effects of the earth's rotation and its oblateness. The angular increment in longitude (in degrees) due to the earth's rotation of $360°$ in one sidereal day of $86,164$ s (23.9345 h) is

$$
\Delta_r = -f_E t = -\left(\frac{360}{86,164}\right) t = -4.178 \times 10^{-3} t
\tag{5.61}
$$

In Eqn (5.61), the negative sign accounts for the fact that the earth is moving in the same sense as the spacecraft for the usual direct, or prograde, orbits thereby reducing the longitudinal angle λ by that angular movement.

For our example cases

- When $t = 488.5$ min $= 29,310$ s, the increment in longitudinal angle is found to be $\Delta_r = -122.5°$.
- For the later time $t = 613.2$ min $= 36,792$ s, the increment in longitudinal angle is found to be $\Delta_r = -153.7°$.

5.4.5 **EFFECT OF REGRESSION OF NODES ON LONGITUDE**

The departure from spherical symmetry, or oblateness, of the earth results in a variation in gravitational acceleration which produces an out of plane force on the spacecraft causing the orbit to precess so that the line of the ascending node rotates slightly in each orbit. Defining the longitudinal angle of the ascending node as $l_{o,1}$ the rate of the movement is given, in degrees per second, by

$$\frac{dl_{o,1}}{dt} = -2.3963 \times 10^9 \frac{\cos i}{r^{3.5}} \tag{5.62}$$

For direct or prograde orbits this effect, called the regression of nodes, acts in the same sense as a speed-up of the earth's rotation. The angular increment in the longitudinal angle due to oblateness is given by

$$\Delta_o = \left(\frac{dl_{o,1}}{dt}\right) t = -2.3963 \times 10^9 \frac{t \cos i}{r^{3.5}} \tag{5.63}$$

In our continuing example case

- the angular increment due to oblateness for $t = 488.5$ min is

$$\frac{dl_{o,1}}{dt} = -2.3963 \times 10^9 \cos 30° \left[\frac{(6378+400)(1+0.3)}{1+0.3 \cos(1139°)}\right]^{-3.5} = -5.343 \times 10^{-5} \text{ deg/s}$$

$$\Delta_o = (-5.705 \times 10^{-5} \text{ deg/s})(29,310 \text{ s}) = -1.672°$$

- for the later time of $t = 613.2$ min, we obtain

$$\frac{dl_{0,1}}{dt} = -2.3963 \times 10^9 \cos 30° \left[\frac{(6378+400)(1+0.3)}{1+0.3 \cos(1367°)}\right]^{-3.5} = (-4.337 \times 10^{-5} \text{ deg/s})$$

$$\Delta_o = (-3.926 \times 10^{-5} \text{ deg/s})(36,790 \text{ s}) = -1.444°$$

5.4.6 **EFFECT OF ROTATION OF APSIDES ON LONGITUDE**

The lack of spherical symmetry of the earth produces another effect similar to regression of nodes. In an orbit over a spherically symmetric planet the argument of the periapsis, ω, remains fixed. However, in practical cases where complete spherical symmetry is lacking, the line of apsides, which defines the major axis of the orbital ellipse, as shown in Figure 5.1, may be caused to rotate. The normal acceleration of an orbiting spacecraft is enhanced when it nears the earth's equatorial bulge and this causes the line of apsides to rotate at the rate

$$\frac{d\omega}{dt} = 1.1943 \times 10^{10} \frac{4 - 5 \sin^2 i}{a^{3.5}(1 - e^2)^2} \tag{5.64}$$

Here the rotation rate is given in degrees per second with the semimajor axis a of the orbit measured in kilometers. There are three interesting points about Eqn (5.64): (i) it applies only to elliptic orbits because in circular orbits the argument of the perigee is undefined: every point on a circular orbit is the perigee (and the apogee); (ii) if $\sin i = (4/5)^{1/2}$, that is, the incidence angle of the plane of the orbit $i = 63.43°$, then $d\omega/dt = 0$ for any eccentricity and this orbit is called the critical, or Molniya, orbit; (iii) the rotation rate is positive for $0 < i < 63.43°$ and $123.43 < i < 180°$ and negative for $63.43° < i < 123.43°$. Note also that the effect is greatest for low orbits (semimajor axis a small) at low inclination (i small). The angular increment in the longitudinal angle due to the rotation of apsides is given by

$$\Delta_a = \frac{d\omega}{dt} t \tag{5.65}$$

Our example case has $e = 0.3$, $a = 9684$ km, and $i = 30°$ so that $d\omega/dt = 4.437 \times 10^{-4}$ deg/s and is positive because the inclination of the plane of the orbit is $30°$. We find that

- at $t = 488.5$ min, $\Delta_\omega = 13.00°$
- at $t = 613.2$ min, $\Delta_\omega = 16.32°$.

5.4.7 THE SPACECRAFT'S LONGITUDE

At this point, we have calculated the proper longitudinal angle λ in the nth orbit and this is measured in an easterly direction from the ascending node of that orbit. To determine the proper earth-based longitude L_o we must include the effects of the angular movement of the earth, the regression of nodes during the time since the first ascending node was passed, and the rotation of apsides. The corrected longitudinal angle is given as follows:

$$L_o = \lambda + \{l_{o,1}(0) + \Delta_o + \Delta_r + \Delta_a\} \tag{5.66}$$

Equation (5.66) states that the correct longitudinal angle of the SSP of the spacecraft on the nth orbit is equal to the longitudinal angle of the spacecraft at the time in question plus a correction. The correction is shown in braces in Eqn (5.66) and includes the longitudinal angle of the ascending node in the first orbit, the angular movement of the ascending node due to the earth's oblateness, the angular movement due to rotation of the earth, and the angular movement of periapsis due to the earth's oblateness, respectively.

There is yet another step in order to put the longitude into the appropriate convention of east and west longitudes. First, if the longitude of the ascending node in the first orbit $L_{o,1}(0)$ is in the eastern hemisphere and given in degrees east longitude, then the corresponding longitudinal angle $l_{o,1}(0) = L_{o,1}$ while if it is in the western hemisphere and given in degrees west longitude, then the corresponding longitudinal angle $l_{o,1}(0) = 360 - L_{o,1}$.

Consider our running example where

- We are in the fourth orbit with $\theta_4 = 59.04°$, latitude $L_a = 25.38°$N, longitudinal angle $\lambda = 55.26°$, increment due to rotation $\Delta_r = -125.3°$, increment due to regression of nodes $\Delta_o = -1.602°$, and increment due to rotation of periapsis $\Delta_a = 13.31°$. Then $\lambda + \Delta_r + \Delta_o + \Delta_a = 55.26° + (-122.50°) + (-1.672°) + (13.00) = -55.91°$. Let us assume that the ascending node in the first orbit is located at $150°$W. In the western hemisphere the longitudinal angle is $l_{o,1} = 360° - 150° = 210°$. Then adding our correction term we get $L_o = -55.91° + 210° = 154.1°$. Because $L_o < 180°$ the spacecraft is in the eastern hemisphere and the longitude is written as $L_o = 154.1°$E.
- At the later time $t = 613.2$ min in our running example, we are further on in the fourth orbit with $\theta_4 = 287.0°$, latitude $L_a = 28.56°$S, longitudinal angle $\lambda = 289.5°$, increment due to rotation

$\Delta_r = -153.7°$, increment due to oblateness $\Delta_o = -1.444°$, and increment due to rotation of periapsis $\Delta_a = 15.97°$. Then $\lambda + \Delta_r + \Delta_o + \Delta_a = (289.5°) + (-153.7°) + (-1.444°) + (16.32°) = 150.7°$. We already assumed the ascending node in the first orbit is located at $150°W$ so that $l_{o,1} = 360° - 150° = 210°$. Then adding our correction term we get $L_o = 150.7° + 210.0° = 360.7°$. Now at the later time $L_o = 0.7°$ and therefore the spacecraft just crossed the prime meridian into the eastern hemisphere with longitude $L_o = 0.7°E$.

5.4.8 A VIEW OF THE GROUND TRACKS OF THE EXAMPLE CASES

Combining all the information on the ground track of the example case we may illustrate the first four orbits on a grid where positive L_a denotes north latitude and negative L_a denotes south latitude, while positive L_o denotes east longitude and negative L_o denotes west longitude. The ground tracks for four orbits for the case where $e = 0.3$, the perigee altitude is $z_p = 400$ km, and the ascending node of the first orbit is located at $L_o = -150°$ or $150°W$ are shown in Figure 5.19. The regression of nodes is clearly illustrated in Figure 5.19 where the ascending node (the point at which the spacecraft passes the equator while moving in a northerly direction) for each successive orbit is seen to move westward.

The case of a circular orbit with an altitude equivalent to the eccentric orbit's perigee of $z_p = 400$ km has been shown to have similar characteristics but with a different time line. The ground tracks of the first two orbits of each are shown in Figure 5.20 where the first ascending node is at $L_o = -150°$ or $150°W$. Note that at the end of the second orbit the next ascending node for the eccentric orbit is at $137°$ or $137°E$ while that of the circular orbit is at $167°$ or $167°E$. The variation of the latitude of the ground tracks for the eccentric and circular orbits is shown as a function of time in Figure 5.21. The period of the eccentric orbit, at 158 min, is almost twice that of the circular orbit, at 92.5 min.

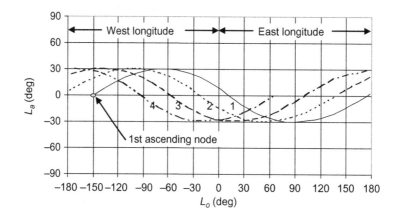

FIGURE 5.19

Ground track of four orbits of the example case where $e = 0.3$, $z_p = 400$ km, and the first ascending node is located at $-150°$ which is equivalent to $150°W$.

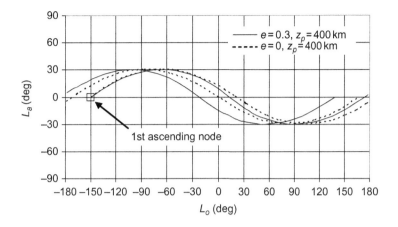

FIGURE 5.20

The ground tracks of the first two orbits of the example cases, one an eccentric orbit where $e = 0.3$, $z_p = 400$ km and one a circular orbit with an altitude equal to z_p are shown. The first ascending node for both is at $-150°$ which is equivalent to $150°$ W.

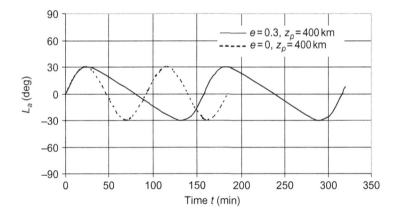

FIGURE 5.21

The spacecraft latitude is shown as a function of time for the first two orbits of the example cases, one an eccentric orbit where $e = 0.3$, $z_p = 400$ km and one a circular orbit with an altitude equal to z_p.

5.5 **THE SPACECRAFT HORIZON**

The photograph in Figure 5.22 shows the launch of the Space Shuttle Atlantis (STS-115) from Cape Canaveral as seen from the ISS. Part of the horizon visible from the ISS is clearly seen across the upper third of the photograph. The locus of all the horizon points visible from an orbiting spacecraft encloses a footprint on the ground as illustrated in Figure 5.23. The footprint constitutes

FIGURE 5.22

The launch of the space shuttle Atlantis (STS-115) from Cape Canaveral as seen from the ISS. Note the horizon which is visible across the upper third of the photograph (NASA).

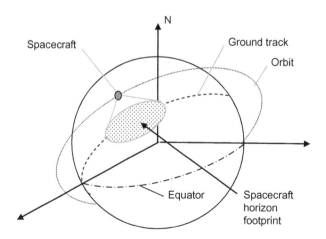

FIGURE 5.23

Illustration of the spacecraft horizon footprint within which line of sight communication with the ground is possible.

the area on the surface of the sphere enclosed by a cone tangent to the sphere when the apex of the cone is at the local radius of the spacecraft. As the spacecraft orbits, this footprint sweeps out a swath within which a line of sight between the ground and the spacecraft is available for two-way communication and observation.

5.5.1 THE HORIZON FOOTPRINT

The geometry of the horizon footprint is schematically illustrated in Figure 5.24 for two orbits. Both have the same perigee altitude $z_p = 400$ km but one is a circular orbit ($e = 0$) and the other is an ellipse with $e = 0.3$. The arc length of the footprint along the surface subtended by the tangent cone is given by

$$C_f = 2\beta R_E = 2R_E \cos^{-1}\left(\frac{1}{1 + \frac{z}{R_E}}\right) \tag{5.67}$$

The arc length of the horizon footprint given by Eqn (5.67) is shown as a function of the spacecraft altitude z in Figure 5.25. For comparison, the altitude for a satellite in GEO and the arc length of half the earth's circumference are indicated in Figure 5.25. A satellite in GEO therefore has a line of sight to almost the entire half-sphere of earth. Because the orbital period of the satellite is equal to that of the rotation of the earth, that almost complete view of the earth remains constant with time, making GEO of such great value for communication and observation.

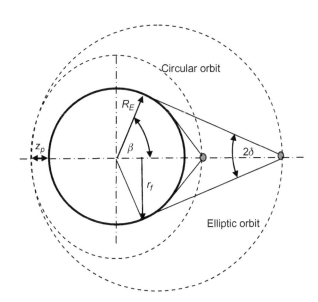

FIGURE 5.24

Schematic diagram of the geometry of the spacecraft horizon footprint radius r_f for an elliptic orbit with $e = 0.3$ and perigee altitude of z_p and a circular ($e = 0$) orbit with an altitude equal to z_p.

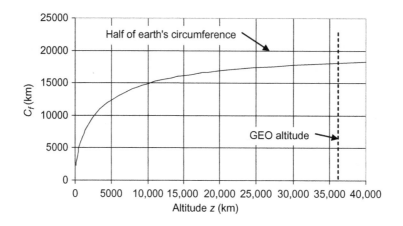

FIGURE 5.25

The arc length of the horizon footprint C_f is shown as a function of the spacecraft altitude z.

The scale of Figure 5.25 obscures any detail regarding the arc length of the footprint at the low altitudes of manned orbital flight, but, for example, at $z = 400$ km the arc length found from Eqn (5.67) is $C_f = 4400$ km, somewhat larger than the great circle arc between New York and San Francisco. Thus a spacecraft, like the ISS, at $z = 400$ km above the center of the United States has a line of sight in any direction that spans the country from coast to coast. However, we must keep in mind that the arc length of this footprint is only about 10% of the earth's circumference and the spacecraft's orbital period is only around 90 min so that line of sight communication between the spacecraft and a point on earth is limited in duration to approximately 9 min or less.

5.5.2 COMMUNICATION WITH THE SPACECRAFT

In practical cases where instrument limitations and atmospheric interference may be encountered communication all the way to the horizon isn't necessarily possible. For example, suppose that the practical line of sight is reduced from the value δ by an incremental angle ε as illustrated in Figure 5.26. The included angle β to the tangent point is reduced to the value β' and the triangle formed by the radial coordinate of the spacecraft, the radius of the earth, and the length of the practical line of sight includes the angle ϕ which may be found from the following trigonometric equation:

$$\frac{R_E}{\sin(\delta - \varepsilon)} = \frac{r}{\sin \phi}$$

Then the angle β' may be found from the relation

$$\beta' = 180° - \phi - (\delta - \varepsilon)$$

In the case of a spacecraft in circular orbit at $z = 400$ km, which was mentioned previously, the angle $\delta = 70.22°$ and $\beta = 19.78°$. If the actual field of view is reduced by $\varepsilon = 3°$ then we find that $\phi = 101.6°$ and $\beta' = 11.23°$. The period of the spacecraft's orbit is 92.52 min so the time for it to travel an angle of $2\beta' = 22.46°$ is $t = (22.46/360)(92.52 \text{ min}) = 5.77$ min, considerably less than the

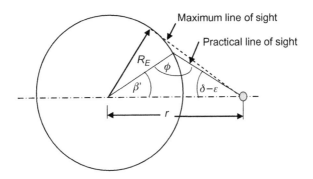

Maximum line of sight

Practical line of sight

R_E

ϕ

β'

$\delta - \varepsilon$

r

FIGURE 5.26

Geometry of the case of reduced line of sight due to instrument limitations or atmospheric interference.

maximum of about 9 min indicated previously. Because this is a rather short time period for direct communication with the spacecraft it is common to uplink to a satellite system which then down-links to an earth station. Indeed, this was the major reason for developing the Tracking and Data Relay Satellite System (TDRSS).

5.6 INTERPLANETARY TRAJECTORIES

Human spaceflight beyond earth orbit inspires ideas for missions of exploration and requires sub-stantially greater expense and planning. However, it is understood that departures for such endea-vors are best carried out from earth orbit, as exemplified by the Apollo program. Although detailed analysis of lunar and interplanetary missions, like those to other planets, their moons, asteroids, and comets, is beyond the scope of this book, we will discuss some simplified notions regarding such efforts.

5.6.1 LUNAR TRAJECTORIES

For simplicity, we assume that the moon follows a circular orbit with radius $r_M = 384,400$ km, a reasonable assumption because its orbital eccentricity is $e \sim 0.055$. Furthermore, we will neglect the influence of the moon's gravity on the motion of the spacecraft. Transfer orbits for a spacecraft from a circular LEO to the moon's orbit are schematically illustrated in Figure 5.27. The minimum energy Hohmann transfer orbit is shown as a solid line and it will take the greatest amount of time to reach the moon's orbit. As the velocity increment ΔV_1 given to the spacecraft in LEO is increased the transfer orbit will widen as shown by the dashed line in Figure 5.27 and the moon's orbit will be reached in shorter and shorter times. Of course if ΔV_1 is too small the spacecraft will never reach the moon's orbit, as indicated by the dotted line.

Assuming that the altitude of the LEO is $z = 400$ km leads to $r_1 = R_E + z = 6778$ km and from Eqn (5.21) we find that the semimajor axis of the Hohmann transfer ellipse is $a = (r_1 + r_2)/2 = (384,400$ km $+ 6771$ km$)/2 = 195,600$ km. Likewise, from Eqns (5.19) and (5.20), we find $\Delta V_1 = 3.082$ km/s and $\Delta V_2 = -0.8287$ km/s, respectively. The spacecraft speed in the 400 km

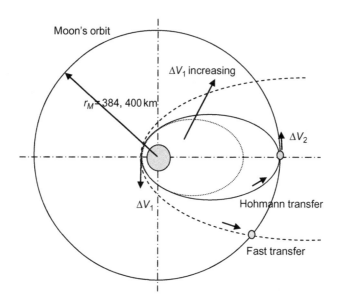

FIGURE 5.27

Various possible transfer orbits between LEO and the moon are shown.

LEO is $V_1 = 7.669$ km/s so the velocity required to impulsively put the spacecraft into the Hohmann transfer orbit is $V = V_1 + \Delta V_1 = 10.751$ km/s. The time required for the transfer is found from Eqn (5.22) to be $\tau_H = 119.6$ h or slightly less than 5 days.

If instead the spacecraft is inserted into a larger orbit, like that shown by the dashed line in Figure 5.27, the moon's orbit could be intercepted earlier than with the Hohmann transfer. In this case, the transit time would be shorter, but the velocity increment needed to place the spacecraft in the moon's orbit would be greater. Consider a simple fast transfer case where the center of the elliptic orbit is located at the distance r_M from the center of the earth. Then the semimajor axis of the transfer orbit is $a = R_E + z + r_M = 391{,}171$ km. In this case, the eccentric anomaly defined by Eqn (5.54) yields $E = 89.13°$ and, using Eqn (5.55), the time to intercept the orbit of the moon is $t = 61.68$ h or 2.57 days, about half the time required by the Hohmann transfer. However, the change in velocity required to put the spacecraft on the moon's orbit is now $\Delta V = 1.435$ km/s, almost twice that required for the Hohmann transfer. In this case, the insertion velocity for the spacecraft in LEO is 10.8 km/s compared to 10.75 km/s for the Hohmann transfer.

Using Eqn (5.51), one may determine the propellant consumption in the two cases. Note that the sign of the velocity increment ΔV is irrelevant as far as fuel consumption is concerned but simply serves to indicate the direction in which the required thrust force must be applied. Assuming a relatively high specific impulse typical of liquid propellant engines, i.e., $I_{sp} = 400$ s, the Hohmann transfer consumes an amount of propellant equal to about 63% of the initial mass of the spacecraft while the fast transfer case consumes about 69%. On the other hand, the longer trip time of the Hohmann transfer consumes more mass in life support material and exposes the crew to greater cosmic radiation exposure. This is a good example of the trade-offs that must be considered in the design of spacecraft, particularly for human flight.

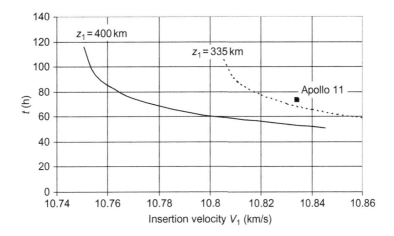

FIGURE 5.28

The calculated time required to intercept the moon's orbit is shown as a function of insertion velocity V_1 for two values of the initial LEO altitude z_1.

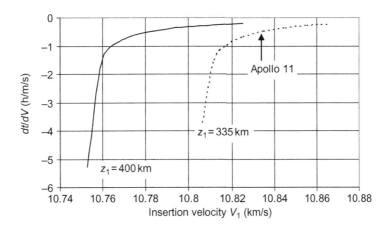

FIGURE 5.29

The sensitivity of the translunar travel time is shown as a function of the insertion velocity for two altitudes in LEO. The insertion velocity for Apollo11, for which $z_1 = 335$ km, is also shown.

In Figure 5.28, the time required to intercept the moon's orbit is shown as a function of the insertion velocity V_1 for two values of the initial LEO altitude z_1. The flight data for Apollo11, for which $z_1 = 335$ km, is also shown. It is clear that the flight time is very sensitive to the insertion velocity as well as to the initial LEO altitude.

The flight data for Apollo11, for which $z_1 = 335$ km, is also shown.

It is instructive to examine the sensitivity of the translunar travel time as a function of the insertion velocity in order to judge the best insertion speed to employ. Figure 5.29 shows the variation of travel time measured in hours with the insertion velocity measured in meter per second. Choosing an insertion velocity just beyond the "knee" of the curves shown in Figure 5.29 ensures an economical approach to translunar insertion. Increasing the insertion velocity much beyond the knee uses additional propellant but provides little additional reduction in transit time.

It should be noted that the escape velocities for the translunar analysis are 10.85 km/s for $z_1 = 400$ km and 10.90 km/s for $z_1 = 335$ km, values only slightly larger than the insertion velocities for the highly eccentric elliptic orbit shown here.

5.6.2 MARTIAN TRAJECTORIES

Human exploration of other planets in the solar system remains a topic of interest, particularly because of the success of the Apollo mission in visiting the moon. The planet Mars is attractive in this regards because it is relatively near, its environment is reasonably hospitable to human presence, and its surface has been safely visited and investigated by unmanned robotic systems. A simple case of a Hohmann transfer between earth and Mars is illustrated in Figure 5.30. For simplicity, we assume the orbits of both planets are circular and coplanar. In actuality $e_E = 0.0167$ and $e_{Mars} = 0.093$ while the orbital inclination of Mars relative to earth is 1.9°. The radii of earth and Mars are given in astronomical units where 1 AU $= 1.496 \times 10^8$ km.

Figure 5.30 shows that the semimajor axis of the Hohmann transfer ellipse is $a = (r_1 + r_2)/2 =$ 1.26 AU $= 1.882 \times 10^8$ km. The Hohmann transfer described previously in Section 5.1.5 was between two orbits around earth. Now the two orbits are around the sun so we must calculate a new value for the central body gravitation constant k/m. Using Eqn (5.6) and Table 5.1, we may calculate

$$\left(\frac{k}{m}\right)_{sun} = \frac{m_{sun}}{m_E} m_E G = (3.329 \times 10^5)(3.986 \times 10^5 \text{ km}^3/\text{s}^2) = 1.327 \times 10^{11} \text{ km}^3/\text{s}^2$$

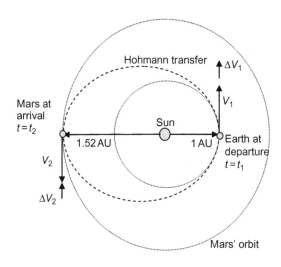

FIGURE 5.30

A Hohmann transfer between Earth and Mars.

Likewise, from Eqn (5.5) we find the spacecraft speed on the Hohmann orbit starting at the earth orbit is $V_1 = 32.64$ km/s while at the Mars orbit it is $V_2 = 21.55$ km/s. However, the orbital velocity of the earth around the sun is calculated from Eqn (5.5) as $V_E = 29.78$ km/s. Similarly, the orbital velocity of Mars around the sun is $V_{Mars} = 24.15$ km/s. The spacecraft velocity increment required to impulsively put the spacecraft into the Hohmann transfer orbit is $\Delta V_1 = V_1 - V_E = 32.71$ km/s $- 29.78$ km/s $= 2.93$ km/s. At the transfer point to the Martian orbit, the spacecraft velocity increment required to impulsively put the spacecraft into the Martian orbit is $\Delta V_2 = V_2 - V_{Mars} = 21.50$ km/s $- 24.15$ km/s $= -2.654$ km/s. Note that the spacecraft in earth orbit must be given excess speed to move onto the Hohmann transfer orbit and then must apply negative thrust to slow down the spacecraft to the orbital speed of Mars. This is illustrated in Figure 5.30.

The time required for the transfer is found from Eqn (5.22) to be

$$\tau_H = \pi \sqrt{\frac{a^3}{m_{sun}G}} = \pi \sqrt{\frac{(1.888 \times 10^8 \text{ km})^3}{1.327 \times 10^{11} \text{ km}^3/\text{s}^2}} = 258.3 \text{ days}$$

The same method may be used to calculate the time for a Hohmann transfer from earth to other planets. Because both planets are in motion it is important to note that planning for any transfer of a manned spacecraft must keep in mind the return trip. For example, if the transfer to Mars is carried out as shown in Figure 5.30, after the 258.3 days the spacecraft would be at the position shown for Mars, but the earth would have moved through 261°. The angular positions of the Earth and Mars, assuming circular orbits, are shown in Figure 5.31 as a function of time. The synodic period, that is, the time lapse between recurring equivalent relative angular positions between two planets is seen to be about 2.13 years for the Earth–Mars system. Therefore, if one wishes to carry out a round-trip minimum energy transfer, about 16 months must be spent on or around Mars until the next favorable alignment occurs.

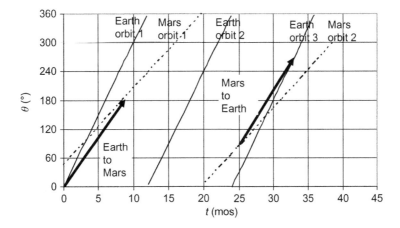

FIGURE 5.31

Orbital positions of Earth and Mars are shown as a function of time in months. Possible Hohmann transfers are shown as bold arrows.

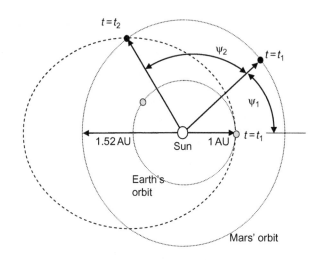

FIGURE 5.32

A transfer orbit between Earth and Mars showing the phase angles between the two planets.

Transfer orbits other than those requiring minimum energy may be used, as illustrated in Figure 5.32. The spacecraft is launched at $t = t_1$ at a point where Mars is ahead of the Earth by an angle ψ_1. The spacecraft arrives at Mars' orbit coincident with the arrival of the planet itself. A transfer orbit with parameters e and p has a calculable time of flight $t_2 - t_1$ to cover the angular distance $\psi_1 + \psi_2$ and this must equal the time required for Mars to travel the angular distance ψ_2. The angle ψ_1 called the phase angle at departure; for the Hohmann transfer considered previously the phase angle at departure is about 45°.

The difficulty of a mission to Mars should not be underestimated. Over the past 50 years there have been 49 attempts, all of course unmanned, of which 26 failed, 2 were partial successes, and 21 were successes.

5.6.3 OTHER PLANETARY TRAJECTORIES

Human exploration of planets of the solar system other than Mars is currently a remote possibility and not only because only Mars is most receptive to human presence on its surface and possibly has supported life at some time. Technical issues are posed by the characteristics of the other planets taken from NSSDC (2013) and listed in Table 5.2. The transit time to the other planets using the minimum energy Hohmann transfer has been calculated using the methods of this chapter and are shown in Table 5.2. The outer planets all require more than about 3 years to reach and protecting crew members from exposure to cosmic radiation poses formidable mass penalties. The synodic period of all the outer planets is just about a year, but the nearest planets, Mars and Venus,

Table 5.2 Some Orbital Characteristics of Planets in the Solar System

Planet	Orbital V (km/s)	Orbital e	Orbital i (degrees)	Orbital r_p (10^6 km)	Synodic Period (days)	Hohmann[a] Transfer (years)	ΔV_1[b] (km/s)
Mercury	47.9	0.205	7.0	46.0	115.9	0.289	−7.55
Venus	35.0	0.007	3.4	107.5	583.9	0.400	−2.51
Earth	29.8	0.0170	0.0	147.1	–	–	–
Mars	24.1	0.094	1.9	206.6	777.9	0.709	2.93
Jupiter	13.1	0.049	1.3	740.5	398.9	2.73	8.78
Saturn	9.7	0.057	2.5	1353	378.1	6.09	10.3
Uranus	6.8	0.046	0.8	2741	369.7	16.1	11.3
Neptune	5.4	0.011	1.8	4445	367.5	30.6	11.6
Pluto	4.7	0.244	17.2	4435	366.7	45.1	11.8

[a]Calculated using methods of this chapter.
[b]Departure speed relative to earth orbital speed for Hohmann transfer.
Data from NSSDC (2013).

have about twice that, making return launch windows more restrictive. The required velocity of insertion into a Hohmann transfer orbit is about three times larger for the outer planets than for the closer planets, thereby making stressing demands on the spacecraft propulsion system. Note that the inner planets require a negative velocity increment relative to the earth's orbital velocity because the orbital transfer is from a larger orbit to a smaller one.

Successful human spaceflight to the moon has been proven by the Apollo program. A manned mission to Mars in the 2030s will require resuming translunar space activities with the Orion space capsule and the Space Launch System (SLS) in the 2014–2020 time period. Then a true step into interplanetary space will be accomplished by Orion's visiting an asteroid around 2025. NASA's Asteroid Redirect Mission (ARM) is aimed at capturing a 7 m to 10 m diameter, approximately 500 ton asteroid and redirecting it to a stable orbit in translunar space. This program is expected to provide a means for protecting the earth from near-earth objects (NEO) that are on a possible collision path and to help develop the technological capabilities for success-ful interplanetary travel. As far as orbital dynamics are concerned, the program will further develop and refine techniques of deep space navigation and rendezvous necessary for further human exploration space.

5.7 CONSTANTS AND CONVERSION FACTORS

Some of the constants and conversions used in this chapter are collected in Table 5.3.

Table 5.3 Selected Constants and Conversion Factors

Parameter	Value	Description
Sidereal day	86,164 s	Time for the earth to make a revolution
AU	1.496×10^8 km	Mean distance from the earth to the sun
f_E	7.292×10^{-5} s^{-1}	Mean rotational speed of the earth about its axis
$m_E G$	3.9860×10^5 km^3/s^2	Gravitational constant for the earth
R_E	6371 km	Mean radius of the earth
V_E	7.909 km/s	Circular orbit velocity evaluated at R_E
r_M	384,400 km	Mean distance from the earth to the moon
$m_{sun} G$	1.327×10^{11} km^3/s^2	Gravitational constant for the sun
G	6.673×10^{11} m^3/kg-s^2	Universal gravitational constant
1 nm	1852 m	Standard distance conversion
1 ft	0.3048 m	Standard length conversion

5.8 NOMENCLATURE

a	semimajor axis of elliptical orbit
c	half the distance between the foci of an elliptical orbit
C_f	arc length of the footprint along the surface of the earth
E	eccentric anomaly
e	orbit eccentricity, see Eqn (5.4)
F	force
f	average angular speed
G	universal gravitational constant $= 6.673 \times 10^{-11}$ m^3/kg-s^2
g	gravitational acceleration
h	angular momentum
I_{sp}	specific impulse
i	incidence angle of orbital plane
k	planetary constant, see Eqn (5.6)
L_a	latitude
L_o	longitude
l_a	latitude angle
l_o	longitudinal angle
$l_{o,1}(0)$	longitudinal angle of the ascending node in the first orbit
m	spacecraft mass
m_p	propellant mass
$\dot{m}_p$	propellant mass flow rate
n	index for number of orbits
p	semilatus rectum of conic section
R_E	radius of the earth
r	radial coordinate
r_M	mean distance of the moon from the earth
t	time
U	energy

V	velocity
ΔV	velocity increment for orbital transfer
x	abscissa
y	ordinate
z	geometric altitude
z_d	drag-limited orbital altitude
z_o	altitude in circular orbit
z_r	radiation effects-limited orbital altitude
α	angle between velocity vectors
β	angle between r and R_E at the maximum line of sight tangent point
β'	angle between r and R_E at the practical line of sight tangent point
Δ_a	correction in longitude due to the rotation of apsides
Δ_o	correction in longitude due to earth's oblateness
Δ_r	correction in longitude due to earth's rotation
δ	angle between radius vector and maximum line of sight
γ	flight path angle
ε	angle between the maximum and practical lines of sight
ϕ	$180 - [\beta' + (\delta - \varepsilon)]$
λ	corrected longitudinal angle
ψ_1	phase angle at departure
θ	true anomaly
τ	orbital period
ω	argument of periapsis

5.8.1 SUBSCRIPTS

a	apoapsis or apogee
E	Earth
e	conditions at atmospheric entry interface
esc	conditions at planetary escape
f	horizon footprint
H	denotes Hohmann transfer orbit
kin	kinetic
P	planet
p	periapsis or perigee
pot	potential
1	denotes orbit 1
2	denotes orbit 2
3	denotes orbit 3

REFERENCES

Bate, R. R., Mueller, D. D., & White, J. E. (1971). *Fundamentals of astrodynamics.* New York, NY: Dover.

Brown, C. D. (1998). *Spacecraft mission design.* Reston, VA: AIAA.

NSSDC (2013). *Planetary fact sheet.* <www.nssdc.gsfc.nasa.gov/planetary/factsheet/>.

Pisacane, V. L. (2005). *Fundamentals of space systems* (2nd ed.). New York, NY: Oxford.

USAF (1965). *Space planners guide.* HQ USAF Systems Command.

ATMOSPHERIC ENTRY MECHANICS

6.1 GENERAL EQUATIONS OF MOTION

We are primarily concerned with the passage of manned vehicles through the atmosphere after carrying out suborbital, orbital, or planetary return missions. For the purposes of preliminary design and analysis, we will consider the trajectories to lie in a plane passing through the center of the planet, so that out-of-plane forces are not considered unless specifically indicated. The general nature of the trajectory and the variables describing it are shown in Figure 6.1. Figure 6.1a is drawn roughly to scale to provide an idea of the relative size of the atmospheric envelope for a typical low earth orbit (LEO) of, say, 400 km altitude. Figure 6.1b, which is not to scale, shows the variables of interest for analysis of a spacecraft following a trajectory that will ultimately bring it down to the surface. The thrust force F is directed at an angle χ to the velocity V of the center of mass of the spacecraft. The velocity V is displaced an angle γ to the local horizon with $\gamma < 0$ as shown for a reentering spacecraft. For ascending craft, like launch vehicles, the velocity would be pointing above the local horizon and $\gamma > 0$. The lift L acts normal to the velocity, the drag D acts in the direction opposite to V, and the weight of the spacecraft mg acts toward the center of the Earth. The geometric altitude of the spacecraft is z and ω is the angle describing the motion of the radius r to the spacecraft. The range x of the spacecraft is measured by the trace of the radius vector on the earth's surface $x = \omega R_E$.

The general equations of motion along and normal to the trajectory are given, respectively, as follows:

$$a_t = \frac{dV}{dt} = \frac{F_t}{m}$$

$$a_n = V\left(\frac{d\gamma}{dt} - \frac{d\omega}{dt}\right) = \frac{F_n}{m}$$

The gravitational acceleration varies with altitude and, according to Newton's law of gravitation, may be written as

$$g = g_E \frac{R_E^2}{r^2} = \frac{k}{r^2} = g_E \frac{1}{\left(1 + \dfrac{z}{R_E}\right)^2}$$

The mean radius of the Earth is taken to be 6371 km and in the case of a typical LEO of $z = 400$ km, the local gravitational acceleration $g = 0.885g_E$. Thus, the gravitational acceleration is still substantial in LEO. At the nominal entry threshold of $z = 100$ km, the gravitational acceleration

Manned Spacecraft Design Principles. DOI: http://dx.doi.org/10.1016/B978-0-12-804425-4.00006-4

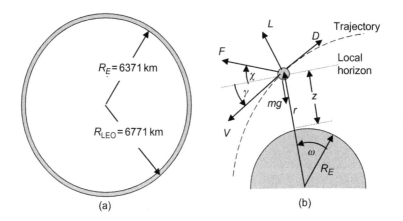

FIGURE 6.1

Schematic diagram of a spacecraft on a planar trajectory around the Earth, (a) relative size of the region of manned LEO operations and (b) notation for trajectory analysis.

$g = 0.969g_E$. At $z = 70$ km, the altitude at which deceleration effects begin to take hold in manned entry missions, $g = 0.978g_E$ making the assumption of $g = g_E$ during reentry a reasonable approximation.

For preliminary design and analysis purposes, we will assume that the Earth is not rotating and that the spacecraft is a point mass. The Earth makes one rotation per day about its axis so that at the equator, the surface velocity is maximum and is given by $2\pi R_E/24 = 1667$ km/h $= 0.463$ km/s. The duration of entry of the Space Shuttle Orbiter starting at $z = 120$ km is about 30 min. Cape Canaveral is located at 28.5°N and 80.5°W and would move about 1400 km to the east during that reentry period. Obviously, for practical mission planning applications the rotation of the Earth must be taken into account. When the stability and control aspects of the spacecraft are considered the vehicle will be treated as a rigid body.

Under the conditions described above, the acceleration of the spacecraft along and normal to the flight path, respectively, may be written in nondimensional form as follows:

$$\frac{1}{g_E}\dot{V} = \frac{F}{mg_E}\cos\chi - \frac{D}{mg_E} - \frac{k}{r^2 g_E}\sin\gamma \tag{6.1}$$

$$\frac{1}{g_E}V\dot{\gamma} = \frac{F}{mg_E}\sin\chi + \frac{L}{mg_E} - \frac{k}{r^2 g_E}\cos\gamma + \frac{V^2}{rg_E}\cos\gamma \tag{6.2}$$

The rates of change of altitude z and of the trajectory angle ω are given, respectively, by

$$\dot{r} = \dot{z} = V\sin\gamma \tag{6.3}$$

$$\dot{\omega} = \frac{\dot{x}}{R_E} = \frac{V}{r}\cos\gamma \tag{6.4}$$

Equations (6.1) and (6.2) show that the spacecraft's motion is characterized by V and γ and control of that motion is achievable only through modulation of the thrust F, the lift L, and the drag D.

If thrust is being produced, the spacecraft will be consuming propellant such that the rate of change of the mass of the spacecraft is given by

$$\dot{m} = -\frac{F}{g_E I_{sp}} \tag{6.5}$$

The entry phase of spaceflight marks the end of a mission and it is generally impractical to carry propellant throughout an entire mission in order to use it to produce thrust for entry trajectory control. Thrust may be employed for spacecraft attitude control during reentry because such use requires relatively small thrusters and correspondingly small propellant expenditure.

It is usually more efficient to use passive aerodynamic means to generate forces of a magnitude sufficient to influence the overall trajectory. Drag is always produced by a spacecraft during reentry and a small degree of asymmetry in shape or angle of attack of the spacecraft is often sufficient to produce a useful lift force. Appropriate location of the center of mass of the spacecraft permits the lift and drag to generate a pitching moment to assist in attitude control. At high altitudes, aerodynamic forces are ineffective and small thrusters are used to produce pitching and rolling moments. Pitch thrusters change the angle of attack and thus the magnitude of the lift vector, while the roll thrusters bank the spacecraft, redirecting the lift vector, as will be discussed subsequently.

6.2 GLIDING ENTRY TRAJECTORIES

For the reasons discussed above, it is typical for entry to be accomplished by gliding (zero thrust) with the flight starting at an entry angle $\gamma < 0$, because at the entry interface $V = V_e$ is directed below the local horizon, and for which the following assumptions apply: $F = 0$ and $m = $ constant. The stipulation of constant mass is based on the premise that because no thrust is being generated the mass of the vehicle is fixed, outside of some possible ablation of heat protection material. Under these assumptions, the equations of motion become

$$\frac{1}{g_E}\dot{V} = -\frac{D}{mg_E} - \sin\gamma \tag{6.6}$$

$$mV\dot{\gamma} = \frac{mV^2}{r}\cos\gamma + L - mg_E\cos\gamma \tag{6.7}$$

The auxiliary relation between altitude and the primary variables is

$$\dot{z} = \dot{r} = V\sin\gamma \tag{6.8}$$

Using Eqn (6.8) we may transform the time derivatives into space derivatives as follows:

$$\frac{d}{dt} = \frac{d}{dz}\dot{z} = V\sin\gamma\frac{d}{dz} \tag{6.9}$$

The lift and drag forces may be written in the conventional aerodynamic forms:

$$L = q(C_L S) = \frac{1}{2}\rho V^2 (C_L S) = \frac{1}{2}\gamma p M^2 (C_L S)$$

$$D = q(C_D S) = \frac{1}{2}\rho V^2 (C_D S) = \frac{1}{2}\gamma p M^2 (C_D S)$$

The quantity q is called the dynamic pressure, C_L and C_D are the lift and drag coefficients, respectively, and S is the reference area on which those coefficients are based. In gliding reentry, the lift and drag provide the only control over the motion of the spacecraft during its descent.

The nominal entry altitude for the Earth is around 100 km and this fact permits us the possibility of making several useful assumptions with no loss in generality. The variation of the gravitational acceleration with altitude is assumed to be zero, such that $g \simeq g_E$ below $h = 100$ km. This has been previously shown to be correct to within an error of about 3%. In the same fashion, we set $r \simeq R_E$ which is correct to within an error of 1.6%. It has also been noticed that, in the development of the model atmosphere in Chapter 2, it was pointed out that below 100 km in altitude the temperature of the atmosphere has relatively little variation and therefore the pressure and density variations with altitude are approximately exponential in shape. It was shown that one may fit an exponential curve that best describes the variation in the region of interest. Using this approach, we may approximate the density profile up to about 100 km by the following expression:

$$\rho \approx \rho_0 \exp(-z/H) \tag{6.10}$$

The quantity H was introduced in Chapter 2 as a scale height of the atmosphere given by

$$H = \frac{RT}{g} = (0.02926 \text{ km}/K)T$$

If the temperature in the atmosphere were constant, then H would be constant, and the equation of hydrostatic equilibrium would result in the atmospheric pressure profile being exponential. The perfect gas equation of state would then further require that the atmospheric density profile be exponential as well. The average scale height from $z = 10$ to 100 km is 6.9 km, which corresponds to an average temperature of 235.87 K. Choosing this scale height yields

$$\frac{p}{p_{sl}} = \delta = \exp\left(-\frac{z}{H}\right)$$

For consistency, with the temperature being assumed constant at 235.8 K, the sea level density would be greater than the standard value and the density profile would have to be given by

$$\frac{\rho}{\rho_{sl}} = \sigma \approx \frac{\rho}{\rho_0} = \exp\left(-\frac{z}{H}\right)$$

Early entry studies by NASA used $\rho_0 = 1.752$ kg/m^3 and $H = 7.16$ km, while some later studies used $\rho_0 = 1.752$ kg/m^3 and $H = 6.7$ km. Here, it is suggested that because $\rho_{sl} = 1.225$ kg/m^3 we should use $\rho_0 = (\rho_0/\rho_{sl})\rho_{sl} = (1.22)(1.225 \text{ kg/m}^3) = 1.5$ kg/m^3 when using the pressure-based scale height $H = 6.9$ km and this should be appropriate for altitudes between 10 and 100 km. Of course, values for H and ρ_0 can be chosen as desired so as to optimize the accuracy of the approximation in specific ranges of the atmosphere.

Introducing Eqns (6.9) and (6.10) into Eqns (6.6) and (6.7) and setting $z/H = s$ as a nondimensional altitude results in the following form of the gliding entry equations:

$$\frac{d}{ds}\left(\frac{V^2}{g_E R_E}\right) + \frac{\rho_0 H C_D S}{m}\frac{e^{-s}}{\sin \gamma}\left(\frac{V^2}{g_E R_E}\right) = -\frac{2H}{R_E} \tag{6.11}$$

$$\frac{d}{ds}\cos \gamma - \frac{H}{R_E}\left(\frac{g_E R_E}{V^2} - 1\right)\cos \gamma = -\frac{1}{2}\left(\frac{L}{D}\right)\frac{\rho_0 H C_D S}{m}e^{-s} \tag{6.12}$$

The independent variable in Eqns (6.11) and (6.12) is $s = z/H$, the nondimensional altitude, while the dependent variables are the specific kinetic energy of the spacecraft $V^2/2$ and the flight path angle γ. The three dimensionless parameters that appear are the spacecraft's lift-to-drag ratio, $L/D = O(1)$, the atmospheric model scale factor $H/R_E = 1.083 \times 10^{-3} \sim 923.3^{-1}$, and the quantity

$$\frac{\rho_0 H C_D S}{m} = (\rho_0 g_E H)\left(\frac{C_D S}{m g_E}\right) = \frac{\rho_0 g_E H}{B} \tag{6.13}$$

6.2.1 THE BALLISTIC COEFFICIENT B

The numerator on the right-hand side of Eqn (6.13) depends upon the values for ρ_0 and H used in the atmospheric model. For the values quoted previously and using $g_E = 9.807$ m/s^2, the product $\rho_0 g_E H = 101.5 \sim 101.3$ kPa $= 1$ atm. Therefore

$$\rho_0 g_E H \frac{C_D S}{m g_E} = \frac{\rho_0 g_E H}{B} \approx \frac{1}{B'}$$

The quantity B, the ratio of the weight of the spacecraft to its drag area, is called the ballistic coefficient and has the units of pressure. Because $\rho_0 g_E H \sim 1$ atm, it is convenient to use B', the normalized ballistic coefficient which is measured in atmospheres. As the ballistic coefficient of the spacecraft becomes larger, the effect of drag on it becomes smaller. The numerical values given here refer to the earth's conditions and other values will arise for other planets.

Note that the parameter defined in Eqn (6.13) is approximately the ratio of sea level atmospheric pressure to the ballistic coefficient (which has the units of pressure). Most manned capsules like Mercury, Gemini, Apollo, Vostok, Soyuz, Shenzhou, and the proposed Orion capsule have values of B in the range of 2−5 kPa, as discussed in Chapter 10. A quick estimate for winged reentry spacecraft like the X-38 crew rescue vehicle suggests a value of $B \sim 3.7$ kPa, while the Space Shuttle has a value of $B \sim 4$ kPa.

However, entry vehicles like those carried by ICBMs have values of B around 10 kPa or more (the GE Mark six reentry vehicle, which was carried aloft by a Titan II launch vehicle and contained a W-53 multi-megaton nuclear warhead, had $B \sim 30$ kPa). Later entry vehicles containing nuclear warheads had even larger values of B. The GE Mark 12 reentry vehicle, which is carried in groups of three aboard the Minuteman III ICBM, is much smaller in size than the Mk 6 but has a ballistic coefficient $B \sim 300$ kPa. This high value is due to the low drag coefficient of its slender and relatively sharp conical shape.

Meteors also have high values of the ballistic coefficient B: common chondrite meteors, stony-iron meteors, and iron meteors have B approximately equal to $22d$, $30d$, and $50d$ kPa, respectively, where d is the meteor diameter (in meters). The Chelyabinsk meteor that entered the earth's atmosphere on February 15, 2013 was a common chondrite having a diameter estimated to be in the range of 17 m $< d < 20$ m, so that the value of its ballistic coefficient was probably in the range 375 kPa $< B < 440$ kPa.

We will see in Chapter 10 that a reasonable correlation for the mass (in kilogram) of a manned spacecraft is given by

$$m \approx 470(d^2 l)^{\frac{2}{3}}$$

Here d and l represent the characteristic cross-sectional and longitudinal dimensions (in meters) of the spacecraft, respectively. Then the ballistic coefficient, measured in kilopascals, is roughly given by

$$B \approx \frac{5.87}{C_D}\left(\frac{l}{d}\right)^{\frac{2}{3}} \qquad (6.14)$$

Thus, for capsule-like spacecraft with $l/d \sim O(1)$ and $C_D \sim O(1)$, the ballistic coefficient will be around 5 kPa, the value already indicated as typical for recent and current spacecraft. The drag coefficient here is based on the maximum cross-sectional area of the vehicle.

6.2.2 THE EQUATIONS IN TERMS OF DENSITY

We may change the independent variable once again by combining the density ratio, $\sigma = \rho/\rho_0$, and the ballistic coefficient B into a new dimensionless variable

$$\eta = \frac{\rho_0 g_E H}{B}\sigma = \frac{\sigma}{B'}$$

Then, we may change variables from s to η using the following relation:

$$\frac{d}{ds} = \frac{d\eta}{ds}\frac{d}{d\eta} = \frac{d}{ds}\left(\frac{e^{-s}}{B'}\right)\frac{d}{d\eta} = -\frac{\sigma}{B'}\frac{d}{d\eta}$$

In addition, we define a nondimensional energy ratio

$$U = \frac{V^2}{g_E R_E} = \left(\frac{V}{V_E}\right)^2 \qquad (6.15)$$

The quantity $V_E = [(6.371 \times 10^6 \text{ m})(9.807 \text{ m/s}^2)]^{1/2} = 7.904$ km/s is the orbital velocity calculated at the surface of the Earth. We may rewrite Eqns (6.11) and (6.12) so that our basic equations for the entry trajectory may be rewritten in the following form:

$$\frac{dU}{d\eta} - \frac{U}{\sin\gamma} - \frac{2}{\eta}\left(\frac{H}{R_E}\right) = 0 \qquad (6.16)$$

$$\frac{d\cos\gamma}{d\eta} + \left(\frac{H}{R_E}\right)\left(\frac{1}{U} - 1\right)\frac{\cos\gamma}{\eta} - \frac{1}{2}\left(\frac{L}{D}\right) = 0 \qquad (6.17)$$

Note that the only vehicle-dependent parameter appearing in this set of equations is the lift-to-drag ratio and the only other parameter is the ratio of the scale height of the atmosphere to the radius of the Earth which we have fixed at $H/R_E = 1/923.3$.

6.3 DECELERATION DURING ENTRY

One of the important design requirements for human spaceflight is limiting the magnitude and duration of spacecraft acceleration to levels the human body can successfully tolerate. Therefore, we first must recognize that the nondimensional acceleration along the flight path is found from Eqn (6.6) to be

$$\frac{1}{g_E}\frac{dV}{dt} = -\left(\frac{C_D S}{m g_E}\right)q - \sin\gamma = -\frac{q}{B} - \sin\gamma \qquad (6.18)$$

Here q is the dynamic pressure

$$q = \frac{1}{2}\rho V^2 = \frac{1}{2}(\rho_0 g_E R_E)\sigma U \tag{6.19}$$

6.3.1 MAXIMUM DECELERATION FOR HUMAN SPACEFLIGHT

According to Eqn (6.18), the entry vehicle is subjected to a deceleration inversely proportional to the ballistic coefficient B because of its drag. At the same time, it is subjected to an acceleration directly proportional to the flight path angle γ because of its weight (recall that γ is negative for entry in our coordinate system). The net effect of the drag and weight effects should be such as to decelerate the vehicle in an acceptable manner.

The typical maximum value of acceleration that can be tolerated by humans is around $9g_E$, the value associated with air superiority fighter aircraft. Early studies on the physiological effects of rapid acceleration were described by Haber (1959). Note that the general acceleration tolerance of human beings depends not only on the magnitude of the acceleration but also on the direction of the acceleration and the duration of exposure. Figure 6.2 illustrates that for an exposure time of more than 100 s, the acceleration should be less than $6g_E$ and considering the incorporation of a reasonable margin of safety, an appropriate value should be less than about $4g_E$. Additional information on the physiological effects of g-loading on humans may be found in Voshell (2004).

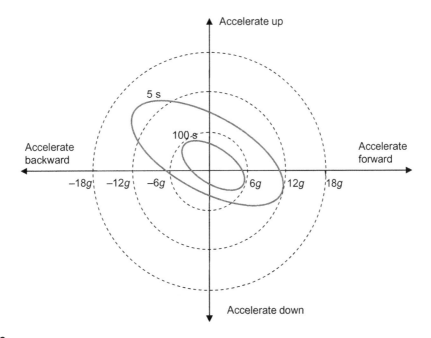

FIGURE 6.2

Duration of exposure to various levels and directions of acceleration that will not impair useful consciousness of crew members.

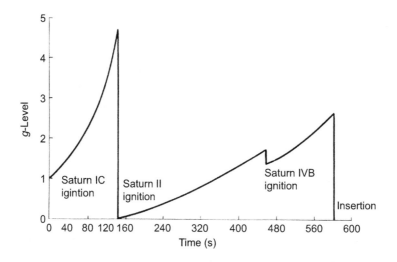

FIGURE 6.3

A typical Apollo launch acceleration profile for the Saturn V launch vehicle is shown.

From NASA (1989).

In light of these effects, the acceleration levels on Apollo flights were generally kept to values within $\pm 8g_E$. Results from NASA (1989) for the acceleration history during boost and reentry for some Apollo missions from NASA (1989) are shown in Figures 6.3–6.5. In addition, Table 6.1 presents maximum decelerations recorded during reentry for Apollo missions 7–17. It is clear that the acceleration levels are kept within $\pm 3g_E$ for the major portion of the operations shown.

From Eqn (6.18) we see that for decelerations no larger than $9g_E$ and because the range of the flight path angle is defined by $-1 < \sin \gamma < 0$, we may assume that for safe human flight the ratio of dynamic pressure to ballistic coefficient is as follows:

$$\frac{q}{B} \le O(10) \qquad\qquad (6.20)$$

This result provides information only on the magnitude of the ratio q/B, but we may use Eqn (6.14) with the stipulation that the fineness ratio of a spacecraft is $l/d = (O)1$ so that $B \sim 5.87/C_D$ with B measured in kilopascals. Detailed spacecraft configurations and mass characteristics are a result of requirements other than merely achieving a certain ballistic coefficient. Indeed, it is just the opposite, with the value of B which arises from satisfying other design issues driving the allowable dynamic pressure levels. For typical manned spacecraft in hypersonic flight, $C_D = O(1)$ so that for $B \sim 5$ kPa the maximum allowable value for the dynamic pressure is around 50 kPa, or 0.5 atm. Keeping the acceleration within safe limits for humans is the reason that the magnitude of the ballistic coefficient for manned entry vehicles is typically $B < 5$ kPa and entry angles $\gamma_e \ll 1$.

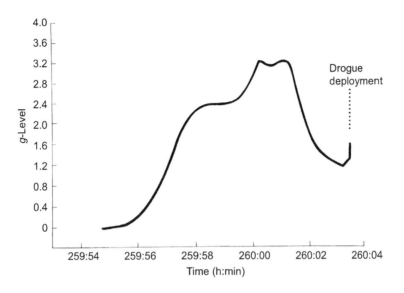

FIGURE 6.4

The Earth orbital reentry deceleration profile for Apollo 7.

From NASA (1989).

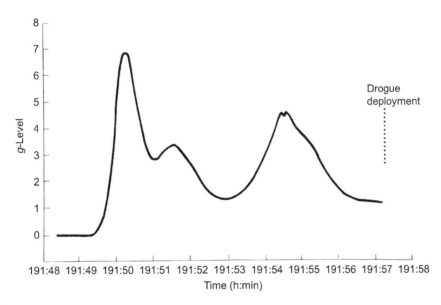

FIGURE 6.5

Reentry deceleration profile for the Apollo 10 Lunar orbital return mission.

From NASA (1989).

Table 6.1 Apollo Manned Spaceflight Reentry g-levels

Flight	Maximum g_E at Entry
Apollo 7	3.33
Apollo 8	6.84
Apollo 9	3.35
Apollo 10	6.78
Apollo 11	6.56
Apollo 12	6.57
Apollo 13	5.56
Apollo 14	6.76
Apollo 15	6.23
Apollo 16	7.19
Apollo 17	6.49

6.3.2 MINIMUM DECELERATION FOR HUMAN SPACEFLIGHT

In order for the entry vehicle to slow down, Eqn (6.18) shows that

$$-\frac{q}{B} - \sin \gamma < 0$$

For typical entry paths, γ is in the range $-90° < \gamma < 0$ so that $-1 < \sin \gamma < 0$, and therefore we conclude that q/B must be at least O(1). Based on the discussion in the previous section that ballistic coefficients for manned spacecraft are likely to lie in the range $1.5 \, \text{kPa} < B < 5 \, \text{kPa}$ we expect the minimum practical value of q to be about 1.5 kPa.

Another consideration relating to the magnitude of the deceleration is the time required for reentry. Integrating Eqn (6.18) from $V = V_e$ at $t = 0$ to $V = 0$ at $t = t_f$ yields

$$-\frac{V_e}{g_E} = -\int_0^{t_f} \left(\frac{q}{B} + \sin \gamma\right) dt = -a_{avg} t_f$$

The variables under the integral sign may vary throughout the flight, but considering a time-averaged value of the deceleration the flight time may be estimated to be

$$t_f \approx \frac{1}{a_{avg}} \frac{V_e}{g_E} \tag{6.21}$$

The entry velocity upon return from LEO is generally around 7.6 km/s; therefore, for average deceleration levels in the range of $0.3g_E$ to $4g_E$, the entry duration lies between about 2600 and 200 s, respectively. The typical duration of the entry of the Space Shuttle Orbiter is about 1820 s with an average deceleration of about $0.4g_E$, while the duration of the entry of the Mercury 7 capsule was about 380 s with an average deceleration of about $2g_E$ and a maximum value of about $8g_E$ for about 30 s. Though the longer reentry time places much milder deceleration levels on the crew, either case poses stresses that are not prohibitive. We will show, however, that a long entry time places thermal constraints on the spacecraft and crew.

6.3.3 DYNAMIC PRESSURE CORRIDOR FOR MANNED SPACECRAFT

The entry corridor of acceptable trajectories for manned spacecraft has been shown to lie in the dynamic pressure range of $1.5\,\text{kPa} < q < 50\,\text{kPa}$. Curves of constant dynamic pressure delineating this safe entry corridor are shown in Figure 6.6. Flight at dynamic pressures greater than 50 kPa would subject to spacecraft to decelerations higher than $10g_E$, while flight at dynamic pressures below 1.5 kPa would result in excessively long flight times and downrange distances. Thus, for safe manned reentry as far as deceleration effects are concerned, the flight trajectory should be kept within the dynamic pressure corridor shown in Figure 6.6.

Consider the hypothetical case of a constant pressure trajectory at a specified design dynamic pressure q_{design} for which Eqn (6.18) may be written as follows:

$$\frac{1}{g_E}\frac{dV}{dt} = -\frac{q_{design}}{B} - \sin\gamma \tag{6.22}$$

Of course, here we require $q_{design} < 10B$ in order to preserve a reasonable deceleration environment for manned reentry.

Because q_{design} is a chosen parameter and $U \sim 1$ at entry, Eqn (6.19) defines an entry altitude that isn't necessarily the conventional nominal value of 100 km. Indeed, we see that the entry altitude

$$z_e = -H\ln\left(\frac{q_{design}}{46,860}\right)$$

For example, when $q_{design} = 10\,\text{kPa}$, which corresponds to a relatively low ballistic coefficient $B \le 1\,\text{kPa}$, the altitude at the start of the constant dynamic pressure trajectory is $z_e = 58.32\,\text{km}$. Therefore, for a smooth trajectory one would need a transitional path between $z = 100\,\text{km}$ and

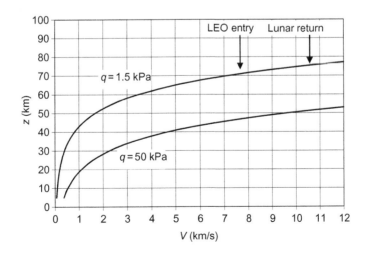

FIGURE 6.6

The flight corridor for tolerable maximum deceleration levels for manned entry. Typical entry velocities for LEO and lunar return missions are shown.

$z = 58.32$ km that would vary in dynamic pressure from $q \sim 0$ up to $q = 10$ kPa, after which one may apply aerodynamic control to stay on $q = 10$ kPa trajectory the rest of the way down. On the other hand, for a high-end design dynamic pressure, say $q_{design} = 40$ kPa, which corresponds to a higher ballistic coefficient $B \leq 4$ kPa, the constant dynamic pressure trajectory would begin at $z = 48.75$ km. Thus, at both ends of the ballistic coefficient spectrum of manned spaceflight we enter the safe descent corridor only after about half of the nominal 100 km of the atmosphere has been traversed.

6.4 HEATING DURING ENTRY

Multiplying Eqn (6.6) by the velocity V yields the rate of energy change of the vehicle

$$m\left(V\frac{dV}{dt} + g_E V \sin\gamma\right) = m\left(\frac{1}{2}\frac{dV^2}{dt} + g_E\frac{dz}{dt}\right) = -\frac{1}{2}C_D\rho SV^3$$

Normalizing this equation results in

$$\frac{d}{dt}\left[\frac{z}{H} + \frac{1}{2g_E H}V^2\right] = -\frac{1}{2H}\left(\frac{C_D S}{mg_E}\right)\rho_0\sigma V^3$$

Rewriting this equation in terms of the dynamic pressure yields

$$\frac{d}{dt}\left[\frac{z}{H} + \frac{1}{2g_E H}V^2\right] = -\frac{q^{3/2}}{BH\sqrt{\rho_0\sigma}} \tag{6.23}$$

Thus, the total energy of the spacecraft diminishes in proportion to the rate at which it does work on the atmosphere. The passage of the spacecraft heats the atmosphere by the essentially adiabatic compression of the bow shock wave and by friction in the boundary layer over the surface of the vehicle. Part of the power expended on heating the atmosphere goes into heating the spacecraft. To ensure the integrity of the spacecraft and its crew, the thermal load generated by reentry must be properly managed. Thermal protection is provided by the heat shield that faces the flow and is subjected to convective heat transfer due to boundary layer friction and radiative heat transfer from the hot gas layer between the bow shock and the heat shield surface.

The details of heat transfer processes in the case of high-speed atmospheric entry are covered in Chapter 9 but for the purposes of our current investigation of trajectory design some general results will be applied. A correlation for the stagnation point heat transfer rate due to convection was given by Sutton and Graves (1971) as

$$q_{t,c} = C\sqrt{\rho}V^3 = C\frac{(2q)^{3/2}}{\rho} \tag{6.24}$$

Note that the functional form of the heat flux in Eqn (6.24) is similar to that of the energy dissipation due to drag given by Eqn (6.23). The coefficient C of this equation is dependent upon the radius of curvature of the heat shield surface R_N (in meters) and is generally given by

$$C = \left(1.7415 \times 10^{-4} \sqrt{kg/m}\right)\frac{1}{\sqrt{R_N}}$$

A correlation for the heat flux at the stagnation point due to thermal radiation from the hot gas in the shock layer was given by Tauber and Sutton (1991) as follows:

$$q_{t,r} = K R_N^{\varpi} \rho^{\xi} f(V) \tag{6.25}$$

Here K and the exponent ξ are constants, while the exponent ϖ and the function $f(V)$ depend upon the altitude and flight velocity considered.

As discussed in detail in Chapter 9, analytic studies of stagnation point heat transfer at hypersonic speeds focused on bodies with spherically symmetric nose caps. Equations (6.24) and (6.25) clearly show that the radius R_N of the nose cap is an important parameter in determining the magnitude of the heat flux at the stagnation point. It is important to note that $q_{t,c} \sim R_N^{-1/2}$ and $q_{t,r} \sim R_N^a$, where $0 < a < 1$, so that a small radius of curvature enhances convective heating while a large radius of curvature enhances radiative heating. It must also be kept in mind that practical spacecraft are generally flown at an angle of attack to provide lift and the resulting three-dimensionality of the flow precludes the use of the local geometric radius of curvature at the stagnation point in heat transfer analyses and an effective radius of curvature must be used. The size and shape of a manned spacecraft is determined by factors other than heat transfer alone, as shown in Chapter 10. For the present purposes, it is sufficient to note that a practical range of radius of curvature may be considered to be $1\,\text{m} < R_N < 4\,\text{m}$.

6.4.1 HEAT TRANSFER CORRIDOR FOR MANNED SPACECRAFT

The stagnation point heat transfer q_t has the units of watt per square meter, although it is more common to quote it in watt per square centimeter. Heat transfer rates on the order of 1 kW/cm² are generally considered to be the upper limit for practical thermal protection systems and 500 W/cm² tends to be the typical value tolerated by existing manned spacecraft for the highly stressful case of lunar reentry. Reentry from LEO generally entails rates of from 100 to 200 W/cm². We may define a general corridor for thermally acceptable flight in the same fashion as was done for the corridor for safe deceleration, that is, in terms of the dynamic pressure. Such an acceptable corridor for convective heat transfer alone is shown in Figure 6.7 for heat shield radii of curvature $R_N = 4$ and 1 m, which are representative of manned spacecraft. For other nose radii, multiply the values of $q_{t,c}$ shown by $R_N^{-1/2}$. A similar acceptable corridor for radiative heat transfer alone is shown in Figure 6.8, also for heat shield radii of curvature $R_N = 4$ and 1 m. For other nose radii at $V < 9$ km/s, multiply the values of $q_{t,c}$ shown by R_N (for $V > 9$ km/s the effect of radius of curvature is nonlinear, as explained in Chapter 9). Note that for speeds typical of LEO entry, the radiative heat transfer is considerably smaller than the convective heat transfer, while for higher entry speeds, like those encountered in lunar return trajectories, the radiative heat transfer is of the same order of magnitude.

There are two factors to consider here: the maximum heat transfer rate to the spacecraft $q_{t,c} + q_{t,r}$ and the maximum heat absorbed by it, $Q_c + Q_r$. The heat transfer rate at the stagnation point is limited by the heat shield material because it can raise the temperature of the heat shield material to unacceptable levels. The total heat absorbed by the spacecraft Q is limited by the ability of the spacecraft's thermal management system to keep the environmental conditions of the crew at a safe level.

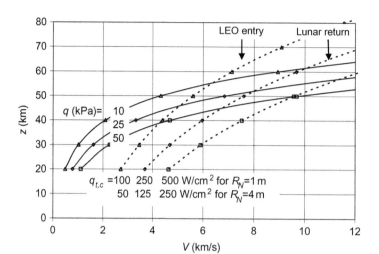

FIGURE 6.7

Acceptable convective heating corridor is shown along with the dynamic pressure corridor for safe deceleration. The convective stagnation point heat transfer rates shown by the dashed lines for $R_N = 1$ and 4 m.

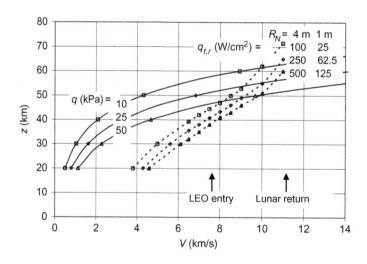

FIGURE 6.8

Acceptable radiative heating corridor is shown along with the dynamic pressure corridor for safe deceleration. The radiative stagnation point heat transfer rates are shown by the dashed lines for $R_N = 1$ and 4 m.

The total heat load $Q = Q_c + Q_r$ depends not only upon the magnitude of the maximum heat transfer rate but also upon the duration of the flight. The total heat transferred to the surface area of the spacecraft wetted by the flow S_w may be written as follows:

$$Q = Q_c + Q_r = \int_0^{t_f} \left[q_{t,c} \iint_{S_w} \left(\frac{q_c}{q_{t,c}} \right) dS_w + q_{t,r} \iint_{S_w} \left(\frac{q_r}{q_{t,r}} \right) dS_w \right] dt$$

The important parameters setting the magnitude of the total heat transferred to the spacecraft are the stagnation heat fluxes due to convection and radiation (which are proportional to $1/R_N^{1/2}$ and R_N, respectively), the surface area of the spacecraft (which is proportional to R_N^2), and the time of flight which was given previously in Eqn (6.21) as

$$t_f \approx \frac{V_e}{a_{avg} g_E}$$

The range of safe acceleration levels in the spacecraft was previously suggested to be in the range $0.3 \le a_{avg} \le 3.0$ so, for example, LEO entry with $V_e \sim 7.6$ km/s the flight duration is in the range $250 \text{ s} < t_f < 2500$ s. Because reasonably fit humans can function easily at the lower range of acceleration, any further reduction serves only to subject the spacecraft to unnecessary additional heating. Detailed discussion of spacecraft heating will be covered in Chapter 9.

6.5 BALLISTIC ENTRY

Equations (6.16) and (6.17) do not have convenient closed form solutions and must be solved numerically or approximately. Unfortunately, the approximate solutions that have been suggested lack either sufficient simplicity or practicality to warrant detailed attention. Though the parameters $H/R_E = 923.3^{-1}$ and $L/D = O(1)$ appearing in Eqns (6.16) and (6.17) are well-defined, the coefficients like $1/\sin \gamma$, $1/U$, and η^{-1} can vary widely over the trajectory path making simple solutions unlikely. As a consequence, numerical solutions are discussed in the following sections.

6.5.1 BALLISTIC OR ZERO-LIFT ENTRY

If the lift-to-drag ratio is zero, which is typical of ballistic missile warheads, early space capsules, and natural objects like meteors, Eqns (6.16) and (6.17) may be rewritten for the earth's atmosphere as follows:

$$\frac{dU}{d\eta} + \frac{U}{\sin \gamma} + \frac{1}{461.6\eta} = 0 \tag{6.26}$$

$$\frac{d\gamma}{d\eta} - \left(\frac{1}{U} - 1 \right) \frac{\cot \gamma}{923.3\eta} = 0 \tag{6.27}$$

The initial conditions are that at the normalized altitude $\eta_e = \sigma_e/B'$, the normalized kinetic energy is $U_e = \frac{V_e^2}{g_E R_E}$, and the reentry angle is γ_e. Typically, the entry density ratio is assumed to be

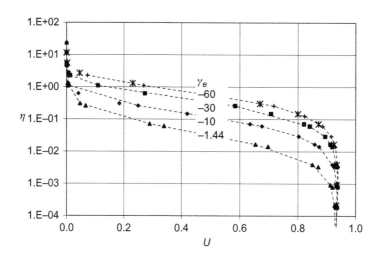

FIGURE 6.9

Numerical solutions for the normalized kinetic energy U as a function of the normalized altitude η are shown for various entry angles in degrees for the case of $L/D = 0$ and $U_e = 0.9266$.

appropriate to around 100 km; for the round number $\sigma_e = 10^{-7}$ the corresponding entry altitude is $z_e = 111$ km. The variable $\eta = \sigma/B'$ incorporates the normalized ballistic coefficient $B' = B/\rho_0 g_E H$ which is essentially the ballistic coefficient measured in atmospheres. The flight path angle γ is measured in radians.

Because there are no free parameters in Eqns (6.26) and (6.27), we may solve them numerically for a typical value of $V_e = 7.6$ km/s, or $U_e = 0.9266$, and several values of the reentry angle γ_e. The decay of the normalized kinetic energy U is shown as a function of the normalized altitude η for $L/D = 0$ in Figure 6.9. Note that small η corresponds to high altitude and increasing η denotes descent toward the surface of the Earth. Thus, the normalized energy U is seen to decrease rapidly as the spacecraft descends (i.e., η increases) and the greater the magnitude of the (negative) reentry angle, the further into the atmosphere that U remains constant, indicating that essentially constant energy is maintained.

The deeper penetration of the spacecraft into the atmosphere as the entry angle γ_e is steepened is seen even more clearly in Figure 6.10. There we also see that the flight path angle γ remains constant until the spacecraft has descended relatively far into the atmosphere. Of course, for any location η the corresponding density ratio $\sigma = B'\eta$ so that the higher the ballistic coefficient B', the higher the value of σ and the lower the geometric altitude z.

As pointed out in previous sections, the most important aspect of reentry performance for manned spacecraft is maintenance of an acceptable level of deceleration. The acceleration of the spacecraft is shown as a function of η in Figure 6.11 for various values of the reentry angle. It is clear that in the absence of any lift, the only way to maintain a safe deceleration level is for the reentry angle to be less than about $-1.5°$.

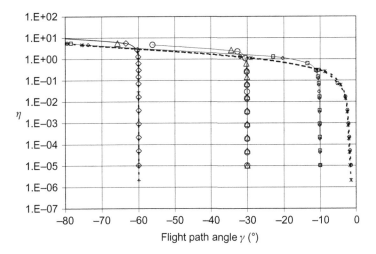

FIGURE 6.10

Numerical solutions for the flight path angle γ as a function of the normalized altitude η are shown for various entry angles in degrees for the case of $L/D = 0$ and $U_e = 0.9266$. The lowest entry angle shown is $\gamma_e = -1.44°$.

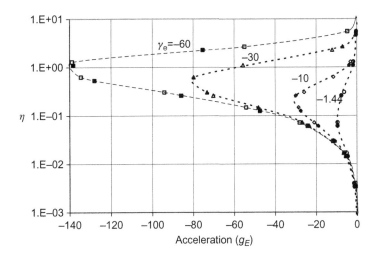

FIGURE 6.11

Numerical solutions for the acceleration in units of g_E as a function of the normalized altitude η are shown for various entry angles in degrees for the case of $L/D = 0$ and $U_e = 0.9266$.

This factor was recognized at the start of the US space program when, in February 1962, the reentry of Col. John Glenn's Friendship 7 in the Mercury program submitted him to about $-7g_E$ to $-8g_E$ for around 30 s. The reentry conditions were $V_e = 7.1$ km/s, or $U_e = 0.8085$, with a reentry angle $\gamma_e = -1.5°$. The Friendship 7 capsule ballistic coefficient was $B' = 0.025$ atm. This reentry

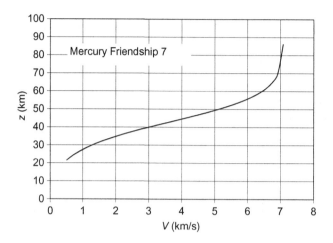

FIGURE 6.12

Numerical solution for the velocity of the Mercury program's Friendship 7 space capsule is shown as a function of altitude.

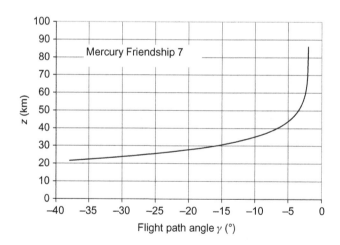

FIGURE 6.13

Numerical solution for the flight path angle of the Mercury program's Friendship 7 space capsule is shown as a function of altitude.

corresponds closely to the case for $\gamma_e = -1.44°$ studied above, so the acceleration history is much like that shown in Figure 6.11. To illustrate the general behavior in terms of physical variables, the velocity and the flight path angle are shown as functions of the geometric altitude in Figures 6.12 and 6.13. The acceleration as a function of the altitude is shown in Figure 6.14.

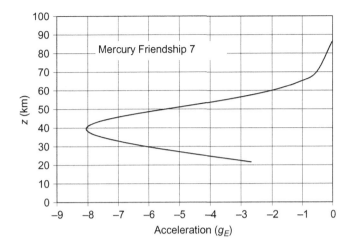

FIGURE 6.14

Numerical solution for the acceleration, in units of g_E, of the Mercury program's Friendship 7 space capsule is shown as a function of altitude.

6.5.2 AN APPROXIMATE SOLUTION FOR STEEP BALLISTIC ENTRY

If, in addition to considering $L/D = 0$, we further assume that the spacecraft enters the atmosphere at a steep negative angle (recall that for reentry the flight path angle $\gamma < 0$), say $-\pi/2 < \gamma < -\pi/4$, then $-1 < \sin \gamma < -0.707$ and $0 > \cot \gamma > -1$. Recall that representative values of the atmospheric parameters were chosen as follows: $\rho_0 g_E H = 1$ atm and $H/R_E = 1.085 \times 10^{-3}$. Using this information, Eqns (6.26) and (6.27) may be written as

$$\frac{dU}{d\eta} - \frac{U}{\sin \gamma} - \frac{1}{461.6\eta} = 0 \tag{6.28}$$

$$\frac{d\cos \gamma}{d\eta} + \left(\frac{1}{U} - 1\right)\frac{\cos \gamma}{923.3\eta} = 0 \tag{6.29}$$

The numerical solutions shown in Figure 6.10 indicate that steep reentry angles tend to remain constant throughout most of the descent. Therefore we assume, for steep reentry, that $\gamma \sim \gamma_e$ and this permits us to integrate Eqn (6.28) leading to the result that

$$U = U_e \left(\frac{\eta}{\eta_e}\right)^{1/461.6U_e} \exp\left(\frac{\eta - \eta_e}{\sin \gamma_e}\right) = U_e \left(\frac{\sigma}{\sigma_e}\right)^{1/461.6U_e} \exp\left[\frac{\sigma - \sigma_e}{\sin \gamma_e}\left(\frac{\rho_0 g_E H}{B}\right)\right] \tag{6.30}$$

This result is compared in Figure 6.15 to the numerical solutions quoted previously and it is clear that quite reasonable agreement is achieved, even for relatively shallow initial reentry angles.

Evaluating Eqn (6.30) at the surface $z = 0$, where $\sigma = 1$, yields the ratio of velocity at the surface V_0 to that at the reentry altitude V_e, where $\sigma_e \sim 10^{-7}$, as follows:

$$\frac{V_0}{V_e} = \sqrt{\frac{U_0}{U_e}} \approx \left(\frac{1}{10^{-7}}\right)^{0.001085} \exp\left(\frac{\rho_0 g_E H}{2B \sin \gamma_e}\right) = 0.98 \exp\left(\frac{50.75}{B \sin \gamma_e}\right) \tag{6.31}$$

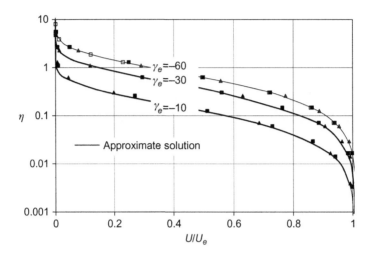

FIGURE 6.15

Comparison of the approximate solutions for U/U_e as a function of η for steep ballistic entry (solid lines) with the numerical results (data symbols).

Note that $\gamma_e < 0$ and therefore $\sin \gamma_e < 0$. The weight of the spacecraft clearly has little effect on the velocity as it provides only the coefficient 0.98 in Eqn (6.31) and therefore may be safely neglected for steep reentry without lift. Thus, the velocity achieved in the vicinity of the surface depends upon the magnitude of the product of the ballistic coefficient and the sine of the entry angle. Bodies with ballistic coefficients $B > 50$ kPa will impact the Earth with substantial velocity, while those with $B < 10$ kPa will slow to very low velocities prior to impact.

Equation (6.29) may be rearranged as follows:

$$d(\ln \cos \gamma) = \frac{-1}{923.3} \left(\frac{1}{U} - 1 \right) d(\ln \eta) \qquad (6.32)$$

The coefficient $(U^{-1} - 1)/921.3 \ll 1$ at least until the normalized kinetic energy drops to very small values. For example, when $U = 0.04$, which corresponds to a velocity $V = 1.579$ km/s, or a Mach number $M < 5$, the coefficient $(U^{-1} - 1)/921.3 = 0.026$. We may make an estimate of the magnitude of the change in flight path angle over the reentry by assuming that the coefficient on the right-hand side of Eqn (6.32) is a constant which leads to the approximate result that

$$\cos \gamma \approx \cos \gamma_e \left(\frac{\eta}{\eta_e} \right)^{-(\frac{1}{U}-1)/923.3} = \cos \gamma_e \left(\frac{\sigma}{\sigma_e} \right)^{-(\frac{1}{U}-1)/923.3}$$

Taking $U = 0.04$ at an altitude of 5 km yields $\cos \gamma \sim 0.7 \cos \gamma_e$; for $\gamma_e = -60°$ we find $\gamma = -69.5°$. Thus, even when evaluating the coefficient at low altitude and energy where its magnitude is greatest results in a change in flight path angle of less than 16°. This corresponding change in $\sin \gamma$ is about 8%. The actual changes would of course be even less than these values Therefore, it is reasonable to assume that Eqn (6.32) shows that γ is approximately constant over most of the reentry flight path with deviations only possible where the kinetic energy decreases to

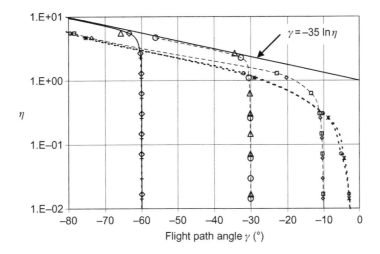

FIGURE 6.16

A logarithmic correlation for steep angle reentry is shown along with numerical solutions for ballistic entry.

very low levels. Figure 6.10 suggests that for steep ballistic reentry the flight path angle remains constant over most of the path and ends with an apparently universal variation with the normalized density coordinate η. Because the scale is semilogarithmic, it may be inferred that for steep reentry angles and low altitudes the flight path angle is correlated to the logarithm of η as follows:

$$\gamma = -35 \ln \eta \tag{6.33}$$

This correlation is shown in Figure 6.16 and demonstrates that the flight path angle for steep ($\gamma_e \sim -30°$ or less) ballistic reentry may be approximated by a constant value $\gamma = \gamma_e$ from the entry point ($\eta \ll 1$) to the corresponding value of η given by Eqn (6.33) and then continuing with $\gamma = -35 \ln \eta$ for the remaining altitude interval.

Results for typical suborbital reentry of a ballistic missile nose cone with the entry conditions $V_e = 6.91$ km/s, or $M = 22$, and $\gamma_e = -60°$, are shown in Figure 6.17. Note that for altitudes above about 50 km, there is no appreciable change in velocity with the vehicle acting as if there were no drag at all. At lower altitudes, the magnitude of the ballistic coefficient determines the behavior of the velocity and the speed reduces from $M = 22$ to only about $M = 12.9$ at ground level for the high ballistic coefficient case, $B' = 1$ atm ~ 100 kPa, whereas the $B' = 0.1$ atm ~ 10 kPa case shows a decay down to zero at ground level.

As mentioned previously, the flight path angle will remain essentially constant until the flight speed is so greatly reduced that $U \ll 1$ and this will occur only for relatively small values of the ballistic coefficient $B \leq O(10$ kPa$)$. Equation (6.29) suggests that as U approaches zero, γ must approach $-\pi/2$. This occurs because once the kinetic energy is depleted the potential energy of the body becomes dominant and it begins to fall vertically. Thus, the flight path angle would rapidly become more negative leading to an almost perpendicular descent in the final stages, for low enough values of the ballistic coefficient. A rough rule of thumb to account for flight path angle changes in the final stage of descent in such cases is to permit the body to decelerate to about $M = 1$, or about 0.34 km/s, at the constant entry trajectory angle and thereupon immediately have

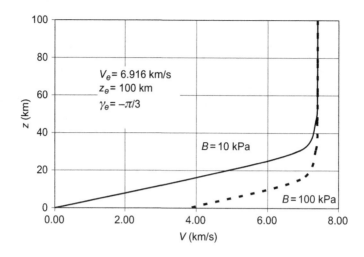

FIGURE 6.17

Approximate velocity–altitude history for steep ballistic entry of a nose cone with reentry conditions $V = 6.92$ km/s ($U_e = 0.88$) and $\gamma_e = -60°$ for low and high ballistic coefficients.

the body drop vertically ($\gamma = -\pi/2$) to the surface. Of course, for higher values of the ballistic coefficient, say $B = O(100 \text{ kPa})$, the kinetic energy would never diminish sufficiently to appreciably change the flight path angle.

It should be clear from Eqn (6.31) that the controlling factor is not B alone, but the product $B \sin \gamma$ which must be large enough to make the retardation due to drag be negligible. If the velocity changes little, Eqn (6.32) shows that the flight path angle will likewise change little. Only meteors entering the earth's atmosphere at shallow angles, if of sufficient size, are likely to have values of $B \sin \gamma$ large enough so that their speed and angle change very little.

6.5.3 TYPICAL BALLISTIC ENTRY CHARACTERISTICS

Solutions for the characteristics of a typical nonlifting entry trajectory are considered in the following discussion. The trajectory is nominally that of a Project Mercury space capsule, like Friendship 7, the first manned space capsule orbited by the United States in February 1962, piloted by John Glenn. The solutions are presented for the purpose of illustrating typical trends. In LEO the orbital velocity is around 98% of $(g_E R_E)^{1/2}$, the orbital velocity at the earth's surface or around 7.74 km/s. Assuming that the retrorocket burn to deorbit the space capsule slows it down such that for a nominal entry altitude of 111 km the entry velocity $V_e = 7.6$ km/s, which corresponds to a normalized kinetic energy $U = 0.9266$, with a reentry angle $\gamma_e = -1.44°$. Solving Eqns (6.26) and (6.27) leads to the velocity–altitude plot shown in Figure 6.18 and the flight path angle–altitude plot shown in Figure 6.19. Because there is little variation above 80 km altitude, that portion of the figures is omitted. Though the velocity changes rapidly below 60 km the flight path angle shows little change from the entry angle until an altitude of about 30 km is reached.

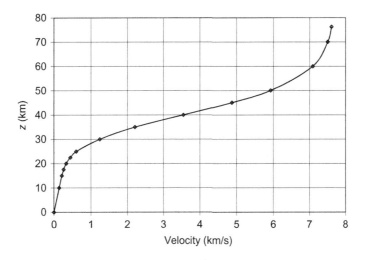

FIGURE 6.18

The entry velocity for a representative trajectory of a Project Mercury space capsule is shown as a function of altitude.

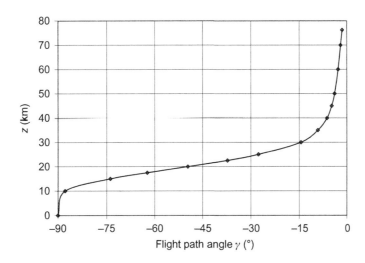

FIGURE 6.19

The entry flight path angle for a representative trajectory of a Project Mercury space capsule is shown as a function of altitude.

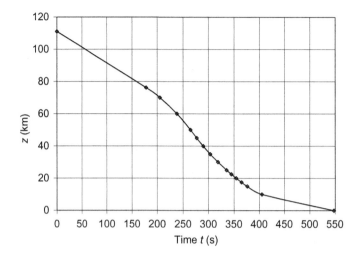

FIGURE 6.20

Entry altitude as a function of time for a representative trajectory of a Project Mercury space capsule.

It is more common to show the important reentry parameters as functions of the flight time from some initial point. Here, we take $t = 0$ at the nominal reentry point where the entry flight conditions are $V_e = 7.6$ km/s and $\gamma_e = -1.44°$ at $z_e = 111$ km. The flight path angle and the velocity define the relationship between time and altitude according to the relation

$$\frac{dz}{dt} = V \sin \gamma$$

The altitude–time history for the trajectory is shown in Figure 6.20. In the same fashion the downrange distance may be found from Eqn (6.4) to be

$$\frac{dx}{dt} = V \cos \gamma$$

The downrange distance–time for the trajectory is shown in Figure 6.21, while the trajectory shape in terms of altitude as a function of downrange distance is shown in Figure 6.22. The velocity and flight path angle are shown as a function of time in Figures 6.23 and 6.24.

The acceleration of the vehicle as it descends may be found by calculating dV/dt along the trajectory and this is shown in Figure 6.25. Here, we see a sharp peak in the deceleration with a value above $7g_E$ for about 40 s between 270 s $< t <$ 310 s which corresponds to the altitude range 35 km $< z <$ 47 km. This level of deceleration is about the limit practical for a manned vehicle, and was one of the reasons only skilled test pilots were selected for NASA's original astronaut program. In order to reduce this deceleration constraint, an alteration of the flight trajectory by generating some lift was deemed necessary. The later Apollo program made use of lift for modulating reentry loads and was brought to greater perfection by the Space Shuttle Orbiter.

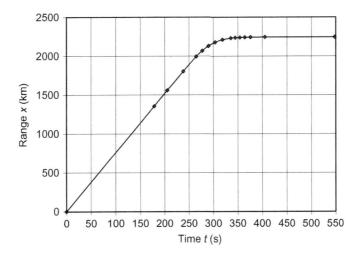

FIGURE 6.21

Downrange distance as a function of time for a representative trajectory of a Project Mercury space capsule.

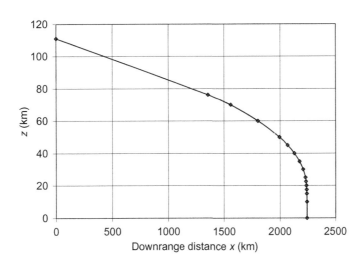

FIGURE 6.22

Entry downrange distance as a function of altitude for the representative trajectory of a Project Mercury space capsule. Note that the altitude scale is stretched compared to the downrange scale.

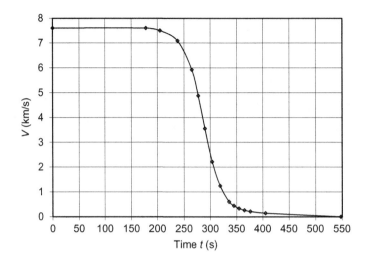

FIGURE 6.23

The entry velocity for a representative trajectory of a Project Mercury space capsule is shown as a function of time from the start of entry.

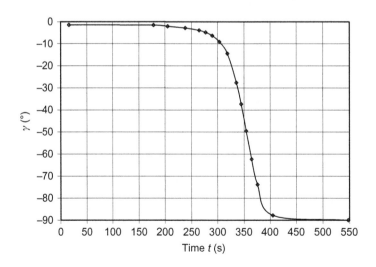

FIGURE 6.24

The entry flight path angle for a representative trajectory of a Project Mercury space capsule is shown as a function of time from the start of entry.

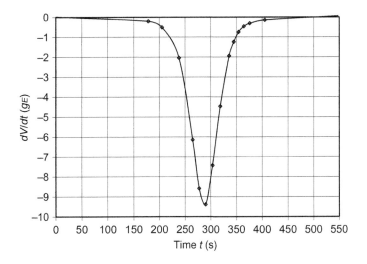

FIGURE 6.25

The deceleration as a function of time for the representative trajectory of a Project Mercury space capsule.

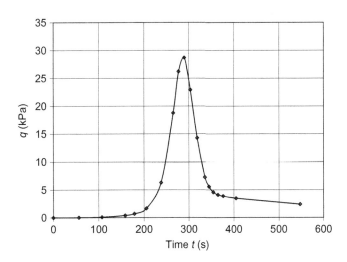

FIGURE 6.26

Variation of the dynamic pressure as a function of time for the representative trajectory of a Project Mercury space capsule.

As shown in Figure 6.26, the dynamic pressure varies in the range of 0 to about 29 kPa, or 0.29 atm, with that maximum occurring at about 40 km altitude. The convective heat transfer rate at the stagnation point was given previously in Eqn (6.24) and is shown in Figure 6.27 with respect to time. The heating characteristics of reentry will be discussed in greater detail in Chapter 9.

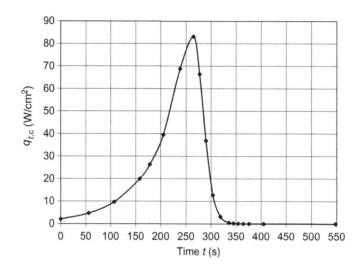

FIGURE 6.27

Variation of the stagnation point convective heat flux as a function of time for the representative trajectory of a Project Mercury space capsule.

6.6 GLIDING ENTRY

Our previous study of ballistic entry showed that in order to keep the acceleration to an acceptable level for manned spaceflight, it was necessary to have very small initial flight path angles. However, we also saw that the flight path angle decreased to more negative values during descent for practical ballistic coefficients and Eqn (6.6) shows that this affects deceleration. Therefore to control deceleration, it is necessary to control the change in the flight path angle and Eqn (6.7) shows that this may be accomplished by applying lift.

6.6.1 LOH'S SECOND-ORDER APPROXIMATE SOLUTION

In general, the governing equations in the form, for example, of Eqns (6.6) and (6.7) must be solved numerically. An approximate solution developed by Loh (1968), who called it a second-order solution, was said to be based on noting the numerical solutions showed that the second term in Eqn (6.17) remains relatively constant throughout trajectories of interest. That is, although U, η, and γ all varied throughout a trajectory, their combination resulted in an approximately constant value. There is no firm theoretical foundation for such an approximation, and the extent of its applicability is not clearly prescribed, even heuristically. In addition, the approximation doesn't lead to a particularly simple solution technique and because numerical solution of systems of ordinary differential equations is now within easy reach of personal computers, we will proceed directly to numerical results to guide the design process. For a comprehensive discussion of Loh's second-order approximate solution, see, for example, Ashley (1992) or Regan and Anandakrishnan (1993).

6.6.2 **ENTRY WITH LIFT**

Because it is necessary to carry out numerical solutions, it is convenient to consider a normalized version of the basic set of time-dependent Eqns (6.6)–(6.8), which may be written as follows:

$$\frac{dv}{d\tau} = -\frac{923.3}{2B'}e^{-s}v^2 - \sin\gamma \tag{6.34}$$

$$\frac{d\gamma}{d\tau} = \left(v - \frac{1}{v}\right)\cos\gamma + \frac{923.3}{2B'}\frac{L}{D}ve^{-s} \tag{6.35}$$

$$\frac{ds}{d\tau} = 923.3v\sin\gamma \tag{6.36}$$

The time t, in seconds, is normalized as follows:

$$\tau = \frac{g_E}{V_E}t \tag{6.37}$$

The velocity, in kilometer per second, is normalized by V_E and is given by

$$v = \frac{V}{V_E} = \frac{V}{7.904} \tag{6.38}$$

The flight path angle γ is, as before, given in radians and the altitude z, in kilometer, is nondimensional by the scale height resulting in

$$s = \frac{z}{H} = \frac{z}{6.9} \tag{6.39}$$

The equations may then be solved for the entry process starting with the initial conditions

$$\begin{aligned} v(0) &= v_e \\ \gamma(0) &= \gamma_e \\ s(0) &= s_e \end{aligned} \tag{6.40}$$

It is useful to note that in Eqns (6.34) and (6.35), the term

$$\frac{923.3}{2B'} = \frac{(R_E/H)\rho_0 g_E H}{2(mg_E/C_D S)} = \frac{1}{2}\frac{\rho_0 V_E^2}{B} \tag{6.41}$$

Here, we have the ratio of dynamic pressure in a circular orbit at the surface of the Earth to the ballistic coefficient, q_E/B.

6.6.3 **LOW *L/D* ENTRY: APOLLO**

The Mercury type space capsules flew ballistic, or nonlifting, trajectories and therefore were limited to small entry angles in order to maintain acceleration levels barely acceptable for manned reentry. As pointed out previously, the only control over the entry trajectory once the spacecraft passes through the entry interface at specific values of V_e and γ_e is through the lift-to-drag ratio L/D. A consequence of the lack of a means to modulate L/D is the inability to fly longer downrange distances over which energy could be dissipated at a slower, more suitable, rate so that the rate of heat transfer to and deceleration of the vehicle could be tolerated.

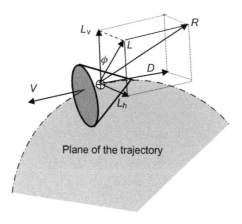

FIGURE 6.28

Schematic diagram of space capsule banked such that the lift vector L is at an angle ϕ to the plane of the trajectory giving rise to components L_v and L_h in and normal to that plane, respectively.

The Apollo space capsule was designed with this feature in mind, as discussed in some detail subsequently in Chapter 8. For the purposes of the discussion of atmospheric entry characteristics, it is sufficient to point out that the Apollo capsule could produce a maximum value of the lift-to-drag ratio in the range of $0.3 < L/D < 0.4$. Though the L/D generated is a function of angle of attack, it was found effective to change L/D by holding a constant angle of attack while banking the capsule such that the lift in the vertical plane of descent, L_v, is reduced by the cosine of the bank angle, as shown schematically in Figure 6.28. Rolling the capsule through positive and negative bank angles results in positive and negative values of the side force L_h so that on the average the capsule remains in the desired trajectory plane.

The Apollo spacecraft was to return from lunar orbit with an entry speed of $V_e \sim 11$ km/s which is about 50% greater than that for entry from low orbit, $V_e \sim 7.7$ km/s. Therefore, more than three times the kinetic energy would have to be dissipated during entry leading to high heat transfer rates, while at the same time keeping deceleration to a manageable level. This prompted the development of control laws to optimize the orientation of the capsule during reentry. For preliminary design purposes and for clarity in treating the nuances of the entry process, we consider a variable L/D case as well as a simpler one in which the lift and drag coefficients are held constant in the governing equations.

For example, we may consider Apollo 4, a case in which the capsule passed the entry interface of $z_e = 120$ km at an earth-relative speed $V_e = 10.74$ km/s and an angle of $\gamma_e = -6.9°$, as reported by Hiltje (1967). The mass of the capsule was 5357 kg and the frontal area, on which the drag coefficient is based, was $S = 12$ m². This leads to a ballistic coefficient $B = W/C_D S = 3.65$ kPa or $B' = 0.036$ atm. The value of the drag coefficient used here is $C_D = 1.2$ which was confirmed by the flight test and is characteristic of Apollo capsules at an angle of attack of 25°, as discussed subsequently in Chapter 8. This angle of attack, and therefore the drag coefficient, was maintained throughout most of the flight.

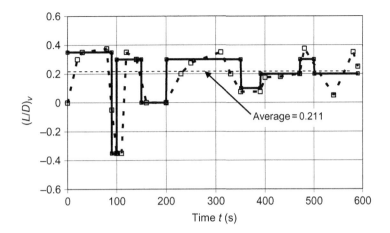

FIGURE 6.29

Approximate variations of the range of $(L/D)_v$ values during Apollo 4 entry and the corresponding stepwise fit used in the numerical calculations. Also shown is the simple average value over the flight.

The effective lift-to-drag ratio in the (vertical) plane of the trajectory, $(L/D)_v$, was varied during the flight by banking the capsule. The effective lift-to-drag ratio varied through the range $-0.35 < L_v/D < 0.375$ during the flight which lasted about 585 s. The reported time history of $(L/D)_v$ for the flight suggested the stepwise approximation for $(L/D)_v$ as shown in Figure 6.29, where the overall average value of about 0.211 is also shown.

Though we may insert a detailed equation for the time or altitude history of (L/D) to use in Eqn (6.35) it seems sufficient for explanatory purposes to use a stepwise approximation for $(L/D)_v$, like that shown in Figure 6.29, to carry out performance calculations. Of course, in designing a trajectory to suit particular needs, modulation of the lift-to-drag ratio (L/D) history must be employed because it is the only control function available for that purpose. It is often considered acceptable to use a constant, average, value of $(L/D)_v$ but we will show that that is not a good choice for typical reentry problems.

In Figure 6.30, the computed altitude–time histories for the stepwise $(L/D)_v$ variation shown in Figure 6.29 and for the constant average value $(L/D)_v = 0.211$ are compared to the flight result reported by Hillje (1967). The histories show an increase in altitude, called a skip, after a descent which is not dependent upon the lift-to-drag ratio. This is largely due to the fact that aerodynamic forces are very small until the altitude drops to about 60 km. The entry angle $\gamma_e < 0$ and the high speed of entry from lunar return ensures that the centripetal acceleration in Eqn (6.35) dominates so that $d\gamma/dt > 0$. At some point, γ passes through zero (yielding the local minimum in z shown in Figure 6.30) and then grows, carrying the spacecraft back to higher altitudes. This lofting of the spacecraft is aided by increasing the lift-to-drag ratio, as is clearly seen in Figure 6.30. However, care must be taken with the trajectory planning to avoid the possibility of having the vehicle skip back out of the atmosphere.

The computational results for the simplified stepwise variation in $(L/D)_v$ provide a reasonable description of the trajectory with the achieving close to the target downrange distance of the actual

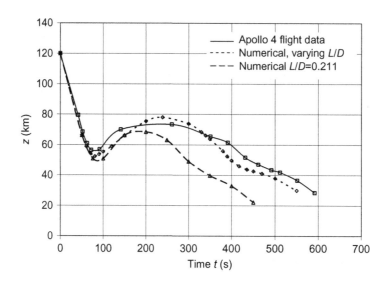

FIGURE 6.30

Computed altitude histories for the stepwise variation of $(L/D)_v$ and for $(L/D)_v = 0.211$ compared to Apollo 4 flight data reported by Hillje (1967).

flight. The constant $(L/D)_v = 0.211$ results are in reasonable agreement with the variable $(L/D)_v$ results and with the actual flight data even into the skip mode, the first third of the flight, but ultimately insufficient lift limits the accuracy of the results in the final two-thirds of the flight.

The actual trajectory flown, that is, the variation of altitude z with downrange distance x, is of operational interest because a recovery team must be placed near the projected terminal point of the flight to pick up the capsule and the crew. The trajectory for the single skip reentry under consideration here is shown as a function of downrange distance in Figure 6.31. The need for lift modulation to control the entry trajectory is clearly illustrated by Figure 6.31 in that the constant average lift case can only reach about two-thirds the downrange distance. This same behavior is exhibited by the flight path angle as can be seen in Figure 6.32, where the flight path angle falls off more rapidly for the average L/D case because insufficient lift is available to keep the spacecraft aloft longer.

The velocity history is shown in Figure 6.33, where using an average value for L/D results in a more rapid deceleration and a deeper penetration into the atmosphere than actually experienced by the spacecraft. The calculation using the stepwise distribution of $(L/D)_v$ displays a similar tendency, but to a lesser degree. Figure 6.34 shows the computed altitude—velocity variation compared to flight data. We will see that the relatively small inaccuracy in predicting accurate velocity—altitude pairs in the range 9 km/s $< V < 11$ km/s has a substantial influence on the accuracy of the prediction of peak values of the other major parameters of the flight, namely, the dynamic pressure, the deceleration, and the maximum heat flux.

Spacecraft structural integrity depends upon the aerodynamic loads, which in turn depend directly on the magnitude of the dynamic pressure $q = \rho V^2/2$. The dynamic pressure history is

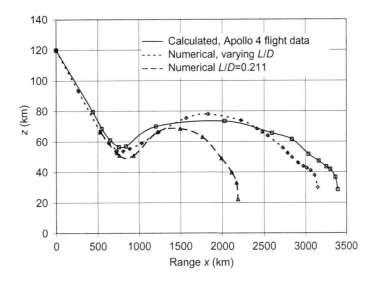

FIGURE 6.31

Computed trajectories for the stepwise variation of $(L/D)_v$ and for $(L/D)_v = 0.211$ compared to that calculated from Apollo 4 flight data reported by Hillje (1967).

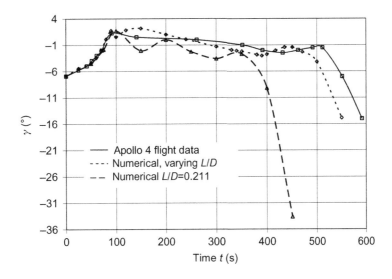

FIGURE 6.32

Computed flight path angle history for the stepwise variation of $(L/D)_v$ and for $(L/D)_v = 0.211$ compared to Apollo 4 flight data reported by Hillje (1967).

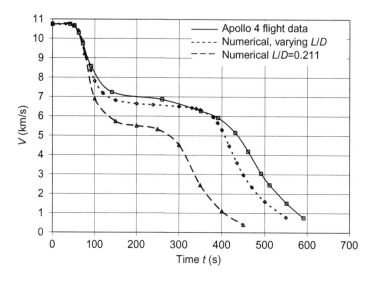

FIGURE 6.33

Computed velocity history for the stepwise variation of $(L/D)_v$ and for $(L/D)_v = 0.211$ compared to Apollo 4 flight data reported by Hillje (1967).

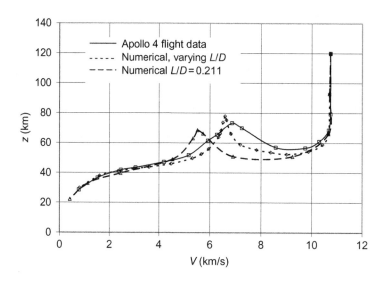

FIGURE 6.34

Computed velocity as a function of altitude for the stepwise variation of $(L/D)_v$ and for $(L/D)_v = 0.211$ compared to Apollo 4 flight data reported by Hillje (1967).

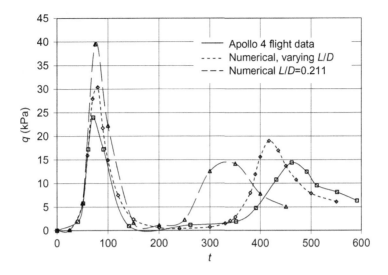

FIGURE 6.35

Computed dynamic pressure history for the stepwise variation of $(L/D)_v$ and for $(L/D)_v = 0.211$ compared to Apollo 4 flight data reported by Hillje (1967).

shown in Figure 6.35, where it is seen that the calculations are conservative, overestimating the magnitude of the maximum value by about 20% when using the stepwise variation of $(L/D)_v$ and by about 50% when using the average value of $(L/D)_v$.

The safety and well-being of the spacecraft crew depends upon maintaining tolerable levels of deceleration during entry. The flight data in Figure 6.36 shows that the Apollo 4 capsule was subjected to a relatively high, but acceptable, maximum deceleration of around $7g_E$. For small flight path angles, Eqn (6.18) demonstrates that the acceleration of the spacecraft, $a = dV/dt$, is directly proportional to the dynamic pressure q. Calculations of the time variation of the normalized acceleration a/g_E are presented in Figure 6.36 and, as in Figure 6.35, the computations seen to be similarly conservative, overestimating the magnitude of the maximum deceleration by about 20% when using the stepwise variation of $(L/D)_v$ and by about 55% when using the average value of $(L/D)_v$.

The degree of thermal protection depends largely on the rate of heat transfer and the stagnation point convective heat transfer portion $q_{t,c}$ is proportional to $\rho^{1/2}V^3$ as discussed in Chapter 9. The computed values for maximum convective heating rate $q_{max,c}$ are conservative by about 20% compared to those calculated using the flight data for Apollo 4, as seen in Figure 6.37.

According to Lee (1972), the calorimeters used to measure convective heat transfer on the heat shield during the Apollo 4 mission failed relatively early in the flight so that maximum values were not available. The Apollo 4 flight data results in Figure 6.37 are calculated from Eqn (6.24) using the reported flight data for altitude and velocity. The convective heat flux calculated with the same equation for the trajectories computed for both the variable L/D and the average L/D cases are shown once again to be conservative compared to the calculated results based on reported flight data.

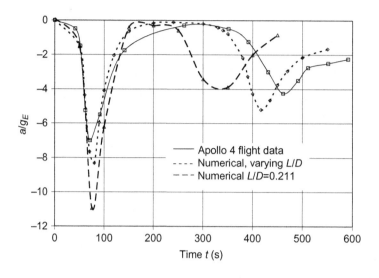

FIGURE 6.36

Computed acceleration history for the stepwise variation of $(L/D)_v$ and for $(L/D)_v = 0.211$ compared to Apollo 4 flight data reported by Hillje (1967).

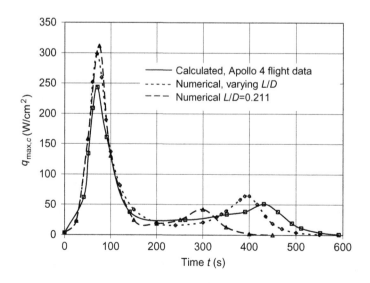

FIGURE 6.37

Computed maximum convective heat flux history for the stepwise variation of $(L/D)_v$ and for $(L/D)_v = 0.211$ compared to results computed from Apollo 4 flight data reported by Hillje (1967).

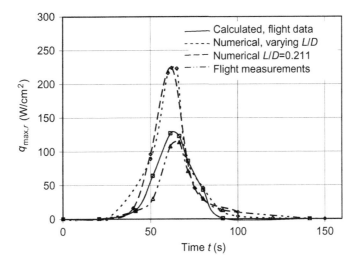

FIGURE 6.38

The computed maximum radiative heat flux history for the stepwise variation of $(L/D)_v$ and for $(L/D)_v = 0.211$ are compared to computed data based on the Apollo 4 trajectory and the measurements reported by Hillje (1967).

The radiometers used on Apollo 4 did collect radiative heat flux data during the high-speed portion of the entry yielding the results shown in Figure 6.38. The radiative heat flux was computed using Eqn (6.25), and for the actual flight trajectory, the results are reasonably close to the measured flight data. However, the maximum radiative transfer rate computed using the variable L/D or the $L/D = 0.211$ trajectory considered here is almost twice as high as that actually measured. This difference is entirely due to the extreme sensitivity of the radiative heat flux to velocity during entry. The total maximum heat flux $q_{\max}$ is shown as a function of time in Figure 6.39 but it must be emphasized that the data shown is derived from, but not actually, flight data. It may be noted that the total maximum heat flux of about 500 W/cm^2 predicted by the computations using variable or constant L/D is close to the prediction of 483 W/cm^2 quoted for the Apollo 4 flight by Pavlosky and St. Leger (1974).

The total heat load Q, which is proportional to the integral of the total heat flux over time and surface area, as discussed in Section 6.4.1, is not calculated here but will be discussed in Chapter 9. However, because the total heat load will be proportional to the maximum heat flux a reference heating load may be defined as

$$Q_{ref} = \int_0^{t_f} (q_{\max,c} + q_{\max,r})dt$$

Integrating the stagnation point heat flux distribution in Figure 6.37 yields the results shown in Figure 6.40.

The predicted heat load for the case of variable $(L/D)_v$, $Q_{ref} = 32$ kJ/cm^2, is 10% higher than that predicted on the basis of the flight trajectory data, $Q_{ref} = 29$ kJ/cm^2. This is due in large part to the difference between the maximum heat flux values during the interval between $t = 50$ s and

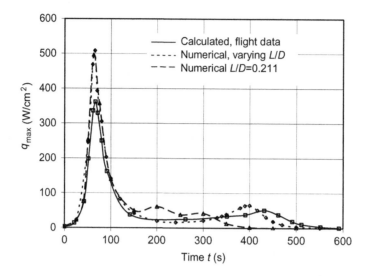

FIGURE 6.39

The computed total heat flux for the stepwise variation of $(L/D)_v$ shown in Figure 6.29 and for $(L/D)_v = 0.211$ are compared to computed data based on the Apollo 4 reported by Hillje (1967) as a function of time.

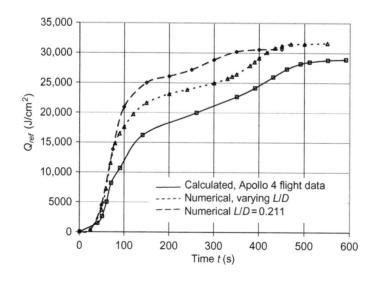

FIGURE 6.40

The computed reference heat loads for the stepwise variation of $(L/D)_v$ and for $(L/D)_v = 0.211$ are compared to computed data based on the Apollo 4 trajectory reported by Hillje (1967) as a function of time.

$t = 70$ s, as seen in Figure 6.39. The heat load prediction of $Q_{ref} = 31$ kJ/m^2 using the constant value of $(L/D)_v = 0.211$ is lower than that for the variable $(L/D)_v$ case largely because the time of flight is likewise underpredicted. Thus, the accuracy of the maximum heat flux and the time of flight is important in determining the total heat load, which must be supported by the spacecraft.

6.6.4 MODERATE *L/D* ENTRY: SPACE SHUTTLE ORBITER

The Space Shuttle Orbiter follows a trajectory that yields a smooth velocity variation with altitude with no skip like that followed by the Apollo capsule. A plot of the trajectory for a typical entry of a Space Shuttle Orbiter from LEO is shown in Figure 6.41. In this case, the nominal trajectory for Space Transportation System STS-5 is depicted. The entry velocity, flight path angle, and altitude are $V_e = 7.7$ km/s, $\gamma_e = -1.5°$, and $z_e = 120$ km, respectively. The altitude drops rapidly at essentially constant velocity until the vehicle reaches an altitude of about 80 km (50 mi) after which time the drop in velocity and altitude is more gradual. During the descent, the Space Shuttle Orbiter is pitched through decreasing angles of attack $\alpha(t)$, while also rolling back and forth through a series of different bank angles $\phi(t)$ as shown in Figure 6.42. These pitching and rolling maneuvers are used to modulate the effective lift and drag so as to control the reentry behavior. Banking back and forth modulates the lift in the vertical plane, while keeping the average side force approximately zero so that the vehicle remains on a trajectory lying within a vertical plane.

Along this path, the dynamic pressure, the deceleration, and the convective heat transfer, as calculated from the STS-5 nominal trajectory data according to the equations of the previous sections, all continually increase from essentially zero to a maximum and then decrease again, as

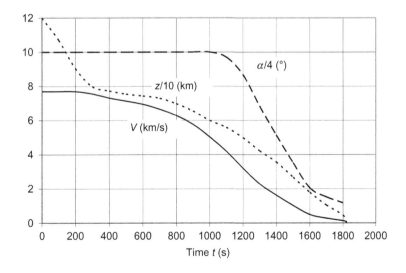

FIGURE 6.41

Plot of a nominal Space Shuttle Orbiter entry trajectory (STS-5) showing the variation of speed, altitude, and angle of attack as a function of time.

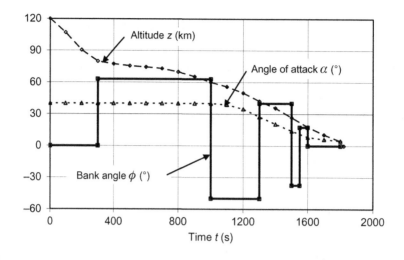

FIGURE 6.42

Plot of a typical Space Shuttle Orbiter entry trajectory showing the nominal variation of Orbiter bank angle ϕ, angle of attack α, and altitude z as a function of time.

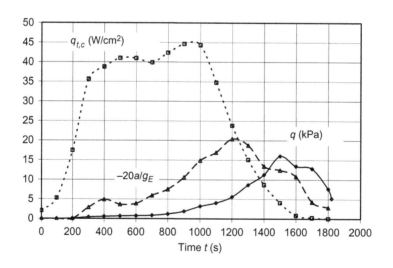

FIGURE 6.43

The nominal variation of stagnation point heat transfer rate, deceleration, and dynamic pressure is shown as a function of time for the typical Space Shuttle Orbiter entry trajectory of Figure 6.41.

shown in Figure 6.43. Note that the three variables reach their maxima at different times and therefore at different altitudes in the typical mission shown. That the nominal entry trajectory of STS-5 shown in Figure 6.41 is one which lies within the acceptable corridor for manned spaceflight is illustrated in Figure 6.44.

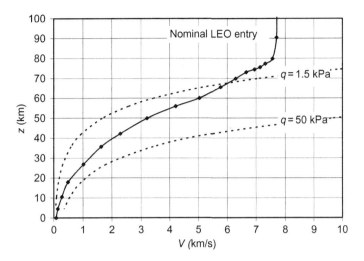

FIGURE 6.44

The nominal trajectory of STS-5 is shown to lie within the dynamic pressure corridor considered acceptable for manned spaceflight.

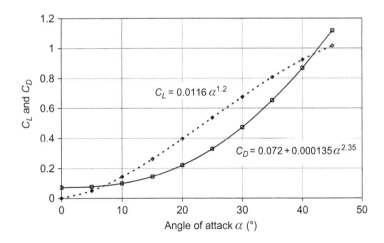

FIGURE 6.45

Approximate lift and drag characteristics for the Space Shuttle Orbiter in hypersonic flight ($M > 5$).

Because the Space Shuttle Orbiter is actually a spaceplane, it has aerodynamic characteristics very much like an airplane. That is, the lift and drag are functions of angle of attack α, and, as mentioned in the previous section, the lift in the vertical plane is a function of the angle of bank ϕ as well. The lift and drag characteristics are described in some detail in Chapter 8 and are summarized for hypersonic flight ($M > 5$) in Figure 6.45. Obviously there is a substantial variation in lift

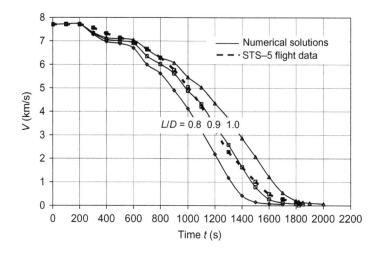

FIGURE 6.46

Comparison of the velocity history of a typical Space Shuttle Orbiter (STS-5) entry trajectory with numerical solutions for constant values of L/D and $B' = 0.0406$ atm.

and drag coefficients over the range of angle of attack covered by STS-5. In addition, the history of the bank angle of STS-5 is not in hand and changes in lift may also occur in addition to those due to angle of attack.

However, we may use the STS-5 trajectory results to assess the behavior of numerical solutions to the equations developed thus far. In Figure 6.46, the nominal velocity history for STS-5 shown in Figure 6.41 is compared to numerical solutions of Eqns (6.34)–(6.36) for the simple case in which constant lift and drag coefficients are assumed to exist throughout the flight. Using a nominal mass $m = 90{,}000$ kg, a reference area $S = 250$ m², and a drag coefficient $C_D = 0.86$ yields a normalized $B' = mg_E/C_DS = 0.0406$ atm which corresponds to flight at $\alpha = 40°$, which is the case for over half the flight time, as can be seen in Figure 6.42. Using this constant value of B' along with an assumed constant value L/D in Eqns (6.34)–(6.36) leads to the results shown in Figures 6.46 and 6.47 where it is seen that the case of $B' = 0.0406$ and $L/D = 0.9$ provides a reasonably good fit to the nominal flight data for velocity and altitude as a function of time from entry.

Note that the computed velocity and altitude histories in Figures 6.46 and 6.47 exhibit a long-period phugoid-type motion at high altitudes. This is characteristic of computations for high L/D (>1) vehicles entering the atmosphere at small angles. Damping due to increasing density reduces these oscillations as lower altitudes are reached, as shown in Figure 6.46. Flight path angle measurements were not available for STS-5 but the general behavior is as illustrated in Figure 6.48.

Although this simple approach is instructive, there is no method for choosing appropriate constant values for B' and L/D. If we use the angle of attack history for STS-5 shown in Figure 6.41, along with the curve fits for the approximate lift and drag characteristics shown in Figure 6.45, in Eqns (6.34)–(6.36), we obtain the velocity and time histories shown in Figures 6.49 and 6.50, respectively. There are two cases considered, one with a zero bank angle and one with a variable bank angle. It is clear from Figures 6.49 and 6.50 that the case with a zero bank angle leads to

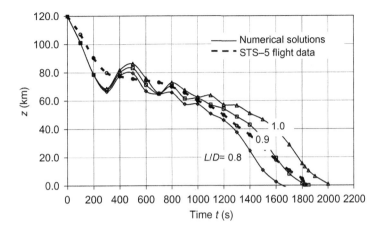

FIGURE 6.47

Comparison of the altitude history of a typical Space Shuttle Orbiter (STS-5) entry trajectory with numerical solutions for constant values of L/D and $B' = 0.0406$ atm.

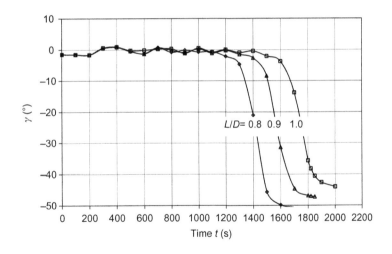

FIGURE 6.48

Comparison of the computed flight path angle history for constant values of L/D and $B' = 0.0406$ atm.

considerably higher velocities and altitudes than those for STS-5. This result suggests that the lift produced is too high leading to excessive entry time and too great range. However, if the lift is reduced by reducing the angle of attack the drag will also be reduced and again the velocity will be increased, so some other means of reducing lift is necessary. Rolling the vehicle such that the lift in the vertical plane of the trajectory is reduced by the cosine of the bank angle provides such lift control. A typical such maneuver for the Space Shuttle Orbiter was shown in Figure 6.42.

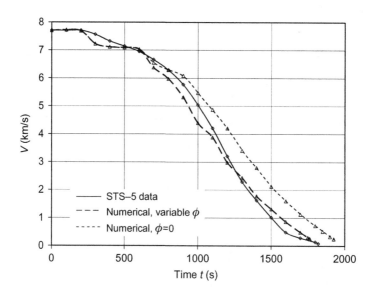

FIGURE 6.49

Computed velocity history for the STS-5 vehicle assuming a variable bank angle and a bank angle of zero compared to the nominal STS-5 data.

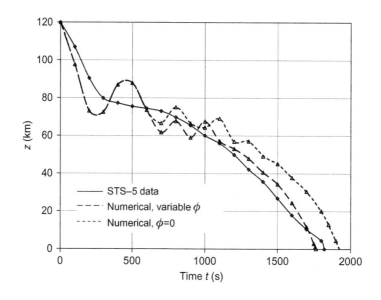

FIGURE 6.50

Computed altitude history for the STS-5 vehicle assuming a variable bank angle and a bank angle of zero compared to the nominal STS-5 data.

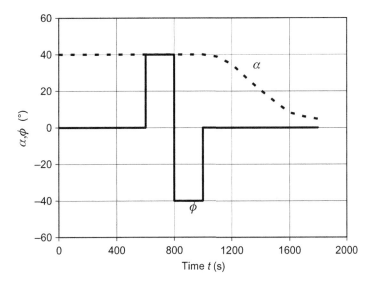

FIGURE 6.51

Notional banking maneuver to reduce lift coefficient at fixed angle of attack.

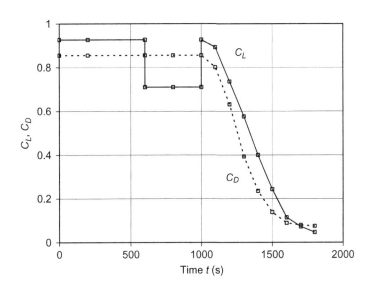

FIGURE 6.52

Variation of C_L and C_D for the notional rolling maneuver shown in Figure 6.51.

For illustration purposes, the simpler maneuver shown in Figure 6.51 is suggested. The corresponding C_L and C_D characteristics are illustrated in Figure 6.52. The spacecraft response to angle of attack variation with and without roll modulation is shown in Figures 6.49 and 6.50. It is clear that the roll maneuver works to reduce the flight time even though the angle of attack variation is

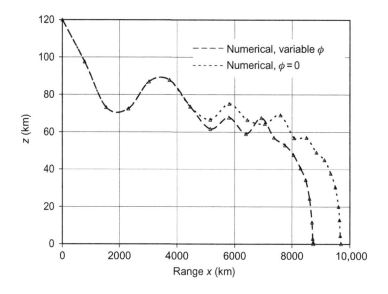

FIGURE 6.53

Range as a function of spacecraft altitude with and without roll modulation of lift.

fixed. It should also be recalled that the STS-5 trajectory is being used for illustration because the angle of attack history is available. The actual roll maneuvers of STS-5 were not available and are not used here, being more complicated, like the typical variation shown in Figure 6.42.

The range covered with and without roll modulation of the lift is shown in Figure 6.53. It appears clear that the roll maneuver has essentially put a time shift into the trajectory without changing its character. This may also be seen in the history of the flight path angles shown in Figure 6.54 which display a behavior similar to that seen for the constant L/D trajectories in Figure 6.48.

Also worth noting again is that the computed trajectories in Figures 6.47, 6.50, and 6.53 display a skip-like behavior unlike the STS-5 flight data in Figure 6.50. As mentioned previously, this arises because Eqns (6.34)−(6.36) involve small differences of small numbers, particularly at high altitudes (the density varies by eight orders of magnitude over the flight path), which results in details of the flight characteristics being magnified. Thus, though the computed flight path angles remain small through most of the flight, Figure 6.54 shows that the sign of the angle fluctuates resulting in the phugoid motion observed in the calculated trajectories. This is best appreciated by first comparing the altitude variation of velocity with and without roll modulation of the lift to that for the STS-5 data shown in Figure 6.55. The computed velocity exhibits the waviness described previously, particularly above an altitude of 60 km. The roll maneuver takes place between 60 and 70 km altitude, and below 60 km the velocity−altitude behavior of the spacecraft with and without roll modulation of the lift is essentially the same. The computed results do not closely match the STS-5 data because, although its angle of attack history was used, its actual roll history is not known and the simple roll maneuver described previously was substituted for illustrative purposes.

This variation of computed altitude coupled to the exponential nature of the atmosphere leads to corresponding oscillations in computed flight characteristics like stagnation point convective heat

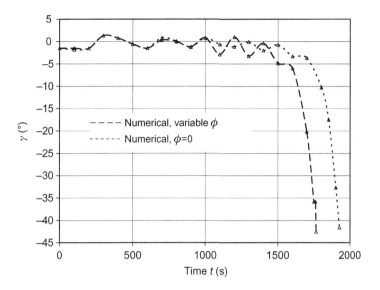

FIGURE 6.54

Comparison of the computed flight path angle history with and without roll modulation of lift.

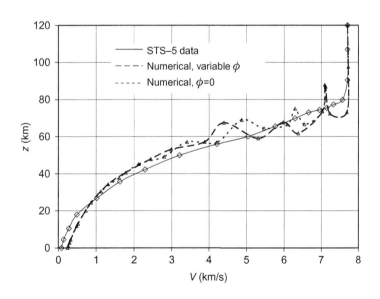

FIGURE 6.55

Altitude variation of spacecraft velocity with and without roll modulation compared to that for the STS-5 data.

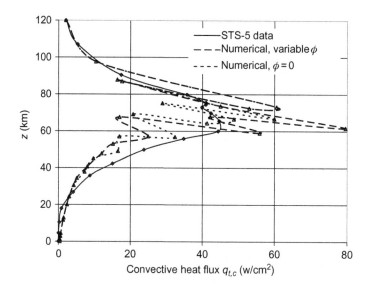

FIGURE 6.56

Altitude variation of stagnation point convective heat flux with and without roll modulation compared to that computed from the STS-5 data.

flux, dynamic pressure, and acceleration level, as illustrated in Figures 6.56–6.58, respectively. The convective heat flux depends upon $\rho^{1/2}V^3$, as described in Section 6.4, and therefore is sensitive to the waviness in the trajectory data. As can be seen in Figure 6.56, the variations in convective heat flux are most noticeable in the altitude range $60\,\text{km} < z < 80\,\text{km}$ and computed peak values can be off by as much as a factor of 2. However, the average value in that altitude range for the calculated trajectories is 43 W/cm² for the case with the roll maneuver and 39 W/cm² without the roll maneuver, compared to a peak value of about 44 W/cm² computed from the STS-5 data. In a similar manner, we may integrate the maximum heat flux over the entire flight and find the reference heat load to be about 44 kW/cm² for the case with the roll maneuver and about 45 kW/cm² without the roll maneuver, compared to 42 kW/cm² computed from the STS-5 data. Thus, using the roll maneuver shortens the flight, increases the peak heat flux, but reduces the total heat load to the spacecraft. The computational results are conservative compared to those for the STS-5 data.

The peak dynamic pressure for the STS-5 data occurs at an altitude well below that for which oscillations in the computed values of q appear, as seen in Figure 6.57. The computed values of q do not show a peak and their magnitude is about three times that of the STS-5 data because the computed velocity is not in very close agreement with the flight data as shown in Figure 6.55. The peak value of dynamic pressure in the STS-5 data occurs in the supersonic range, $M \leq 3$, where the lift and drag coefficients depart from the hypersonic values used in the computations. In Figure 6.58, the computed acceleration levels show the altitude variations discussed previously but are all within the range of the STS-5 data which has a maximum of $a/g_E \sim -1$. The Space Shuttle Orbiter was designed to carry scientific specialists who were not necessarily pilots trained for and capable of tolerating the relatively high deceleration levels common to previous flights with space capsules. The lifting capability of a spaceplane was crucial to providing such a benign environment

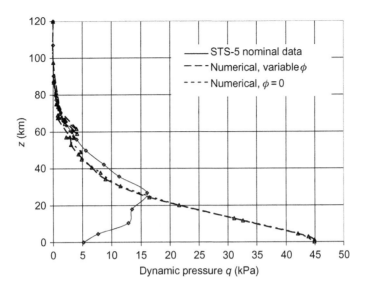

FIGURE 6.57

Altitude variation of dynamic pressure with and without roll modulation compared to that computed from the STS-5 data.

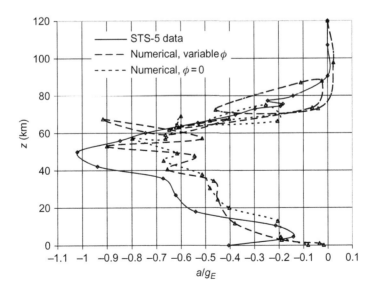

FIGURE 6.58

Altitude variation of normalized acceleration with and without roll modulation compared to that computed from the STS-5 data.

for the crew, although very small entry angles were still necessary. As a result, the calculation of the entry trajectory of a spaceplane is dependent upon accurate lift and drag information for use throughout the flight.

6.7 LOW-SPEED RETURN AND RECOVERY: PARACHUTES

Capsule-type entry vehicles lack sufficiently high L/D to avoid unacceptably high terminal speed and therefore require the assistance of an additional aerodynamic decelerator. Parachute recovery systems have the light weight, simplicity, and reliability to provide safe recovery of space capsules. The typical sequence of events for the landing of the Gemini capsule, for example, taken from Vincze (1966), is shown in Figure 6.59. The nominal deployment conditions for the recovery are shown in Table 6.2.

The Gemini recovery system used a high-altitude drogue parachute to deploy the pilot parachute pack. The use of a drogue parachute in tandem with a pilot parachute for deploying a main landing parachute was shown to work well. The use of the pilot and drogue parachutes also served to prevent recontact of the recovery and rescue (R&R) section with the main parachute canopy. The concept of landing shock attenuation by water entry of the cabin section at the corner of the heat shield, as illustrated by the last sketch in the deployment sequence in Figure 6.59, thus eliminating the additional weight and complexity of shock absorption equipment, proved successful.

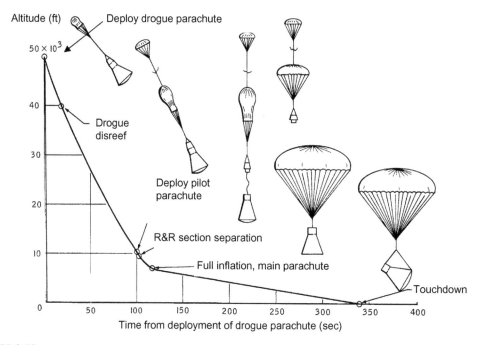

FIGURE 6.59

Deployment sequence of the parachute landing system for the Gemini spacecraft.

From Vincze (1966).

Table 6.2 Nominal Deployment Conditions for Gemini Parachute System from Vincze (1966)		
Parameter	**Deploy Reefed Drogue at an Altitude of 15.24 km (50,000 ft)**	**Drogue Disreef at an Altitude of 12.19 km (40,000 ft)**
Mach number M	0.84	0.57
Dynamic pressure q	57.45 kPa (120 lb/ft^2)	4.692 kPa (98 lb/ft^2)
Flight path angle γ	$-65°$	$-89°$

FIGURE 6.60

Parachute recovery of Apollo 15. Note that one of the parachutes has not inflated properly. Courtesy NASA.

For heavy payloads a cluster of several parachutes may be used. Clustering does slightly reduce the drag performance compared to that of single canopy, but this approach offers a degree of redundancy which takes into account the possibility that one parachute may not deploy successfully. The typical three-parachute cluster used, as shown in Figure 6.60 for the Apollo 15 landing, clearly illustrates this safety factor in practice.

Because parachutes may be folded and packed in a fashion which requires little volume makes them particularly suited to the tight quarters of spacecraft. When they are deployed they are not generally free to open fully, but instead are initially in a reefed configuration. Reefing a parachute, like reefing a sail, means restricting the amount of surface area exposed and therefore the force produced. Thus, reefing permits control of the opening-shock load on the parachute. Disreefing is the process of severing (usually pyrotechnically) the reefing line, which initially restrains the opening of the parachute thereby allowing the parachute to fully open and provide the desired drag force. Reefing may be carried out in several stages so as to more precisely control the deceleration process.

6.7.1 AERODYNAMICS OF PARACHUTES

The parachute is a simple light-weight aerodynamic decelerator operating primarily through a pure drag force. In general, the drag force produced by the parachute in a steady vertical descent in still air may be written as follows:

$$D = q_v C_D S \tag{6.42}$$

Under the assumed flight condition, the drag force on the parachute D is in the vertical direction and in the opposite sense of the descent rate V_v and the dynamic pressure experienced is based on the equivalent airspeed of the descent $q_v = \rho_{sl}(V_{E,v})^2/2$, where the general definition of an equivalent airspeed is $V_E = \sigma^{1/2} V$; at sea level $V_E = V$. Here, C_D and S are grouped together because there are several definitions used for the reference area S and therefore C_D depends upon the definition chosen for S:

1. Nominal area S_0: the actual three-dimensional canopy constructed surface area computed as the sum of the gore areas inclusive of vent area, slots, and other openings within the gore outline. Note that a gore is one of the many triangular segments of cloth which are joined together to make the canopy. There may be anywhere from two to twelve dozen gores used to fashion a parachute, depending upon the inflated diameter of the parachute. Areas of surfaces such as ribs, flares, panels, or additional fullness to the cloth are also included. The nominal diameter D_0 is the diameter of a circle whose area is S_0 and the associated drag coefficient is denoted by $C_{D,0}$.
2. Constructed area S_c: the area of a circle whose diameter is D_c, where the constructed diameter D_c is the diameter of the parachute (measured along the radial seam) when projected on a planar surface. The nominal diameter D_0 is a hypothetical diameter while D_c is an actual dimension of the parachute. Except for flat circular parachutes, D_c differs from D_0.
3. Projected area S_p: the area of the vertically descending inflated canopy projected onto the horizontal plane; basically, the planform area of the inflated parachute. The projected diameter D_p is defined as the diameter of a circle whose area is S_p and the associated drag coefficient is denoted by $C_{D,p}$.

6.7.2 PARACHUTE DESIGN PARAMETERS

The porosity of a parachute canopy includes both the small interstices between the threads of the parachute fabric and the larger vents and slots specifically incorporated in the canopy design. Porosity and its distribution over the canopy affect the drag coefficient, the characteristic time for canopy inflation, the shock loads, and the static stability of the parachute system. The constructed profile (flat, conical, or spherical) and the planform shape (circular, square, triangular, or cruciform) are important factors in parachute performance as are the relative length of the suspension lines (l/D_0). The unit-canopy loading, W/C_DS, that is, the ballistic coefficient of the parachute, determines the equilibrium rate of descent in still air. Changes in unit-canopy loading affect the operating characteristics of the parachute in ways that cause the drag coefficient to vary with rate of descent. Canopies constructed of annular rings of fabric, such as the ringslot and ringsail types typically used for spacecraft recovery, have little variation of drag coefficient with unit loading. There are a number of other factors which are important in determining the dynamics of parachute deployment that are beyond the scope of the preliminary design analysis pursued here; for further details see Cockrell (1987), Maydew and Peterson (1991), and Knacke (1992).

6.7.3 PARACHUTE MATERIALS

Fabrics used in parachute construction include nylon, Dacron, Nomex, and Kevlar. Solid nylon has a density of 1138 kg/m^3, while Dacron and Nomex have densities around 1377 kg/m^3 and Kevlar has a density of 1440 kg/m^3. Different types of packing for actual parachutes lead to the following densities (in kilogram per cubic meter): manual, 350−450; vacuum, mechanical, or pneumatic press, 480−580; hydraulic press, 640−740. The nylon fabrics used in the Gemini recovery parachute used as an example had areal densities of 0.038 and 0.075 kg/m^2. Assuming that the canopy is constructed solely of a single piece of cloth of area S_0, the areal density for the canopies of Table 6.3 lies in the range of $0.077−0.123 \text{ kg/m}^2$ with an average of 0.098 kg/m^3.

6.7.4 RINGSAIL PARACHUTES

Many designs for parachutes and other aerodynamic deceleration systems have been used, as described by Ewing (1971), Cockrell (1987), Maydew and Peterson (1991), and Knacke (1992). However, the ringsail parachute, invented by E.G. Ewing in 1953 proved to be the mainstay for the US manned spacecraft program. This parachute, which is described in detail by Ewing (1972), has a unique shape which is illustrated in Figure 6.61. The upper canopy is constructed of wide concentric cloth strips separated by gaps to provide porosity, just like the ringslot parachute. However, over the remainder of the canopy there are no gaps but porosity is obtained instead from crescent-shaped slots, which result from the flare produced by the difference in length between the leading edge of each sail and the trailing edge of the sail below it.

   The characteristics of the ringsail design as used in spacecraft recovery applications are shown in Table 6.3 and include data reported by Ewing (1972) and by Delurgio (1999). The vertical velocity is in terms of the equivalent airspeed $V_{E,v}$, and the calculated values of C_{D0} are obtained using Eqn (6.42) with the reported values of S_0 and $V_{E,v}$. The ratio of the length of the suspension lines to the nominal diameter of the parachute l/D_0 also has an influence on C_{D0}, as will be discussed

Table 6.3 Ringsail Characteristics for Various Recovery Applications

Parachute	Mass (kg)	No. of Canopies	V_v (m/s)[a]	D_0 (m)	S_0 (m²)	C_{D0} Calculated	C_{D0} Reported	l/D_0
Apollo	2152	1	8.47	26.9	566.0	0.848	0.85	1.4
Apollo	4304	2	8.50	26.9	566.0	0.842	0.845	1.4
Apollo	4304	3	7.01	26.9	566.0	0.826	0.825	1.4
Apollo	4960	3	8.08	26.1	534.4	0.760	0.798	1.4
Apollo	5853	3	9.39	26.1	534.4	0.664	0.698	1.4
Apollo	6455	3	8.99	26.1	534.4	0.798	0.85	1.4
Gemini	1993	1	9.02	25.7	517.0	0.759	0.76	0.94
Mercury	978	1	8.41	19.2	289.5	0.765	0.76	0.97
Mercury	1060	1	7.32	19.2	289.5	1.097	0.91	0.97
ASSET	492	1	16.8	9.0	63.90	0.439	0.67	0.93
E-6	589	1	10.7	12.5	122.6	0.676	0.68	0.93
E-5 Samos	770	1	6.28	22.6	401.5	0.780	0.78	0.94
Century	4422	1	8.50	39.3	1210	0.810	0.9	1.15
EELV	9060	3	6.10	41.5	1349	0.965	1.03	1.15
K-1Orbiter	12,231	3	5.97	47.5	1775	1.031	1.1	1.15
K-1Launch	20,385	6	5.58	47.5	1775	0.986	1.1	1.15
20 K	9060	1	8.38	56.0	2464	0.839	0.84	1.18
20 K	9060	1	7.77	57.8	2622	0.917	0.92	1.23

[a]*Equivalent airspeed.*

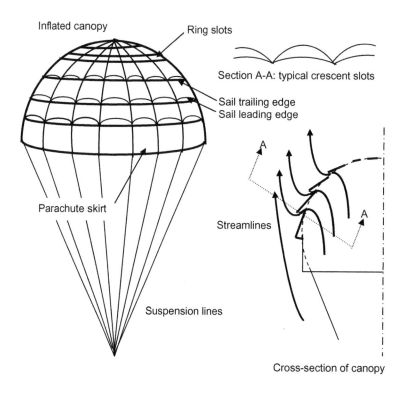

FIGURE 6.61

Schematic diagram of the main design features of the ringsail parachute.

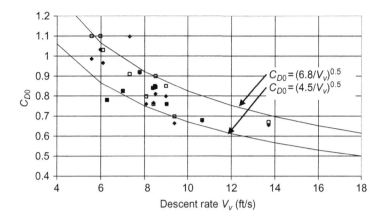

FIGURE 6.62

Variation of C_{D0} as a function vertical equivalent airspeed. Data is from Table 6.2 with closed and open symbols denoting calculated and reported values, respectively.

subsequently. The drag coefficient data of Table 6.3, both calculated and reported, is shown in Figure 6.62 along with a correlation of the form

$$C_{D0} = \sqrt{\frac{K_D}{V_v}} \tag{6.43}$$

Here, K_D is a coefficient with a range of $4.5 \le K_D \le 6.8$ and curves based on Eqn (6.43) using the two extreme values of K_D are also shown on Figure 6.62. There is scatter in the data but the trend seems sufficiently accurate for preliminary design purposes. The complex design and construction of the parachutes and their naturally flexible structure make for a reasonable degree of variability in their performance characteristics.

The variation of C_{D0} with relative suspension line length is shown in Figure 6.63. Ewing (1972) suggests that the drag coefficient is optimum at about $l/D_0 = 1.15$. The correlation shown in Figure 6.63 is

$$C_{D0} = 1 - K_l \left[\left(\frac{l}{D_0} \right)^2 - 1.17 \right] \tag{6.44}$$

Here, we see that the trend is for the drag coefficient to drop off from a maximum when the relative suspension line length is larger or smaller than 1.17 rather than 1.15. The coefficient K_l has a range of $4 \le K_l \le 5$ and curves based on Eqn (6.44) using the two extreme values of K_l are also shown on Figure 6.63. Once again, there is scatter in the data but the trend seems sufficiently accurate for preliminary design purposes.

The mass of a single canopy m_c is shown in Figure 6.64 as a function of the nominal diameter D_0. The data does follow the expected quadratic growth as described by the approximation

$$m_c = K_m D_0^2 \tag{6.45}$$

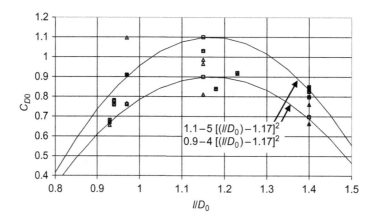

FIGURE 6.63

Variation of C_{D0} as a function of relative suspension line length. Data is from Table 6.2 with closed and open symbols denoting calculated and reported values, respectively.

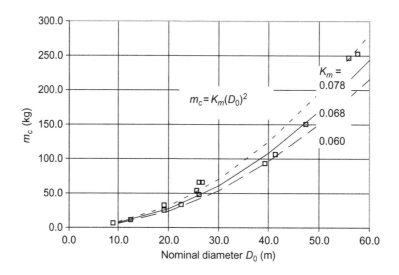

FIGURE 6.64

Variation of canopy mass m_c as a function of canopy nominal diameter D_0. Data shown is from Table 6.2 and curves are correlations for heavy, medium, and light construction.

The coefficient K_m varies depending on construction details with $K_m = 0.060$, 0.068, and 0.078 kg/m^2 being considered reasonable approximations for light, medium, and heavy construction, respectively. These values appear to embrace the range of canopy mass in Figure 6.64. However, there are often additional items contained in the parachute landing system like reserve

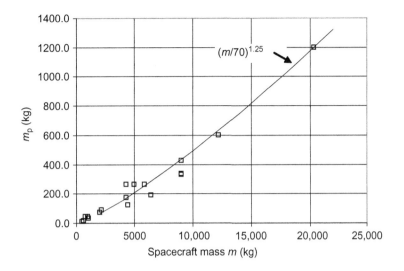

FIGURE 6.65

Variation of parachute recovery system mass m_p as a function of the returned spacecraft mass m. Data shown is from Table 6.2 using the correlation $m_p = 1.25m_c$.

parachutes, capsule slings, flotation devices for water recovery, pressure switches to activate devices at specific altitudes, etc., and these add to the weight of the complete recovery system. Using information reported by Ewing (1972) for the Mercury, Gemini, and Apollo landing systems indicates that the mass of the parachute recovery system, excluding any reserve parachutes, is proportional to the mass of the main canopy $m_p = K_p m_c$ where $1.3 < K_p < 1.45$. Applying this approximation leads to the results shown in Figure 6.65 for the complete parachute recovery system mass m_p (exclusive of a reserve parachute) as a function of the mass of the returned spacecraft m.

From knowledge of the mass of a single canopy m_c based on the above approximations, one may determine a prospective value for the mass of the complete parachute recovery system m_p. This value and the spacecraft mass m may be entered into Figure 6.65 to examine its consistency with the current database.

Although a single canopy is recommended for the aerial recovery of entry capsules, operational and safety requirements may make it necessary to select a cluster of identical canopies. Normally, parachute clusters are avoided, if possible, because they tend to open nonuniformly and they have a diminished drag coefficient compared to a single canopy. Some data for ringsail parachutes presented by Ewing (1972) suggests that the ratio of drag coefficient of a cluster of n parachutes ($n \le 4$) to that of a single parachute may be approximated by

$$\frac{C_{D,c}}{C_{D0}} = 1 - 0.027(n - 1)$$

The parachute design information gathered here provides input for weight and volume requirements for the next level of detail design of the spacecraft to be recovered.

6.8 LOW-SPEED RETURN AND RECOVERY: SPACEPLANES

The use of space capsules was and remains a successful approach for manned spaceflight because of their relative simplicity and reliability. However, the objective of reducing the expense of space access brought forward the concept of a spacecraft that could function like a conventional airplane, that is, a reusable vehicle. The space shuttle was the first and still the only spaceplane, one that could be launched into orbit as is a space capsule, but, upon completion of its mission could deorbit, enter the atmosphere, and fly down to a horizontal landing on a runway. It was expected that this feature would permit the manned vehicle to be refurbished and turned around for another trip to space. In addition to being reusable, the spaceplane's lift-to-drag ratio, several times greater than that of a capsule, would permit substantial flexibility in terms of downrange and cross-range performance. The typical return of a shuttle from space is illustrated in Figure 6.66 in which it is seen that the reaction control system (RCS) used at the highest altitudes converts to aerodynamic control of pitch and roll at intermediate altitudes and finally the yaw system changes from thruster force to aerodynamic force application below $z = 30$ km. Below this altitude the shuttle is flying supersonically ($1 < M < 4$) for which there is reasonable experience garnered from SR-71 operations. The shuttle can be flown in the same fashion and brought down to the subsonic speed regime at about the levels flown by commercial aircraft. In this section, we concentrate on the final phase of return and recovery which is shown on Figure 6.66 as the approach, flare, and land segment.

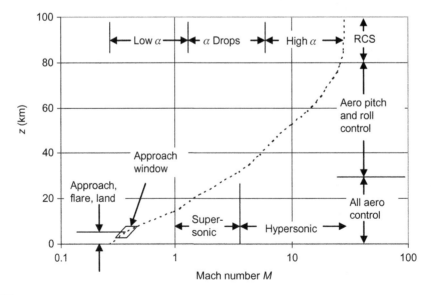

FIGURE 6.66

The approach and landing phase of a spaceplane (M decreasing) like the shuttle is shown in relation to the overall entry process.

6.8.1 **LOW-SPEED CHARACTERISTICS OF SPACEPLANES**

Kempel, Painter, and Thompson (1994) describe the development of the HL-10, a wingless vehicle called a lifting body. Interest in such lifting bodies as practical candidates for manned entry from space with airplane-like landing capability grew rapidly in the 1960s and several that were flight-tested extensively by NASA and the US Air Force are shown in Figure 6.67. The experience gained with flying the lifting bodies and the X-15 hypersonic research airplane was the basis for the successful development of the Space Shuttle vehicle.

The landing procedure consisted of three parts: final approach, flare, and postflare deceleration, as illustrated in Figure 6.68. The pilots first establish a preflare aim point on the ground when at an altitude of about 1200 m above ground level while flying the final approach at 140−156 m/s (270−300 kts) along the −16° to −18° (nose down) steep glide path. The flare is initiated at about 300 m above ground level with a load factor around $n = 1.5$ and serves to bring the vehicle onto the shallow glide path angle of about $\gamma_o = -1.5°$. This maneuver reduces the vehicle velocity to about 115−125 m/s (220−240 kts) at which point the landing gear is lowered. The continued deceleration along the shallow glide path permits touchdown at about 81−116 m/s (155−223 kts). Kempel et al. (1994) point out that "The steep approaches were never a problem for the pilots, although they were breathtaking to watch. The M2-F2 descents were particularly spectacular in their steepness."

It has been pointed out by Hoey (1963) that values of $L/D < 3$ during the landing flare is considered unacceptable for piloted landing, independent of the landing wing loading $(W/S)_l$. The L/D during flare is essentially equivalent to $(L/D)_{max}$. The X-15 and the Space Shuttle Orbiter have $(W/S)_l$ in the range of 3350−3830 N/m^2 (70−80 lb/ft^2) and $(L/D)_{max}$ in the range of 4.15−4.75. The HL-10 and the M2-F2 have L/D in the range of 3.5−4 but with $(W/S)_l$ about half those of the

FIGURE 6.67

Three early lifting body designs at NASA Dryden Research Center. From left to right, the X-24A, the M2-F3, and the HL-10. Courtesy of NASA.

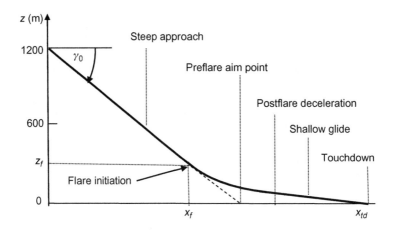

FIGURE 6.68

Landing approach for high-speed vehicles with low L/D.

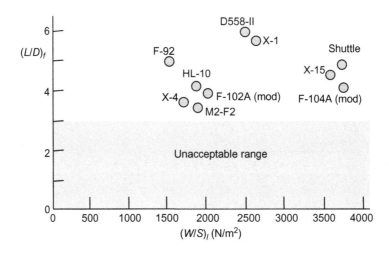

FIGURE 6.69

Lift-to-drag ratio during flare as a function of wing loading for several high-speed aircraft and lifting bodies.

X-15 and the shuttle. Power-off landings are readily achievable with vehicles whose $(L/D)_{max}$ is in the range of 3.5–5. Factors in the configuration design which contribute to such landings are good visibility, good handling qualities, effective speed brakes, and minimal trim changes. Some representative values for lift-to-drag ratio in flare as a function of wing loading for high-speed aircraft in landing are shown in Figure 6.69.

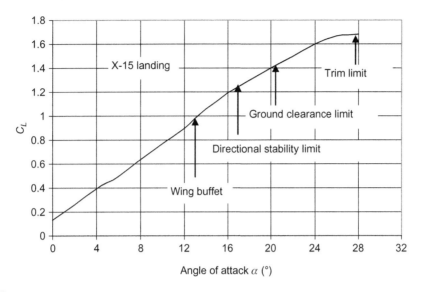

FIGURE 6.70

Lift coefficient of the X-15 as a function of angle of attack. Shown on the figure are the limiting operational factors in landing.

Approach and landing may be considered to be quasi-static processes so lift is approximately equal to landing weight. As discussed by Sforza (2014) for commercial aircraft, the landing speed is generally taken as $V_l = 1.2V_s$ to account for gusts during landing while the approach speed $V_a > 1.3V_s$, where V_s is the stall speed given by

$$V_s = \sqrt{\frac{(W/S)}{\frac{1}{2}\rho_{sl}\sigma(C_L)_{max}}} \qquad (6.46)$$

The stall speed may be considered the minimum usable speed and $C_{L,max}$ the maximum usable lift coefficient. Because high-speed aircraft fuselages are often long, the true stall angles of attack are never reached in landing because of the danger of tail strikes and other controllability issues. For example, Johnston and Gaines (1963) point out that the maximum usable angle of attack for the X-15 is about 13°. The range of limitations they found on angle of attack of the X-15 is shown in Figure 6.70, and these are representative of most high-speed vehicles. Nominal lift curves for several representative hypersonic vehicles are shown in Figure 6.71. They cover the range of from a pointed slender body (X-15) to a delta-winged bluff body (Shuttle) to a pure lifting body (HL-10).

A comparison of the variation of the lift-to-drag ratio as a function of angle of attack for the representative space access vehicles are shown in Figure 6.72. For a compilation and assessment of flight-derived lift and drag data for spaceplanes see Saltzman, Wang, and Iliff (2002). In order to attain the necessary landing L/D of around 4, Figure 6.72 shows that the lifting body shape must

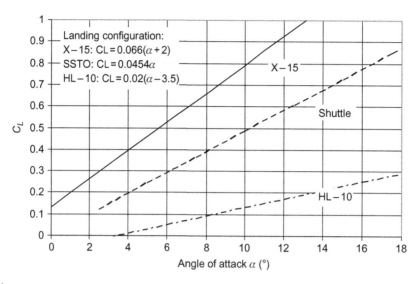

FIGURE 6.71

Nominal lift curves as a function of angle of attack in the landing configuration for several space access vehicles.

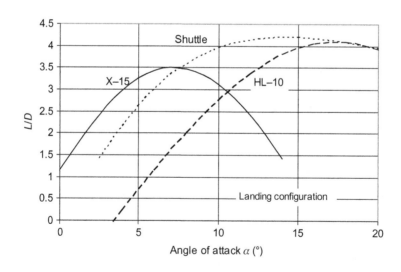

FIGURE 6.72

The variation of L/D as a function of angle of attack in the landing configuration for several space access vehicles.

fly at much larger angles of attack than the more slender bodies. This is illustrated in Figure 6.73, where the HL-10 is shown landing accompanied by the NASA F-104 chase plane. On the other hand, the X-15 lands at a more moderate angle of attack, as shown in Figure 6.74. In between the two extremes is the landing attitude of the Shuttle which is shown in Figure 6.75.

FIGURE 6.73

Landing of the HL-10 lifting reentry body and the F-104 chase plane at NASA Dryden Research Center (NASA).

FIGURE 6.74

The XD-15 is shown landing accompanied by the F-104 chase plane. Note moderate angle of attack of the X-15 and its main landing gear which uses skids rather than wheels (NASA).

FIGURE 6.75

STS-46 Shuttle Atlantis lands on Runway 33 at the Kennedy Space Center (NASA).

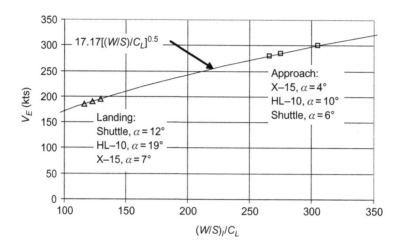

Nominal landing and approach equivalent speeds (in knots) for the representative spaceplanes. The associated angle of attack is also shown.

The equivalent airspeed, which is generally used since it is effectively the indicated airspeed monitored by the pilot, is given by

$$V_E = V\sqrt{\sigma}$$

In practice, the equivalent speed is usually measured in knots and therefore for equilibrium flight at a unit load factor the equivalent speed may be written as

$$V_E = 17.17\sqrt{\frac{(W/S)}{C_L}} \tag{6.47}$$

Recall that the speed in knots is about twice that in meter per second, that is, V_E (kts) = $1.943 V_E$ (m/s).

As discussed previously in connection with the equation for the stall velocity, spaceplanes rarely operate near their maximum lift coefficient and land at rather high speeds. Furthermore, the nominal approach speed for spaceplanes like those we have been considering is generally in the range $1.4 < V_a/V_l < 1.6$, substantially larger than in commercial aircraft where it is typically $V_a/V_l = 1.25$. The landing and approach speeds for our representative spaceplanes are shown in Figure 6.76. Equation (6.47) has been used to retrieve the associated lift coefficients and the equations in Figure 6.71 have been used to extract the corresponding angles of attack and they are also shown on Figure 6.76.

Although the nominal assumption here is that the approach and landing process is quasi-static in nature, this is more accurate for conventional aircraft than for high performance aircraft. The latter tend to flare more aggressively, with the load factor n in the vertical direction being closer to $n = 1.4$ rather than $n \sim 1$, which is the case for commercial and transport aircraft (Sforza, 2014). The typical variation of the load factor with the flare altitude for the X-15 is shown in Figure 6.77. The knee of the curve in that figure defines the altitude for which executing the flare is the easiest. At lower load factors, the flare must be initiated at higher altitudes which requires increased for completing the flare

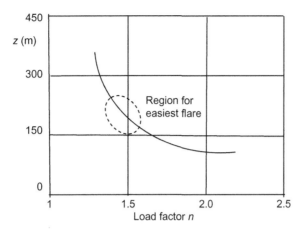

FIGURE 6.77

Typical variation of the flare initiation altitude as a function of load factor n for X-15 approaches.

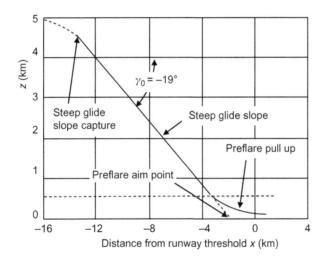

FIGURE 6.78

A steep glide path approach for a spaceplane is illustrated here for conditions typical of a space shuttle flight.

while at higher load factors the flare initiation altitude is low resulting in reduced time for the landing approach thereby leading to greater stress on piloting technique.

6.8.2 STEEP GLIDE FROM 5000 TO 500 M

The final stage of the descent to the surface has the spaceplane approaching on a steep glide to facilitate a rapid landing. The steep glide path is illustrated in Figure 6.78 for conditions typical of a space shuttle flight. The steep glide path is captured at an altitude of about $z = 5$ km and at a distance of about 12 km from the runway threshold, $x = 0$ on Figure 6.78. The pilot locates a preflare aim point on the ground and uses the speed brake to control speed so as to maintain constant equivalent flight speed V_E until an altitude of around $z = 500$ m is reached, whereupon a pull-up is executed in order to prepare for the final flare prior to landing.

During the steep glide at a constant flight path angle $\gamma = \gamma_0$, the equilibrium conditions require that

$$\frac{1}{g}\frac{dV}{dt} = -\frac{D}{W} - \sin\gamma_0 \tag{6.48}$$

$$mV\frac{d\gamma}{dt} = 0 = L - mg\cos\gamma_0 \tag{6.49}$$

Combining these equations yields

$$\tan\gamma_0 = -\frac{D}{L} - \frac{W}{L}\left(\frac{1}{g}\frac{dV}{dt}\right) \tag{6.50}$$

In descent along the constant flight path angle, the acceleration is

$$\frac{dV}{dt} = \frac{dV}{dz}\frac{dz}{dt} = V\sin\gamma_0\frac{dV}{dz} = \frac{1}{2}\sin\gamma_0\frac{d}{dz}\left(\frac{V_E^2}{\sigma}\right)$$

For a constant equivalent airspeed descent, this becomes

$$\frac{dV}{dt} = \frac{1}{2}V_E^2\sin\gamma_0\left(-\frac{1}{\sigma}\frac{d\sigma}{dz}\right) \tag{6.51}$$

The steep glide takes place typically altitudes less than 5000 m, where temperature in the first layer of the standard atmosphere ($z < 11{,}000$ m) is characterized by a constant lapse rate, as pointed out in Chapter 2. Using Eqn (2.4) for the pressure variation with altitude, the data given in Table 2.1, and the state Eqn (2.1), one may obtain the following expression for the density ratio:

$$\sigma = 1.225(1 - 2.2557 \times 10^{-5}z)^{4.256} \tag{6.52}$$

Then

$$-\frac{1}{\sigma}\frac{d\sigma}{dz} = \frac{9.6 \times 10^{-5}}{1 - 2.2557 \times 10^{-5}z} \approx 9.6 \times 10^{-5} \tag{6.53}$$

The error incurred in choosing $z = 0$ in the approximation shown above is about 1% in flight path angle or velocity at $z = 4$ km. Then, using Eqns (6.51) and (6.53) in Eqn (6.50) yields

$$\tan\gamma_0 = -\frac{D}{L} - \frac{W}{L}\frac{9.6 \times 10^{-5}\sin\gamma_0}{g}\left(\frac{1}{2}V_E^2\right) \tag{6.54}$$

Equation (6.49) requires that

$$\frac{W}{L} = \frac{1}{\cos\gamma_0}$$

Using this information in Eqn (6.54) transforms it into

$$\tan\gamma_0 = -\frac{D}{L} - \frac{9.6 \times 10^{-5}V_E^2\tan\gamma_0}{2g} \tag{6.55}$$

We will assume that the drag coefficient depends only on the profile drag, the lift-induced drag, and the drag caused by deployment of the speed brake. The inverse of L/D is given by

$$\left(\frac{D}{L}\right) = \frac{C_{D,0} + kC_{L,t}^2 + \left(\frac{\partial C_D}{\partial \delta_{sb}}\right)\delta_{sb}}{C_{L,t}} = \frac{C_{D,0} + \left(\frac{\partial C_D}{\partial \delta_{sb}}\right)\delta_{sb}}{C_{L,t}} + kC_{L,t} \tag{6.56}$$

The trimmed lift coefficient is

$$C_{L,t} = \frac{L}{qS} = \frac{W\cos\gamma}{qS} = \left(\frac{W}{S}\right)\frac{\cos\gamma}{q} \tag{6.57}$$

Substituting Eqn (6.57) into Eqn (6.56) leads to

$$\left(\frac{D}{L}\right) = q\frac{C_{D,0} + \left(\frac{\partial C_D}{\partial \delta_{sb}}\right)\delta_{sb}}{\left(\frac{W}{S}\right)\cos\gamma} + \frac{k\cos\gamma}{q}\left(\frac{W}{S}\right) \tag{6.58}$$

Then, after substituting Eqn (6.58), with $\gamma = \gamma_0$, into Eqn (6.55) and using the relation $q = \rho_{sl}V_E^2/2$ we may rearrange terms to arrive at the following relation:

$$q\sin\gamma_0\left[1 + \frac{9.6\times10^{-5}q}{\rho_{sl}g}\right] = -\left[C_{D,0} + \left(\frac{\partial C_D}{\partial\delta_{sb}}\right)\delta_{sb}\right]\frac{q^2}{\left(\frac{W}{S}\right)} - k\cos^2\gamma_0\left(\frac{W}{S}\right)$$

This is a quadratic equation of the form $Aq^2 + Bq + C = 0$ where

$$A = \sin\gamma_0\left[\frac{9.6\times10^{-5}}{\rho_{sl}g}\right] + \left[C_{D,0} + \left(\frac{\partial C_D}{\partial\delta_{sb}}\right)\delta_{sb}\right]\left(\frac{W}{S}\right)^{-1}$$

$$B = \sin\gamma_0$$

$$C = k\cos^2\gamma_0\left(\frac{W}{S}\right)$$

The solution of the quadratic equation for q is

$$q = \frac{-B \pm \sqrt{B^2 - 4AC}}{2A}$$

We may use the nominal conditions of equilibrium glide for the Space Shuttle shown in Table 6.4.

The results for the equilibrium glide are shown in Figure 6.79, which also includes a particular operating point for STS-4. Though the data point shown is for $\delta_{sb} = 15°$, the operating range for deflection angle of the speed brake is up to $25°$ for subsonic flight. A scale for velocity in knots is presented on the figure and it may be noted that velocity in knots is just about twice the velocity in meter per second: V (kts) $= 1.943V$ (m/s).

With an equilibrium glide slope of $-19°$ and a constant equivalent speed $V_E = 150$ m/s (291 kts) the time elapsed from an altitude of about 4500 to 500 m is given by

$$\Delta t = t_2 - t_1 = \frac{1}{\sin\gamma_0}\int_{z_1}^{z_2}\frac{dz}{V} = \frac{1}{V_E\sin\gamma_0}\int_{z_1}^{z_2}\sqrt{\sigma}dz \tag{6.59}$$

Table 6.4 Nominal Equilibrium Glide Conditions for the Space Shuttle Orbiter		
Landing weight W_s	934,000 kN	210,000 lb
Wing area S	250 m²	2690 ft²
Landing wing loading $(W/S)_l$	3740 N/m²	78.1 lb/ft²
Air weight density $\rho_{sl}g$	12 kN/m³	0.0765 lb/ft³
Induced drag factor k	0.173	0.173
Zero-lift drag coefficient $C_{D,0}$	0.067	0.067
Speed brake effectiveness $\frac{\partial C_D}{\partial \delta_{sb}} = 0.00068/\text{degree}$	0.00068/degree	0.00068/degree
Speed brake deployment angle δ_{sb}	0–87.5°	0–87.5°

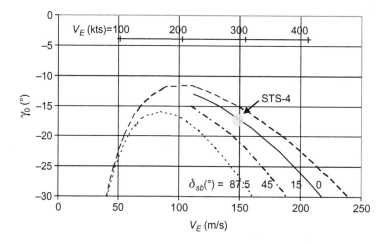

FIGURE 6.79

Equilibrium glide slope as a function of equivalent speed for the conditions appropriate to the space shuttle. An operating point for the flight of STS-4 is indicated.

Using Eqn (6.59) results in the elapsed time given by

$$\Delta t = \frac{-1.4173 \times 10^3}{V_E \sin \gamma_0} \left[(1 - 2.2557 \times 10^{-5}z_2)^{3.128} - (1 - 2.2557 \times 10^{-5}z_1)^{3.128} \right] \tag{6.60}$$

For the space shuttle example chosen previously, the elapsed time is about 72.4 s and the horizontal distance traveled, $\Delta x = \Delta z/\tan \gamma_0$, is about 11.6 km. Therefore, at the end of the equilibrium glide for the typical space shuttle case shown in Figure 6.78 the spacecraft would be at an altitude of around 500 m and about 2000 m from the runway threshold with an equivalent speed of 150 m/s or a true speed of 153.7 m/s. At this stage, a pull-up maneuver is executed to bring the vehicle into an attitude appropriate for landing.

6.8.3 **PREFLARE PULL-UP FROM 500 TO 150 M**

At a distance of around 2—3 km from the runway threshold, the spaceplane is flying at around 500 m altitude with an equivalent speed of about 125—150 m/s (250—300 kts) and a steep equilibrium glide path angle of around $-15°$ to $-20°$. The steep glide terminates in a pull-up maneuver to bring the flight path angle up to around $-1°$ to $-3°$ in preparation for the final flare and landing. This maneuver is initiated at around 500 m altitude so it is reasonable to consider density to be constant. We assume that the pull-up is carried out at a constant load factor $n = L/W$. The applicable equations of motion for this case are

$$\frac{dV}{dt} = -g\left[\left(\frac{D}{W}\right) + \sin \gamma\right] = -g\left[n\left(\frac{D}{L}\right) + \sin \gamma\right] \tag{6.61}$$

$$\frac{d\gamma}{dt} = \frac{g}{V}\left[\left(\frac{L}{W}\right) - \cos \gamma\right] = \frac{g}{V}[n - \cos \gamma] \tag{6.62}$$

The auxiliary equations for the spacecraft position in the trajectory are given by

$$\frac{dz}{dt} = V \sin \gamma \tag{6.63}$$

$$\frac{dx}{dt} = V \cos \gamma \tag{6.64}$$

If we introduce new nondimensional variables $X = V/V_{ref}$, $Y = \gamma$, and $T = t/t_{ref}$ with the normalizing quantities given by $V_{ref} = V_0$ and $t_{ref} = V_0/g$, where V_0 is the initial velocity at the start of the pull-up, then the equations of motion become

$$\frac{dX}{dT} = n\left(\frac{D}{L}\right) + \sin Y \tag{6.65}$$

$$\frac{dY}{dT} = \frac{n - \cos Y}{X} \tag{6.66}$$

The initial conditions are as follows:

$$X(0) = 1 \text{ and } Y(0) = \gamma(0)$$

The flight path angle is not in general very small until the bottom of the flight path and, as pointed out previously in the analysis of the entry phase, there are no simple closed form solutions for this equation set so that numerical solutions are required.

Flying at a constant load factor requires increasing lift as the flight speed is reduced through drag. This requires some increase in the angle of attack of the vehicle during the maneuver. However, if the pull-up is carried out at constant L/D, where the angle of attack is held constant the results are close to those for constant load factor. This is because the pull-up is carried out at high L/D where the variation of L/D with angle of attack is small.

Consider the typical case of a lifting entry vehicle carrying out a pull-up maneuver at a constant load factor with initial conditions of $V_0 = 150$ m/s (291 kts), $\gamma_0 = -19°$, and $z_0 = 500$ m. Two cases are considered: $n = 1.4$ with a lift-to-drag ratio $L/D = 4$ and $n = 1.2$ with $L/D = 3$. The flight path angle variation with time during the pull-up is virtually linear, as shown in Figure 6.80. The object of the maneuver is to increase the flight path angle to about $\gamma = -2°$ so the duration of the

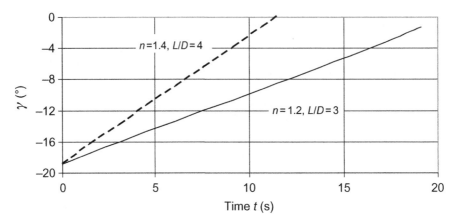

FIGURE 6.80

Variation of the flight path angle during $n = 1.4$ and $n = 1.2$ pull-up maneuvers, both starting at $V_0 = 150$ m/s and $z = 500$ m.

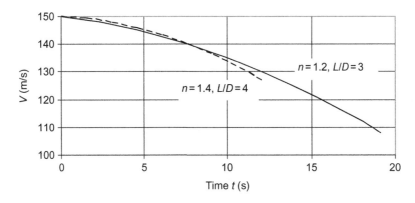

FIGURE 6.81

Variation of the flight speed during $n = 1.4$ and $n = 1.2$ pull-up maneuvers, both starting at $V_0 = 150$ m/s and $z = 500$ m.

maneuver is about 10 s for $n = 1.4$ and almost twice that for $n = 1.2$. Note that if the pull-up is continued the flight path angle becomes positive and the vehicle would be climbing. During the maneuver the flight speed decreases as shown in Figure 6.81 and the altitude decreases as shown in Figure 6.82.

The trajectory of the vehicle during the pull-up is depicted in Figure 6.83. The downrange distance x shown here is measured from the start of the pull-up maneuver. The $n = 1.4$ pull-up reaches an appropriately small flight path angle at around $x = 1600$ m, but this is achieved at an altitude $z = 230$ m which is higher than that required for the glide path for final approach. An

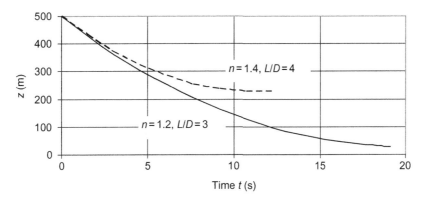

FIGURE 6.82

Variation of the altitude during $n = 1.4$ and $n = 1.2$ pull-up maneuvers, both starting at $V_0 = 150$ m/s and $z = 500$ m.

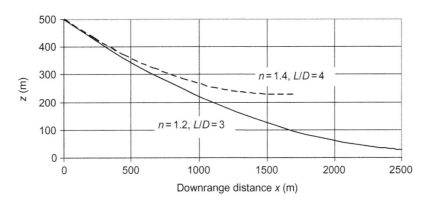

FIGURE 6.83

Variation of the flight path angle during $n = 1.4$ and $n = 1.2$ pull-up maneuvers, both starting at $V_0 = 150$ m/s and $z = 500$ m.

additional flare to reduce altitude sufficiently to capture the final landing glide path would be required. The $n = 1.2$ pull-up reaches the final glide path angle of about $-1.5°$ at $z = 30$ m but 800 m further downrange, at about $x = 2400$ m.

6.8.4 THE LANDING AIR RUN

The landing process follows the sequence of events described in the previous sections with the final stage of the landing occurs at the runway as shown in Figure 6.84. There the landing field length is denoted by $x_l = x_a + x_g$, that is, the sum of an air run and a ground run. These two portions of the landing process are treated sequentially below.

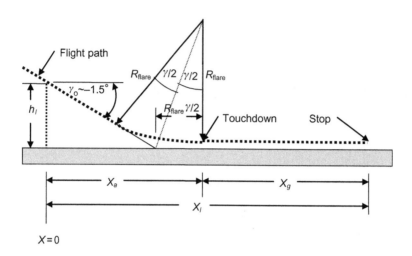

FIGURE 6.84

Stages of the landing process showing shallow glide, final flare, and actual landing. Flight path angles and all distances are exaggerated for clarity.

The air run may be considered to start at the $h_l = 15.24$ m (50 ft) obstacle, so that in an x–z coordinate system with $x = 0$ at $z = h_l$, the approach speed equals V_a, which for spaceplanes is generally in the range $1.4 < V_l < 1.6$ whereas for commercial airliners the value is lower, around $1.25V_l$. The vehicle executes a gradual final flare of large radius R, increasing the flight path angle from γ_0 of around $-1.5°$ to $-3°$ up to zero at which time the vehicle has slowed to V_l, that is, to a value around $1.2V_s$ to $1.25V_s$ at an acceptable sink rate and touched down on the runway, ending the air run at $x = x_a$.

A simplified analysis of the air run considers the flight path angle, which is negative in the landing phase, to change so slowly that the angular acceleration may be ignored in the force balance, leading to the following approximate equations:

$$-W \sin \gamma - D = m\frac{dV}{dt} \tag{6.67}$$

$$L - W \cos \gamma = m\frac{d}{dt}\left(R\frac{d\gamma}{dt}\right) \approx 0 \tag{6.68}$$

During the flare, the deceleration along the trajectory may be written as $\frac{dV}{dt} = V\frac{dV}{dx}$. The distance traveled during the air run from $V = V_a$ at $x = 0$, to $V = V_l$ at $x = x_a$, may be found by integrating the equations

$$-D - mg\gamma = mV\frac{dV}{dx} = \frac{1}{2}m\frac{dV^2}{dx} \tag{6.69}$$

$$L - W = 0 \tag{6.70}$$

In these equations, it was assumed that the flight path angle is small enough to properly replace the sine and cosine functions with the leading terms of their expansions. This leads to the following equation describing the change in the total, kinetic plus potential, energy of the vehicle:

$$d\left(\frac{1}{2}mV^2\right) + mg\gamma dx = -Ddx \tag{6.71}$$

The change in total energy is equal to the work done by the external forces acting on the vehicle. The weight is approximately equal to the lift during the flare (the load factor $n \sim 1$) so that $mg \sim L$ and may be considered constant during the flare. Then Eqn (6.71) becomes

$$d\left(\frac{1}{2g}V^2\right) = -\left(\gamma + \frac{D}{L}\right)dx \tag{6.72}$$

If it is assumed that the actual, slowly varying, small flight path angle may be replaced by an average value, say γ_a, then Eqn (6.72) may be integrated to yield

$$V_l^2 - V_a^2 = -2g\left[\left(\frac{D}{L}\right)_l + \gamma_a\right]x_a \tag{6.73}$$

Because the lift during the landing air run was assumed to be approximately equal to the weight and the drag is proportional to the square of the lift, the quantity $(D/L)_l$ must also be considered approximately constant and equal to an effective value which includes the ground effect, which may be appreciable when the vehicle is within one span height above the ground.

Equation (6.73) may be rearranged to yield the air run as follows:

$$x_a = \frac{1}{2g}\left[\frac{V_a^2 - V_l^2}{\left(\frac{D}{L}\right)_l + \gamma_a}\right]$$

In order to provide a stall margin upon landing, the landing speed is taken as $V_l = 1.2V_s$ where the stall speed is given by

$$V_s = \sqrt{\frac{2(W/S)}{\rho_{sl}\sigma C_{L,\mathrm{max}}}}$$

Then, in terms of the stall characteristics of the vehicle, the air run equation becomes

$$x_a = 1.44\left[\left(\frac{V_a}{V_l}\right)^2 - 1\right]\frac{\left(\frac{W}{S}\right)_l}{8\rho_{sl}\sigma C_{L,l,\mathrm{max}}}\left[\left(\frac{D}{L}\right)_l + \gamma_a\right]^{-1} \tag{6.74}$$

As mentioned previously, the ratio of approach speed to landing speed lies in the range of $1.25 < V_a/V_l < 1.6$. The lower end of the range is typical of commercial airliners (Sforza, 2014), while the upper end of the range typical of high performance aircraft. Note that the effective glide ratio in landing $(D/L)_l$ may be approximated by

$$\left(\frac{D}{L}\right)_l \approx \frac{C_{D,l}}{C_{L,l}} = \frac{\left(C_{D,0} + \frac{C_L^2}{\pi eA}\right)_l}{C_{L,l}} = 1.44\frac{C_{D,0,l}}{C_{L,\mathrm{max},l}} + \frac{C_{L,\mathrm{max},l}}{1.44\pi eA}$$

In the absence of lift and drag characteristics, as in the first stages of preliminary design, one may approximate the effective drag to lift ratio to be around $(D/L)_l = 1/4$ for spaceplanes in the landing configuration, as depicted in Figure 6.72, while the average flight path angle γ_a may be taken to be around $-1.5°$ or -0.026 rad; note that the value of x_a should be positive. For example, assuming $V_a/V_l = 1.5$, the total landing air run would be approximated by

$$x_a = 8 \frac{\left(\dfrac{W}{S}\right)_l}{g\rho_{sl}\sigma C_{L,\max,l}}$$

(6.75)

For typical spaceplane, we may take $(W/S)_l = 3.6$ kPa and $C_{L,\max,l} = 1$ so that at sea level the air run would measure about 2400 m.

6.8.5 THE LANDING GROUND RUN

On the simplest level, one may assume that the average deceleration during the ground run is

$$\frac{dV}{dt} = \frac{d}{dx}\left(\frac{1}{2}V^2\right) = -a_{avg} = \text{const.}$$

Integrating from the $V = V_l$ (the landing speed) at $x = 0$ to $V = 0$ at the end of the landing where $x = x_g$ yields $x_g = \frac{V_l^2}{2a_{avg}}$. For a case of $V_l = 100$ m/s (196 kts) and $a_{avg} = 0.31$ g, we find the ground run to be $x_g = 1645$ m. The distance covered during the ground run may be put in terms of the stall characteristics of the aircraft by setting $V_l = 1.2V_s$ resulting in the following relation:

$$x_g = \frac{1.44\left(\dfrac{W}{S}\right)_l}{a_{avg}\rho_{sl}\sigma C_{L,\max,l}}$$

(6.76)

At the next level of approximation, the dynamics of the situation require that the normalized acceleration during the ground run be given by

$$\frac{1}{g}\left(\frac{dV}{dt}\right)_g = \frac{1}{2g}\left(\frac{dV^2}{dx}\right)_g = \left(\frac{F}{W}\right)_g - \left(\frac{D}{W}\right)_g - \left(\frac{F_b}{W}\right)_g$$

For aircraft in general, the thrust F at landing is essentially idle thrust so we may take $(F/W)_g = $ constant. During the ground run however, the use of thrust reversers provides a negative thrust level which aids in braking. So far, spaceplanes are unpowered in landing and in that case the thrust during the ground run will be zero, $(F/W)_g = 0$. However, it may be necessary to provide negative thrust, that is, additional drag, through the use of a drag parachute. In that case, the thrust force during the ground run is actually a drag force which may be written as

$$\left(\frac{F}{W}\right)_g = -\frac{D_p}{W} = -\frac{1}{2W}C_{D,p}\rho_{sl}\sigma V^2 S_p$$

Here, the parachute drag is characterized by a drag coefficient $C_{D,p}$ based on an appropriate parachute area S_p, typically the frontal area of the inflated parachute.

The drag of the spaceplane during the ground run may be written as

$$\left(\frac{D}{W}\right)_g = \frac{C_{D,g}\rho_{sl}\sigma V^2}{2\left(\frac{W}{S}\right)_l}$$

Here, the landing drag coefficient during the ground run $C_{D,g}$ will depend on the specific landing configuration like flap and slat deflection and speed brake setting.

The braking force term may be written in terms of the normal force the wheels exert on the runway as follows:

$$\frac{F_b}{W} = \mu_b\left(1 - \frac{L}{W}\right)$$

The braking coefficient of friction lies in the range of $0.4 < \mu_b < 0.6$ for concrete runways. The use of skids rather than wheels on some high performance research aircraft like the X-15 results in a braking coefficient of about 0.33 on dry lake bed runways according to measurements reported by McKay and Scott (1961).

Expanding the lift L in terms of landing conditions leads to the following result:

$$\left(\frac{F_b}{W}\right)_g = \mu_b\left(1 - \frac{L}{W}\right)_g = \mu_b\left[1 - \frac{C_{L,g}\rho_{sl}\sigma V^2}{2\left(\frac{W}{S}\right)_l}\right] \tag{6.77}$$

The deceleration equation then becomes

$$a = \frac{1}{2}\frac{dV^2}{dx} = -g\left[\frac{C_{D,p}\rho_{sl}\sigma V^2 S_p}{2W} + \mu_b\right] - g\left[1 - \mu_b\left(\frac{C_L}{C_D}\right)_g\right]\frac{C_{D,g}\rho_{sl}\sigma V^2}{2\left(\frac{W}{S}\right)_l} \tag{6.78}$$

The lift-to-drag ratio $L/D = C_L/C_D = 0$ during the ground run because as the aircraft touches down it levels out, reducing the angle of attack, while it is also slowing down. Indeed, touchdown of an aircraft generally automatically initiates deployment of spoilers to dump the lift and let the aircraft settle down on the ground thereby increasing the rolling resistance which helps slow the aircraft down.

Equation (6.78) states that the acceleration $a = a_1 + a_2$, where a_1 is the deceleration due to the drag chute, if any, and braking, and a_2 is the deceleration due to aerodynamic drag of the spaceplane less the effect of aerodynamic lift, if any, on reducing the normal force between the wheels and the runway. Equation (6.78) is of the form

$$\frac{dV^2}{dx} + AV^2 + B = 0 \tag{6.79}$$

With $(L/D)_g = 0$, the coefficients A and B are given by

$$A = \frac{\rho_{sl}\sigma g}{\left(\frac{W}{S}\right)_l}\left[C_{D,p}\left(\frac{S_p}{S}\right) + C_{D,g}\right]$$

$$B = 2\,g\mu_b$$

The solution in terms of the ground distance is

$$V^2 = \left[V_l^2 + \frac{B}{A} \right] e^{-Ax} - \frac{B}{A} \tag{6.80}$$

The total ground distance is then determined by setting $V = 0$ and is given by

$$x_g = \frac{1}{A} \ln \left(1 + \frac{A}{B} V_l^2 \right) \tag{6.81}$$

This equation may be put in the form

$$x_g = \frac{V_l^2}{2.88\, g} \left[\frac{C_{L,l,\mathrm{max}}}{C_{D,p}(S_p/S) + C_{D,g}} \right] \ln \left[1 + \frac{1}{2\mu_b C_{L,l,\mathrm{max}}} \left(\frac{V_l}{V_s} \right)^2 \right] \tag{6.82}$$

With a landing speed $V_l = 100$ m/s, $C_{L,l,\mathrm{max}} = 1$, $\mu_b = 0.5$, and $(V_l/V_s) = 1.2$, the ground run is

$$x_g = \frac{316}{C_{D,p}(S_p/S) + C_{D,g}} \tag{6.83}$$

The spaceplane drag coefficient with full speed brake deployment might be around $C_{D,g} = 0.125$ which would lead to $x_g = 2530$ m without deployment of a drag parachute.

A drag parachute can provide an effect approximately equal in magnitude to that provided by the aerodynamic drag of the spaceplane alone. For example, in describing the development of a drag parachute for the space shuttle orbiter, Meyerson (2001) shows that parachute drag can be written as

$$D_p = f(t) K_w C_{D,p,0} S_p q_\infty = C_{D,p} S_p q_\infty$$

The effective drag coefficient $C_{D,p}$ includes transient effects of deployment expressed by $f(t)$ and blockage effects due to the vehicle wake expressed by K_w. For the space shuttle, these factors have experimentally determined ranges of $0 < f(t) < 1.05$ and $0.65 < K_w < 0.9$. However, it must be noted that the drag parachute for the space shuttle orbiter is to be deployed only between 118 (230 kts) and 72 m/s (140 kts) and then jettisoned between 41 (80 kts) and 27 m/s (50 kts). Therefore the parachute drag is applied for only part of the ground run duration. To keep the relatively simple relation for the ground run given by Eqn (6.80), an effective parachute drag applied throughout the ground run is required. One reasonable approach is to consider the ratio of the impulse produced during actual deployment to that which would have been produced had the parachute been deployed for the whole landing time. This ratio may be considered to represent $f(t)$ yielding

$$f(t) \approx \frac{\displaystyle\int_{t_1}^{t_2} D_p dt}{\displaystyle\int_{0}^{t_f} D_p(t) dt}$$

Here t_1 and t_2 represent the times of full inflation and jettisoning of the drag parachute, respectively, while $t = 0$ and t_f represent the touchdown time and the time to final stop, respectively. Using the flight data for parachute drag as a function of time presented by Meyerson (2001) leads to the approximate result that $f(t) = 0.62$. With this result and the drag produced at the reported speed of parachute deployment, one may estimate the wake blockage factor to be around $K_w = 0.8$. Then, the parachute drag coefficient in the case of the space shuttle is $C_{D,p} = (0.62)(0.8)(0.575) = 0.285$. In the case of the shuttle $S_p/S = 0.467$ so that the drag contribution of the parachute, $C_{D,p}(S_p/S) = 0.133$, is about the same as that of the spacecraft itself, $C_{D,g} = 0.125$.

The stopping distance requirement for a space shuttle orbiter with a mass of 112,344 kg (248,000 lb) is reported by Meyerson (2001) to be 2438 m (8000 ft). This is about the same as the value of $x_g = 2530$ m obtained above for the case where the parachute is not deployed. Using the parachute according to the assumptions above reduces the ground run to 1225 m, well within the specified requirement.

6.9 SUMMARY OF CONSTANTS AND PARAMETERS

In this section we summarize some of the data used throughout this chapter.

6.9.1 ATMOSPHERIC ENTRY PARAMETERS

1. *Basic constants*
   $R_E = 6371$ km
   $g_E = 9.807$ m/s^2
   $V_E = (g_E R_E)^{1/2} = 7.904$ km/s, the orbital velocity at the surface of the Earth
   $H = 6.9$ km for exponential fit to atmospheric pressure profile
   $H/R_E = 1.083 \times 10^{-3} = 923.3^{-1}$
2. *Exponential density atmospheric model*
   $\rho_0 = 1.5$ kg/m^3 (this is a reference value for the $z = 0$ value of the exponential approximation for ρ, that is, $\sigma = \rho/\rho_0 = \exp(-z/H)$, and is not accurate at low altitudes; e.g., the actual standard sea level density is $\rho_{sl} = 1.225$ kg/m^3).
3. *Scale height and atmospheric pressure*
   $\rho_0 g_E H = (1.500$ kg/m$^3)(9.807$ m/s$^2)(6.900$ km$) = 101.5$ kPa ~ 1 bar ~ 1 atm. This says that a column of air of height H and density ρ_0 exerts essentially the same pressure on the surface of the Earth as does the entire atmosphere.
4. *Dynamic pressure for orbit at the earth's surface*
   $\rho_0 g_E R_E = (1.500$ kg/m$^3)(9.807$ m/s$^2)(6371$ km$) = 93.72 \times 10^6$ Pa $= 93.72$ MPa $= 925.1$ atm. Note also that $\rho_0 g_E R_E = \rho_0 V_E^2 = 2q_E$ twice the dynamic pressure corresponding to orbit at the at the earth's surface, where $q_E = 46,300$ kPa.

6.9.2 FLARE AND LANDING PARAMETERS

Typical values for the flare and landing characteristics of several spaceplanes is provided in Table 6.5. Some additional details and references are provided as follows:

1. *Altitude, speed, and flight path angle at flare initiation*:
   a. Average flare initiation altitude for the X-15 (Matranga, 1961) is $z_f = 244$ m (800 ft) with a range of 122 m (400 ft) to 488 m (1600 ft), while the average flare initiation velocity $V_f = 154.5$ m/s (300 kts) with a range of 124.6 m/s (242 kts) to 174.6 m/s (339 kts) although touchdown speeds for both extremes were found to be close to the average V_{td}.

Table 6.5 Typical Flare and Landing Characteristics of Several Spaceplanes

Performance Characteristic	X-15	Lifting Bodies[a]	Space Shuttle
Flare altitude (m)	244	223	518
Flare velocity (m/s)	155	125	146
Flare time (s)	<30	14	>25
Flare γ (°)	−18 to −20	−8 to −25	−19
Touchdown velocity (m/s)	95−98	98	100
Sink rate (m/s)	1.22	1.5[b]	0.46−0.76
Max sink rate (m/s)	2.74	2.74[b]	2.74
Touchdown α (°)	7	10	10−12
Normal acceleration (g)	1	1.15−2.37	1
Ground run (m)	1642	2446	1520

[a]*The lifting bodies tested included the HL-10, the M2-F2, and the X-24A.*
[b]*Estimated.*

   b. The typical altitude and speed for the M2-F2 lifting body spaceplane is given as $z_f = 223$ m (730 ft) and $V_f = 125$ m/s (240 kts) by Drake (1964). Kempel et al. (1994) report that both the HL-10 lifting body spaceplane and the M2-F2 initiated the flare at a speed between 141 m/s (270 kts) and 156 m/s (300 kts) and an altitude above the ground of about 305 m (1000 ft). The flight path angle was reported by Kempel et al. (1994) to be $\gamma_f = -16°$ to $-18°$.

   c. Myers, Johnston, and McRuer (1983) indicate that the space shuttle starts the flare at about $z_{fi} = 520$ m with $V_f = 146$ m/s (280 kts) at a flight path angle of $\gamma_f = -19°$.

2. *Time for flare and touchdown*:

   a. Average time to complete the flare and touchdown for the X-15 (Matranga, 1961) is $t_f = 27$ s with t_f typically less than 30 s.

   b. Drake (1964) reports that the time for flare and touchdown of the M2-F2 is about 14 s.

   c. The shuttle flare completion time is greater than 25 s.

3. *Touchdown velocity*

   a. Average value for the X-15 (Matranga, 1961) is $V_{td} = 95.3$ m/s (185 kts). Average values for the X-15 (Wilson, 1967) is 98.1 m/s (190.5 kts).

   b. Average value for three lifting bodies (Larsen, 1972) the M2-F2, HL-10, and the X-24A is 97.8 m/s (190 kts) with a range of 82.9 m/s (161 kts) to 118 m/s (229 kts). Kempel et al. (1994) report touchdown between 81 m/s (155 kts) and 116 m/s (223 kts).

   c. Nominal value for the Space Shuttle (Myers et al., 1983) is 100 m/s (195 kts) with a maximum of 116 m/s (225 kts).

4. *Sink rate*

   a. Average value for the X-15 (Matranga, 1961) is $V_{v,td} = 1.22$ m/s (4 ft/s). The design limit is 2.74 m/s (9.0 ft/s) for $0° < \alpha_{td} < 10°$ with a linear decay to zero for $10° < \alpha_{td} < 13°$. In one of the first four flights, this value was exceeded with $V_{v,td} = 9.5$ ft/s and $\alpha_{td} = 11°$ resulting in severe damage to the aircraft.

b. No specific data for the design limits on the sink rate were found for the lifting bodies. However, the HL-10, for example, uses the landing gear of the T-38 Talon trainer which has a maximum sink rate of 1.73 m/s (5.67 ft/s). Therefore, it is likely that the maximum sink rate at landing for a vehicle like the HL-10 is set at about the same value as that of the X-15 and the space shuttle, 2.74 m/s (9 ft/s). This appears reasonable because the 3926 kg (6460 lbs) landing mass of the HL-10 is somewhat more than half of that of the T-38.

c. Design sink rate for the Space Shuttle (Myers et al., 1983) is in the range 0.46 m/s (1.5 ft/s) $< V_{v,td} <$ 0.76 m/s (2.5 ft/s). The design limit is 2.74 m/s (9 ft/s).

5. *Landing angle of attack*
   a. Average value for the X-15 (Matranga, 1961) is $\alpha_{td} = 7°$ with a range of $4-11°$.
   b. The angle of attack of the HL-10 in landing is about 10° while the M2-F2 is about $-2°$ according to Kempel et al. (1994). Pyle (1971) shows that although the lift curve slopes are approximately equal for the HL-10 and the M2-F2, the zero-lift angle of attack is 3.6° for the HL-10 and $-9.7°$ for the M2-F2.
   c. Nominal angle of attack in landing is around $10-12°$ for the space shuttle.

6. *Normal acceleration in landing flare*
   a. The average value of the normal acceleration component for the X-15 during the landing flare is approximately 1g (Matranga, 1961).
   b. The average value of the normal acceleration component for the three lifting bodies is 1.53g with a range from 1.15g to 2.37g (Larson, 1972). Kempel et al. (1994) report the flare is carried out at about 1.5g.
   c. The normal acceleration of the shuttle is suggested by Myers et al. (1983) to be approximately 1g.

7. *Ground run distance*
   a. Average value for the X-15 (McKay & Scott, 1961) is $x_g = 1642$ m (53,865 ft). The X-15 main landing gear has skids, not wheels, and the effective friction coefficient was determined to be around 0.33.
   b. Average value for the three lifting bodies (Larson, 1972) is $x_g = 2446$ m (8026 ft) with a range of 1286 m (4220 ft) to 3885 m (12,750 ft).
   c. Nominal value for the Space Shuttle (Myers et al., 1983) is $x_g = 2440$ m (8000 ft) with a minimum of approximately 1520 (5000 ft).

6.10 NOMENCLATURE

a	acceleration
B	ballistic coefficient $W/C_D S$
B'	normalized ballistic coefficient $B/\rho_0 g_E H$
C	constant, Eqn (6.24)
C_D	drag coefficient
$C_{D,0}$	parachute drag coefficient based on S_0
C_L	lift coefficient
D	drag
D_0	nominal diameter of parachute

d	characteristic cross-sectional dimension of spacecraft
F	thrust
F_b	braking force
g	gravitational acceleration
g_E	gravitational acceleration at the earth's surface
H	atmospheric scale height
I_{sp}	specific impulse
K	constant in Eqn (6.25)
K_D	coefficient, Eqn (6.43)
K_l	coefficient, Eqn (6.44)
K_m	coefficient, Eqn (6.45)
K_w	factor expressing effect of vehicle weight on drag chute performance
k	gravitational constant equal to $g_E R_E^2$
L	lift
l	characteristic length of spacecraft
M	Mach number or pitching moment
m	mass
n	load factor in pull-up maneuver
p	pressure
Q	heat transfer
Q_{ref}	reference heating load
q	dynamic pressure
q_c	convective heat flux
q_r	radiative heat flux
q_v	dynamic pressure in vertical descent
R	gas constant or resultant force
R_E	mean radius of the Earth
R_{LEO}	radius of low earth orbit
R_N	vehicle nose radius
r	radius measured from the earth's center
S	spacecraft reference area
S_p	parachute reference area
S_w	spacecraft wetted surface area
S_0	parachute nominal area
s	normalized altitude z/H
T	temperature
t	time
t_f	flight duration
U	$V^2/g_E R_E$
V	velocity
V_E	orbital velocity at the earth's surface or equivalent airspeed, Eqn (6.47)
V_s	stall speed
v	velocity ratio V/V_e
W	weight
X	V/V_{ref}, Eqn (6.65)
x	downrange distance
Y	flight path angle
z	geometric altitude

α angle of attack
χ angle between thrust vector and local horizon
δ atmospheric pressure ratio p/p_{sl}
ϕ bank angle
γ flight path angle between velocity vector and local horizon or specific heat ratio
η dimensionless variable σ/B'
μ_b braking coefficient of friction
ω trajectory angle, Figure 6.1
ρ atmospheric density
ρ_0 reference atmospheric density at sea level
σ atmospheric density ratio ρ/ρ_{sl}
τ normalized time $t(g_E/V_E)$
ϖ exponent in Eqn (6.25)
ξ exponent in Eqn (6.25)

6.10.1 SUBSCRIPTS

a   approach conditions
c   convective
design   design condition
e   conditions at the entry into the atmosphere
l   landing condition
max   maximum condition
n   normal direction
p   conditions with parachute deployed
r   radiative
s   stall conditions
sl   sea level
t   tangential direction or stagnation condition
v   vertical component

REFERENCES

Ashley, H. (1992). *Engineering analysis of flight vehicles.* New York, NY: Dover.

Cockrell, D. J. (1987). *The aerodynamics of parachutes.* NATO Advisory Group For Aerospace Research And Development. AGARDograph No. 295.

Delurgio, P. R. (1999). *Evolution of the ringsail parachute.* AIAA 99-1700.

Drake, H. M. (1964). *Aerodynamic testing using special aircraft.* NASA TM-X-51605.

Ewing, E. G. (1971). *Deployable aerodynamic deceleration systems.* NASA-SP-8066.

Ewing, E. G. (1972). *Ringsail parachute design.* Air Force Flight Dynamics Laboratory Technical Report AFFDL-TR-72-3.

Haber, H. (1959). The physical factors of the space environment. In H. Seifert (Ed.), *Space technology* (pp. 27-01−27-40). New York, NY: John Wiley & Sons.

Hillje, E. R. (1967). *Entry flight aerodynamics from Apollo Mission AS-202.* NASA TN D-4185.

Hoey, R. G. (1963). *Horizontal landing techniques for hypersonic vehicles.* AGARD 428.

Johnston, E. W., & Gaines, L. M. (1963). Low speed characteristics of the X-15. In S. Scala, A. Harrison, & M. Rogers (Eds.), *Dynamics of manned lifting planetary entry* (p. 668). New York, NY: Wiley & Sons.

Kempel, R. W., Painter, W. D., & Thompson, M. O. (1994). *Developing and flight testing the HL-10 lifting body: A precursor to the space shuttle.* NASA Reference Publication 1332.

Knacke, T. W. (1992). *Parachute recovery systems design manual.* Santa Barbara, CA: Para Publishing.

Larson, R. R. (1972). *Statistical analysis of landing contact conditions for three lifting body research vehicles.* NASA TN D-6708.

Lee, D. B. (1972). *Apollo experience report—aerothermodynamic evaluation.* NASA TN D-6843.

Loh, W. H. T. (1968). *Re-entry and planetary physics and technology.* New York, NY: Springer-Verlag.

Matranga, G. J. (1961). *Analysis of X-15 landing approach and flare characteristics determined from the first 30 flights.* NASA TN D-1057.

Maydew, R. C., & Peterson, C. W. (1991). *Design and testing of high-performance parachutes.* NATO Advisory Group For Aerospace Research And Development. AGARDograph No. 319.

McKay, J. M., & Scott, B. J. (1961). *Landing gear behavior during touchdown and runout for 17 landings of the X-15 research airplane.* NASA TM X-518.

Meyerson, R. E. (2001). *Space shuttle orbiter drag parachute design.* AIAA 2001–2051; also NASA Conference Paper JSC-CN-6790.

Myers, T. T., Johnston, D. E., & McRuer, D. T. (1983). *Space shuttle flying qualities and flight control system assessment study—phase II.* NASA CR-170406.

NASA (1989). *Biomedical results of Apollo.* NASA SP-368.

Pavlosky, J. E., & St. Leger, L. G. (1974). *Apollo experience report—thermal protection system.* NASA TN D-7564.

Pyle, J. S. (1971) "Lift and Drag Characteristics of the HL-10 Lifting Body During Subsonic Gliding Flight", NASA TN D-6263.

Regan, F. J., & Anandakrishnan, S. M. (1993). *Dynamics of atmospheric re-entry.* American Institute of Aeronautics and Astronautics, 1993.

Saltzman, E. J., Wang, K. C., & Iliff, K. W. (2002). *Aerodynamic assessment of flight-determined subsonic lift and drag characteristics of seven lifting-body and wing-body reentry vehicle configurations.* NASA/TP-2002-209032.

Sforza, P. M. (2014). *Commercial airplane design principles,* Waltham, MA: Elsevier.

Sutton, K., & Graves, R. A. (1971). *A general stagnation-point convective-heating equation for arbitrary gas mixtures.* NASA TR R-376.

Tauber, M., & Sutton, K. (1991). Stagnation-point radiative heating relations for Earth and Mars entries. *Journal of Spacecraft and Rockets, 28*(1), 40–42.

Vincze, J. (1966). *Gemini spacecraft parachute landing system.* NASA TN D-3496.

Voshell, M. (2004). *High acceleration and the human body.* November in <http://csel.eng.ohio-state.edu/voshell/gforce.pdf>.

Wilson, R. J. (1967). *Statistical analysis of landing contact conditions of the X-15 airplane.* NASA TN D-3801.

LAUNCH MECHANICS

7.1 GENERAL EQUATIONS FOR LAUNCH VEHICLES

The boost phase of an orbital space mission is critical to the design process because the initial weight of the vehicle, which is generally related to its cost, is at its maximum at liftoff. The payload of the launch system is typically a small fraction of the total weight. Figure 7.1 is a photograph of the launch of the space shuttle STS-115 from Cape Canaveral on September 9, 1996, as seen from the International Space Station (ISS). Note that the exhaust plume column appears almost vertical at the launch point and then begins to curve as STS-115 rises. The Florida coastline is evident at the bottom of the photograph and the distant horizon appears at the top of the photograph.

The boost trajectory and associated nomenclature is shown in Figure 7.2. The velocity V is tangent to the trajectory and the thrust F is inclined at an angle χ to V. The angle between the tangent to the trajectory and the local horizon is the flight path angle γ. The lift is normal to the velocity and the drag is in the same direction as V, but in the opposite sense. The weight of the spacecraft mg is directed toward the center of the earth.

The equations of motion for flight of a point mass m over a windless nonrotating planet were given previously in Chapter 6 and are repeated below:

$$\dot{V} = \frac{F}{m}\cos\chi - \frac{D}{m} - \frac{k\sin\gamma}{r^2} \tag{7.1}$$

$$\dot{\gamma} = \frac{V\cos\gamma}{r} + \frac{F\sin\chi}{mV} + \frac{L}{mV} - \frac{k\cos\gamma}{Vr^2} \tag{7.2}$$

$$\dot{r} = \dot{z} = V\sin\gamma \tag{7.3}$$

$$\dot{\omega} = \frac{\dot{x}}{r} = \frac{V\cos\gamma}{r} \tag{7.4}$$

Here $k = 398{,}600 \text{ km}^3/\text{s}^2$ is the gravitational constant for the earth and the local gravitational acceleration is given by

$$g = \frac{g_E R_E^2}{r^2} = \frac{k}{r^2} = \frac{g_E}{\left(1 + \dfrac{z}{R_E}\right)^2} \tag{7.5}$$

The gravitational acceleration and mean radius of the earth at the equator are $g_E = 9.80665 \text{ m/s}^2$ and $R_E = 6378 \text{ km}$, respectively. The thrust, lift, and drag forces are discussed subsequently. Equations (7.1)–(7.4) are sufficient to determine the performance of a launch vehicle considered as a point mass moving in a plane. When the stability and control characteristics of the vehicle are

Manned Spacecraft Design Principles. DOI: http://dx.doi.org/10.1016/B978-0-12-804425-4.00007-6

FIGURE 7.1

Launch of STS-115 from Cape Canaveral on September 9, 1996, as seen from the ISS.

Courtesy: NASA.

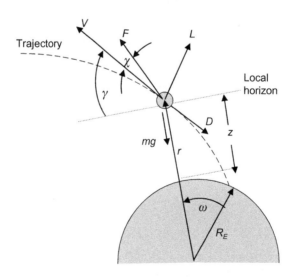

FIGURE 7.2

Schematic diagram of a boost trajectory with zero lift.

desired it must be considered at least to be a rigid body and additional equations are required. In the simplest case of rigid body motion in a plane, a pitching moment equation is required and the moment of inertia of the vehicle must be known. This aspect of launch performance is deferred to a later section.

7.2 THRUST, LIFT, AND DRAG FOR A SIMPLIFIED BOOST ANALYSIS

The three external forces acting on the vehicle, considered here as a point mass, are the thrust, lift, and drag. The thrust is provided by the propulsion system while the lift and drag are dependent upon the geometrical configuration of the launcher. The importance of each of these terms in the preliminary design phase is discussed subsequently.

7.2.1 ROCKET ENGINE THRUST

Jet propulsion is the motive force for the spacecraft launch process so we must be able to characterize the thrust in terms of other variables in the system of Eqns (7.1)–(7.4). As shown by Sforza (2012), the thrust of the rocket engine is given by the sum of the momentum of the exhaust gas and the pressure imbalance at the nozzle exit plane:

$$F = \dot{m}_p V_e + A_e(p_e - p_\infty) = \dot{m}_p V_e \left[1 + \frac{1}{\gamma_e M_e^2} \left(1 - \frac{p_\infty}{p_e} \right) \right] \tag{7.6}$$

The thrust may be characterized by the mass flow of propellant and an effective exhaust velocity

$$V_{eff} = V_e \left[1 + \frac{1}{\gamma_e M_e^2} \left(1 - \frac{p_\infty}{p_e} \right) \right] \tag{7.7}$$

The effective exhaust velocity depends solely upon the conditions at the exit plane of the nozzle and the pressure of the ambient p_∞ into which it exhausts. Note that when the exit pressure matches the ambient pressure, $p_e = p_\infty$, the exit velocity $V_e = V_{eff}$. Under this condition, the nozzle is said to be "matched" and it can be shown (Sforza, 2012) that the thrust is a maximum for this case. Because the fuel flow rate was usually measured in terms of weight rather than mass, the traditional figure of merit for jet engines is the specific impulse I_{sp}, measured in seconds and defined by the following equation:

$$I_{sp} = \frac{F}{\dot{m}_p g_E} = \frac{V_{eff}}{g_E} \tag{7.8}$$

In the English system of units, I_{sp} describes the number of pounds of thrust per pound of propellant consumed per second so that the units of I_{sp} are seconds. Using the SI system of units, one may equivalently consider the measure of thrust production efficiency to be the force, in newtons, produced per kilogram per second of propellant consumed, which is equivalent to the effective exhaust velocity $V_{eff} = g_E I_{sp}$, measured in m/s.

For fixed conditions in the combustion chamber, the nozzle effective exhaust velocity in Eqn (7.7) depends solely on the exhaust gas composition and the geometry of the nozzle. The only other factor in Eqn (7.7) is p_∞, the pressure of the ambient into which the nozzle exhausts. Because the launch vehicle ascends through the atmosphere the ambient pressure for the rocket nozzle is $p_\infty = p_\infty(z)$. Therefore, for a fixed geometry nozzle operating at fixed chamber conditions p_c and T_c, the pressure at the exit plane can match the ambient pressure at only one altitude. This means that the thrust, and therefore I_{sp}, or V_{eff}, will also vary with altitude even though the propellant mass flow is kept constant.

An unambiguous value for the specific impulse of a rocket nozzle occurs for operation in a vacuum where the pressure outside the nozzle is zero. The specific impulse in this case is a maximum and is denoted by $I_{sp,vac}$. The specific impulse quoted by rocket engine manufacturers is usually $I_{sp,vac}$ unless specifically noted otherwise. Sforza (2012) shows that the altitude variation of the specific impulse for a fixed geometry nozzle may be reasonably approximated by the following expression:

$$\bar{I}_{sp} = \frac{I_{sp}}{I_{sp,vac}} = 1 - \Gamma \frac{\varepsilon \exp\left(-\dfrac{z}{H}\right)}{p_c} \tag{7.9}$$

The combustion chamber pressure p_c must be specified in units of atmospheres in Eqn (7.9). The ratio of the nozzle exit area to the nozzle throat area $\varepsilon = A_e/A_t$ is called the expansion ratio of the nozzle and is generally specified by the engine manufacturer. For rocket engines designed to operate within the atmosphere, the quantity ε/p_c lies in the range of $0.15 < \varepsilon/p_c < 0.5$ and Eqn (7.9) is appropriate. However, rocket engines designed to operate outside the atmosphere have a range of $2 < \varepsilon/p_c < 6$ and because they are always operating very close to vacuum conditions Eqn (7.9) is not applicable.

The factor H is the scale height for the exponentially modeled pressure variation with altitude described in Section 2.2.3, where the value $H = 7.16$ km is often used. The quantity Γ is a function of the approximate value for the ratio of specific heats γ in the nozzle and is defined by

$$\Gamma = \frac{1}{\gamma}\left(\frac{\gamma-1}{2}\right)^{\frac{\gamma}{\gamma-1}}\left(\frac{\gamma+1}{\gamma-1}\right)^{\frac{\gamma+1}{2(\gamma-1)}} \approx 0.753(\gamma-1)^{0.33} \tag{7.10}$$

The approximate form for Γ shown in Eqn (7.10) is accurate to within about $\pm1\%$ for the typical values of γ in rocket nozzles, say $1.15 < \gamma < 1.4$, over which range $0.4 < \Gamma < 0.55$. Using the quoted range of the variables, we find that the parameter of interest in Eqn (7.9) has a maximum range of $0.06 < \Gamma\varepsilon/p_c < 0.275$. Then according to Eqn (7.9), the maximum reduction in I_{sp} occurs at ground level and $0.94 < I_{sp}(0)/I_{sp,vac} < 0.725$.

Thus the correction term in Eqn (7.9) is basically set by the term $\exp(-z/H)$ showing that the specific impulse rises as the altitude increases and reaches essentially the maximum value of $I_{sp,vac}$ at sufficiently high altitudes, say $3H$ or $4H$, or about $20-30$ km. The vacuum-specific impulse and associated expansion ratio is shown in Figure 7.3 for rockets using either liquid or solid propellants. Rocket engines with values of $\varepsilon > 100$ are only practical for operation at very high altitudes and relatively small levels of thrust, as illustrated by the data in Figure 7.4 which is for the same rocket engines shown in Figure 7.3. This limitation is due to the fact that the weight of the rocket nozzle, which is basically the frustum of a shallow angle thin-walled cone, is proportional to the area of its surface. It can be shown that the surface area of the nozzle is approximately proportional to $(F/p_c)\varepsilon$. For example, to produce a given thrust level at a specified chamber pressure the weight of the nozzle with $\varepsilon = 400$ will be 10 times that of a nozzle with $\varepsilon = 40$.

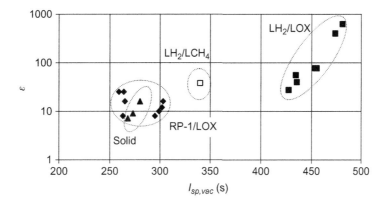

FIGURE 7.3

The vacuum-specific impulse and associated expansion ratio is shown for rockets using either liquid or solid propellants.

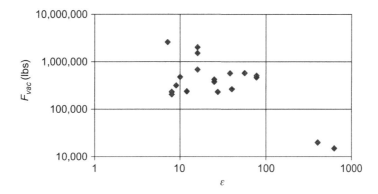

FIGURE 7.4

The vacuum thrust and associated expansion ratio is shown for the rocket engines depicted in Figure 7.3.

The thrust F may be controlled by changing the mass flow of propellant and/or the effective exhaust velocity of the nozzle, as can be seen from Eqn (7.8). The mass flow of propellant may be controlled during flight by means of valves in liquid rocket engines and by appropriate propellant grain design in solid rocket engines, as described by Sforza (2012). The effective exhaust velocity V_{eff} may be controlled during flight by varying the nozzle geometry so as to adjust the exit pressure appropriately. However, such complexities are typically ignored in preliminary design studies and relegated to later detailed design phases. Therefore, for the development of preliminary design techniques we assume that the mass flow of propellant is constant and the nozzle geometry is fixed so that the thrust is a function only of altitude through the specific impulse and may be written as follows:

$$F(z) = \dot{m}_p(g_E I_{sp,vac})\bar{I}_{sp}(z) = F_{vac}\bar{I}_{sp}(z) \tag{7.11}$$

Here F_{vac} is the vacuum thrust of the rocket engine, and, like $I_{sp,vac}$, it is a value typically supplied by manufacturers. The mass of the space launch system will decrease as it ascends because of the consumption of propellant so that

$$\frac{dm}{dt} = \dot{m} = -\dot{m}_p \tag{7.12}$$

Because the propellant consumption rate remains constant, the mass of the launch vehicle is given by

$$m(t) = m_0 + \int_0^t \dot{m}\,dt = m_0 - \int_0^t \dot{m}_p\,dt = m_0 - \dot{m}_p t \tag{7.13}$$

Using Eqn (7.11) in Eqn (7.13), in which m_0 denotes the initial mass of the launcher, yields

$$m(t) = m_0 - \frac{F_{vac}}{g_E I_{sp,vac}} t$$

Obviously, the burn time of the engine is limited by the total mass of propellant available m_p so that the mass of the launch system at the time of burn-out t_{bo} is

$$m_{bo} = m_0 - m_p = m_0 - \frac{F_{vac}}{g_E I_{sp,vac}} t_{bo} \tag{7.14}$$

From Eqn (7.14), the burn-out time is obtained as

$$t_{bo} = \left(\frac{m_p}{m_0}\right) \left(\frac{F_{vac}}{m_0 g_E}\right)^{-1} I_{sp,vac} = \frac{\left(\frac{W_p}{W}\right)_0}{\left(\frac{F_{vac}}{W}\right)_0} I_{sp,vac} \tag{7.15}$$

Thus the time to burn-out of the propellant is directly proportional to the amount of propellant available at launch and the efficiency with which it is used, but inversely proportional to the take-off value of the vacuum thrust to weight ratio. In general, $(m_p/m)_0 = O(1)$ and $(F_{vac}/W)_0 = O(1)$ so that $t_{bo}/I_{sp,vac} = O(1)$. A reasonable rule of thumb is that the burn-out time for a rocket-powered launcher is about equal to the vacuum-specific impulse because typically $m_p/m_0 \sim 0.9$ and $(F_{vac}/W)_0 \sim 1.3$. Three recent and three earlier launch vehicles for manned spacecraft along with their launch mass m_0 and vacuum thrust to weight ratio $(F_{vac}/W)_0$ are shown in Figure 7.5.

7.2.2 LAUNCH VEHICLE DRAG AND LIFT

The lift and drag forces in Eqns (7.1) and (7.2) are defined as follows:

$$D = C_D q S \tag{7.16}$$

$$L = \frac{L}{D} C_D q S \tag{7.17}$$

(a) Space Shuttle
m_0 = 2,580,000 kg
$(F_{vac}/W)_0$ = 1.27

(b) Soyuz
m_0 = 330,000 kg
$(F_{vac}/W)_0$ = 1.42

(c) Long March
m_0 = 464,000 kg
$(F_{vac}/W)_0$ = 1.43

(d) Mercury Atlas 9
m_0 = 120,000 kg
$(F_{vac}/W)_0$ = 1.58

(e) Gemini Titan II
m_0 = 154,200 kg
$(F_{vac}/W)_0$ = 1.52

(f) Apollo Saturn V
m_0 = 3,000,000 kg
$(F_{vac}/W)_0$ = 1.34

FIGURE 7.5

Recent and past launchers for manned spacecraft and their initial mass m_0 and initial vacuum thrust to weight ratio $(F_{vac}/W)_0$.

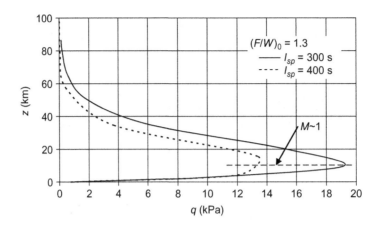

FIGURE 7.6

Dynamic pressure as a function of altitude for a typical launcher with $(F/W)_0 = 1.3$ and $I_{sp} = 300$ and 400 s. Dashed line denotes the altitude where $M \sim 1$.

Thus the contribution of lift and drag to the acceleration of the vehicle depends on the dynamic pressure and the lift to drag ratio L/D of the ascent vehicle. The dynamic pressure is given by

$$q = \frac{1}{2}\rho_{sl}\sigma V^2 \approx \frac{1}{2}\rho_{sl}\exp\left(-\frac{z}{H}\right)V^2$$

or

$$q = \frac{1}{2}p_{sl}\delta M^2 \approx \frac{1}{2}p_{sl}\exp\left(-\frac{z}{H}\right)M^2$$

(7.18)

In Eqn (7.18), we again use the idea of an exponential atmosphere, with H being the scale height of the atmosphere, $\delta = p/p_{sl} = \sigma = \rho/\rho_{sl}$, and $T = \text{constant} = 215$ K, as described in Chapters 2 and 6.

During ascent the density ratio σ drops exponentially while the velocity increases and therefore the dynamic pressure tends to be relatively small in magnitude throughout this flight regime. For example, the maximum dynamic pressure during ascent for the Space Shuttle Orbiter typically occurs around $M = 1$ and is approximately 30 kPa. Typical curves of dynamic pressure as a function of altitude for the typical cases we have been considering are shown in Figure 7.6.

The effects of drag may be assessed by considering the drag to thrust ratio of the launcher (D/F) noting that the thrust of a rocket engine is independent of the flight velocity while the drag is proportional to the square of the flight velocity through the dynamic pressure, as given in Eqn (7.18). The drag to thrust ratio may be written as

$$\frac{D}{F} = \frac{C_D q S}{F} = \left[\frac{1}{2}\frac{\gamma p_{sl}S}{W_0\left(\frac{F_{vac}}{W_0}\right)}\right]\left[C_D M^2 \frac{\exp\left(-\frac{z}{H}\right)}{1 - \frac{\varepsilon\Gamma}{p_c}\exp\left(-\frac{z}{H}\right)}\right]$$

(7.19)

Here we have used the exponential atmosphere approximation to define the altitude variation of q according to Eqn (7.18) and of I_{sp} according to Eqn (7.9). The first term in square brackets in Eqn (7.19) is constant while the second term in square brackets varies during ascent. We have shown in Figure 7.6 that the maximum dynamic pressure occurs early in the ascent at about $M = 1$ and an altitude of one or two scale heights. It is well known that the maximum drag coefficient of a body also occurs around $M = 1$; see, e.g., Ashley (1974) or Hoerner (1958). At one scale height, $z = H$, and above, the fraction in the second square bracket has a value less than 0.4 because $0.06 < \Gamma \varepsilon / p_c < 0.275$ for typical engines. For slender missiles, Stoney (1961) shows that the drag coefficient C_D, based on cross-sectional area S, rises rather abruptly from about 0.15 to about 0.4 or more through the Mach number range $0.85 < M < 1.3$. However, experimental results collected by Walker (1968) on models of the Saturn V launch vehicle show values about two or three times that. Thus, using a nominal value of $C_{D,max} = 1.2$, the maximum value of the second square bracket in Eqn (7.19) may be approximated as follows:

$$\left[C_D M^2 \frac{\exp\left(-\dfrac{z}{H}\right)}{1 - \Gamma \dfrac{\varepsilon}{p_c}\exp\left(-\dfrac{z}{H}\right)} \right]_{max} \approx [(1.2)(1)(0.4)] = 0.48$$

To make an order of magnitude estimate of the term in the first square bracket of Eqn (7.19), we may consider the launch vehicle to be a long cylinder with length to diameter ratio l/d. Assume that the initial mass of the launcher is essentially the initial mass of the propellant $m_p = \rho_p v$, where ρ_p is the propellant density and the volume $v = \pi d^2 l/4$. For the atmosphere $\gamma = 1.4$ and the initial thrust to weight ratio may be taken as $(F_{vac}/W)_0 = 1.4$. Then approximate magnitude of the term in the first square bracket of Eqn (7.19) may then be estimated as follows:

$$\left[\frac{1}{2} \frac{\gamma p_{sl}}{\left(\dfrac{F_{vac}}{W}\right)_0} \frac{S}{W_0} \right] = \frac{1}{2} \frac{\gamma p_{sl}}{\left(\dfrac{F_{vac}}{W}\right)_0} \frac{S}{\rho_p g_E S l} \approx \frac{1}{2} \frac{(1.4)}{(1.4)} \frac{\rho_w g_E l_w}{\rho_p g_E l} \approx \frac{1}{2} \frac{l_w}{l} \tag{7.20}$$

The density of various propellants, both liquid and solid, is shown in Table 7.1 (Sforza, 2012) from which we assume ρ_p to be around 1000 kg/m³, or approximately the density of water so we've taken $\rho_p \sim \rho_w$. Furthermore, in Eqn (7.20), the sea level atmospheric pressure has been expressed in terms of the height of a water column, l_w. The length of typical launchers is several tens of meters, say 50 m, and l_w is approximately 10 m so that l_w/l is on the order of 0.2. Thus the maximum value of the drag to thrust ratio given in Eqn (7.19) may be then approximated as follows:

$$\left(\frac{D}{F}\right)_{max} \approx \left[\frac{1}{2}\frac{l_w}{l}\right][0.4 C_{D,M=1}] \approx 0.2 C_{D,M=1}\frac{l_w}{l} \approx 0.24\frac{l_w}{l} \approx 0.048 \tag{7.21}$$

Therefore, $(D/F)_{max} = O(10^{-2})$ and we may expect a maximum error of about 5% by neglecting drag in preliminary design studies of launch performance. Of course, as we shall see subsequently, there is no special difficulty in including the drag term in numerical calculations, but the drag coefficient is configuration dependent as well as a function of flight Mach number and Reynolds number, $C_D = C_D(M, Re)$, and an accurate description of it is required for each launch vehicle geometry.

Table 7.1 Propellant Densities for Several Fuel and Oxidizer Combinations[a]

Oxidizer	Oxidizer Density (kg/m³)	Fuel	Fuel Density (kg/m³)	O/F Ratio	Propellant Mixture Density (kg/m³)
LOX (liquid oxygen)	1141 ($T = 90.3$ K)	LH$_2$ (liquid Hydrogen)	69.5 ($T = 20.8$ K)	4	926
				6	987
LOX (liquid oxygen)	1141 ($T = 90.3$ K)	RP1 (rocket propellant-1)[b]	800 ($T = 298$ K)	2.2	1034
				2.6	1046
LOX (liquid oxygen)	1141 ($T = 90.3$ K)	LCH$_4$ (liquid methane)	415 ($T = 109$ K)	2.7	944
N$_2$O$_4$ (nitrogen tetroxide)	1450 ($T = 293$ K)	UDMH (unsymmetrical dimethyl hydrazine)[c]	791 ($T = 295$ K)	2.8	1277
Ammonium perchlorate	1950 (composite solid)	PBAN (polybutadiene-acrylonitrite and aluminum)	1360 (composite solid)	2.3	1771

[a]*Densities of cryogenic propellants are typically for temperatures just below the boiling point while those of other propellants are for near standard atmospheric temperature.*
[b]*Kerosene fraction with empirical chemical formula $CH_{1.953}$.*
[c]*Hydrazine derivative $(CH_3)_2NNH_2$.*

The lift force acts normal to the velocity so the effects of lift arise indirectly through the flight path angle as given by Eqn (7.2). The lift to thrust ratio may be written as

$$\frac{L}{F} = \left(\frac{L}{D}\right)\frac{D}{F} \tag{7.22}$$

We have just demonstrated that the drag to thrust ratio D/F is $O(10^{-2})$ for long slender missiles at small angles of attack. As we will show later, slender body theory predicts the lift to drag ratio to be $L/D = 2\alpha/C_D$ when α is in radians or $L/D = 0.034\alpha/C_D$, when α is measured in degrees. Because $C_D = O(10^{-1})$ at small angles of attack we find that $L/F < O(10^{-2})$. It is desired that L/D remain small during flight because the fragile nature of lightweight, thin-walled missile fuselages cannot support the bending moments that would develop concurrently with the lift. As a consequence, it is reasonable to also neglect the lift of conventional missile launch vehicles during ascent without serious impact on the accuracy of the calculations. As before, there is no special difficulty in including the lift term in numerical calculations but the lift coefficient is also configuration dependent as well as a function of flight Mach number and Reynolds number, $C_L = C_L(M, Re)$, and an accurate description of it is required for each launch vehicle geometry.

There is continuing interest in launch systems that take advantage of lift to fly to orbit, rather than depending on thrust alone to boost the spacecraft, but no such systems are yet under serious development. Based on the above analysis, we assume that for preliminary design purposes the aerodynamic lift and drag are negligible and set $L = D = 0$, that is, the ascent trajectory of the vehicle, considered as a point mass, depends only on the applied thrust F. When we subsequently investigate the stability of the launch vehicle, we must treat it at least as a rigid body and the lift and drag will need to be revisited insofar as they affect the pitching moment.

7.3 **THE NONDIMENSIONAL EQUATIONS OF MOTION**

A nondimensional representation of Eqns (7.1)–(7.4) which describes the motion of the launch vehicle may be given as follows:

$$\frac{d\bar{V}}{d\bar{t}} = \frac{G_0\bar{I}_{sp}\cos\chi - G_1 e^{-\bar{z}/\bar{H}}\bar{V}^2}{1 - G_0\bar{t}} - \frac{\sin\gamma}{(1+\bar{z})^2} \tag{7.23}$$

$$\frac{d\gamma}{d\bar{t}} = \frac{\cos\gamma}{(1+\bar{z})^2}\left[G_2(1+\bar{z})\bar{V} - \frac{1}{\bar{V}}\right] + \frac{1}{1 - G_0\bar{t}}\left[\frac{G_0\bar{I}_{sp}\sin\chi}{(1+\bar{z})^2\bar{V}} + G_3 e^{-\bar{z}/\bar{H}}\bar{V}^2\right] \tag{7.24}$$

$$\frac{d\bar{z}}{d\bar{t}} = G_2\bar{V}\sin\gamma \tag{7.25}$$

$$\frac{d\bar{x}}{d\bar{t}} = G_2\bar{V}\cos\gamma \tag{7.26}$$

The equations include the effects of variation of the gravitational acceleration, thrust vectoring, drag, and lift. The nondimensional variables are as follows:

$$\bar{t} = \frac{t}{I_{sp,vac}} \tag{7.27}$$

$$\bar{V} = \frac{V}{g_E I_{sp,vac}} \tag{7.28}$$

$$\bar{z} = \frac{z}{R_E} \tag{7.29}$$

$$\bar{x} = \frac{x}{R_E} \tag{7.30}$$

Definitions of the nondimensional parameters are given as follows:

$$G_0 = \frac{\dot{m}_p I_{sp,vac}}{m_0} \tag{7.31}$$

$$G_1 = \frac{\rho_{sl}(g_E I_{sp,vac})^2 C_D S}{2m_0 g_E} \tag{7.32}$$

$$G_2 = \frac{g_E I_{sp,vac}^2}{R_E} \tag{7.33}$$

$$G_3 = \left(\frac{C_L}{C_D}\right)G_1 \tag{7.34}$$

Note that, in general, $C_D = C_D(M,Re)$ and $C_L = C_L(M,Re)$ while $M = M(z,V)$ and $Re = Re(z,V)$ so that appropriate relations must be provided to achieve the desired accuracy. The only restriction on the initial conditions is that $\cos\gamma(0)$ and $V(0)$ are not both zero. Aside from the initial conditions, the only other external control over the motion as posed here is through the thrust vector angle $\chi(t)$ which is assumed to lie in the plane of the motion. As pointed out previously, even if no thrust vector control (TVC) is applied the vehicle will perform a gravity turn as it ascends. Furthermore, the contributions of the drag and lift to the performance of launch vehicles have been shown to be secondary and may be neglected, at least for preliminary design purposes.

7.4 SIMPLIFIED BOOST ANALYSIS WITH CONSTANT THRUST AND ZERO LIFT AND DRAG

For simplicity, we assume that the boost phase is carried out with constant thrust $F = F_{vac}$ and with $\chi = 0$, that is, with the thrust vector aligned with the velocity vector. Under these conditions the quantity

$$G_0 = \frac{\dot{m}_p I_{sp}}{m_0} \approx \left(\frac{F_{vac}}{W}\right)_0$$

The drag and lift are assumed to be negligible so we set $G_1 = G_3 = 0$. Furthermore we make the approximation that the altitude during the launch phase will always be very much less than the radius of the earth so that

$$r = R_E + z = R_E\left(1 + \frac{z}{R_E}\right) \approx R_E \tag{7.35}$$

The altitude for low earth orbit (LEO) is typically $z \leq 400$ km so that the maximum value of the error incurred is 6.3% which is often considered reasonable for preliminary design purposes. Under these assumptions, the basic equations (7.23)–(7.26) may be simplified to the following set of dimensional equations:

$$\frac{1}{g_E}\dot{V} = \frac{\left(\dfrac{F_{vac}}{W}\right)_0}{1 - \left(\dfrac{F_{vac}}{W}\right)_0 \dfrac{t}{I_{sp}}} - \sin\gamma \tag{7.36}$$

$$\dot{\gamma} = \cos\gamma\left(\frac{V}{R_E} - \frac{g_E}{V}\right) \tag{7.37}$$

$$\dot{r} = \dot{z} = V\sin\gamma \tag{7.38}$$

$$\dot{x} = V\cos\gamma \tag{7.39}$$

Equation (7.36) shows that the acceleration of the vehicle along the trajectory is dependent on the flight path angle only through the component of gravitational acceleration in that direction. Equation (7.37) shows that the rate of change of the flight path angle, that is, the degree of turning of the velocity vector is proportional to the component of the gravitational acceleration normal to the trajectory. This gives rise to the so-called gravity turn that launch vehicles undergo without any external control application.

7.4.1 APPROXIMATE SOLUTION FOR SMALL TIMES

In order to get an idea of how the flight path angle will change with altitude in a typical trajectory without fully solving the set of Eqns (7.36)–(7.39), we may consider the solution for small values of time that might allow us to solve the equations analytically. We know that launches ideally start at $t = 0$ with $V(0) = 0$ and $\gamma(0) = \pi/2$. Enforcing these conditions would make the right-hand side of Eqn (7.37) indefinite so that the flight path angle development could not be calculated. If we assume that the liftoff proceeds vertically for a short period of time so that $V \ll 1$, but not zero, but retain the requirement that $\gamma(0) = 90°$, then Eqn (7.37) requires that the flight path angle remains constant at $90°$ and the vehicle will ascend vertically throughout its flight. Though such a trajectory may be acceptable for a sounding rocket, it is inadequate for achieving orbit because at some point the flight path angle should become small, indeed zero, for a circular orbit. If we instead assume that $\gamma < 90°$ but keep $V(0) = 0$, then Eqn (7.37) will have a singularity and again a solution cannot be obtained.

Instead we assume that during the early stages of launch the flight path angle is slightly different from $90°$ such that $\gamma = \pi/2 - \varepsilon$ where $\varepsilon \ll 1$ and therefore $\sin \gamma = 1$ to $O(\varepsilon^2)$. Furthermore we assume that at $t = 0$ the velocity $V(0) = V_0$ where V_0 is small, say several tens of meters per second. For constant values of initial thrust to weight ratio $(F_{vac}/W)_0$ and specific impulse $I_{sp,vac}$, we may solve Eqn (7.36) to obtain

$$V = -g_E I_{sp,vac}\left\{\ln\left[1 - \left(\frac{F_{vac}}{W}\right)_0 \frac{t}{I_{sp,vac}}\right] + \frac{t}{I_{sp,vac}}\right\} + V_0 \tag{7.40}$$

Note that $g_E I_{sp,vac}$ is a characteristic velocity of the launch process while $I_{sp,vac}$ is a characteristic time. The parameters in Eqn (7.40) have the following magnitudes: $g_E = O(10 \text{ m/s}^2)$, $I_{sp,vac} = O(10^2 \text{ s})$, $(F_{vac}/W)_0 = O(1)$, and the characteristic velocity $g_E I_{sp,vac} = O(10^3 \text{ m/s})$. The velocity given in Eqn (7.40) is shown as a function of time for $(F_{vac}/W)_0 = 1.3$ and various values of $I_{sp,vac}$ in Figure 7.7, along with numerical solutions of Eqns (7.15) and (7.16) for comparison. The approximate results, represented by data symbols, begin to underestimate the exact results around 70–80 s into the flight because the effects of $\sin \gamma$, which were neglected in arriving at Eqn (7.40), begin to become increasingly important.

For small values of time, Eqn (7.40) may be expanded to the following form:

$$V = -g_E I_{sp,vac}\left\{-\left(\frac{F_{vac}}{W}\right)_0 \frac{t}{I_{sp,vac}} - \frac{1}{2}\left[-\left(\frac{F_{vac}}{W}\right)_0 \frac{t}{I_{sp,vac}}\right]^2 + \cdots + \frac{t}{I_{sp,vac}}\right\} + V_0 \tag{7.41}$$

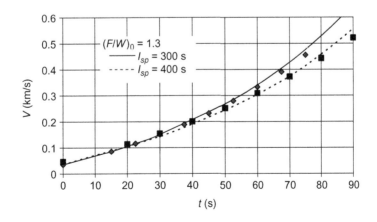

FIGURE 7.7

Velocity of launch vehicle as a function of time for a given initial vacuum thrust to weight ratio and two values of vacuum-specific impulse. Curves are numerical solutions of the equations while symbols refer to the approximate result given in Eqn (7.40).

We see that at very early times $V = V_0 + [(F_{vac}/W)_0 - 1]g_E\, t$ illustrating that the initial thrust level must be greater than the weight and that the magnitude of $(F_{vac}/W)_0$ determines the acceleration of the launch vehicle off the pad. This equation is an approximation for small times and increasingly underestimates the more accurate result given by Eqn (7.40) with an error on the order of 10% at $t = 75$ s. Under the assumption of small values of time, V is likewise small and we see that $V/R_E \ll g_E/V$ so Eqn (7.37) may be approximated as follows:

$$\dot\gamma = -\frac{g_E \cos\gamma}{V} \tag{7.42}$$

Substituting the solution for V given by Eqn (7.41), which is valid to second order in $\bar{t} = t/I_{sp,vac}$, into Eqn (7.42) yields

$$\frac{d\gamma}{\cos\gamma} = -\frac{d\bar{t}}{V_0 + \left[\left(\dfrac{F_{vac}}{W}\right)_0 - 1\right]\bar{t} + \dfrac{1}{2}\left(\dfrac{F_{vac}}{W}\right)_0^2 \bar{t}^2} \tag{7.43}$$

This equation may be integrated with the following result:

$$\ln\left[\tan\left(\frac{\pi}{4} + \frac{\gamma}{2}\right)\right] = -\frac{1}{\sqrt{Q}}\ln\left\{\frac{\left(\dfrac{F_{vac}}{W}\right)_0^2 \bar{t} + \left(\dfrac{F_{vac}}{W}\right)_0 - 1 + \sqrt{Q}}{\left(\dfrac{F_{vac}}{W}\right)_0^2 \bar{t} + \left(\dfrac{F_{vac}}{W}\right)_0 - 1 - \sqrt{Q}}\right\} + \ln C \tag{7.44}$$

In Eqn (7.44), C is a constant of integration and the parameter Q is given by

$$Q = \left[\left(\frac{F_{vac}}{W}\right)_0 - 1\right]^2 - 2\frac{V_0}{g_E I_{sp,vac}}\left(\frac{F_{vac}}{W}\right)_0^2 \tag{7.45}$$

It is obvious that for $t = 0$ where ideally $\gamma(0) = \pi/2$, the left-hand side of Eqn (7.44) is unbounded. This singularity is always present and can present difficulties in starting numerical solutions to Eqn (7.37). However, here we assume that at $t = 0$ the initial flight path angle is slightly less than 90° with $\gamma(0) = \gamma_0 = \pi/2 - \varepsilon$ and $\varepsilon \ll 1$. We may simplify Eqn (7.44) for γ close to $\pi/2$ by making the following approximation:

$$\tan\left(\frac{\pi}{4} + \frac{\gamma}{2}\right) \approx \frac{2}{\cos\gamma} \tag{7.46}$$

This approximation is reasonably accurate, overpredicting γ by about 1.5% for $\gamma = 75°$. If we use Eqn (7.46) and confine our attention to flight path angles not very much smaller than 90°, Eqn (7.44) may be simplified, after some manipulation, to

$$\cos\gamma = \cos\gamma_0 \left\{\frac{\left(\dfrac{F_{vac}}{W}\right)_0^2 \bar{t} + \left(\dfrac{F_{vac}}{W}\right)_0 - 1 + \sqrt{Q}}{\left(\dfrac{F_{vac}}{W}\right)_0^2 \bar{t} + \left(\dfrac{F_{vac}}{W}\right)_0 - 1 - \sqrt{Q}}\left[\frac{\left(\dfrac{F_{vac}}{W}\right)_0 - 1 + \sqrt{Q}}{\left(\dfrac{F_{vac}}{W}\right)_0 - 1 - \sqrt{Q}}\right]\right\}^{\frac{1}{\sqrt{Q}}} \tag{7.47}$$

Using Eqns (7.38) and (7.39), we may determine the early portion of the trajectory of a launch vehicle. The approximate trajectories of launch vehicles with an initial thrust to weight ratio of $(F_{vac}/W)_0 = 1.3$ and with specific impulse values of $I_{sp,vac} = 300$ s and 400 s for the first 90s of the

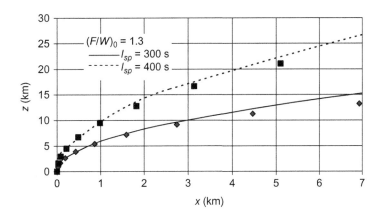

FIGURE 7.8

The trajectories of launchers with $(F_{vac}/W)_0 = 1.3$ and $I_{sp,vac} = 300$ and 400 s are shown for the first 90 s of the flight. Curves are numerical solutions of the complete equations while symbols refer to the approximate results. Note that the scale used exaggerates the turn of the trajectory.

flight are compared to corresponding numerical solutions in Figure 7.8. We note again that the second-order velocity approximation of Eqn (7.41) used for the development increasingly underestimates the actual velocity as was previously shown in Figure 7.7. As a result, the decrease in flight path angle with time as given by Eqn (7.42) is increasingly exaggerated. However, this degree of accuracy is sufficient to illustrate the development of the gravity turn that occurs. The effect of increasing the specific impulse is also clearly illustrated in Figure 7.8. Higher values of $(F_{vac}/W)_0$ have an effect similar to that of higher values of $I_{sp,vac}$.

The solution to Eqn (7.37) for $\gamma(t)$ is very sensitive to the assumed value for the initial flight path angle $\gamma(0)$. This is because at small times $d\gamma/dt \sim -\cos\gamma/V$ and this can quickly force γ to small or even negative values. In general, to achieve orbit at a specified altitude z with no TVC, the initial conditions at launch must be adjusted until the proper orbital values are obtained at the end of the launch process. Therefore dependence on a gravity turn alone for orbital insertion requires "shooting" for the correct final conditions. This sensitivity of the flight path angle evolution as a function of $\gamma(0)$ is shown in Figure 7.9 where the values of $\gamma(0)$ are very close to 90° and are only 0.001 rad (0.057°) apart. The curves represent the numerical solutions of the complete equations while the symbols refer to the corresponding approximate results. It is apparent again that the approximate results are only adequate for times less than about 70 s.

7.4.2 NUMERICAL SOLUTIONS FOR ZERO LIFT AND DRAG

Equations (7.36)–(7.39) may be readily solved with standard numerical integration schemes for simultaneous ordinary differential equations using essentially arbitrary initial conditions. Of course, small changes in initial conditions generally lead to substantial changes in the overall trajectory. Because the velocity equation (7.36) involves $\sin\gamma$, it isn't sensitive to the initial value $\gamma(0)$ which is close to 90°. Thus the initial velocity $V(0)$ is generally taken to be several tens of meters per

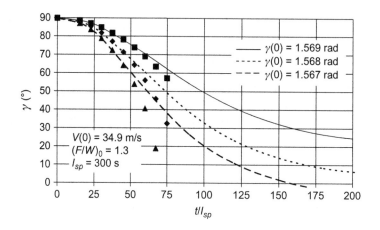

FIGURE 7.9

Flight path angle of a launcher with a given value of $(F_{vac}/W)_0$ and $I_{sp,vac}$ for three values of γ (0). Curves are numerical solutions of the complete equations while symbols refer to the approximate results.

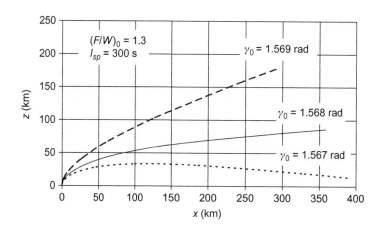

FIGURE 7.10

The trajectory of a launcher with $(F_{vac}/W)_0 = 1.3$ and $I_{sp,vac} = 300$ s for three different initial flight path angles.

second. Note that too small a value of V(0) may lead to an excessively large initial value for $d\gamma/dt$ and subsequent numerical difficulties. We have already shown the substantial change in flight path angle with time for small differences in γ(0) in Figure 7.9. The cases illustrated there used V(0) = 34.9 m/s (approximately 100 ft/s) and the full trajectories corresponding to these conditions are shown in Figure 7.10. Using Eqn (7.14) under the simplifying assumptions of this section shows that the mass ratio of the launch vehicle at burn-out, that is when all the propellant is consumed, is

$$\left(\frac{m}{m_0}\right)_{bo} = 1 - \left(\frac{W_p}{W_0}\right) = 1 - \left(\frac{F_{vac}}{W}\right)_0 \frac{t_{bo}}{I_{sp,vac}} \tag{7.48}$$

Therefore the time to burn-out given by Eqn (7.15) becomes

$$t_{bo} = I_{sp,vac} \frac{\left(\dfrac{W_p}{W_0}\right)}{\left(\dfrac{F_{vac}}{W}\right)_0} \tag{7.49}$$

The maximum burn time would correspond to the launch vehicle being comprised completely of propellant and therefore completely consumed leading to a burn-out time $t_{bo} = I_{sp,vac}/(F_{vac}/W)_0$. Because in practice $I_{sp} = O(10^2)$ and $(F_{vac}/W)_0 = O(1)$ the maximum burn time is confined to several hundreds of seconds. For the cases shown in Figure 7.10 the maximum burn time, according to Eqn (7.49), is 230 s.

The launch vehicle performance illustrated in Figure 7.10 is clearly dependent upon the initial flight path angle as suggested by the results for the history of the flight path angle shown previously in Figure 7.9. The higher value of γ_0 takes the vehicle closer to the higher altitudes needed for a practical orbit for manned spacecraft but the flight path angle is considerably higher than needed for a nearly circular orbit. Conversely the lowest value of γ_0 shown doesn't permit the launch vehicle to reach high enough altitude and actually begins to descend prior to burn-out. The intermediate value of γ_0 results in a flight path angle close to zero which is needed for a circular orbit but doesn't reach quite high enough altitude for a suitable orbit.

Using a launch vehicle with the same initial thrust to weight ratio but a more effective engine having an $I_{sp,vac} = 400$ s produces the trajectories shown in Figure 7.11. Here it is seen that this launch vehicle can reach higher altitudes as well as approach smaller final flight path angles at altitudes close to those suitable for orbit. It is worth noting the large difference in trajectories for initial flight path angles that differ only by about 0.5°. Because of the higher specific impulse the maximum burn time for this launch vehicle is about 308 s.

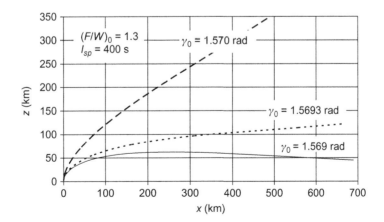

FIGURE 7.11

The trajectory of a launch vehicle with $(F_{vac}/W)_0 = 1.3$ and $I_{sp,vac} = 400$ s for three different initial flight path angles.

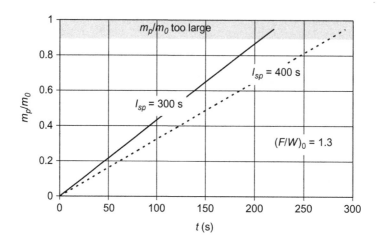

FIGURE 7.12

Propellant fraction as a function of time for a launcher with $(F_{vac}/W)_0 = 1.3$ and $I_{sp,vac} = 300$ and 400 s. Shaded area at top of figure denotes region where structural and payload requirements preclude carriage of propellant.

We now have the opportunity to consider some of the shortcomings of a launch vehicle that must meet practical requirements. Using Eqn (7.11), we may find the propellant consumption as a function of time and this is shown for the cases considered in Figure 7.12. As mentioned previously, the launch vehicle cannot be entirely made of propellant because the launcher weight includes structural weight, including engines, and payload weight. A rough rule of thumb is that about 10% of the initial weight is needed for these categories. As a result the maximum propellant weight fraction is considered to be about 90% of the initial weight so the upper 10% of the propellant fraction is highlighted in Figure 7.12 as being unusable in practice. This cuts the burn time to 210 and 275 s for the $I_{sp,vac} = 300$ and 400 s cases, respectively. The effect of this limited propellant capacity on velocity is illustrated in Figure 7.13. Because orbital velocity for manned spacecraft is around 7.9 km/s, as described in Chapter 5, the propellant limitation makes achievement of orbital velocity possible only for the $I_{sp,vac} = 400$ s case for the stipulated value of $(F/W)_0$.

Added to these difficulties for achieving orbit is the excessive acceleration loads imposed on the launch vehicle late in the flight, as shown in Figure 7.14. Accelerations above around $4g_E$ to $5g_E$ are unacceptable for human spaceflight, as described in Chapter 6. It therefore seems quite unlikely that a rocket-powered single stage to orbit vehicle would be desirable for launching manned spacecraft into orbit.

7.4.3 AN APPROXIMATION FOR BURN-OUT VELOCITY

It is interesting to note that although the flight path angle history is sensitive to velocity and initial flight path angle and that a simple analytic solution is not achievable, such is not the case for the velocity history. If in Eqn (7.36), which models the velocity change achieved by the application of

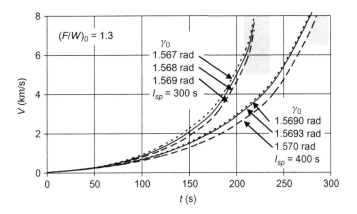

FIGURE 7.13

Velocity as a function of time for a launch vehicle with $(F_{vac}/W)_0 = 1.3$ and $I_{sp,vac} = 300$ and 400 s. Shaded areas at top of figure denote regions where structural and payload requirements preclude carriage of propellant.

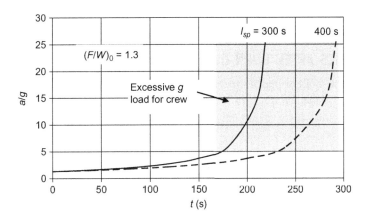

FIGURE 7.14

Acceleration history for a launcher with $(F/W)_0 = 1.3$ and $I_{sp} = 300$ and 400 s. Shaded area denotes region where acceleration levels are unacceptable for human spaceflight.

constant thrust $F = F_{vac}$ under the action of gravity, we assume that the $\sin \gamma$ term may be replaced by an average value, then the velocity is given by

$$V = -g_E I_{sp} \left\{ \left[1 - \left(\frac{F}{W} \right)_0 \frac{t}{I_{sp}} \right] + \sin \gamma_{avg} \frac{t}{I_{sp}} \right\} + V_0 \tag{7.50}$$

In Figure 7.7, we showed that choosing $\sin \gamma_{avg} = 1$ yields results close to the numerical solutions for small values of time. For late times in the flight, choosing $\sin \gamma_{avg} = 0.5$ provides quite accurate values compared to the numerical solutions, as demonstrated in Figure 7.15 for the case where the propellant weight is 90% of the initial weight of the vehicle. The velocity is over-estimated at small times and becomes quite close to the numerical result at larger times.

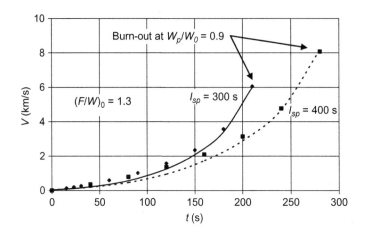

FIGURE 7.15

Velocity history up to burn-out corresponding to $W_p/W_0 = 0.9$ for $(F_{vac}/W)_0 = 1.3$ and $I_{sp,vac} = 300$ and 400 s. Numerical solutions are shown as curves and approximate solutions given by Eqn (7.50) are shown as data symbols.

7.4.4 TRAJECTORIES AFTER BURN-OUT

After all the propellants are consumed, or the engine thrust is cut off, the launch vehicle will coast along a trajectory under the influence of gravity alone. The velocity at burn-out may be approximated using Eqn (7.49) in Eqn (7.50) to obtain

$$V_{bo} = - g_E I_{sp,vac} \left[\ln\left(1 - \frac{W_p}{W_0} \right) + \sin \gamma_{avg} \frac{W_p}{W_0} \left(\frac{F_{vac}}{W} \right)_0^{-1} \right] \tag{7.51}$$

The effect of propellant fraction W_p/W_0 on the burn-out velocity for the case where we assume $\sin \gamma_{avg} = 0.5$ is shown in Figure 7.16. It is clear that the specific impulse has a substantial effect on the magnitude of the velocity at burn-out. The effect of increasing the initial thrust to weight ratio $(F_{vac}/W)_0$ to 1.4 from 1.3 on the burn-out velocity is negligible and therefore is not included in Figure 7.16.

Once the propellant is consumed and we have burn-out, or engine cut-off, Eqn (7.36) becomes

$$\dot{V} = - g_E \sin \gamma \tag{7.52}$$

We may transform from time to vertical distance as the independent variable by introducing

$$\frac{d}{dt} = V \sin \gamma \frac{d}{dz} \tag{7.53}$$

Applying Eqn (7.53) to Eqns (7.52) and (7.37) leads to the following equations:

$$V \frac{dV}{dz} = - g_E \tag{7.54}$$

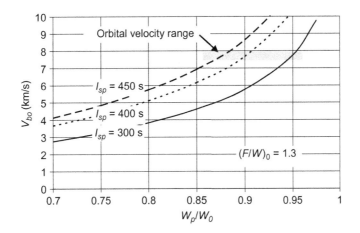

FIGURE 7.16

Velocity at burn-out is shown as a function of propellant fraction W_p/W_0 for $(F_{vac}/W)_0 = 1.3$ and $I_{sp,vac} = 300$, 400, and 450 s. The velocity range required to achieve LEO is shown as a shaded region.

$$\tan \gamma \frac{d\gamma}{dz} = \frac{1}{R_E} - \frac{g_E}{V^2} \tag{7.55}$$

If we set the initial conditions to those at burn-out, i.e., at $z = z_{bo}$ the velocity $V(z_{bo}) = V_{bo}$ and the flight path angle $\gamma(z_{bo}) = \gamma_{bo}$ we may integrate Eqns (7.54) and (7.55) to obtain

$$V^2 = V_{bo}^2 - 2g_E(z - z_{bo}) \tag{7.56}$$

$$\cos \gamma = \cos \gamma_{bo} \frac{\exp\left[-\dfrac{z - z_{bo}}{R_E}\right]}{\sqrt{1 - 2\dfrac{g_E R_E}{V_{bo}^2}\dfrac{z - z_{bo}}{R_E}}} \tag{7.57}$$

The relationship between downrange distance and altitude is

$$\frac{dz}{dx} = \tan \gamma \tag{7.58}$$

We may use Eqns (7.56)–(7.58) to calculate the trajectory of the launch vehicle subsequent to burn-out or engine cut-off. Although we can calculate the velocity at burn-out reasonably accurately using Eqn (7.51), as shown by the results in Figure 7.13, there is no simple method to obtain γ_{bo} and therefore z_{bo}, as illustrated by the results in Figures 7.10 and 7.11, so that numerical solutions must be employed to determine the trajectory up to burn-out.

The trajectories for two cases of a launch vehicle with $(F_{vac}/W)_0 = 1.3$ and $I_{sp,vac} = 400$ s discussed in Section 7.2.2 are shown after engine cut-off in Figures 7.17 and 7.18. The former case burns out at $V_{bo} = 7.39$ km/s and $\gamma_{bo} = 28.99°$ while the latter has $V_{bo} = 7.39$ km/s and $\gamma_{bo} = 1.577°$. In both cases, the velocity is insufficient for orbit at the altitudes shown and the

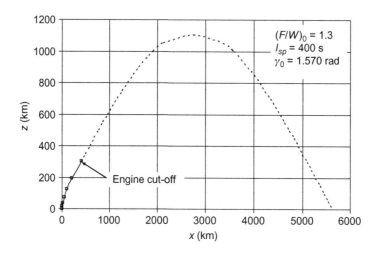

FIGURE 7.17

Approximate trajectory of a given launch vehicle after engine cut-off for a relatively high flight path angle at cut-off. The initial powered trajectory and engine cut-off point are shown for $V_{bo} = 7.39$ km/s corresponding to $W_p/W_0 = 0.9$.

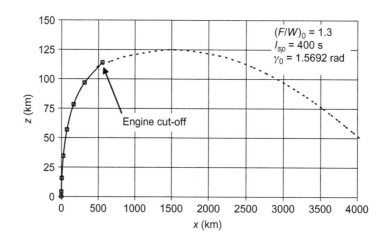

FIGURE 7.18

Approximate trajectory of a given launch vehicle after engine cut-off for a relatively low flight path angle at cut-off. The initial powered trajectory and engine cut-off point are shown for $V_{bo} = 7.39$ km/s corresponding to $W_p/W_0 = 0.9$.

launch vehicle would enter the atmosphere and descend to earth as shown. Figure 7.17 shows that the launch vehicle with the relatively large flight path angle at burn-out will continue to ascend to very high altitude before entering the atmosphere and returning to earth. On the other hand, the launch vehicle with the very small flight path angle at burn-out will rise only somewhat higher than the burn-out altitude before descending to earth, as shown in Figure 7.18. Both cases have the same burn-out velocity so it is the flight path angle that determines the nature of the trajectory.

7.4.5 EFFECTS OF EARTH'S ROTATION

The dynamics of the launch vehicle developed in this chapter are based on a nonrotating earth. The earth rotates toward the east at the rate of one revolution in 23 h 56 m so the linear speed of rotation at the surface of the earth is $V_\omega = (465.1 \text{ m/s}) \cos L_a$ where L_a is the magnitude of the latitude angle. The major spaceports and their location are listed in Table 7.2.

A launch vehicle on the pad possesses the earth's eastward velocity and this additional velocity aids in achieving orbit. A NASA launch from Cape Canaveral provides about 5% of the orbital velocity required while an ESA launch from Kourou provides about 5.9%. Thus eastward, or prograde, launches from sites close to the equator are most attractive. The total elapsed time from liftoff to orbital insertion is typically less than 10 min during which the launch site on the earth moves eastward a distance of only $279 \cos L_a$ km. Earth's rotation is important in detailed trajectory determination, which is outside the scope of this book. For more on this aspect of launch mechanics see, e.g., Bate, Mueller, and White (1971) and Wertz and Larson (1999).

Sometimes reported trajectory characteristics are based on an inertial coordinate system with the origin at the center of the earth rather than the so-called topocentric-horizon coordinate system we have been using which has an origin on the earth ($x = y = 0$, $z = 0$) and rotates with it. We have considered velocities to be measured with respect to a fixed position on the earth and flight path angles to be measured with respect to a local horizon ($z = $ constant). Using the development presented by Bate et al. (1971), one may show that the magnitude of the inertial velocity V_i is given in terms of the velocity, V, the flight path angle, γ, and the azimuth angle, β, all measured relative to a nonrotating earth, and the linear velocity due to rotation, V_ω, as follows:

$$V_i = \sqrt{V^2 + V_\omega^2 + V V_\omega \cos \gamma \sin \beta}$$

The velocity vector diagram is shown in Figure 7.19. The earth rotates eastward and the azimuth angle β is measured clockwise from north. Eastward ($0° < \beta < 180°$) launches gain an assist from the earth's rotation while westward launches ($180° < \beta < 360°$) lose. For a nonrotating earth where $V_\omega = 0$ we recover the result $V_i = V$. However, at liftoff on a rotating earth, where $V = 0$ and $\gamma = 90°$, we see that the inertial velocity $V_i = V_\omega$. As $V \gg V_\omega$ during ascent, the effect of

Table 7.2 Major Spaceports and Their Locations			
Country	**Spaceport**	**Latitude**	**Longitude**
China	Jiuquan	40.6°N	99.9°E
China	Xichang	28.25°N	102.0°E
European Union	Kourou (French Guiana)	5.2°N	52.8°W
India	Sriharikota Island	13.9°N	80.4°E
Japan	Tanegashima	30.4°N	131.0°E
Russia	Baikonur	45.6°N	63.4°E
Russia	Plesetsk	62.8°N	40.1°E
United States	Cape Canaveral	28.5°N	81.0°W
United States	Vandenberg AFB	34.4°N	120.35°W

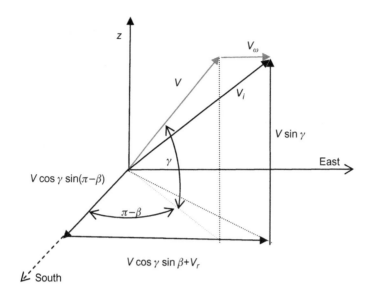

FIGURE 7.19

Velocity diagram for flight over a rotating earth.

earth's rotation diminishes and V_i approaches V. Therefore the reported inertial velocity history of a launch vehicle will show a value equal to V_r at liftoff and as the launch progresses the difference in true and relative velocity would diminish to near zero.

The inertial flight path angle is related to the local horizon-based flight path angle by

$$\sin \gamma_i = \frac{V \sin \gamma}{\sqrt{V^2 + V_\omega^2 + VV_\omega \cos \gamma \sin \beta}}$$

For a nonrotating earth where $V_\omega = 0$ we recover the result $\gamma_i = \gamma$. However, at liftoff on a rotating earth, where $V = 0$ and $\gamma = 90°$, we find $\sin \gamma_i = 0$ which means that $\gamma_i = 0$ although $\gamma = 90°$. Therefore when the inertial flight path angle history of a launch vehicle is reported it will start at zero and grow to some maximum before decreasing because as $V \gg V_\omega$ the difference between true and local horizon-based flight path angle would diminish to near zero.

7.5 STAGING OF ROCKETS

The weight of the launch vehicle decreases as the propellant burns, but the structure and tankage weight remains fixed as the vehicle rises. This weight is parasitic, so it is wasteful to continue carrying this portion of the vehicle mass against the pull of gravity. Ideally this mass should be jettisoned continuously during ascent. Such a procedure is not practical, but an approximation to this may be accomplished by jettisoning discrete portions of the structure and tankage along the flight path when the propellant in them is completely consumed. This concept is known as staging. Stages and stacks are different divisions of a launch vehicle and this difference is illustrated for the Saturn V vehicle in Figure 7.20.

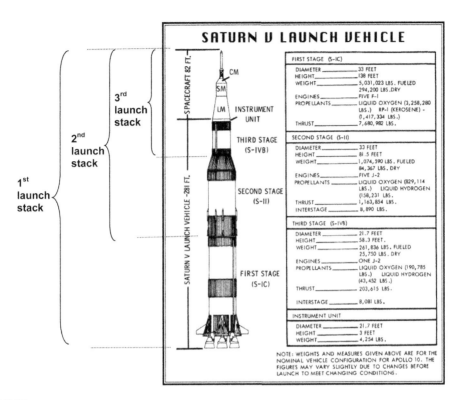

SATURN V LAUNCH VEHICLE

SPACECRAFT 82 FT.

SATURN V LAUNCH VEHICLE -281 FT.

CM

SM

LM   INSTRUMENT
      UNIT

THIRD STAGE
(S-IVB)

SECOND STAGE
(S-II)

FIRST STAGE
(S-IC)

FIRST STAGE (S-IC)	
DIAMETER	33 FEET
HEIGHT	138 FEET
WEIGHT	5,031,023 LBS. FUELED
	294,200 LBS.DRY
ENGINES	FIVE F-1
PROPELLANTS	LIQUID OXYGEN (3,258,280
	LBS.) RP-1 (KEROSENE) -
	(1,417,334 LBS.)
THRUST	7,680,982 LBS.

SECOND STAGE (S-II)	
DIAMETER	33 FEET
HEIGHT	81.5 FEET
WEIGHT	1,074,590 LBS. FUELED
	84,367 LBS. DRY
ENGINES	FIVE J-2
PROPELLANTS	LIQUID OXYGEN (829,114
	LBS.) LIQUID HYDROGEN
	(158,231 LBS.)
THRUST	1,163,854 LBS.
INTERSTAGE	8,890 LBS.

THIRD STAGE (S-IVB)	
DIAMETER	21.7 FEET
HEIGHT	58.3 FEET.
WEIGHT	261,836 LBS. FUELED
	25,750 LBS. DRY
ENGINES	ONE J-2
PROPELLANTS	LIQUID OXYGEN (190,785
	LBS.) LIQUID HYDROGEN
	(43,452 LBS.)
THRUST	203,615 LBS.
INTERSTAGE	8,081 LBS.

INSTRUMENT UNIT	
DIAMETER	21.7 FEET
HEIGHT	3 FEET
WEIGHT	4,254 LBS.

NOTE: WEIGHTS AND MEASURES GIVEN ABOVE ARE FOR THE
NOMINAL VEHICLE CONFIGURATION FOR APOLLO 10. THE
FIGURES MAY VARY SLIGHTLY DUE TO CHANGES BEFORE
LAUNCH TO MEET CHANGING CONDITIONS.

3rd
launch
stack

2nd
launch
stack

1st
launch
stack

FIGURE 7.20

The Saturn V launch vehicle showing the three stages and the corresponding three stacks.

A stage is a propulsion unit consisting of the engine, its propellant and the associated tankage, structure, and controls. This unit carries an inactive payload and the combination is called a launch stack. For example, the first stage shown in Figure 7.20 is a propulsion unit which carries the remainder of the vehicle. The combination is called the first launch stack. The payload of the first launch stack is the second launch stack which is comprised of the second stage propulsion unit and its inactive payload. In turn, the payload of the second launch stack is the third launch stack, which is comprised of the third stage propulsion unit and its inactive payload. In this case, the inactive payload of the third launch stack is the true payload of the launch vehicle.

The equations of motion given by Eqns (7.23)–(7.26) may be applied to the operation of any given number of stages because staged operation may be considered to be a sequence of initial value problems. For simplicity in illustrating the characteristics of staged operation, we may instead use the approximate Eqns (7.36)–(7.39) developed previously. To achieve orbit the appropriate speed and flight path angle must be achieved at some specified altitude. We note that, as described previously and illustrated in Figure 7.15, the velocity of the launch vehicle depends mainly on the rocket engine performance and is not very sensitive to the flight path angle. The flight path angle is generally controlled by appropriate variation of the thrust angle χ during flight. Thus the important factor in achieving orbit is the velocity of the vehicle.

7.5.1 GENERAL RELATIONS FOR STACKS AND STAGES

We shall consider a three-stage vehicle suitable for safe human spaceflight. One-stage and two-stage vehicles are easily considered within the framework presented for the three-stage launcher. Schematic diagrams of one-, two-, and three-stage vehicles are shown in Figure 7.21. Clearly defined on the figure are the corresponding stacks for the three different launch vehicles.

The total initial mass of a given stage $i = 1, 2, 3$, etc., which is part of a given stack $J = I, II, III$, etc., is

$$m_0^J = m_0^{J+1} + m_{str,i} + m_{eng,i} + m_{p,i} = m_0^{J+1} + m_{s,i} + m_{p,i} \tag{7.59}$$

The total mass of the propellant in the ith stage is denoted by $m_{p,i}$. Using Eqn (7.14), an approximate relation for the total propellant mass in a given stage with constant thrust and specific impulse is

$$\frac{m_{p,i}}{m_0^J} = \frac{1}{I_{sp,i}} \left(\frac{F}{W_0} \right)^J (t_{bo,i} - t_{ig,i}) = k_{p,i} \Delta t_i \tag{7.60}$$

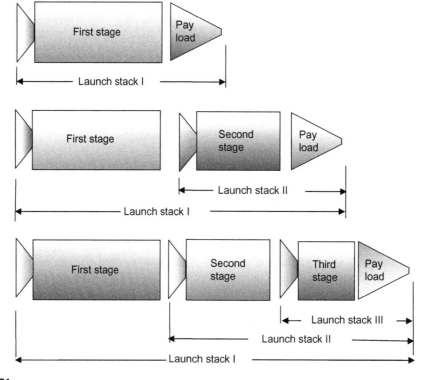

FIGURE 7.21

Schematic diagrams of one-stage, two-stage, and three-stage launch vehicles are shown, starting from the top. The launch stacks comprising each launch vehicle are defined.

Because the initial propellant fraction must be less than unity and the thrust to weight ratio must be greater than unity, the burn time of any stage must be less than the specific impulse provided by that stage's engines, that is, less than several hundred seconds.

We assume that $m_{s,i}$, the empty mass of the stage, is proportional to $m_{p,i}$, the mass of propellant carried by the stage. The adequacy of this assumption will be discussed subsequently. Thus the sum of the structural and engine masses $m_{s,i}$ of a given stage i within a given stack J is

$$\frac{m_{s,i}}{m_0^J} = \frac{m_{str,i} + m_{eng,i}}{m_0^J} = k_{s,i}\frac{m_{p,i}}{m_0^J} \tag{7.61}$$

For a three-stage launch vehicle, the mass balances for each of the stacks $J = I, II,$ and III are as follows:

$$m_0^I = \{[m_{pay} + m_{s,3} + m_{p,3}] + [m_{s,2} + m_{p,2}]\} + [m_{s,1} + m_{p,1}] \tag{7.62}$$

$$m_0^{II} = \{[m_{pay} + m_{s,3} + m_{p,3}]\} + [m_{s,2} + m_{p,2}] \tag{7.63}$$

$$m_0^{III} = \{m_{pay}\} + [m_{s,3} + m_{p,3}] \tag{7.64}$$

The initial mass of each stack may be calculated sequentially, starting with the specified payload mass m_{pay}, from the following equations:

$$m_0^{III} = \frac{m_{pay}}{1 - (1 + k_{s,3})k_{p,3}\Delta t_3} \tag{7.65}$$

$$m_0^{II} = \frac{m_0^{III}}{1 - (1 + k_{s,2})k_{p,2}\Delta t_2} \tag{7.66}$$

$$m_0^I = \frac{m_0^{II}}{1 - (1 + k_{s,1})k_{p,1}\Delta t_1} \tag{7.67}$$

We may combine these equations to show that the initial mass of the launch vehicle is

$$m_0^I = \frac{m_{pay}}{[1 - (1 + k_{s,3})k_{p,3}\Delta t_3][1 - (1 + k_{s,2})k_{p,2}\Delta t_2][1 - (1 + k_{s,1})k_{p,1}\Delta t_1]} \tag{7.68}$$

The payload mass is determined by the mission specification but there appears to be nine parameters which may be varied so as to achieve the minimum initial mass for the launch vehicle. If each of the terms in the square brackets in Eqn (7.68) are independently maximized then the initial launch mass will be minimized. This requires that the structural parameters $k_{s,i}$, the propellant parameters $k_{p,i}$, and the burn times Δt_i be independently minimized, so that the denominator of Eqn (7.68) would be maximized. However, there are limitations to this scheme. In practice, the selection of these parameters is quite limited so they can be neither arbitrarily nor independently chosen.

Let us consider each of these factors separately:

• The thrust to initial weight ratio of a stack may be set arbitrarily (within a rather limited range) but because the initial weight of any stack is an outcome of the calculations the engine thrust then becomes fixed. Unless an engine is going to be designed to meet this required thrust level it becomes necessary to choose from (relatively few) existing engines, and it is unlikely that any will closely match the specified thrust level.

- The specific impulse depends mainly on the propellant mix selected and limited to several narrow ranges of values of around 425−450 s for LH_2−LOX propellants and 260−300 s for RP1−LOX propellants. Obviously, the choice of propellant also further limits the choice of engines.
- The structural factors $k_{s,i}$ likewise cannot be selected arbitrarily but instead are limited to values between about 0.05 to about 0.15. Existing systems suggest that the lowest values of 0.05−0.08 are achievable for booster and first stages while values of 0.1−0.15 seem to be the norm for second and upper stages.
- The burn time for each stage is equally constrained because, as pointed out previously, it must be less than the specific impulse of the engines used in the stage. However, the vehicle must be accelerated to orbital velocity, here assumed to be approximated by $V_E = 7.909$ km/s, but at a rate suitable for safe human spaceflight. The average acceleration in achieving $V_E = 7.909$ km/s in a total burn time of 300 s is about $2.6g_E$ while for a total burn time of 450 s it is about $1.8g_E$. Because the instantaneous acceleration increases very rapidly near the end of the burn it can easily become unacceptably large.

For simplicity, we will use the approximation of Eqn (7.50) to determine the velocity of a given stage. Assuming that $V(0) = 0$ the velocity of the first stack, up to the burn-out time of the first stage $t = t_{bo,1}$ at which point $V = V(t_{bo,1})$ is given by

$$V_1(t) = - g_E I_{sp,1} \ln[1 - k_{p,1}t] - g_E t \sin \gamma_{avg,1} \tag{7.69}$$

If there is only one stack, the first stage is dropped at $t = t_{bo,1}$ and the payload coasts at $V = V_{bo,1}$. However, if there is a second stack the second stage is ignited as the first stage is jettisoned at $t = t_{bo,1}$ and propels the second stack at a velocity given by

$$V_2(t) = - g_E I_{sp,2} \ln[1 - k_{p,2}(t - t_{bo,1})] - g_E(t - t_{bo,1})\sin \gamma_{avg,2} + V_{bo,1} \tag{7.70}$$

If there are only two stacks, the second stage is dropped at $t = t_{bo,2}$ and the payload is free to coast along at $V = V_{bo,2}$. However, if there is a third stack the third stage is ignited as the second stage is jettisoned at $t = t_{bo,2}$ and propels the third stack at a velocity given by

$$V_3(t) = - g_E I_{p,3} \ln[1 - k_{p,3}(t - t_{bo,2})] - g_E(t - t_{bo,2})\sin \gamma_{avg,3} + V_{bo,2} \tag{7.71}$$

The propellant mass of each stage as a fraction of its stack's initial mass is given by

$$
\begin{aligned}
\frac{m_{p,1}}{m_0^I} &= k_{p,1}(t_{bo,1} - 0) = k_{p,1}\Delta t_1 \\
\frac{m_{p,2}}{m_0^{II}} &= k_{p,2}(t_{bo,2} - t_{bo,1}) = k_{p,2}\Delta t_2 \\
\frac{m_{p,3}}{m_0^{III}} &= k_{p,3}(t_{bo,3} - t_{bo,2}) = k_{p,3}\Delta t_3
\end{aligned}
\tag{7.72}
$$

The acceleration of each stage and its corresponding stack is given by

$$
\frac{\dot{V}_1}{g_E} = \frac{\left(\dfrac{F}{W_0}\right)^I}{1 - k_{p,1}t} - \sin \gamma_{avg,1}
$$

$$
\frac{\dot{V}_2}{g_E} = \frac{\left(\dfrac{F}{W_0}\right)^{II}}{1 - k_{p,2}(t - t_{bo,1})} - \sin \gamma_{avg,2} \tag{7.73}
$$

$$
\frac{\dot{V}_3}{g_E} = \frac{\left(\dfrac{F}{W_0}\right)^{III}}{1 - k_{p,3}(t - t_{bo,2})} - \sin \gamma_{avg,3}
$$

The mass of the payload is to be placed in a circular orbit at a speed appropriate to the orbit altitude, given by

$$
V = \sqrt{gr} = \sqrt{\frac{g_E R_E}{1 + \frac{z}{R_E}}} \approx \sqrt{g_E R_E}\left(1 - \frac{z}{2R_E}\right) \approx \sqrt{g_E R_E} = V_E \tag{7.74}
$$

Note that for LEO $z \ll R_E$ so that the final approximation for the appropriate orbital velocity is simply V_E, the orbital velocity calculated at the earth's surface. This approximation yields a velocity about 3% higher than the exact value at $z = 200$ km and is acceptable for preliminary design purposes. In this simple approach, the engines are assumed to have constant thrust F making the initial thrust to weight ratio of each stack, $(F/W_0)^J$, likewise constant. It is also assumed that the specific impulse I_{sp} is constant and therefore $k_{p,i} = (F/W_0)^J/I_{sp,i} =$ constant and that the average flight path angles $\gamma_{avg,i}$ are specified. It is often convenient to use the vacuum values of thrust and specific impulse because these values are most often quoted in the literature. Then Eqns (7.69)–(7.71) may be used to solve for the burn time t_{bo} required to accelerate from $V = V(0)$ to $V = V_E(t_{bo,i}) = 7.909$ km/s and Eqn (7.72) to solve for the fuel fraction needed in each stage. The maximum acceleration in any stage occurs at burn-out, $t = t_{bo,J}$ and may be found using Eqn (7.73).

To aid in selecting engines for a particular launch vehicle design a list of liquid propellant rockets including their vacuum thrust and specific impulse is provided in Table 7.3.

7.5.2 SINGLE STAGE TO ORBIT

Let us assume a hypothetical single stage rocket having a typical initial thrust to weight ratio of $(F/W)_0 = 1.3$ and a LH_2–LOX propellant generating $I_{sp} = 450$ s, so that the coefficient $k_{p,1}$ is given by

$$
k_{p,1} = \frac{\left(\dfrac{F}{W_0}\right)^I}{I_{sp,1}} = \frac{1.3}{450 \text{ s}} = 0.002889 \text{ s}^{-1}
$$

Table 7.3 Specific Impulse and Thrust in Vacuum for a Variety of Rocket Engines in Ascending Order for Each Propellant

Propellant	Engine	$I_{sp,vac}$ (s)	Engine	F_{vac} (lbs)
LH_2-LOX	RS-68	409	HM7-B	14,568
LH_2-LOX	J-2	425	CECE	15,000
LH_2-LOX	Vulcain	431	RS-44	15,000
LH_2-LOX	Vulcain 2	433	RL10B-2	25,000
LH_2-LOX	STME	434	Vinci	40,468
LH_2-LOX	J-2S	435	J-2	232,000
LH_2-LOX	CECE	445	J-2S	256,000
LH_2-LOX	HM7-B	446	Vulcain	256,295
LH_2-LOX	RS-53	454	Vulcain 2	303,507
LH_2-LOX	SSME	454	RS-53	470,000
LH_2-LOX	RL10B-2	462	SSME	512,300
LH_2-LOX	Vinci	465	STME	580,000
LH_2-LOX	RS-44	481	RS-68	768,000
$RP1-LOX$	MA-5	259	H1	205,000
$RP1-LOX$	H1	263	RD-107	223,000
$RP1-LOX$	MA-5A	264	RS-27	231,700
$RP1-LOX$	RS-27	295	RS-27A	237,000
$RP1-LOX$	RSX	299	NK33	339,000
$RP1-LOX$	RS-27A	302	MA-5	377,500
$RP1-LOX$	F1-A	303	NK43	395,000
$RP1-LOX$	F1	304	MA-5A	423,500
$RP1-LOX$	RD-107	310	RSX	480,000
$RP1-LOX$	NK33	331	RD-180	933,400
$RP1-LOX$	RD-180	338	F-1	1,748,000
$RP1-LOX$	NK43	346	F-1A	2,020,000
$UMDH-N_2O_4$	YF-20B	260	YF-20B	166,000
$UMDH-N_2O_4$	Viking 5C	278	Viking 5C	169,065
$MMH-N_2O_4$	Aestus II	340	Aestus II	12,456
CH_4-LOX	STBE	340	CECE	15,000
CH_4-LOX	CECE	340	STBE	575,000

Using Eqns (7.69) and (7.72), we find

$$V_1(t) = -4.413 \ln(1 - 0.002889t) - 0.009807t \sin \gamma_{avg,1} \tag{7.75}$$

$$\frac{m_{p,1}}{m_0^I} = k_{p,1} \Delta t_1 = 0.002889 \ s^{-1} \Delta t_1 \tag{7.76}$$

The velocity histories for different values of γ_{avg} are shown in Figure 7.22 illustrating the effects of gravity as given by $\sin \gamma_{avg}$. The time necessary to reach $V_E = 7.909$ km/s for

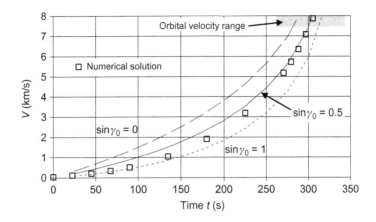

FIGURE 7.22

Velocity history of a single stage rocket with $(F/W)_0 = 1.3$ and $I_{sp} = 450$ s for three values of $\sin \gamma_{avg}$. The circular velocity $V_E = 7.909$ km/s lies in the shaded area. Data symbols represent the numerical solution with varying γ.

$\sin \gamma_{avg} = 0$, 0.5, and 1.0 is 287 s, 306 s, and 316 s, respectively. The two extremes are only 29 s apart and because neither is completely accurate we assume that the middle value, $\sin \gamma_{avg} = 0.5$, may be selected as a reasonable estimate for the gravitational effect. The numerical solution also shown in Figure 7.22 reinforces this conclusion, at least for preliminary design studies.

Using $\sin \gamma_0 = 0.5$ in Eqn (7.75) yields the result that the approximate orbital velocity V_E is reached after a burn of about 305 s. The amount of propellant may now be calculated from Eqn (7.76) using the first stage burn-out time $\Delta t_1 = t_{bo} - t_{ig} = 305$ s $- 0 = 305$ s. This operation yields the propellant fraction required, $m_{p,1}/m_0^I = 0.8811$. Because this is a single stage vehicle $k_{p,2} = k_{p,3} = 0$ and Eqn (7.68) reduces to

$$m_0^I = \frac{m_{pay}}{1 - (1 + k_{s,1})(0.002889 \text{ s}^{-1})(305 \text{ s})} - \frac{m_{pay}}{1 - 0.8811(1 + k_{s,1})} \tag{7.77}$$

The maximum acceleration occurs at the end of the burn when the vehicle is lightest and is given by

$$\frac{\dot{V}_{max}}{g_E} = \frac{1.3}{1 - k_{p,1}t_{bo}} - \sin \gamma_{avg} = \frac{1.3}{1 - (0.002889 \text{ s}^{-1})(305 \text{ s})} - \frac{1}{2} = 10.43 \tag{7.78}$$

Thus the acceleration level near the end of the burn is unacceptably high for human spaceflight. The time at which the acceleration is $6g_E$ is about $t = 277$ s so that there would be a period of around 28 s during which the acceleration would be in excess of $6g_E$. The likelihood of excessive acceleration is one of the limitations in using single stage to orbit launch vehicles for human spaceflight.

With the payload mass specified, Eqn (7.77) provides an equation for the initial mass of the entire single stage stack. It will be shown in Chapter 10 that for a four person crew one may expect a spacecraft mass of about 10,000 kg. Adding about 170 kg per crew member to account for their weight and that of their spacesuits and related personal equipment results in a total spacecraft weight of about 10,680 kg.

The solution to Eqn (7.77), when using an intermediate value of $k_{sp,1} = 0.062$ for first stage, is $m_0 = 166,200$ kg. In this simplified analysis, we assumed that the thrust is constant and that $(F/W_0)^I = 1.3$. Therefore the thrust required for this one-stage launch vehicle is about $F = 477,000$ lbs. A list of liquid propellant rockets including their vacuum thrust and specific impulse was given in Table 7.3. The required thrust level is about that developed by one RS-53 engine, which has a specific impulse of 453.5 s and a vacuum thrust of 470,000 lbs or a Space Shuttle Main Engine (SSME) which also has a specific impulse of 453.5 s with a vacuum thrust of 512,000 lbs. In practice, the thrust of a rocket engine at sea level is generally about 10–20% less than at vacuum conditions, but as mentioned previously, rapidly rises to about the vacuum thrust level during ascent. Using the result that $m_p/m_0 = 0.8811$, we find the mass of propellant required to be $m_p = 146,400$ kg.

The total mass m_0 in this analysis is quite sensitive to the choice of $k_s = m_s/m_p$ and this must be kept in mind when carrying out a preliminary design. Using the often quoted value of $k_s = 0.1$ would lead to $m_0 = 346,900$ kg, the thrust required would be $F = 995,500$ lbs and the mass of propellant $m_p = 305,700$ kg. Obviously, the structural mass coefficient plays a major role in determining the initial mass of a one-stage rocket.

7.5.3 TWO-STAGE VEHICLE TO ORBIT

Now suppose that the ascent vehicle is made up of two stages, each with engines of identical specific impulse $I_{sp,1} = I_{sp,2} = 450$ s. Further assume that each stack has the same initial thrust to weight ratio $(F/W_0)^I = (F/W_0)^{II} = 1.3$. As a result $k_{p,1} = k_{p,2} = 0.002889$ s^{-1} and we continue with the same average value of $\sin \gamma_{avg} = 0.5$ to account for the effect of gravitational attraction. The first stage velocity history would be exactly the same as in the single stage vehicle treated in the previous section, but because we are now using two stages, the first stage has a propellant supply sufficient to provide thrust only up to some intermediate time, say $t_{bo,1} = 200$ s. As pointed out in the discussion of Eqn (7.68), the choice of burn time influences the total mass of the launch vehicle. Here, a burn time of the first stage $t_{bo,1} = 200$ s is selected to maintain a reasonably low acceleration level during the first stage of the flight. At $t_{bo,1} = 200$ s, the velocity of the complete vehicle has been calculated from Eqn (7.75) to be $V_1(200) = 2.825$ km/s (9250 fps). The fuel fraction consumed during this 200 s burn may be determined from Eqn (7.60) as follows:

$$\frac{m_{p,1}}{m_0^I} = k_{p,1} \Delta t_1 = 0.002889 \text{ s}^{-1}(200 \text{ s}) = 0.5778 \tag{7.79}$$

The sequence of events is schematically illustrated in Figure 7.23. Using Eqn (7.68) and assuming that $k_{s,1} = 0.062$ as in the single stage case, we may rewrite the mass of the entire two-stage vehicle as follows:

$$m_0^I = \frac{10,680 \text{ kg}}{[1 - 1.062(0.002889 \text{ s}^{-1})(200 \text{ s})][1 - (1 + k_{s,2})(0.002889 \text{ s}^{-1})\Delta t_2]} \tag{7.80}$$

In order to solve Eqn (7.80), we must assume a value for $k_{s,2}$ and the burn-out time $t_{bo,2}$. Recall that the second stage ignites at time $t_{bo,1} = 200$ s with $V_{bo,1} = 2.82$ km/s. If we use these conditions in Eqn (7.70), we obtain

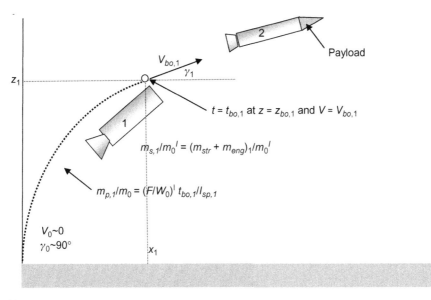

FIGURE 7.23

Notional illustration of the first stage burn-out and initiation of the second stage.

$$V_2 = 2.82 \text{ km/s} - (4.413 \text{ km/s})\ln\left[1 - (0.002889 \text{ s}^{-1})(t - 200 \text{ s})\right]$$
$$- \frac{0.009807 \text{ km/s}^2}{2}(t - 200 \text{ s}) \tag{7.81}$$

Solving Eqn (7.81) shows that the approximate orbital velocity $V_E = 7.909$ km/s is achieved at a burn-out time $t_{bo,2} = 465$ s. Note that the fuel fraction burned during operation of the second stage is then given by

$$\frac{m_{p,2}}{m_0^{II}} = k_{p,2}\Delta t_2 = (0.002889 \text{ s}^{-1})(465 \text{ s} - 200 \text{ s}) = 0.7656 \tag{7.82}$$

In keeping with results for second/upper stages to be discussed in a subsequent section, we use $k_{sp,2} = 0.12$ in Eqn (7.80) which then yields

$$m_0^I = \frac{10,680 \text{ kg}}{(0.3863)(0.1425)} = 193,900 \text{ kg}$$

The mass of the second stack is

$$m_0^{II} = \frac{10,680 \text{ kg}}{1 - 1.12(0.7656)} = 74,930 \text{ kg}$$

The thrust required for the first stage is therefore $F^I = 556,700$ lbs, which is about the thrust level of two J-2S engines ($I_{sp,vac} = 435$ s and $F_{vac} = 265,000$ lbs) as listed in Table 7.2. The thrust required by the second stage to propel the second stack is $F^{II} = 215,000$ lbs, which can readily be satisfied by a single J2-S engine. To launch the same payload into LEO using these assumptions, we see that this two-stage system weighs 12% more than a single stage system and requires the same number of equivalent engines. The propellant mass required for the first stage is $m_{p,1} = (0.5778)(194,000 \text{ kg}) = 112,100 \text{ kg}$ and for the second stage $m_{p,2} = (0.7656)(74,930 \text{ kg}) = 57,370$ kg. In addition to the increased total weight, the two-stage system is somewhat more complex suggesting that reliability may become an issue. There are advantages to staging and we will discuss them subsequently, after we evaluate one more staged launcher: a three-stage vehicle. On the other hand, the maximum acceleration with the two-stage vehicle is around $5g_E$ and this occurs at the end of the second burn. This gentler acceleration is one of the main advantages of staging for human spaceflight.

Note that had we used the conventional assumption $k_{s,2} = k_{s,1} = 0.10$ in determining the mass of the second stack we would calculate a value of 67,660 kg. The mass of the entire vehicle would have been only 185,700 kg, a value about 3.3% less than that calculated previously using the perhaps more appropriate values. The thrust and propellant mass required is decreased proportionately. The important technical factor to keep in mind here is that the uncertainty in the magnitude of the structural weight fraction has a significant impact on the total vehicle mass.

7.5.4 THREE-STAGE VEHICLE TO ORBIT

The same process may be used to account for a vehicle with three stages. For simplicity, we again assume all stages have $(F/W_0)' = 1.3$ and $I_{sp,I} = 450$ s. Assuming again that the first stage burns for 200 s results in a velocity at first stage burn-out of $V_{bo,1} = 2.825$ km/s. We shall again assume that $k_{s,1} = 0.062$ and that $k_{s,2} = k_{s,3} = 0.12$. The fuel fraction burned is as it was in the two-stage case given in Eqn (7.79), i.e., $m_{p,1}/m_0' = 0.5778$ and from Eqn (7.68)

$$m_0' = \frac{m_{pay}}{[1 - (1.062)(0.002889 \text{ s}^{-1})(200 \text{ s})][1 - 1.12k_{p,2}\Delta t_2][1 + 1.12k_{p,3}\Delta t_3]} \qquad (7.83)$$

When the first stage drops off and the second stage rocket fires, the velocity is given approximately by

$$V_2(t) = 2.825 \text{ km/s} - 4.413 \ln\left[1 - (0.002889 \text{ s}^{-1})(t - 200 \text{ s})\right]$$
$$- \frac{0.009807 \text{ km/s}^2}{2}(t - 200 \text{ s}) \qquad (7.84)$$

Assuming the burn-out time for the second stage to be $t_{bo,2} = 345$ s, Eqn (7.70) yields $V_{bo,2} = 4.504$ km/s and

$$\frac{(m_{p,2})_{bo}}{m_0^{II}} = (0.002889 \text{ s}^{-1})(t_{bo,2} - 200 \text{ s}) = 0.4189 \qquad (7.85)$$

Now the third stage fires and accelerates the vehicle to the circular velocity, $V_c = 7.909$ km/s. One may determine the burn-out time by using the velocity relation

$$V_3 = 4.504 \text{ km/s} - 4.413 \log_e[1 - (0.002889 \text{ s}^{-1})(t - 430 \text{ s})]$$
$$- \frac{0.009807 \text{ km/s}^2}{2}(t - 345 \text{ s}) \qquad (7.86)$$

The resulting time is $t_{bo,3} = 566$ s and $m_{p,3}/m_0^{III} = 0.002889(566 - 345) = 0.6385$. Equation (7.83) yields

$$m_0^I = \frac{10,680 \text{ kg}}{[0.3863][1 - 1.12(0.002889 \text{ s}^{-1})(145 \text{ s})][1 - 1.12(0.002889 \text{ s}^{-1})(221 \text{ s})]}$$

$$m_0^I = 182,800 \text{ kg}$$

Using Eqns (7.65) and (7.66), we find the mass of the third stack is $m_0^{III} = 37,490$ kg and that of the second stack $m_0^{II} = 70,620$ kg. The mass of the individual stages are as follows:

$$m_1 = m_0^{III} - m_{pay} = 37,490 \text{ kg} - 10,680 \text{ kg} = 26,810 \text{ kg}$$
$$m_2 = m_0^{II} - m_0^{III} = 70,620 \text{ kg} - 37,480 \text{ kg} = 33,140 \text{ kg}$$
$$m_1 = m_0^I - m_0^{II} = 182,800 \text{ kg} - 70,620 \text{ kg} = 112,180 \text{ kg}$$

From the mass of each stack and the assumed thrust to weight ratio of each stage is $(F/W_0)^J = 1.3$ we determine that $F^I = 524,000$ lbs, $F^{II} = 202,700$ lbs, and $F^{III} = 107,600$ lbs. As shown in Table 7.2, the requirement for the first stack can easily be accommodated by three J-2 engines, the second stack by one J-2 but that for the third stack is not readily met by existing LOX−LH$_2$ engines. The third stage could be powered by 2 Pratt & Whitney RL-60 engines (currently under development) providing a total of about 100,000−130,000 lbs of thrust.

Note that had we used the conventional assumption $k_{s,3} = k_{s,2} = k_{s,1} = 0.10$ in determining the mass of the third stack would be $m_0^{III} = 35,880$ kg and that of the second stack would be $m_0^{II} = 66,530$ kg. The mass of the entire vehicle would be $m_0^I = 182,600$ kg, a value very close to that calculated previously using the somewhat more appropriate values.

The velocity histories for the three different configurations are shown in Figure 7.24 and the acceleration histories in Figure 7.25 for one-, two-, and three-stage launch vehicles with $(F/W_0) = 1.3$ and $I_{sp} = 450$ s for all stages. Using a staged launch vehicle provides control over the

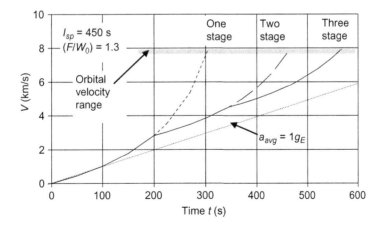

FIGURE 7.24

Velocity histories for one-, two-, and three-stage rockets with constant specific impulse of 450 s and a payload weight of 10,680 kg. The range of orbital velocities is shown by the shaded region and the line of constant average acceleration equal to $1g_E$ is also indicated.

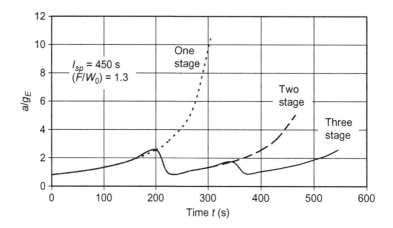

FIGURE 7.25

Acceleration histories for one-, two-, and three-stage rockets with constant specific impulse of 450 s and a payload weight of 10,680 kg.

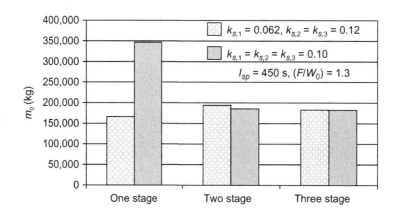

FIGURE 7.26

Comparison of launch weights for one-, two-, and three-stage rockets with constant values of $(F/W_0) = 1.3$ and $I_{sp} = 450$ s and a payload mass of 10,680 kg but different values of the structural coefficient $k_{s,i}$.

acceleration levels, an important factor for manned spacecraft. The advantage of three stages in terms of acceleration is clear in Figure 7.25 where we see that the acceleration remains below $3g_E$ for the entire flight.

Figure 7.26 compares the initial mass m_0 of the launch vehicles whose velocity and acceleration histories were compared in the previous figures. Two cases for the structural parameter $k_{s,i}$ are illustrated. In one case, the first stage structural parameter $k_{s,1} = 0.062$ while the second and third stages have $k_{s,2} = k_{s,3} = 0.12$. These values appear to be reasonably consistent with current practice. The second case sets $k_{s,1} = k_{s,2} = k_{s,3} = 0.10$ and represents a common simplification used for preliminary studies. Only the single stage mass is dramatically affected because the two choices for $k_{s,1}$ are quite different. However, the mass of two- and three-stage vehicles are seen to be not nearly so sensitive. The reduction in launch mass achieved by using three stages instead of two is about 6%. On the other hand, the difference in mass of a multiple stage vehicle compared to a one-stage vehicle depends entirely on the choice of k_s for the first stage. In the case shown in Figure 7.26, using $k_{s,1} = 0.62$ results in a single stage vehicle mass about 90% that of the multiple stage vehicle, while using $k_{s,1} = 0.10$ results in a single stage vehicle with a mass about twice that of a multiple stage vehicle.

One may change the burn-out times to alter the velocity and acceleration characteristics of the launch vehicle. If in the cases treated previously we increase $t_{bo,1}$ to 225 s from 200 s and $t_{bo,2}$ to 400 s from 345 s we find relatively small changes in the overall characteristics as shown in Table 7.4. The velocity history shown in Figure 7.27 exhibits slightly higher final velocities for the first and second stages because of the longer burns in each and the orbital velocity is reached in 560 s rather than 566 s. The acceleration history in Figure 7.28 shows somewhat higher accelerations throughout the flight as compared to the previous cases with earlier burn-out times. The launch masses for the different stage combinations are depicted in Figure 7.29. Because the one-stage launch vehicle burns out when orbital velocity is reached its initial mass is the same as in the previous cases. However, the two-stage vehicles with $t_{bo,1} = 225$ s have slightly reduced masses compared those with $t_{bo,1} = 200$ s studied previously: 9.3% less for the cases with $k_{s,1} = 0.062$ and $k_{s,2} = k_{s,3} = 0.12$ and 3.3% less for the cases with $k_{s,1} = k_{s,2} = k_{s,3} = 0.10$. For three-stage vehicles, the increase in burn-out times of the first and second stages produces the opposite result: a 2% increase in initial mass for the cases with $k_{s,1} = 0.062$ and $k_{s,2} = k_{s,3} = 0.12$ and a 0.3% increase initial mass for the cases with $k_{s,1} = k_{s,2} = k_{s,3} = 0.10$.

Having assessed the effects of burn-out time on staged launch vehicle performance, we turn to the effect of propellants. We may compare the performance of launchers using either of two common liquid propellant combinations, LH_2–LOX or RP1–LOX. We will again assume a constant ratio of thrust to initial weight, $(F/W)_0 = 1.3$, and a constant specific impulse for all stages with $I_{sp} = 450$ s for the LH_2–LOX combination and $I_{sp} = 300$ s for the RP1–LOX combination. The methodology is exactly the same as in all the previous cases, therefore we will proceed directly to the results.

The velocity and acceleration history for launchers using either of the two propellants is shown for one-, two-, and three-stage vehicles in Figures 7.30–7.32, respectively. Figure 7.30 shows that the acceleration is so high in the one-stage launcher that the time to orbit is only about 220 s for the RP1–LOX launcher and, as before, about 305 s for the LH_2–LOX launcher. Clearly the acceleration levels are in excess of that suitable for manned spacecraft for the one-stage launcher. For the two-stage launchers shown in Figure 7.31 the acceleration levels are still quite high, particularly for the RP1–LOX vehicle, although it is clear that the velocity history is being smoothed out by the staging process. In the case of three-stage launch vehicles shown in Figure 7.32, the acceleration levels appear to be acceptable now that the time to orbit is around 500 or more seconds. Indeed, as the number of stages increases the velocity and acceleration histories for the two propellant combinations are increasingly similar.

However, there are substantial differences in mass between launchers using different propellant combinations. First, it is impossible to determine initial mass values for orbit-capable one- and two-stage RP1–LOX vehicles if practical values for the structural coefficients $k_{s,I}$ are used.

Table 7.4 Mass and Thrust Characteristics for Launch Vehicles with Equal I_{sp} and $(F/W)_0$ but Varying Number of Stacks, Burn-Out Times, and Structural Mass Coefficients

Case Considered	Number of Stages	Stack Number	m_0 (kg)	F (lbs)	m_p (kg)	t_{bo} (s)
$(F/W)_0 = 1.3$	1	1	166,200	477,000	146,400	305
$I_{sp} = 450$ s						
$k_1 = 0.062$	2	1	193,900	556,400	112,100	200
$k_2 = 0.12$		2	74,930	215,000	57,370	465
$k_3 = 0.12$						
	3	1	182,800	524,600	105,600	200
		2	70,620	202,700	29,580	345
		3	37,490	107,600	23,940	566
$(F/W)_0 = 1.3$	1	1	346,900	995,500	305,700	305
$I_{sp} = 450$ s						
$k_1 = 0.10$	2	1	185,700	532,900	107,300	200
$k_2 = 0.10$		2	67,660	194,200	51,800	465
$k_3 = 0.10$						
	3	1	182,600	524,000	105,500	200
		2	66,540	191,000	27,870	345
		3	35,880	103,000	22,910	566
$(F/W)_0 = 1.3$	1	1	166,200	477,000	146,400	305
$I_{sp} = 450$ s						
$k_1 = 0.062$	2	1	177,400	509,000	115,300	225
$k_2 = 0.12$		2	54,970	157,800	39,550	474
$k_3 = 0.12$						
	3	1	186,500	535,100	121,200	225
		2	57,750	165,700	34,200	430
		3	18,440	52,900	11,980	560
$(F/W)_0 = 1.3$	1	1	346,900	995,500	305,700	305
$I_{sp} = 450$ s						
$k_1 = 0.10$	2	1	179,600	515,400	116,700	225
$k_2 = 0.10$		2	51,180	146,900	36,820	474
$k_3 = 0.10$						
	3	1	183,100	525,500	119,000	225
		2	52,190	149,800	30,910	430
		3	18,190	52,200	11,380	560

However, as shown in Figure 7.33, we are able to find initial mass values of a three-stage RP1−LOX launcher using reasonable values for the $k_{s,i}$. This discrepancy arises because the propellant consumption efficiency of the RP1−LOX engine is only three-quarters that of the LH$_2$−LOX engine. One result of this disparity is that the mass of the RP1−LOX powered vehicle is more than three times that of the LH$_2$−LOX powered vehicle. This is a major reason for the interest in high specific impulse engines.

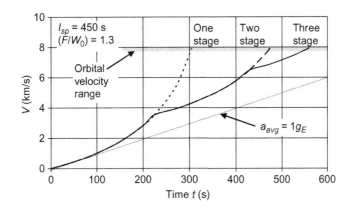

FIGURE 7.27

Velocity histories for one-, two-, and three-stage rockets with constant specific impulse of 450 s and a payload weight of 10,680 kg. The burn-out times are later than those shown in Figure 7.24.

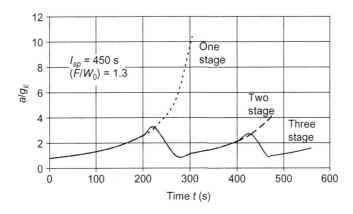

FIGURE 7.28

Acceleration histories for one-, two-, and three-stage rockets with constant specific impulse of 450 s and a payload weight of 10,680 kg. The burn-out times are later than those shown in Figure 7.24.

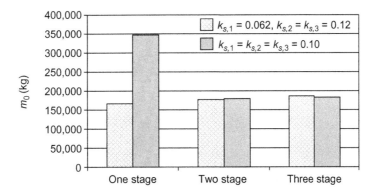

FIGURE 7.29

Comparison of launch mass for one-, two-, and three-stage rockets with constant values of $(F/W_0) = 1.3$ and $I_{sp} = 450$ s and a payload mass of 10,680 kg but different values of the structural coefficient $k_{s,i}$. The burn-out times are later than those shown in Figure 7.24.

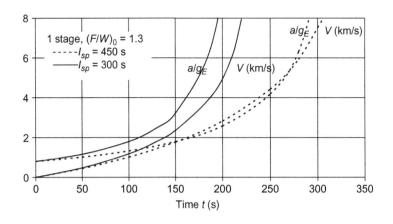

FIGURE 7.30

Velocity and acceleration history for one-stage launchers using either LH_2–LOX or RP1–LOX as propellants.

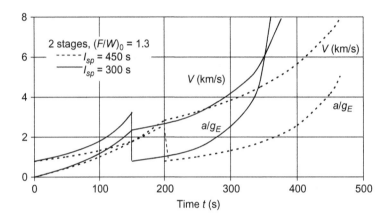

FIGURE 7.31

Velocity and acceleration history for two-stage launchers using either LH_2–LOX or RP1–LOX as propellants.

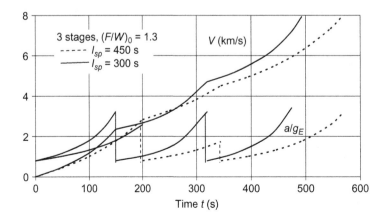

FIGURE 7.32

Velocity and acceleration history for three-stage launchers using either LH_2–LOX or RP1–LOX as propellants.

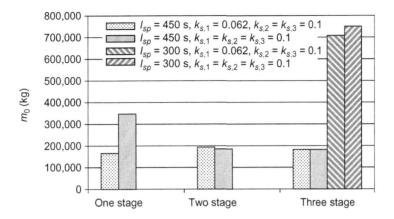

FIGURE 7.33

Comparison of launch mass for one-, two-, and three-stage rockets with constant values of $(F/W)_0 = 1.3$ and a payload mass of 10,680 kg but different constant values of I_{sp} and the structural coefficient $k_{s,i}$.

Another factor to consider is the burn-out time to be selected for each stage. Of course, for a one-stage launcher the burn-out time cannot be arbitrarily chosen but instead must be that at which orbital velocity is achieved. However for a multi-stage launcher the burn-out time for any stage but the last may be set by the mass of propellant carried. The last stage must burn as long as is necessary to achieve orbital velocity. We have been selecting burn-out times of the intermediate stages so as to avoid excessive acceleration. It is clear from Eqn (7.73) that for a given set of performance parameters the maximum acceleration occurs at the burn-out time. Similarly, the average acceleration is

$$a_{avg} = \int_{t_{ig,i}}^{t_{bo,i}} \dot{V}_i dt = \frac{V_i(t_{bo,i}) - V(t_{ig,i})}{\Delta t_i}$$

The normalized average acceleration over the total burn time Δt may be written as

$$\frac{a_{avg}}{g_E} = \frac{V_E}{g_E \Delta t} = \frac{807}{\Delta t}$$

As may be seen from Figures 7.24 and 7.25, for example, the average acceleration lies in the range $1 < a_{avg}/g_E < 2$ so that total burn times are in the range of 450–750 s. One may carry out a series of calculations of the initial mass of the launch vehicle as a function of the average acceleration for various values of individual stage burn times and structural coefficients $k_{s,i}$. Results for two- and three-stage rockets with identical values for stage thrust to weight ratio $(F/W_0) = 1.3$ and specific impulse $I_{sp} = 450$ s, representative of LH_2–LOX powered engines, appear in Figure 7.34; similar results but with specific impulse $I_{sp} = 300$ s, representative of RP1–LOX powered engines are shown in Figure 7.35.

Note that the two-stage rockets with $k_{s,1} = k_{s,2} = k_{s,3} = 0.1$ achieve minimum initial mass at the minimum average acceleration while those with $k_{s,1} = 0.062$ and $k_{s,2} = k_{s,3} = 0.12$ show a continually decreasing initial mass as the acceleration increases from its minimum value. At the minimum

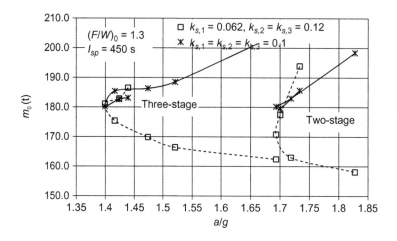

FIGURE 7.34

Initial mass in metric tons (1 t = 1000 kg) of two- and three-stage rockets with m_{pay} = 10.68 t as a function of average overall acceleration for different values of the structural coefficients $k_{s,i}$. All stages have $(F/W)_0$ = 1.3 and I_{sp} = 450 s.

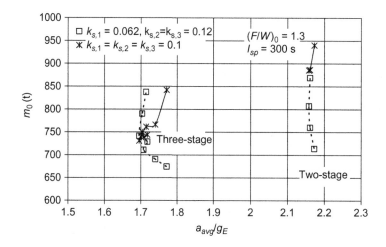

FIGURE 7.35

Initial mass in metric tons (1 t = 1000 kg) of two- and three-stage rockets with m_{pay} = 10.68 t as a function of average overall acceleration for different values of the structural coefficients $k_{s,i}$. All stages have $(F/W)_0$ = 1.3 and I_{sp} = 300 s.

average acceleration, the two-stage rockets with $k_{s,1}$ = 0.062 and $k_{s,2}$ = $k_{s,3}$ = 0.12 show an initial mass 6% less than those with $k_{s,1}$ = $k_{s,2}$ = $k_{s,3}$ = 0.1. A similar behavior is shown by three-stage rockets except that at the minimum average acceleration those with $k_{s,1}$ = 0.062 and $k_{s,2}$ = $k_{s,3}$ = 0.12 show about the same initial mass as those with $k_{s,1}$ = $k_{s,2}$ = $k_{s,3}$ = 0.1.

Table 7.5 Breakdown of Weights for Four Practical Systems

	Space Shuttle (LH$_2$/LOX and Solid Rocket Booster)	Soyuz (RP1/LOX)	Apollo–Saturn V (RP1/LOX and LH$_2$/LOX)	Gemini–Titan II (UDMH/N$_2$O$_4$)	Mercury–Atlas (LOX/RP1)
m_{pay} (t)	134[a]	6.80	42.9	3.85	1.40
$m_{str+eng}$ (t)	273	13.0	235	6.62	5.39
$m_{pay}/m_{str+eng}$	0.492	0.523	0.182	0.582	0.259
Stages	1.5	2	3	2	1.5

[a]*Includes Orbiter vehicle.*

Thus with equal structural coefficients for each identically powered stage, it is possible to make a good approximation to the minimum initial mass by designing for minimum average acceleration. That is, design for maximum total burn-out time for achieving orbital velocities which, in turn, means equal burn time for each stage. When the first stage structural coefficient $k_{s,1}$ is substantially less than that for the other stages, it is acceptable to design for minimum average acceleration. Such an approach will lead to equal or lower initial mass compared to the case where the $k_{s,i}$ are all equal. Then, from that point one may consider designing for a somewhat larger average acceleration so as to reap the reward of further reduced initial mass. Structural and engine mass characteristics of some operational vehicles are shown in Table 7.4. Note that the payload mass tends to be about 20–50% of the launch vehicle structural and engine mass. More detailed information about structural and engine mass characteristics appears in the following section (Table 7.5).

7.5.5 SOME COMMENTS ON ENGINE AND STRUCTURE WEIGHT

The payload mass is part of the mission specification, but the engine weight (in lbs) is reasonably well-correlated to the engine thrust (in lbs) as follows:

$$W_{eng} = a_1 + a_2(F_{vac})^{a_3}$$

It is often convenient to work with the thrust to weight ratio so that the mass fraction of the engines may be written as follows:

$$\frac{m_{eng}}{m_0} = \frac{a_1 + a_2(F_{vac})^{a_3}}{W_0} = \frac{a_1}{m_0 g_E} + \frac{a_2}{(m_0 g_E)^{1-a_3}}\left[\left(\frac{F_{vac}}{W}\right)_0\right]^{a_3} \tag{7.87}$$

Data from eight LH$_2$–LOX engines, nine RP1–LOX engines, one UMDH–N$_2$O$_4$ engine, and one CH4–LOX engine were taken from various sources and the engine weight as a function of vacuum thrust is shown in Figures 7.36 and 7.37. The coefficients in Eqn (7.87) are listed in Table 7.6. Thrust force is given here in pounds and therefore the weight in pounds is also used rather than mass. Note that the correlation given in Eqn (7.87) represents the LH$_2$–LOX engine data in Figure 7.36 to within about $\pm 10\%$ except for one case, the R-68, a relatively new engine first flown in 2002. This engine embodies a simplified design philosophy which reduces the parts count and results in lowered development and production costs. Consequently it is heavier relative to previous engines for which weight reduction was a major factor. Similarly, Eqn (7.87) represents

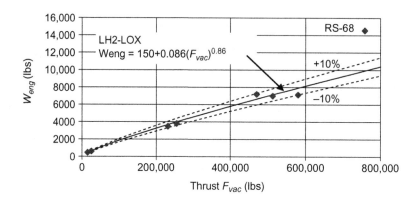

FIGURE 7.36

Engine weight as a function of vacuum thrust for eight LH_2-LOX engines. The LH_2-LOX correlation underestimates the 14,650 lb weight of the RS-68 engine (designed for reliability and cost rather than weight) by 32%.

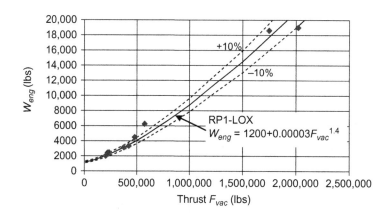

FIGURE 7.37

Engine weight as a function of vacuum thrust for nine LH_2-LOX engines. The $RP1-LOX$ correlation underestimates the 6250 lb weight of the CH_4-LOX engine by 25% and overestimates the 1570 lb weight of the N_2O_4-UMDH engine by 15%.

the $RP1-LOX$ engine data in Figure 7.37 to within about $\pm 10\%$ except for two cases which uses hydrocarbon fuels other than RP1. The weight of the STBE, an engine which uses liquid methane CH_4 rather than the more complex hydrocarbon fuel RP1 (essentially kerosene) as the fuel is under-estimated by about 25%. The weight of the YF-20B engine, which uses unsymmetrical dimethyl hydrazine (UDMH) as the fuel and nitrogen tetroxide (N_2O_4) as the oxidizer is overestimated by about 15%. This engine is used in China's Long March vehicle shown in Figure 7.5(c).

Table 7.6 Coefficients for Engine Weight (in lbs) Correlation

Propellant	a_1	a_2	a_3
LH$_2$–LOX	150	0.086	0.86
RP1–LOX	1200	0.00003	1.4

The structural mass of the vehicle, on the other hand, is much more difficult to estimate, but it should depend on the mass of fuel carried. The ratio of structural mass, including engines, to propellant mass as a function of propellant mass for 26 different stages of launch vehicles was determined from various sources, including Isakowitz (1995) and SSRM (1988) and the results are tabulated in Table 7.7. The ratio of structural mass to propellant mass is shown as a function of propellant mass in Figure 7.38. The trends of the data in Figure 7.38 are illustrated by means of a correlation based on propellant mass (in kg) given by

$$\frac{m_s}{m_p} = \frac{W_s}{W_p} = \frac{k_s}{m_p^{0.15}} \tag{7.88}$$

The data for 34 rocket stages with different propellant combinations and with propellant mass varying from 6000 kg to over 2,000,000 kg shows an appreciable degree of scatter. On the other hand, the range of the ratio of structural mass (including engines) to propellant mass for first and booster stages generally lies below that for second and upper stages. Figure 7.38 shows that there is reasonable differentiation between first/booster stages and second/upper stages. There are two first stage data points which fall out of range on the high side while the remainder falls in a broad swath which shows a distinct reduction in k_s as the propellant mass increases. For simplicity, we will use a constant value for the ratio $m_s/m_p = k_s$ to permit bracketing the range of values realized in practice as well as to aid in illustrating the importance of this parameter. In preliminary design discussions, it is common practice to assume $k_s = m_s/m_p = 0.10$ for any stage whereas the practical evidence suggests lower values for lower stages and higher values for higher stages. We have shown that as the number of stages increases the importance of the detailed choice of k_s diminishes. This should be expected because $k_s = 0.10$ would overestimate the mass of the lower stages and underestimate the weight of the upper stages and these difference tend to cancel one another as far as the overall mass of the launch vehicle is concerned. Indeed, $k_s = 0.10$ is the average value for the 34 liquid propellant stages studied. However, $k_s = 0.077$ is the average for the 16 liquid propellant booster/first stages studied, while $k_s = 0.115$ is the average for the 13 liquid propellant second stages studied and $k_s = 0.11$ is the average for the five third stages considered.

7.6 LONGITUDINAL STABILITY OF LAUNCH VEHICLES

An idealized launch vehicle is shown in Figure 7.39 in which the important angles are identified: the flight path angle γ measured between the vehicle velocity vector and the local horizon, the pitch angle θ measured between the launcher axis and the local horizon, and the angle of attack α

Table 7.7 Ascent Vehicle Stages and the Associated Propellant and Structural Mass Including Engines

Vehicle	Stage	Propellant	$k_s = m_s/m_p$	m_p (kg)	m_s (kg)
L.March 2F	Booster	N_2O_4-UMDH	0.0794	37,800	3000
Soyuz	First	RP1−LOX	0.0965	39,200	3784
Delta 7925	First	RP1−LOX	0.0591	96,120	5680
H-II A	First	LH_2−LOX	0.1277	101,000	12,900
Atlas E	Booster	RP1−LOX	0.0720	112,700	8110
Titan II	First	N_2O_4−UDH	0.0569	117,800	6700
Titan IV	First	UDMH−N_2O_4	0.0516	155,000	8000
Atlas IIA	First	RP1−LOX	0.0658	156,260	10,282
Ariane-5	First	LH_2−LOX	0.0772	158,000	12,200
L.March 2F	First	N_2O_4-UMDH	0.0718	174,000	7851
H-II B	First	LH_2−LOX	0.1361	177,800	24,200
Ariane	First	UH25−N_2O_4	0.0646	226,000	14,600
Atlas V	First	RP1−LOX	0.0741	284,089	21,054
Proton	First	N_2O_4−UDH	0.0742	419,400	31,100
Shuttle E.T.	First[a]	LH_2−LOX	0.0493	718,000	34,880
Saturn S-IC	First	RP1−LOX	0.0621	2,145,700	133,270
Delta 7925	Second	N_2O_4−UDH	0.1583	6000	950
Ariane-5	Second	LH_2−LOX	0.1227	9700	1190
H-IIB	Second	LH_2−LOX	0.2048	16,600	3400
H-IIA	Second	LH_2−LOX	0.1834	16,900	3100
Atlas IIA	Second[b]	LH_2−LOX	0.1087	16,930	1840
Atlas V	Second[b]	LH_2−LOX	0.1077	20,830	2243
Titan II	Second	N_2O_4−UDH	0.0830	28,900	2400
Ariane	Second	UH25−N_2O_4	0.0966	34,000	3,285
Titan IV	Second	UDMH−N_2O_4	0.1286	35,000	4,500
L.March 2F	Second	N_2O_4−UDH	0.0640	86,000	5,500
Soyuz	Second	RP1−LOX	0.0721	95,400	6,875
Proton	Second	N_2O_4−UDH	0.0751	156,000	11,720
Saturn S-II	Second	LH_2−LOX	0.0712	456,100	32,500
Ariane	Third	LH_2−LOX	0.1121	10,700	1,200
Titan IV	Second[b]	LH_2−LOX	0.1315	21,100	2,775
Soyuz	Third	RP1−LOX	0.1070	22,000	2,355
Proton	Third	N_2O_4−UDH	0.0898	46,600	4,185
Saturn IVB	Third	LH_2−LOX	0.1097	106,300	11,660

[a]Shuttle ET has no engines and had three versions: standard weight tank (1981−83) = 34,880 kg, lightweight tank (1983−98) = 29,900 kg, and the super lightweight tank (post-1998) = 26,500 kg.
[b]Centaur upper stage.

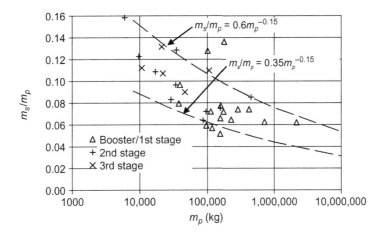

FIGURE 7.38

The ratio of structural mass (including engines) to propellant mass as a function of propellant mass for 34 different stages.

measured between the relative velocity of the vehicle and the vehicle axis. The flight path angle describes the motion of the center of gravity of the vehicle with respect to the local horizon while the pitch angle describes the rigid body motion of the vehicle about its center of gravity. Finally, the angle of attack describes the presentation of the vehicle to the relative wind and thereby determines the aerodynamic forces experienced by the vehicle.

The idealized launch vehicle shown in Figure 7.39 can deflect the thrust vector through an angle δ in order to stabilize and control the vehicle in pitch. In practice this is generally accomplished either by swiveling the entire engine about a top-mounted universal joint called a gimbal or by deflecting the nozzle alone about joints having flexible seals. The former is common on large liquid propellant engines because the combustion chamber is relatively small and flexible propellant lines may be used to connect to the propellant tanks. Because in solid propellant rockets, the propellant and nozzle form an integral unit which is necessary to deflect the nozzle alone to achieve TVC. Other methods which are sometimes used rely upon disturbing the flow in the nozzle either mechanically, by using guidance vanes, or by asymmetrical injection of fluid into the nozzle.

The thrust F is assumed to act through the engine gimbal or nozzle deflection axis while the aerodynamic force acts through the center of pressure CP. The aerodynamic force is defined by the components A and N, denoting the axial and normal components, respectively. The definition of the center of pressure is that point at which the aerodynamic force acts with the moment about that point equal to zero. Taking moments about the center of gravity CG leads to the following result:

$$I\ddot{\theta} = \sum M_{CG} = F(l - x_{CG})\sin\delta + N(x_{CG} - x_{CP}) \tag{7.89}$$

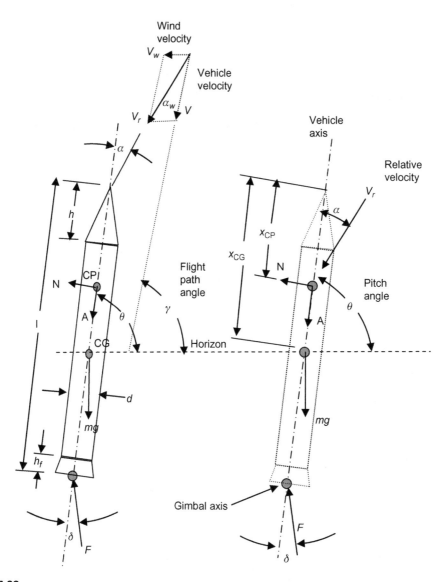

FIGURE 7.39

Schematic diagram of an idealized launcher in the pitch plane showing thrust deflected through an angle δ and a general horizontal wind component V_w.

The normal and axial forces are typically defined in terms of their variation with angle of attack as follows:

$$N = \left(\frac{\partial C_N}{\partial \alpha} \right) \alpha qS = C_{N,\alpha} \alpha qS$$

$$A = \left(\frac{\partial C_A}{\partial \alpha} \right) \alpha qS = C_{A,\alpha} \alpha qS$$

For generality the launch vehicle shown in Figure 7.39 is depicted with a slender flare at the base. Note that the dynamic pressure q is based on the relative velocity V_r and the reference area S for force and moment coefficients may be taken either as the nominal cross-sectional area of the first stage or of the base itself area so care must be exercised in applications. The rate of change of the pitching velocity given in Eqn (7.89) may be written as

$$\ddot{\theta} = \frac{F(l - x_{CG})}{I} \sin \delta + \frac{C_{N,\alpha} qS(x_{CG} - x_{CP})}{I} \alpha \tag{7.90}$$

7.6.1 MOMENTS OF INERTIA OF LAUNCH VEHICLES

Consider the idealized three-stage launch vehicle shown in Figure 7.40. It is assumed to be comprised of cylindrical first and second stages of radius $R_{c1} = R_{c2} = R$, an interstage section in the shape of the frustum of a cone with smaller radius r_f, a cylindrical third stage with radius $R_{c3} = r_f$, and a conical payload section with base radius $r_{co} = r_f$. The length of each segment is indicated in Figure 7.40 and the overall length, measured from the line passing through the gimbal axes of the engines, is denoted by l.

The center of gravity of the idealized launch vehicle may be determined by taking moments about the baseline, that is, the line through the nozzle gimbal axes which leads to

$$m(l - x_{CG}) = m_{co}h_{co} + m_{c3}h_{c3} + m_f h_f + m_{c2}h_{c2} + m_{c1}h_{c1} \tag{7.91}$$

If, for this idealized case, we assume each segment is a solid of uniform density then we may easily find the location of its center of gravity and define the individual distances from the gimbal axis baseline as follows:

$$h_{c1} = \frac{l_{c1}}{2}$$

$$h_{c2} = l_{c1} + \frac{l_{c2}}{2}$$

$$h_f = l_{c1} + l_{c2} + \frac{l_f}{4} \left[\frac{R^2 + 2Rr_f + 3r_f^2}{R^2 + Rr_f + r_f^2} \right]$$

$$h_{c3} = l_{c1} + l_{c2} + l_f + \frac{l_{c3}}{2}$$

$$h_{co} = l_{c1} + l_{c2} + l_f + l_{c3} + \frac{l_{co}}{4}$$

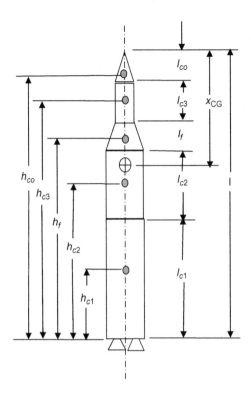

FIGURE 7.40

Schematic diagram of a generic launch vehicle showing center of gravity locations for each segment of the launcher as well as the center of gravity for the complete vehicle.

The moment of inertia of the individual segments about a horizontal axis through their centers of gravity are as follows:

$$I_{co} = \frac{3m_{co}l_{co}^2}{80}\left[1 + 4\left(\frac{r_{co}}{l_{co}}\right)^2\right] \tag{7.92}$$

$$I_{c,i} = \frac{m_{ci}l_{ci}^2}{12}\left[1 + 3\left(\frac{R_{ci}}{l_{ci}}\right)^2\right] \tag{7.93}$$

$$I_f = \frac{3m_f(R^5 - r_f^5)}{10(R^3 - r_f^3)} - \frac{m_f l_f^2}{16}\left[\frac{R^2 + 2Rr_f + 3r_f^2}{R^2 + Rr_f + r_f^2}\right]^2 \tag{7.94}$$

Table 7.8 gives values of the ratios (B/A), (D/A), and (F/A) defined in Eqn (7.95) as a function of the ratio of the frustum radius to the base radius r_f/R. Figure 7.41 illustrates the behavior of these ratios as one proceeds from a right circular cone $(r_f = 0)$ through cone frustums up to a right circular cylinder $(r_f = R)$.

Table 7.8 Coefficients for Moments of Inertia and Center of Gravity of Uniform Solid Cylinders, Cones, and Cone Frustums

r_f/R	B/A	D/A	F/A
1.00	2.000	1.667	3.333
0.95	1.966	1.586	3.248
0.90	1.930	1.511	3.159
0.85	1.892	1.442	3.065
0.80	1.852	1.378	2.967
0.75	1.811	1.319	2.865
0.70	1.767	1.266	2.758
0.65	1.721	1.219	2.647
0.60	1.673	1.176	2.531
0.55	1.623	1.139	2.410
0.50	1.571	1.107	2.286
0.45	1.517	1.080	2.157
0.40	1.462	1.057	2.026
0.35	1.404	1.039	1.891
0.30	1.345	1.025	1.755
0.25	1.286	1.015	1.619
0.20	1.226	1.008	1.484
0.15	1.166	1.003	1.352
0.10	1.108	1.001	1.225
0.05	1.052	1.000	1.107
0	1.000	1.000	1.000

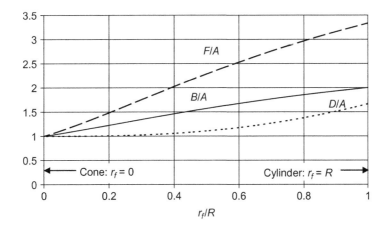

FIGURE 7.41

Variation of the ratios (B/A), (D/A), and (F/A) as a function of the ratio of the frustum radius to the base radius r_f/R proceeding from a right circular cone ($r_f = 0$) through cone frustums up to a right circular cylinder ($r_f = R$).

$$A = 1 + \left(\frac{r_f}{R}\right) + \left(\frac{r_f}{R}\right)^2$$

$$B = 1 + 2\left(\frac{r_f}{R}\right) + 3\left(\frac{r_f}{R}\right)^2$$

$$D = 1 + \left(\frac{r_f}{R}\right) + \left(\frac{r_f}{R}\right)^2 + \left(\frac{r_f}{R}\right)^3 + \left(\frac{r_f}{R}\right)^4 \tag{7.95}$$

$$F = 1 + 3\left(\frac{r_f}{R}\right) + 6\left(\frac{r_f}{R}\right)^2$$

Moments of inertia for cylinders, right circular cones and frustums of such cones with height l, base radius R, and frustum radius r_f about an axis parallel to their bases and passing through their centers of gravity are given by

$$I = \frac{l^2}{10}\left[\left(\frac{F}{A}\right) - \frac{5}{8}\left(\frac{B}{A}\right)^2\right] + \frac{3R^2}{20}\left(\frac{D}{A}\right) \tag{7.96}$$

The centers of gravity are located a distance x_{CG} from their bases given by

$$h_{CG} = \frac{l}{4}\left(\frac{B}{A}\right) \tag{7.97}$$

Approximations for these ratios accurate to within about 3.5% for all r_f/R are as follows:

$$\frac{B}{A} \simeq 1 + \left(\frac{r_f}{R}\right)^{0.87}$$

$$\frac{D}{A} \simeq 1 + \frac{2}{3}\left(\frac{r_f}{R}\right)^{2.5}$$

$$\frac{F}{A} \simeq 1 + \frac{7}{3}\left(\frac{r_f}{R}\right)^{0.94}$$

If the shape of the payload segment is more accurately described by a paraboloid of revolution than a cone, the moment of inertia I_p and center of gravity location $h_{p,CG}$, measured from the base, are as follows:

$$I_p = \frac{m}{18}\left(3R^2 + l_p^2\right)$$

$$h_{p,CG} = \frac{l_p}{3} \tag{7.98}$$

A compilation of mass and area properties of a wide variety of bodies of engineering interest is presented, for example, by Myers (1962).

The moment of inertia of a body of mass m about an axis parallel to one passing through the center of gravity of the body and separated from it by a distance d may be found by using the parallel axis theorem according to the following relation:

$$I = I_{CG} + md^2 \tag{7.99}$$

Then the moment of inertia of the launch vehicle depicted in Figure 7.40 may be built up as a sum of the moments of inertia of the individual elements about the center of gravity of the launch vehicle. Then for each launch stack we have

$$
\begin{aligned}
I_{CG}^{I} = {} & I_{c1} + I_{c2} + I_f + I_{c3} + I_{co} + m_{c1}(l - h_{c1} - x_{CG}^{I})^2 \\
& + m_{c2}(l - h_{c2} - x_{CG}^{I})^2 + m_f(l - h_f - x_{CG}^{I})^2 + m_{c3}(l - h_{c3} - x_{CG})^2 + m_{co}(l - h_{co} - x_{CG}^{I})^2
\end{aligned}
\tag{7.100}
$$

$$
\begin{aligned}
I_{CG}^{II} = {} & I_{c2} + I_f + I_{c3} + I_{co} + m_{c2}(l - h_{c2} - x_{CG}^{II}) \\
& + m_f(l - h_f - x_{CG}^{II}) + m_{c3}(l - h_{c3} - x_{CG}^{II}) + m_{co}(l - h_{co} - x_{CG}^{II})^2
\end{aligned}
\tag{7.101}
$$

$$
I_{CG}^{III} = I_f + I_{c3} + I_{co} + m_f(l - h_f - x_{CG}^{III}) + m_{c3}(l - h_{c3} - x_{CG}^{III}) + m_{co}(l - h_{co} - x_{CG}^{III})^2
\tag{7.102}
$$

Here we have assumed that the frustum is part of the third stage. The center of gravity of each launch stack is given by

$$
x_{CG}^{I} = m^{I} l - [m_{co}h_{co} + m_{c3}h_{c3} + m_f h_f + m_{c2}h_{c2} + m_{c3}h_{c3}]
\tag{7.103}
$$

$$
\begin{aligned}
x_{CG}^{II} = {} & m^{II}(l_{c2} + l_f + l_{c3} + l_{co}) \\
& - [m_{co}(h_{co} - l_{c1}) + m_{c3}(h_{c3} - l_{c1}) + m_f(h_f - l_{c1}) + m_{c2}(h_{c2} - l_{c1})]
\end{aligned}
\tag{7.104}
$$

$$
\begin{aligned}
x_{CG}^{III} = {} & m^{III}(l_f + l_{c3} + l_{co}) \\
& - [m_{co}(h_{co} - l_{c1} - l_{c2}) + m_{c3}(h_{c3} - l_{c1} - l_2) + m_f(h_f - l_{c1} - l_{c2})]
\end{aligned}
\tag{7.105}
$$

The mass of each stack is given by

$$
\begin{aligned}
m^{I} &= m_{c_1} + m_{c2} + m_f + m_{c3} + m_{co} \\
m^{II} &= m_{c2} + m_f + m_{c3} + m_{co} \\
m^{III} &= m_f + m_{c3} + m_{co}
\end{aligned}
$$

The three stacks of the notional three-stage vehicle depicted in Figure 7.21 are shown schematically in Figure 7.42 and include the general layout of propellant tanks and engines. Note that the mass of each stack will decrease with time as the propellant contained within them is consumed. This change in mass causes the moment of inertia to also decrease with time, i.e., $I = I(t)$. In addition, the propellant distribution in the vehicle's tanks continually changes with time causing the center of gravity of the launch vehicle to also change with time, i.e., $x_{CG} = x_{CG}(t)$. In liquid propellant rockets, the level of both fuel and oxidizer is continually dropping while in solid rockets the propellant grain is continually being hollowed out as it is being consumed. In Eqn (7.90), we see that the pitch acceleration is directly proportional to the ratio x_{CG}/I so that in more detailed design phases it becomes important to account for center of gravity travel and moment of inertia changes.

7.6.2 FORCE AND MOMENT ESTIMATION FOR LAUNCH VEHICLES

Launch vehicles may be generally considered slender bodies with fineness ratios $l/d = O(10)$, as shown in the photographs of practical vehicles presented previously. Engineering analyses made use of this attribute of launch vehicles when developing estimation methods for the forces and moments experienced during flight. Early studies applied the ideas of potential flow theory for the evaluation of missile-like bodies with high fineness ratios. However, the accuracy of such an approach is limited to bodies at very small angles of attack. Allen and Perkins (1951) supposed that the flow over a slender body could be considered the superposition of the potential flow and a correction arising from the cross-flow experienced by the body. At angle of attack α, the body is

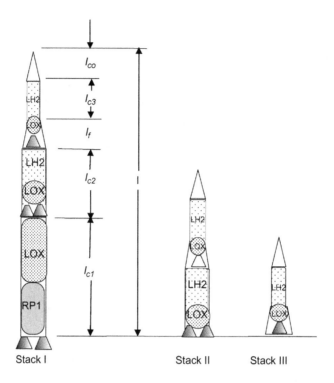

FIGURE 7.42

The three stacks of a notional three-stage vehicle similar to the Saturn V illustrating the general layout of propellant tanks and engines. The fineness ratios of the stacks are not to scale.

subjected to a component of the flow of magnitude $V_r \cos \alpha$ aligned with its axis and a component of magnitude $V_r \sin \alpha$ normal to its axis, where V_r is the velocity relative to the vehicle as illustrated in Figure 7.39. The cross-flow normal to the axis may separate from the body resulting in a drag force normal to the body that adds to the normal force induced by the potential flow. Thereafter, many semiempirical procedures were developed based on this concept and incorporated a wide variety of experimental data (Hamner and Leff, 1966; Muraca, 1966; Teren, Davidson, Borsody, & Daniele, 1968; Jorgensen, 1973a, 1973b).

Although launch vehicles tend to be slender bodies of revolution incorporating several simple shapes like cones, ogives, cone frustums, cylinders, and planar fins they produce relatively complicated transient flow fields. For the purposes of assessing the forces and moments of a launch vehicle in a preliminary design effort, we will employ the relatively simple approach developed by Jorgensen (1973a). This approach assumes that the actual vehicle geometry may be approximated by a slender cone–cylinder–flare or ogive–cylinder–flare body. The normal and axial force components, $C_N = N/qS$ and $C_A = A/qS$, and the pitching moment coefficient about the center of gravity of the vehicle, $C_m = M/qSd$, are given as follows:

$$C_N = \frac{S_b}{S} \sin 2\alpha \cos\frac{\alpha}{2} + \eta C_{D,N} \frac{S_p}{S} \sin^2 \alpha \qquad (7.106)$$

$$C_A = C_{D,0} \cos^2 \alpha \tag{7.107}$$

$$C_m = \left[\frac{\upsilon}{Sd} - \frac{S_b}{S}\left(\frac{l}{d} - \frac{x_{CG}}{d}\right)\right]\sin 2\alpha \cos\frac{\alpha}{2} + \eta C_{D,N}\frac{S_p}{S}\left(\frac{x_{CG}}{d} - \frac{x_p}{d}\right)\sin^2 \alpha \tag{7.108}$$

The first term on the right-hand side of Eqns (7.106) and (7.108) arises from classical slender body theory (Heaslet & Lomax, 1954) while the second term on the right-hand side of Eqns (7.106) and (7.108) arises from the inclusion of the effects of the cross-flow component $V_r \sin \alpha$. Here the ratio S_b/S is present because we are choosing $S = \pi d^2/4$ to be the reference area whereas slender body theory provides force and moment results in terms of the cross-sectional area of the vehicle base. Note that the axial force coefficient C_A in Eqn (7.107) is assumed not to be affected by the cross-flow and involves only the zero-lift drag coefficient of the body.

This formulation of the governing equations requires only the geometry of the vehicle, its center of gravity location, and the nature of the cross-flow term, as defined by the product $\eta C_{D,N}$, which is a function of cross-flow Reynolds and Mach numbers. Launch vehicle trajectories limit the magnitude of those parameters to practical ranges thereby providing a means for selecting reasonable values for them. The Reynolds number normal to the body may be written in terms of the Mach number normal to the body as follows:

$$\mathrm{Re} = \frac{Vd}{\nu} = \left(\frac{a}{\nu}\right)Md \tag{7.109}$$

Using Eqns (2.16) and (2.21), we may approximate the unit Reynolds number, i.e., the Reynolds number per unit diameter, as follows:

$$\frac{\mathrm{Re}}{d} \approx 2.56 \times 10^7 \exp\left(-\frac{z}{7.16}\right)M \tag{7.110}$$

For launches from the surface of the earth the variation of the unit Reynolds number, the Mach number, and the dynamic pressure is shown in Figure 7.43 for a typical flight profile. It is clear that the maximum value of the Reynolds number is around $10^7 d$ and it occurs at subsonic speed

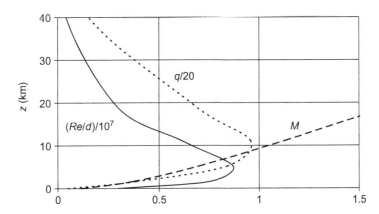

FIGURE 7.43

Variation of the unit Reynolds number (in m^{-1}), the Mach number, and the dynamic pressure (in kPa) for a typical flight profile.

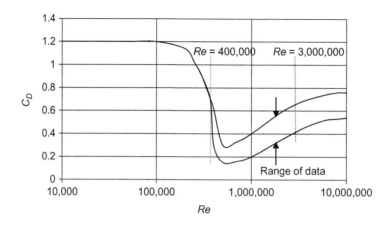

FIGURE 7.44

The drag coefficient for a circular cylinder placed normal to the oncoming flow. At $Re > 400{,}000$ measured data fall in the range shown. The range above 3 million is typical for the cross-flow Re of manned launch vehicles.

while the maximum dynamic pressure q occurs at about the sonic speed. The diameter of manned spacecraft launch vehicles is typically in the range $3 \text{ m} < d < 10 \text{ m}$ so that the maximum Reynolds number is likely to be in the range 30 million to 100 million. For small angles of attack ($\alpha \le 10°$), the cross-flow Mach number is around 10% of the flight Mach number, $M \sin \alpha = O(10^{-1})$, and therefore the cross-flow Reynolds number based on body diameter would be around 3 million to 10 million.

Jones, Cincotta, and Walker (1969) present data for the drag coefficient $C_D = C_D q(\pi d^2/4)$ measured on large circular cylinders ($d = 0.914$ m) mounted normal to the flow at Reynolds numbers based on cylinder diameter up to 10^7 for $M \le 0.2$ and compare it to those measured in other experiments. The reported drag coefficients are found to have a consistent value for $Re < 400{,}000$ while beyond that the data is spread over a band as shown in Figure 7.44. The range of data shown, which represents results from several different investigators, is ascribed to various factors including surface roughness, local subsonic Mach numbers above 0.2, etc. The region likely to be applicable to manned spacecraft launch vehicles, $3 \times 10^6 < Re < 10^7$, is also shown in Figure 7.44. The drag coefficient in this region exhibits a range of $0.4 < C_D < 0.76$. In spite of this fairly wide spread in the $C_{D,N}$, it seems prudent to use $C_{D,N} = 0.53$ which corresponds to the average value measured by Jones et al. (1969) for the Reynolds number range $3 \times 10^6 < Re < 10^7$ ($C_D = 0.54$ at $Re = 3$ million and $C_D = 0.52$ at $Re = 10$ million).

Note that this approach assumes that the drag coefficient depends solely on the Reynolds number based on the normal component of the free stream velocity. Bursnall and Loftin (1951) present results suggesting that this is accurate only for small angles of attack ($\alpha < 20°$) for flows in the supercritical Reynolds number range. Launch vehicles are controlled so as to minimize the angle of attack to reduce bending moments on the airframe making the suggested approach for determining cross-flow drag coefficients acceptable.

The term η is the ratio of the cross-flow drag coefficient of a circular cylindrical body of fineness ratio l/d to that of an infinite cylinder of equal diameter and is of order unity. Though

Jorgensen (1973a) includes η to modify the cross-flow drag coefficient for the finite fineness ratio of the launch vehicle, what little data there for η has been measured at very much lower values of cross-flow Reynolds number than is expected here: 8.8×10^4 compared to 3×10^6. One may approximate the data presented in Jorgensen (1973a) by the following correlation:

$$\eta = 0.05 \left(\frac{l}{d}\right) + 0.52 \tag{7.111}$$

He reports that using these data for subsonic flows provided useful results, although at supersonic and hypersonic speeds using $\eta = 1$ appears yield more accurate results.

7.6.3 STATIC LONGITUDINAL STABILITY

The normal and axial forces N and A may be considered to act at the center of pressure, where, by definition, no moment exists. Symmetry suggests that the axial force contributes no moment about the center of gravity of the vehicle while the normal force produces the following moment:

$$M = N(x_{CG} - x_{CP}) \tag{7.112}$$

From Eqn (7.112) and the definition of the moment and normal force coefficients we find the location of the center of pressure CP to be given by

$$\frac{x_{CP}}{d} = \frac{x_{CG}}{d} - \frac{C_m}{C_N} \tag{7.113}$$

From Figure 7.39, where the CG is shown located aft of the CP, we see that the angle of attack α causes the moment produced by the normal force about the CG to increase. This increase in moment tends to further increase the angle of attack and the body is said to be statically unstable in pitch. We may rearrange Eqn (7.113) to read

$$\frac{x_{CG}}{d} - \frac{x_{CP}}{d} = \frac{C_m}{C_N} \tag{7.114}$$

When $x_{CG} = x_{CP}$, the moment coefficient about the center of gravity is zero and the vehicle is said to be statically neutral in pitch. When $x_{CG} < x_{CP}$, the moment coefficient about the center of gravity is negative and the vehicle is said to be statically stable in pitch. From Figure 7.39, which is drawn with $x_{CP} < x_{CG}$, we see that the normal force will act to increase the angle of attack demonstrating static instability. However, if the center of gravity is placed ahead of the center of pressure the normal force will act to reduce the angle of attack demonstrating static stability.

Therefore, in order to maintain static stability the CG must be located ahead of the CP or a restoring moment must be provided by other means. Deflecting the thrust vector F through an angle δ as shown in Figure 7.39 or using fins placed near the base of the launch stack, like tail surfaces on an airplane, can produce the required restoring effect. TVC is the means by which the launch vehicle can both be guided in trajectory as well as kept stable in pitch. There is a cost to using TVC because it is necessarily accompanied by a small loss in thrust in the flight direction. The use of aft fins is typically confined to stability augmentation alone and does incur a drag penalty throughout the flight of the stack on which they are used. Because of the symmetry of the notional vehicle its stability in yaw can be treated in exactly the same fashion as its stability in pitch.

Consider the case where the angle of attack $\alpha \ll 1$ such that $\sin \alpha = \alpha$ and $\cos \alpha = 1$ to $O(\alpha^2)$. Using this approximation in Eqns (7.106) and (7.108) permits us to write the ratio C_m/C_N as follows:

$$\frac{C_m}{C_N} = \left[\frac{v}{S_b d} - \left(\frac{1 - x_{CG}}{d} \right) \right] + O(\alpha) \tag{7.115}$$

Combining Eqn (7.115) with Eqn (7.114) yields the following expression for the location of the center of pressure of the vehicle:

$$\frac{x_{CP}}{d} = \frac{l}{d} - \frac{v}{S_b d} \tag{7.116}$$

Thus, within the slender body approximation, the center of pressure may be moved aft by biasing the volume aft. The afterbody region might be a simple flare, that is, the frustum of a cone, or it may be more elaborate, as on the Soyuz or Saturn V launch vehicles shown in Figure 7.5. We assume then that the volume of the afterbody may be expressed as $h_f S_m$, where $S_m = \pi d_m^2/4$ is the mean cross-sectional area of the afterbody, d_m is the associated mean radius, and h_f is its axial extent, as illustrated in Figure 7.39. The volume of the launch vehicle may then be expressed as follows:

$$\frac{v}{S_b d} = \left[\frac{l}{d} - \frac{2h}{3d} - \frac{h_f}{d} \left(1 - \frac{d_m^2}{d^2} \right) \right] \left(\frac{S}{S_b} \right) \tag{7.117}$$

For a simple cone–cylinder–flare combination of total length l where the cone portion is of length h and the flare portion is of length h_f the nondimensional volume becomes

$$\frac{v}{S_b d} = \left[\frac{l}{d} - \frac{2h}{3d} + \frac{1}{3} \frac{h_f}{d} \left(2 - \frac{d_b^2}{d^2} - \frac{d_b}{d} \right) \right] \frac{S}{S_b}$$

Note that to maintain the adequacy of the slender body theory the general slope of the afterbody must be small such that $(d_b - d)/2h_f \ll 1$. Thus, when there is no flare, i.e., $d_b = d$, the third term within the square brackets is zero. Combining Eqns (7.116) and (7.117) leads to

$$\frac{x_{CP}}{d} \approx \frac{2h}{3d} \frac{S}{S_b} + \left(1 - \frac{S}{S_b} \right) \frac{l}{d} - \frac{h_f}{d} \left(\frac{d_m^2}{d^2} - 1 \right) \frac{S}{S_b} \tag{7.118}$$

First we note that if there is no aft flare, i.e., $d_m = d$ and $S = S_b$, then the simple result is that $x_{CP} = 2h/3d$. In this cone–cylinder case, as the nose cone length increases the center of pressure moves aft until the cylindrical portion vanishes and $h = l$. It is instructive to note that the slender body theory attributes no effect to the cylindrical afterbody at $\alpha \sim 0$. Indeed, in the case where $h = 0$, i.e., for a flat-faced cylinder, the center of pressure is predicted to be located at the leading edge, $x_{CP} = 0$. Of course, the slender body approximation breaks down near the flat face of the cylinder because the body slope there is no longer small. However, the local effects die out rather quickly and Jorgensen (1973a) shows that the slender body results are reasonably accurate, when compared to the experiments, for example, of Jernell (1968).

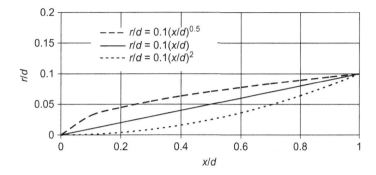

FIGURE 7.45

Slender body shapes of fineness ratio $l/d = 10$ and decreasing values of v/Ad.

Next, we consider the case of a cone–cylinder–flare body like that shown in Figure 7.39. Then the second term in brackets on the right-hand side of Eqn (7.118) is positive and we see that adding the flare to the aft end of the launch vehicle moves the CP aft, thereby serving to enhance its stability.

Several slender bodies of fineness ratio $l/d = 10$ and decreasing values of v/Ad are shown in Figure 7.45. Equation (7.116) then shows that the center of pressure of the bodies moves aft as the body shape changes from convex to flat to concave. This figure illustrates why the addition of a flared base aids in stabilizing slender missiles.

7.6.4 FIN-STABILIZED LAUNCH VEHICLES

The use of a flared base for stabilization of a launch vehicle has the disadvantage of generating high drag because of the increased frontal area presented by that base. It is more effective to use rear-mounted tail surfaces, like that on an airplane, to provide the stabilizing moment required. One pair of tail surfaces generates a stabilizing moment in the pitch plane while another pair generates a stabilizing moment in the yaw plane. We may consider this to represent weather-vane stability in each plane. Because of the rotational symmetry typical of launch vehicle it is sufficient to deal only with the pitch plane. Thus we consider a simple cone–cylinder body with rear-mounted tail surfaces as shown in an elevation, or pitch plane, view in Figure 7.46.

The tail surfaces lying in the pitch plane produce only axial forces and these are negligible insofar as pitching moments are concerned. However, the tail surfaces lying in the yaw plane add a normal force given by

$$N_f = \left(\frac{dC_{N_f}}{d\alpha}\right)\alpha q S_f$$

Here S_f is the planform area of the tail surface which will be defined in more detail subsequently. The tail normal force produces a negative pitching moment

$$M_f = -N_f l_f$$

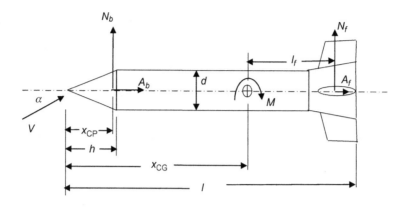

FIGURE 7.46

A slender cone–cylinder missile with a rear-mounted horizontal and vertical tail shown in the pitch plane at angle of attack α.

The moment arm l_f is generally taken to be the distance between the center of pressure of the tail and the center of gravity of the complete launch vehicle. We may put these results in coefficient form consistent with those in Eqns (7.106) and (7.108) as follows:

$$C_{Nf} = \frac{N_f}{qS} = \left(\frac{dC_{Nf}}{d\alpha}\right)\alpha\frac{S_f}{S}$$

$$C_{mf} = \frac{M_f}{qSl} = \left(\frac{dC_{Nf}}{d\alpha}\right)\alpha\frac{S_f l_f}{Sl}$$

For small angles of attack, we assume that the contribution of the rear-mounted fins may be superimposed on the basic force structure acting on the cone cylinder alone so that adding C_{Nf} to Eqn (7.106) and C_{mf} to Eqn (7.108) leads to the following:

$$\frac{C_m}{C_N} = \frac{2\left[\dfrac{v}{Sd} - \dfrac{S_b}{S}\left(\dfrac{l}{d} - \dfrac{x_{CG}}{d}\right)\right] + \eta C_{D,n}\dfrac{S_p}{S}\left(\dfrac{x_{CG}}{d} - \dfrac{x_{CP}}{d}\right)\alpha - \left(\dfrac{dC_{Nf}}{d\alpha}\right)\left(\dfrac{S_f l_f}{Sd}\right)}{2\dfrac{S_b}{S} + \eta C_{D,n}\dfrac{S_p}{S}\alpha + \left(\dfrac{dC_{Nf}}{d\alpha}\right)\left(\dfrac{S_f}{S}\right)} \tag{7.119}$$

Substituting Eqn (7.119) into Eqn (7.114) and evaluating it at zero angle of attack yields the center of pressure as follows:

$$\frac{x_{CP}}{d} = \frac{x_{CG}}{d} - \frac{\left[\dfrac{v}{S_b d} - \left(\dfrac{l}{d} - \dfrac{x_{CG}}{d}\right) - \dfrac{1}{2}\left(\dfrac{dC_{Nf}}{d\alpha}\right)\left(\dfrac{S_f l_f}{S_b d}\right)\right]}{1 + \dfrac{1}{2}\left(\dfrac{dC_{Nf}}{d\alpha}\right)\left(\dfrac{S_f}{S_b}\right)} \tag{7.120}$$

Assuming the simple shape of a cone–cylinder–flare combination, as shown in Figure 7.46, and using Eqn (7.117), we write

$$\frac{v}{S_b d} - \left(\frac{l}{d} - \frac{x_{CG}}{d}\right) = \frac{x_{CG}}{d} - \frac{2}{3}\frac{h}{d}\frac{S}{S_b} + \frac{h_f}{d}\left(\frac{d_m^2}{d^2} - 1\right)\frac{S}{S_b} - \frac{l}{d}\left(1 - \frac{S}{S_b}\right) \tag{7.121}$$

The contribution of the tail fins to the normal force experienced by the vehicle may be written as

$$\frac{1}{2}\left(\frac{dC_{N_f}}{d\alpha}\right)\frac{S_f}{S_b} = K_f \tag{7.122}$$

Using Eqns (7.121) and (7.122) in Eqn (7.120) leads to

$$\frac{x_{CP}}{d} = \frac{\frac{2}{3}\left(\frac{h}{d}\right)\frac{S}{S_b} + K_f\left[\left(\frac{x_{CG}}{d}\right) + \left(\frac{l_f}{d}\right)\right] + \frac{l}{d}\left(1 - \frac{S}{S_b}\right) - \frac{h_f}{d}\left(\frac{d_m^2}{d^2} - 1\right)\frac{S}{S_b}}{1 + K_f} \tag{7.123}$$

As mentioned previously, we take l_f to be the distance from the center of pressure of the tail to the CG of the vehicle. However, because the tail is typically located at the aft end of the vehicle it is sufficient, for preliminary design purposes, to consider the sum of the distance from the nose to the CG and the distance from the CG to the fin normal force to be approximately equal to the complete length of the vehicle, i.e., $x_{CG} + l_f \approx l$. Therefore the approximate location of the center of pressure according to Eqn (7.123) is

$$\frac{x_{CP}}{d} \approx \frac{\frac{2}{3}\left(\frac{h}{d}\right)\left(\frac{S}{S_b}\right) + \left(K_f + 1 - \frac{S}{S_b}\right)\left(\frac{l}{d}\right) - \frac{h_f}{d}\left(\frac{d_m^2}{d^2} - 1\right)\left(\frac{S}{S_b}\right)}{1 + K_f} \tag{7.124}$$

When the normal force contribution of the tail vanishes ($K_f = 0$) the previous result for the center of pressure location, Eqn (7.118) is recovered. We may write the normal force per radian term of the tail as follows:

$$\frac{dC_{N_f}}{d\alpha} \approx C(2\pi) \tag{7.125}$$

Here the value 2π corresponds to the ideal lift curve slope of an infinite wing in incompressible flow and for small angles of attack the lift coefficient and the normal force coefficient are approximately equal, $C_L \sim C_N$. The proportionality factor C in Eqn (7.125) is a function of the aspect ratio and sweepback of the tail fins as well as the free stream Mach number, as discussed by Sforza (2014). For small aspect ratio tails typical of missile applications, the factor $C \sim 1/2$ and the total tail area normal to the plane of the free stream velocity vector may be taken to be equal to about one-third the cylinder or base cross-sectional area, $S_f/S_b \sim 1/3$. Thus a reasonable approximation for illustrative purposes is that

$$K_f = \frac{1}{2}\frac{dC_{N_f}}{d\alpha}\frac{S_f}{S_b} \approx \frac{1}{2}(2C\pi)\left(\frac{1}{3}\right) \approx \frac{\pi}{6} \approx \frac{1}{2} \tag{7.126}$$

To get an idea of the effect of the addition of fins to a cone–cylinder or a cone–cylinder–flare missile we may consider the case of an idealized Saturn V launch vehicle approximated as a cone–cylinder with $h/d = 4.32$, $l/d = 10.39$, and rear-mounted fins as shown in Figure 7.47. The four discrete engine shrouds on the Saturn V (Figure 7.20) provide an equivalent flare with $h_f/d = 0.806$, $S_f/S = 0.78$, and $d_m/d = 1.047$. The variation in the center of pressure location given by Eqn (7.124) is shown in Figure 7.48 as a function of the tail fin contribution K_f. It is clear that even for $K_f = 1/2$, as suggested by Eqn (7.126), the center of pressure is shifted rearward by about two calibers or diameters. The incorporation of the flare moves the center of pressure aft yet another caliber.

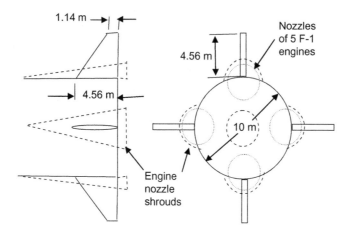

FIGURE 7.47

Schematic diagram of the aft end of the idealized Saturn V launch vehicle showing the stabilizing fins. The Saturn V has shrouds, shown as dashed lines, around the nozzles of its five F-1 engines, indicated here as dotted lines.

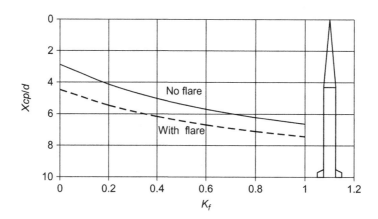

FIGURE 7.48

Effect of rear-mounted fins on center of pressure location at zero angle of attack with and without an aft flare. The parameter K_f is proportional to the tail lift curve slope $dC_{L_t}/d\alpha$ and therefore $K_f = 0$ corresponds to the vehicle without fins.

Walker (1968) reports results of experiments on models of the Saturn V launch vehicle carried out at five different wind tunnel facilities. Six models were used: one faithfully replicating all protuberances, fins, and shrouds, four others each having a different mix of features to permit comparisons of their individual and combined effects, and one with the entire launch escape system removed as if a launch abort sequence had occurred. Mach numbers covered the range $0.5 < M < 8$ and

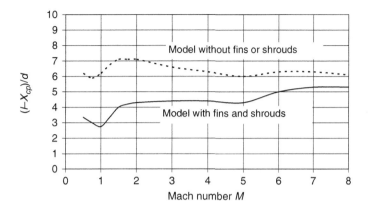

FIGURE 7.49

Experimental results for the location of the center of pressure of a Saturn V model with and without fins and shrouds as a function of flight Mach number as reported by Walker (1968).

Reynolds numbers covered the range 0.5 million to 8 million based on the vehicle reference diameter. The center of pressure location for the completely detailed model compared to that for the same model with neither tail fins nor shrouds is shown as a function of Mach number in Figure 7.49. It is clear that the tail fins and shrouds, though relatively small compared to the remainder of the vehicle, provide a dramatic rearward shift in the center of pressure location, which, as discussed previously, is necessary for improving the static longitudinal stability of the vehicle.

For long slender bodies we typically deal with the components of the resultant aerodynamic force resolved along and normal to the body axis denoted by the symbols A and N. The relationship between these components and the usual lift and drag components, L and D, is given by the following equations:

$$C_N = C_L \cos \alpha + C_D \sin \alpha$$
$$C_A = - C_L \sin \alpha + C_D \cos \alpha$$

For small angles of attack these equations become, to $O(\alpha^2)$, as follows:

$$C_N \approx C_L + C_D\alpha$$
$$C_A \approx C_D - C_L\alpha$$

Then for $\alpha \ll 1$ and $C_D/C_L \ll 1$, the normal force slope is equal to the lift curve slope or

$$\frac{\partial C_N}{\partial \alpha} \approx \frac{\partial C_L}{\partial \alpha}$$

Basic wing theory and analysis is discussed by Sforza (2014) and the empirical approach for subsonic flows presented there is followed here. For subsonic Mach numbers, the lift curve slope of a conventional wing is given, per radian, by the following equation:

$$\frac{\partial C_L}{\partial \alpha} = \frac{A}{2 + \sqrt{\left(\frac{A\beta}{\kappa}\right)^2 \left[1 + \left(\frac{\tan \Lambda_{c/2}}{\beta}\right)^2\right] + 4}} \; 2\pi \tag{7.127}$$

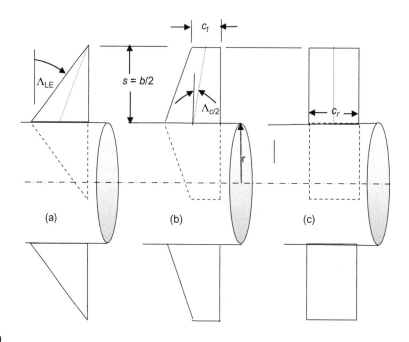

FIGURE 7.50

Some simple planforms for tail fins: (a) delta, (b) cropped delta, and (c) rectangular wing. The dashed lines illustrate how fin planforms are joined to form wings used in the lift calculations while the dotted lines denote the mid-chord line.

A wing may be formed from simple tail fin planforms by joining them at their root chords as depicted by the dashed lines in Figure 7.50. Equation (7.127) shows that the lift curve slope for a wing differs from the theoretical value of 2π for an airfoil by a factor which is a function of wing aspect ratio A, mid-chord sweep angle $\Lambda_{c/2}$, and the parameters β and κ. For the special case of delta wings in subsonic flow, Polhamus (1971) provides charts which are useful for design purposes and a simple curve fit to the data in those charts yields

$$\frac{\partial C_L}{\partial \alpha} \approx \frac{4}{(\beta A)^{0.1}} \sin\left(\frac{\beta A}{\pi}\right)$$

The aspect ratio A of the wing is the ratio of the area of a square whose side is the span b to the actual planform area of the wing, denoted by S:

$$A = \frac{b^2}{S} \tag{7.128}$$

The Mach number enters Eqn (7.127) through the subsonic form of the Prandtl−Glauert factor

$$\beta = \sqrt{1 - M^2} \tag{7.129}$$

If the sweepback angle of the leading edge is known the following equation may be used to find the mid-chord sweep for any straight-tapered wing:

$$\tan \Lambda_{c/2} = \tan \Lambda_{LE} - \frac{2}{A}\left(\frac{1-\lambda}{1+\lambda}\right) \tag{7.130}$$

The quantity λ is the taper ratio, c_t/c_r. For trapezoidal wings, the sweepback angle of the leading edge may be found from that at any other constant percent chord line ($\nu = \%c/100\%$), according to the following:

$$\tan \Lambda_{LE} = \tan \Lambda_{\nu c} + \frac{4n}{A}\left(\frac{1-\lambda}{1+\lambda}\right) \tag{7.131}$$

For example, if the quarter chord sweepback angle is known, ($\nu = 25\%/100\% = 1/4$) the sweepback angle of the leading edge is easily determined. In a similar fashion, once the sweepback angle of the leading edge is known, the sweepback angle of any other constant percent chord line can be easily found:

$$\tan \Lambda_{\nu c} = \tan \Lambda_{LE} - \frac{4n}{A}\left(\frac{1-\lambda}{1+\lambda}\right) \tag{7.132}$$

The factor κ in Eqn (7.127) depends on the lift curve slope of the airfoil section (parallel to the free stream) used in the wing and is the ratio of the experimental two-dimensional (i.e., airfoil) lift curve slope (per radian) at the appropriate Mach number $(c_{la})_M$ to the theoretical value at that Mach number, $2\pi/\beta$, or $\kappa = (c_{la})_M/(2\pi/\beta)$. Note that the Prandtl–Glauert correction for subsonic compressibility is $(c_{l\alpha})_M = c_{l\alpha}/\beta$, so in the absence of an experimental value for $(c_{l\alpha})_M$ one may use $\kappa = c_{l\alpha}/2\pi$, that is, the ratio of the actual low-speed airfoil lift curve slope to that of the airfoil in ideal incompressible flow. If no airfoil information is available one may take $\kappa = 1$ with little loss in accuracy.

Graphical representations of experimental data at subsonic speeds for many NACA airfoils are collected by Abbott and Von Doenhoff (1959) and tabulated experimental and theoretical data for other NACA airfoils is presented by Hoak et al. (1978). Loftin and Bursnall (1948) carried out experiments on selected NACA airfoils at Reynolds numbers up to 25×10^6. The main conclusion of that research was that the airfoil lift curve slope was essentially unaffected by the increase in Reynolds number, remaining remarkably close to the theoretical incompressible value $c_{l,\alpha} = 2\pi$ per radian or $a = 0.11$ per degree.

In the case of supersonic flight, the lift curve slope of a wing becomes considerably more dependent on the details of wing planform. The theory of wings in supersonic flow is covered in some depth by Jones and Cohen (1960). Piland (1949) presents a summary of lift and center of pressure characteristics of a variety of different planforms using linearized supersonic flow theory. The lift curve slope depends upon the value of βA where β is now the supersonic form of the Prandtl–Glauert factor

$$\beta = \sqrt{M^2 - 1} \tag{7.133}$$

In the case of a delta wing alone, as shown in Figure 7.50(a), the theoretical lift slope is given, along with the region of applicability, by Piland (1949) as follows:

$$\beta A \leq 4 : \beta\left(\frac{\partial C_L}{\partial \alpha}\right)_W = \frac{1}{2}\pi\beta A[E''(\beta \cot \Lambda_{LE})] \approx \frac{1}{2}\pi\beta A(1 - 0.903\beta A) \tag{7.134}$$

$$\beta A > 4 \text{:} \beta \left(\frac{\partial C_L}{\partial \alpha}\right)_W = 4 \tag{7.135}$$

The quantity $E''(\beta \cot \Lambda_{LE})$ in Eqn (7.134) is the elliptic integral of the second kind with modulus $\beta \cot \Lambda_{LE}$ which we approximate here by the quantity $(1 - 0.903\beta A)$.

In the case of a rectangular wing alone, as shown in Figure 7.50(c), the theoretical lift slope depends upon the value of βA and is given, along with the region of applicability, by Piland (1949) as follows:

$$\beta A \geq 1 \text{:} \beta \left(\frac{\partial C_L}{\partial \alpha}\right)_W = 4 - \frac{2}{\beta A} \tag{7.136}$$

For other wings with sweepback and taper the theoretical lift slope becomes a complicated expression. Pitts, Nielsen, and Kaattari (1957) show a graphical representation for a wing with taper ratio $\lambda = 0.5$ and Λ_{LE}, Λ_{TE}, or $\Lambda_{c/2}$ equal to zero. We present a simple approximate fit for the case $\lambda = 0.5$, shown in Figure 7.50(b) as a planform half-way between the delta wing and rectangular wing, as follows:

$$\beta A \geq 2, \lambda = 0.5 \text{:} \beta \left(\frac{\partial C_L}{\partial \alpha}\right)_W \approx 4 \left[1 - \frac{1}{(1.55\beta A)^{4/3}}\right] \tag{7.137}$$

The lift curve slopes discussed above are for wings alone. However, when wings are mounted on a body there may be interference between the two. A fundamental study of lift on wing–tail–body combinations and the interference effects between those components of a flight vehicle is presented by Pitts et al. (1957). They show that under the assumptions of slender body theory the lift coefficient of a wing mounted on a body, $C_{L,W(B)}$, may be approximated by

$$C_{L,W(B)} = K_{W(B)} \left(\frac{\partial C_L}{\partial \alpha}\right)_W \alpha \tag{7.138}$$

Here $\left(\frac{\partial C_L}{\partial \alpha}\right)_W$ is the lift curve slope for the wing alone. Note that the basic assumption underlying the theory is that the wing alone, denoted by the subscript W, refers to the situation where the two wing halves are joined at their root chord as illustrated by the added dashed lines in Figure 7.50. The expression for the constant $K_{W(B)}$ given by Pitts et al. (1957) may be approximated by the following:

$$K_{W(B)} \approx 1 + \left(\frac{r}{s}\right)^{1.15}; \quad 0 \leq \frac{r}{s} \leq 1 \tag{7.139}$$

Clearly the factor $K_{W(B)}$ indicates that wing mounted on a body has a greater lift slope than it would if it were acting alone. As the half-span of the wing $s = b/2$ grows much larger than the radius of the body, the effect of the body diminishes and in the case of conventional commercial aircraft where $r/s = O(10^{-1})$ the effect is small. Once the radius becomes greater than the half-span the body acts like a reflection plane and $K_{W(B)} = 2$.

7.6.5 LAUNCH VEHICLE DIAMETER ESTIMATION

Once the payload mass and its orbit, along with acceleration constraints, are specified one may use the methods discussed in Section 7.5 to begin developing the general configuration of the launch vehicle that will be required. Using the initial thrust to weight ratio and specific impulse as parameters, the propellant type, total mass, and number of engines and stages of candidate launch vehicles may be assembled. Then the general geometric configuration may be established. The major function of the stages is to house the necessary propellant safely and effectively. We start by considering the launch vehicle to be cylindrical and the payload mass to be a small fraction of the liftoff mass so that we may approximate the liftoff mass as follows:

$$m_0 \approx (1 + k_s)m_p = (1 + k_s)\rho_p k_p \frac{\pi}{4} d^3 \left(\frac{l}{d}\right) \tag{7.140}$$

The volume fraction of the launch vehicle occupied by the propellant is given here as k_p. Then the diameter of the launch vehicle may be written as

$$d \approx \left[\frac{4m_0}{\pi(1+k_s)k_p\rho_p(l/d)}\right]^{1/3} \tag{7.141}$$

Data for the maximum diameter of several launch vehicles, manned and unmanned, are shown in Figure 7.51 as a function of liftoff mass. Equation (7.141) is also shown in Figure 7.51 in which $k_s = 0.1$ is used, as discussed in Section 7.5, along with the following density approximation based on the vehicle data:

$$k_p\rho_p = k_p\rho_w = 650 \, \text{kg/m} \tag{7.142}$$

The correlation used in Figure 7.51 (with m_0 in metric tons) is

$$d = 1.22 \left[\frac{m_0}{l/d}\right]^{\frac{1}{3}} \tag{7.143}$$

Note that some data in Figure 7.51 fall outside the band of reasonable l/d values. The overall density of the Saturn V is only about half that given by Eqn (7.142) leading to a large diameter. The use

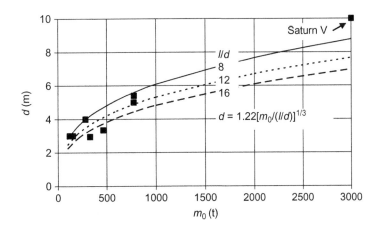

FIGURE 7.51

Variation of maximum diameter of launch vehicles with initial mass. Symbols denote actual flight vehicle data.

of five F-1 engines, four outboard and one center, likely forced a somewhat larger diameter than predicted by Eqn (7.143). On the other hand, the two smallest diameters also fall outside the given l/d band. Those data are for the Soyuz and Long March launch vehicles have large fineness ratios with l/d more than 16 and 17, respectively. However, those vehicles have an overall density almost that of water yielding a density about 50% greater than that assumed by Eqn (7.142). These differences are essentially packaging issues that are typically resolved later in the design process. Thus it is suggested that a diameter falling within the range of the curves in Figure 7.51 for a given value of m_0 be used as the starting point of the configuration design. Selecting a larger diameter in the reasonable range increases drag penalties, but the smaller l/d associated with that choice increases structural integrity. The opposite will be true for selecting a smaller diameter in the range.

7.6.6 LAUNCH VEHICLE CONFIGURATION DESIGN

Having selected a candidate set of values for d and l/d, one may choose a geometry for further development. A simple cone—cylinder model can emulate a relatively complex launch vehicle, like the Saturn V, for example, and permit us to evaluate performance in a manner acceptable for preliminary design purposes. This configuration requires only an assumption of the nose length h in order to proceed. The three-stage Saturn V launch vehicle was shown previously in Figure 7.4 to illustrate the concepts of stacking and staging. In Figure 7.52, it is shown to scale along with a simplified cone—cylinder model of it. We shall use this model to illustrate the design procedures for a three-stage launch vehicle. Thus we base the model launch vehicle on the dimensions of the Saturn V Apollo 10 launch vehicle using the data shown in Table 7.9.

The mass of the spacecraft to be launched was specified and the subsequent trajectory study yielded the required mass and type of propellant along with the particular engine and the mass of the associated structure for each of the three stages. At the simplest level, one may proceed by assuming that the density of the entire vehicle is uniform and, at the next level of accuracy, that the density of each stage and of the payload is different but uniform in each component. Here we assume that the spacecraft and the three stages each have uniform density so that their masses, listed in Table 7.10, are acting through their centers of gravity, as illustrated in Figure 7.53. The densities of these different components of the Saturn V and of the model launch vehicle are also given in Table 7.10 and illustrated in Figure 7.54 along with the density of the vehicle as a whole and, for comparison purposes, the density of water. Recall that the design model has been taken to be a cone of base diameter d serving as an envelope for the combination of the spacecraft and the third stage while the second and third stages are assumed to be cylinders of that same diameter d. The actual dimensions appear in subsequent tables.

The approximate methods developed previously for assessing vehicle stability require knowledge of the stack planform area, the axial location of the centroid of that area, and the associated volume. The data for the three stacks are presented in Table 7.11. The center of gravity location for each of the stacks is also required and these quantities are presented in Table 7.12. Inclusion of the actual cone—cylinder—frustum details of the Saturn V launch vehicle doesn't materially alter the centroid and center of gravity locations although they would unnecessarily complicate the aerodynamic calculations appropriate to preliminary design without providing significantly improved accuracy.

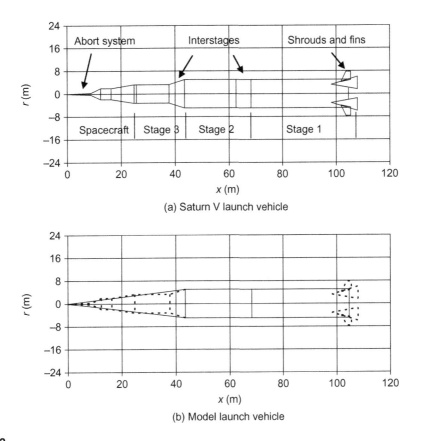

FIGURE 7.52

Schematic diagrams of (a) the Saturn V launch vehicle and (b) the model launch vehicle. Dashed lines in (b) illustrate the actual outline of the Saturn V.

7.6.7 LAUNCH VEHICLE CENTER OF PRESSURE

The center of pressure location for a slender body is given by Eqn (7.118) and for the basic cone–cylinder configuration of the model launch vehicle shown in Figure 7.52 is given by

$$\frac{x_{CP}}{d} = \frac{2}{3}\frac{h}{d} = \frac{2}{3}\frac{43.47 \text{ m}}{10.06 \text{ m}} = 2.88 \tag{7.144}$$

The four engine shrouds of the actual launch vehicle illustrated by the dashed lines in Figure 7.47 were shown previously to be approximated by an equivalent flare with $h_f/d = 0.81$, $S_f/S = 0.78$, and $d_m/d = 1.047$ leading to

$$\frac{x_{CP}}{d} = \frac{2}{3}\frac{h}{d}(0.78) + (1 - 0.78)\frac{l}{d} - 0.806(1.047^2 - 1)(0.78) = 4.47$$

Table 7.9 Approximate Dimensions of the Components of the Saturn V Apollo 10 and the Model Launch Vehicle

Component	x (m)	Saturn V r (m)	Saturn Model r (m)
Launch escape system	0	0	0.00
Command module	8.85	0.33	1.02
Service module	12.40	1.95	1.43
Lunar module	16.31	1.95	1.88
Instrumentation unit	24.79	3.28	2.85
Stage 3	25.70	3.28	2.96
Interstage	37.72	3.28	4.34
Stage 2	43.47	5.03	5.03
Interstage	62.64	5.03	5.03
Stage 1	68.17	5.03	5.03
Gimbal plane	104.6	5.03	5.03

Table 7.10 Mass Distribution for the Saturn V Apollo 10 Launch Vehicle and the Model Launch Vehicle

Segment	Saturn V Apollo 10 Mass (t)[a]	Model Launcher Mass (t)	Model Launcher Density (t/m³)
Spacecraft	72.02	72.07	0.252
Third stage	116.8	124.2	0.166
Second stage	480.4	490.8	0.249
First stage	2276	2279	0.682
Total	2945	2966	0.467

[a]*Metric ton (t) = 1000 kg.*

However, if the stability is to be augmented by the addition of fins it is shown in Eqn (7.124) that the center of pressure location would be given by

$$\frac{x_{CP}}{d} \approx \frac{\frac{2}{3}\left(\frac{h}{d}\right)(0.78) + (K_f + 1 - 0.78)\left(\frac{l}{d}\right) - 0.806(1.047^2 - 1)(0.78)}{1 + K_f}$$

$$\frac{x_{CP}}{d} \approx \frac{2.246 + (K_f + 0.22)(10.39) - 0.06048}{1 + K_f} \tag{7.145}$$

The tail contribution denoted by K_f was defined in Eqn (7.122) and depends upon the lift curve slope $dC_{L_f}/d\alpha$ of the tail as well as on the area ratio S_f/S_b. Assuming the nominal value $K_f = 0.5$ suggested by Eqn (7.126) leads to $x_{CP}/d = 6.44$, more than double that for the case with no fins or

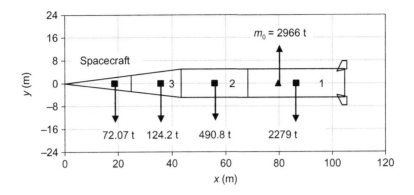

FIGURE 7.53

Schematic diagram of model launch vehicle illustrating the masses of the different segments acting at their respective centers of gravity.

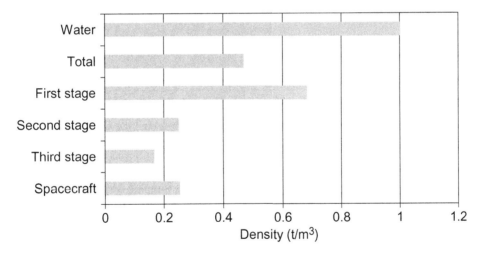

FIGURE 7.54

Approximate densities (metric tons per cubic meter) of the components of the Saturn V Apollo 10 launch vehicle. The density of water is shown for comparison.

flare-like shrouds. The theoretical results developed here for cone–cylinder–flare bodies are constant while the experimental results vary with Mach number. This is because slender body theory for the cone–cylinder–flare admits no dependence on Mach number. Here, when we added fins a constant nominal value for K_f was selected. More accurate analyses like those discussed in the previous section would reveal Mach number dependence for K_f, because $dC_{L,f}/d\alpha$ is a function of Mach number.

Table 7.11 Planform Area, Centroid Locations, and Volumes for the Model Launcher

Item i	$r_{i,min}$ (m)	$r_{i,max}$ (m)	l_i (m)	A_p (m²)	v (m³)	h'_i (m)	h_i (m)	$A_{p,i}h_i$ (m³)	$l-x_p$ (m)
Stack I									
Spacecraft	0	2.855	24.84	70.9	212.0	8.28	88.00	6241	
Stage 3	2.874	5.029	18.63	147.2	936.5	8.46	69.56	10,241	
Stage 2	5.029	5.029	24.70	248.4	1963	12.35	48.74	12,109	
Stage 1	5.029	5.029	36.39	366.0	2891	18.20	18.20	6660	
Sums			104.6	832.5	6002			35,250	42.34
Stack II									
Spacecraft	0	2.855	24.84	70.92	212.0	8.280	51.61	3660	
Stage 3	2.874	5.029	18.63	147.2	936.5	8.468	33.17	4883	
Stage 2	5.029	5.029	24.70	248.4	1963	12.35	12.35	3068	
Sums			68.17	466.5	3112			11,611	24.88
Stack III									
Spacecraft	0	2.855	24.84	70.92	212.0	8.280	26.91	1908	
Stage 3	2.874	5.029	18.63	147.2	936.5	8.468	8.468	1247	
Sums			43.47	218.1	1149			3155	14.46

Table 7.12 Center of Gravity Locations for the Model Launcher

Item i	$r_{i,min}$ (m)	$r_{i,max}$ (m)	l_i (m)	m_i (kg)	h''_i (m)	$l-x_{CG,I}$ (m)	$m_i x_{CG,I}$ (kg-m)	$l-x_{CG}$ (m)
Stack I								
Spacecraft	0	2.855	24.84	72.07	6.210	85.93	6193	
Stage 3	2.874	5.029	18.63	124.2	7.663	68.75	8536	
Stage 2	5.029	5.029	24.70	490.8	12.35	48.74	23,922	
Stage 1	5.029	5.029	36.39	2279	18.20	18.20	41,467	
Sums			104.6	2966			80,118	27.01
Stack II								
Spacecraft	0	2.855	24.84	72.07	6.210	49.54	3570	
Stage 3	2.874	5.029	18.63	124.2	7.663	32.36	4027	
Stage 2	5.029	5.029	24.70	490.8	12.35	12.35	6061	
Sums			68.17	687.1			13,658	19.88
Stack III								
Spacecraft	0	2.855	24.84	72.07	6.210	13.87	1790	
Stage 3	2.874	5.029	18.63	124.2	7.663	7.663	951.8	
Sums			43.47	196.3			2742	13.97

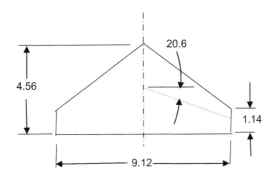

FIGURE 7.55

Fins joined at root to form single wing for lift slope calculation. Linear dimensions are in meters and the angle is in degrees.

We may consider using the fins shown in Figure 7.47 and applying the approach due to Pitts et al. (1957) discussed in the previous section in order to determine the lift of the fins when in the presence of the launch vehicle body. This method first considers the wing alone, where the wing is formed by joining the two opposing fins at their roots as illustrated in Figure 7.50. The fins in Figure 7.47 are close in configuration to those shown in Figure 7.50(b) and are shown as a wing alone in Figure 7.55.

The area of the two fins joined as one wing as shown in Figure 7.55 is $S_f = 26 \text{ m}^2$ and from Eqn (7.128) the aspect ratio is $A = b^2/S_f = 3.2$. The sweepback of the mid-chord is shown in Figure 7.55 to be $\Lambda_{c/2} = 20.6°$ and the taper ratio $\lambda = c_t/c_r = 0.25$. From Eqn (7.122) we find

$$K_f = \frac{1}{2}\left(\frac{dC_{Nf}}{d\alpha}\right)\frac{S_f}{S} = \frac{1}{2}\left(\frac{dC_{Nf}}{d\alpha}\right)\frac{26 \text{ m}^2}{79.45 \text{ m}^2} = 0.163\left(\frac{dC_{Nf}}{d\alpha}\right) \tag{7.146}$$

For small α the wing normal force and lift curve slopes are equal

$$\left(\frac{dC_N}{d\alpha}\right)_f = \left(\frac{dC_L}{d\alpha}\right)_w$$

For subsonic flight, the lift curve slope of the wing may be calculated using Eqn (7.127). For supersonic flight, Eqns (7.136)–(7.139) may be applied to determine a reasonable estimate for the lift curve slope of the wing. Using these equations for fins with the same aspect ratio $A = 3.2$ as those shown in Figure 7.55 for the model vehicle leads to the normal force slope variation with Mach number shown in Figure 7.56. For $M = 1$ the estimate provided by McDevitt (1955) for rectangular wings, $dC_L/d\alpha = \pi A/2$, was used. As can be seen, there is not a great deal of variation between the results for the different geometries. Using the result for K_f in Eqn (7.146) and the normal force slope for a taper ratio $\lambda = 0.5$, which should be adequate to represent the actual taper ratio $\lambda = 0.25$, leads to the results shown in Figure 7.57 for the center of pressure location derived from Eqns (7.144) and (7.145). The wind tunnel test results on a complete Saturn V model and one with the base shrouds and fins removed are also shown in Figure 7.57 for comparison. The results

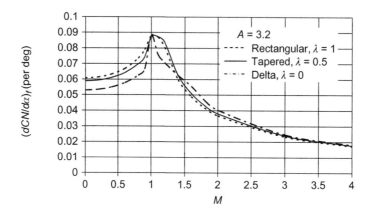

FIGURE 7.56

Normal force slope for the fins of aspect ratio $A = 3.2$ and different planform geometries when mounted on the model launch vehicle.

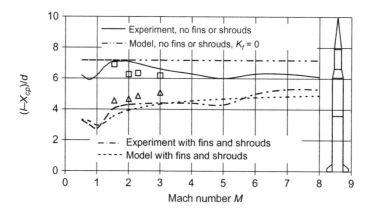

FIGURE 7.57

Experimental center of pressure locations for the Saturn V launch vehicle compared to the theoretical results for the cone–cylinder–flare–tail model. The actual vehicle is shown to scale to illustrate the CP location on the body. Data symbols represent results of some other empirical methods described in the text.

appear to be quite reasonable considering the relative simplicity of the analysis. It is interesting to note that the center of pressure location moves little outside the transonic region. The data symbols in Figure 7.57 represent results obtained using experimentally based empirical models developed for cone cylinders by Hamner and Leff (1966) and Muraca (1966) which will be discussed subsequently.

One may calculate the movement of the center of gravity during flight by first considering the reduction in mass of the launch vehicle because of propellant consumption in the first stage according to Eqn (7.13):

$$m(t) = m_0 \left[1 - \left(\frac{F_{vac}}{W} \right)_0 \frac{t}{I_{sp,vac}} \right]$$

The center of gravity location measured from the gimbal axis is

$$l - x_{CG} = \frac{m_{0,1}(l - x_{CG,1}) \left[1 - \left(\frac{F_{vac}}{W} \right)_0 \frac{t}{I_{sp,vac}} \right] + \sum_{i=2}^{4} m_i(l - x_{CG,i})}{m_{0,1} \left[1 - \left(\frac{F_{vac}}{W} \right)_0 \frac{t}{I_{sp,vac}} \right] + \sum_{i=2}^{4} m_i} \qquad (7.147)$$

Here $i = 2$ and 3 denote stages 2 and 3, while $i = 4$ denotes the spacecraft being launched. The center of pressure location is a function of Mach number, as shown in Figure 7.57, and therefore recourse to the previous trajectory calculations for the launcher is needed to obtain the Mach number as a function of time, $M = M(t)$. Assuming typical values of $(F_{vac}/W)_0 = 1.4$ and $I_{sp,vac} = 260$ s leads to the results shown in Figure 7.58.

The CP and CG locations as a function of time for a Saturn V launch are shown as solid and dotted lines in Figure 7.58 and are constructed from data presented by Haeussermann (1965), Pinson (1971), and Walker (1968). The calculations based on the approaches developed here are shown as open symbols and the shaded area denotes the region of maximum dynamic pressure. It is clear from Figure 7.58 that the launch vehicle is unstable during the 160-s first stage burn, except perhaps for a few seconds at around $t = 60$ s where $M \sim 1$. The calculated values for center of gravity and center of pressure are in good agreement with the flight data during the entire burn.

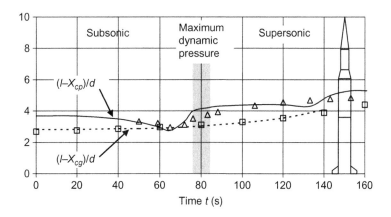

FIGURE 7.58

CP and CG locations as a function of time for a Saturn V launch are shown as solid and dotted lines, respectively. Calculated values are shown as open symbols and the shaded area denotes the region of maximum dynamic pressure.

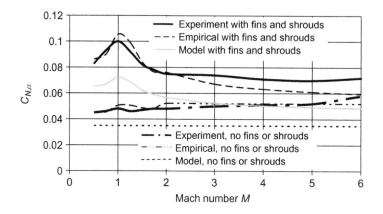

FIGURE 7.59

Comparison of theoretical and empirical results for the normal force coefficient slope on a launch vehicle similar to the Saturn V.

Although slender body theory provides a reasonable estimate of CP location, it underestimates the normal force coefficient slope $C_{N,\alpha}$ by a substantial amount. This is because it doesn't properly account for the contribution of the cylindrical afterbody to the normal force. This is clearly demonstrated by Sherer (1966) through comparisons of slender body theory to the method of characteristics. Other empirical approaches do provide a more accurate estimation of the normal force slope.

Experimental results for the Saturn V wind tunnel model with and without tail fins and engine shrouds compiled by Walker (1968) are shown in Figure 7.59. Also presented in that figure are results for the approximate Saturn V model based on the slender body theory and tail fin estimates described previously. It is obvious that the experimental data for $C_{N,\alpha}$ is substantially underestimated by this approach. Hamner and Leff (1966) present carpet plots of $C_{N,\alpha}$ and $C_{m,\alpha}$ based on experimental data for a wide variety of cone cylinders in the Mach number range of $0.7 < M < 2$, cone angles $10° < \theta < 40°$, and cylindrical afterbody lengths with $(l - h)/d$ up to 6. In applying this approach, we chose a cone half-angle $\theta = 10°$ and an afterbody cylinder length $(l - h)/d = 6$ yielding an overall $l/d = 8.83$ and thereby providing a configuration reasonably close to the Saturn V dimensions. Muraca (1966) gives an empirical estimation method for cone–cylinder–frustum combinations in the speed range $0.8 < M < 7$. Calculations for the cone–cylinder approximation to the Saturn V ($\theta = 6.56°$ and $l/d = 10.4$) using Muraca's method yield numbers consistent with those of Hamner and Leff (which are confined to $M \leq 2$) and both are used in constructing the curves labeled empirical in Figure 7.59. It is apparent that these two methods yield results in much better agreement with the experimental data for the case without fins and shrouds than that those of the slender body theory. When the tail contribution presented previously is combined with the empirical cone–cylinder methods good agreement with experiment is again achieved.

7.6.8 TVC DEFLECTION REQUIREMENTS

As has been demonstrated, the typical launch vehicle is unstable in pitch, and, by similar considerations, also in yaw because the CP is usually forward of the vehicle CG. Therefore, disturbances
which cause the angle of attack to change from zero will be amplified by the induced normal force.
The most important disturbances arise from the ambient wind field, although other factors like
thrust misalignment, aerodynamic bias, etc., may also contribute. Deflection of the thrust vector is
the only practical control mechanism for such vehicles and it is important to assess the extent to
which the thrust vector must be deflected.

A relatively simple analytic approach to evaluating TVC requirements for launch vehicles is
presented by Teren et al. (1968). They carry out a linearization of the equations of motion about a
nominal trajectory for which the nominal angle of attack $\alpha_n = 0$, the nominal angle of attack due to
wind $\alpha_{w,n} = 0$, and the deflection angle $\delta_n = 0$. From Figure 7.60, we see that the angle of attack
may be written in terms of the pitch angle θ, the flight path angle γ, and the wind angle of
attack α_w as follows:

$$\alpha = \theta - \gamma - \alpha_w \tag{7.148}$$

Then for the nominal trajectory

$$\theta_n = \gamma_n \tag{7.149}$$

$$\alpha_n = \alpha_{w,n} = 0 \tag{7.150}$$

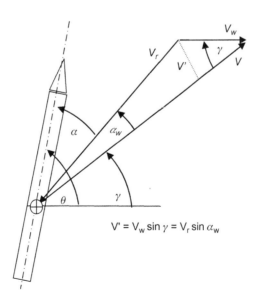

FIGURE 7.60

General velocity relationships for a launch vehicle in a crosswind.

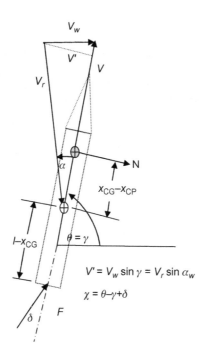

FIGURE 7.61

Velocity relationships for a nominal launch vehicle trajectory with $\theta = \gamma$ and a crosswind showing that $\alpha = -\alpha_w$.

The vehicle is assumed to encounter a wind disturbance V_w and the linearized analysis carried out by Teren et al. (1968) shows that for most cases the perturbed angle of attack is given by

$$\alpha \approx -\alpha_w \tag{7.151}$$

They point out that this result is equivalent to assuming a "zero drift" trajectory where the flight path angle perturbation during the wind encounter remains zero. The nominal case is shown in Figure 7.61 where we see that Eqns (7.149)–(7.151) are satisfied.

If the launch vehicle is trimmed so that there is no pitch change, then according to Eqn (7.90) the thrust deflection required is

$$\sin \delta \approx -\frac{C_{N,\alpha}qS(x_{CG}-x_{CP})}{F(l-x_{CG})}\alpha = \frac{C_{N,\alpha}qS(x_{CG}-x_{CP})}{F(l-x_{CG})}\alpha_w \tag{7.152}$$

Recall, from Figure 7.39, that the moment arm from the gimbal axis to the CG is $l - x_{CG}$. If we allow for the presence of a tail, the equilibrium equation(7.152) becomes

$$\sin \delta \approx \frac{qS}{F(l-x_{CG})}\left[C_{N,\alpha}(x_{CG}-x_{CP}) - C_{Nf,\alpha}\frac{S_f l_f}{Sd}\right]\alpha_w \tag{7.153}$$

In Figure 7.61, it is noted that the following relationship between the velocity components exists:

$$\sin \alpha_w = \frac{V_w}{V_r} \sin \gamma_n \tag{7.154}$$

Substituting Eqn (7.154) into Eqn (7.153) under the assumption that $\alpha_w \ll 1$ yields

$$\sin \delta \approx \frac{qS}{F\left(\frac{l}{d} - \frac{x_{CG}}{d}\right)} \left[C_{N,\alpha}\left(\frac{x_{CG}}{d} - \frac{x_{CP}}{d}\right) - C_{N_f,\alpha}\frac{S_f(l - x_{CG})}{Sd}\right] \frac{V_w}{V_r} \sin \gamma_n \tag{7.155}$$

Note that in Eqn (7.155), we have again used the approximation that $l_f = (l - x_{CG})$. To develop an estimate of the TVC requirement, we rewrite Eqn (7.155) as follows:

$$\sin \delta \approx \frac{\left(\dfrac{qS}{W_0}\right)}{\left(\dfrac{F}{W_0}\right)} \left[C_{N_f}\left(\frac{\dfrac{l}{d} - \dfrac{x_{CP}}{d}}{\dfrac{l}{d} - \dfrac{x_{CG}}{d}} - 1\right) - C_{N_f,\alpha}\frac{S_f}{S}\right] \frac{V_w}{V_r} \sin \gamma_n \tag{7.156}$$

We may discuss the nature and magnitude of the various terms in Eqn (7.156) serially. First let us consider the ratio of the dynamic pressure force to the initial weight of the launch vehicle

$$\frac{qS}{W_0} = \frac{\pi d^2 q}{4 m_0 g_E} \tag{7.157}$$

Using the correlation for launch vehicle diameter based on Eqns (7.142) and (7.143) and shown in Figure 7.51 (note that m_0 in the correlation is measured in metric tons) Eqn (7.157) becomes (with q in kPa)

$$\frac{qS}{W_0} \approx \frac{\pi q}{4 m_0 g_E}\left\{1.22\left[\frac{m_0}{(l/d)}\right]^{\frac{1}{3}}\right\}^2 = 0.1191 q\left[m_0\left(\frac{l}{d}\right)^2\right]^{-\frac{1}{3}} \tag{7.158}$$

Equation (7.158) shows that increased mass and fineness ratio l/d, which is equivalent to increased moment of inertia around the pitch axis, will reduce the TVC requirement posed by Eqn (7.156). For the typical range of vehicle mass and dimensions, such as those listed in Table 7.8, we find the following approximate range (with q in kPa):

$$0.0015q \le \frac{qS}{W_0} \le 0.006q \tag{7.159}$$

The maximum dynamic pressure experienced by launch vehicles is in the range of 20–40 kPa so that Eqn (7.159) suggests that the maximum value of qS/W_0 is less than about 0.25 and probably closer to 0.1.

Within our slender body assumption, Eqn (7.106) shows that the normal force slope for a cone−cylinder−flare alone, evaluated at a small angle of attack, is given by

$$C_{N,\alpha} = \frac{dC_N}{d\alpha} \approx 2\frac{S_b}{S} \tag{7.160}$$

For configurations with tails, Eqn (7.125) and the corresponding discussion suggests that

$$C_{N_f,\alpha} \frac{S_f}{S} = \frac{dC_{N_f}}{d\alpha} \frac{S_f}{S} \approx \pi \frac{S_f}{S} \tag{7.161}$$

Similarly, the results for CP and CG positions shown in Figure 7.58 for the Saturn V suggest that

$$\frac{\frac{l}{d} - \frac{x_{CP}}{d}}{\frac{l}{d} - \frac{x_{CG}}{d}} - 1 \approx \frac{4}{3} - 1 = \frac{1}{3} \tag{7.162}$$

Using the previous estimates for the terms appearing in Eqn (7.155) along with the assumption of a modest value of $F/W_0 = 1.3$ and $\sin \gamma_n \sim 1$, we may estimate a nominal value of Eqn (7.156) as follows:

$$\sin \delta \approx \frac{0.1}{1.3} \left(\frac{2 S_b}{3 S} - \pi \frac{S_f}{S} \right) \frac{V_w}{V_r} \sin \gamma_n \approx 0.05 \frac{S_b}{S} \left(1 - \frac{3\pi}{2} \frac{S_f}{S_b} \right) \frac{V_w}{V_r} (1) \tag{7.163}$$

From Eqn (7.163) we see that when no tail is used ($S_f = 0$) the thrust vector deflection necessary for a wind speed $V_w/V_r = 0.1$ and $S_b/S = 1.3$ is about 0.4°. However, using a tail with $S_f/S_b = 0.2$ would reduce the required thrust deflection to a negligible value. Thus Eqn (7.163) may be used as an aid in sizing a tail for mitigating the thrust deflection requirement. It must be recognized that in reality most of the terms in the basic TVC control equation (7.156) vary with the trajectory variables so that the thrust deflection needed will vary during the ascent.

An illustration of thrust deflection requirements during ascent is provided by considering a launch vehicle like the Saturn V. Using trajectory data provided by Pinson (1971) and the aerodynamic characteristics given by Walker (1968) we may consider one of the wind profiles proposed by Daniele and Teren (1968), shown in Figure 7.62. Applying the data for the Saturn V vehicle reported previously in this chapter to Eqn (7.156) leads to the thrust vector deflection history shown in Figure 7.63. Here the slender body theory was used to account for the cone–cylinder–flare

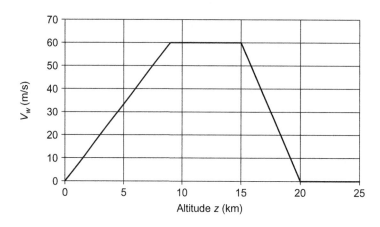

FIGURE 7.62

A typical profile of wind velocity as a function of altitude as proposed by Daniele and Teren (1968).

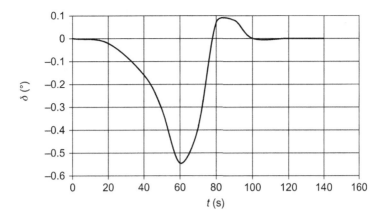

FIGURE 7.63

Time history of thrust vector deflection angle for a typical Saturn V launch.

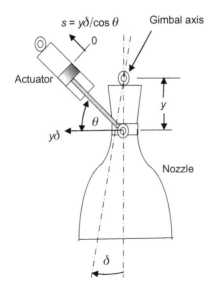

FIGURE 7.64

Thrust deflection system using a hydraulic actuator to pivot nozzle about gimbal axis.

contribution to the normal force coefficient slope. If the more elaborate empirical methods discussed at the end of Section 7.6.6 are applied, the thrust vector deflections would be approximately 30% larger than indicated in Figure 7.63.

Although movements of a rocket engine nozzle through fractions of a degree appear to be difficult to achieve one must keep in mind that the machinery is quite large in scale. The simple arrangement shown in Figure 7.64 is representative of actual TVC technology. Choosing, as nominal dimensions in Figure 7.64, $y = 1$ m, $\theta = 45°$, and $\delta = 1°$, we find $y\delta = 1.75$ cm and $s = 2.46$ cm, displacements that are readily achievable and controllable.

7.7 GENERAL LAUNCH VEHICLE DESIGN CONSIDERATIONS

A useful reference for Space Shuttle operations and specifications may be found in the Space Shuttle Reference Manual, SSRM (1988). A general description of the launch vehicle design process has been presented by Blair, Ryan, Schutzenhofer, and Humphries (2001).

7.7.1 LIQUID PROPELLANT TANKS

At this point in the design process the general characteristics of the launch vehicle have been established: the number of stages, the structural mass, number, and type of engine for each stage, the selection of propellants for each stage, the characteristics of the motion along the trajectory and general mass properties such as center of gravity, center of pressure, mass distribution, moment of inertia, and total mass. The preliminary design then moves on to more detail design to define tank shape and size, tank internal pressure requirements, based on propellant type and required volume. Additional constraints applied include factor of safety, criteria for failure modes, methods for combining loads, and ability to stand on the launch pad (unpressurized) with any or all other stages full or empty. A general analysis of propellant tanks suitable for preliminary design is presented here. For additional details, consult the design guidelines for propellant tanks presented by Huzel and Huang (1967, 1992) and Wagner (1974).

The general configuration of the tank system of a large liquid bipropellant rocket is shown in Figure 7.65. In such applications, the fuel and oxidizer are typically fed to the powerhead, which contains the combustion chamber, by means of turbopumps. The fuel and oxidizer tanks are pressurized by external means to ensure the pressure level at the turbopump inlet is sufficiently high to avoid cavitation within the turbopump, typically $200-700$ kPa. Separate external tanks for pressurizing the main tanks are illustrated in Figure 7.65 although cryogenic propellants, for example, may be tapped off the main tanks, passed through a heat exchanger to gasify the fuel and/or the oxidizer and increase its pressure. The fuel and/or oxidizer gas may then be ducted back to the appropriate main tank to maintain the proper ullage pressure. The thrust produced by accelerating the products

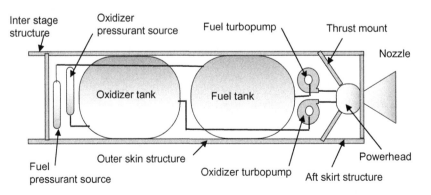

FIGURE 7.65

General layout of propellant tank system for a large lower stage liquid bipropellant rocket.

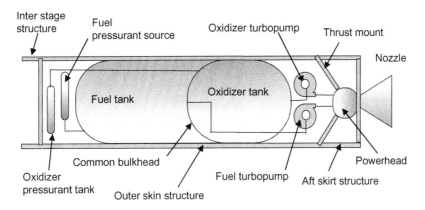

FIGURE 7.66

General layout of propellant tank system for upper stages of a large liquid bipropellant rocket.

of propellant combustion through the nozzle is transferred through the thrust mount and into the outer skin structure to which the propellant tanks are attached. This layout is representative of the Saturn V first stage (S-IC) and the external tank (ET) of the Space Shuttle. In the S-IC, the propellant was LOX/RP1 with approximately equal volumes for the oxidizer and fuel tanks. The SSME propellant was LOX/LH$_2$ resulting in an oxidizer tank about one-third the volume of the fuel tank because of the very low density of hydrogen.

A variation of this tandem tank configuration is one in which the fuel and oxidizer tanks share a common bulkhead, as shown in Figure 7.66. Such a system was employed on the second stage (S-II) and third stage (S-IVB) of the Saturn V launch vehicle, both of which used LOX/LH$_2$ as the propellant. In these cases, the oxidizer tank is again about one-third the volume of the fuel tank because of the very low density of hydrogen.

For lower thrust applications a purely pressure-fed system is used, as shown in Figure 7.67. The helium tank provides pressurization in the range of 1.4–2.8 MPa for both propellants, shown here as a nitrogen tetroxide (N$_2$O$_4$) oxidizer and a monomethyl hydrazine (MMH) fuel. This combination is hypergolic, meaning that the propellants react on contact so there is no need for an ignition system. This approach was used as the auxiliary propulsion system (APS) on the Saturn V S-IVB stage for vehicle attitude control as well as on the Space Shuttle Orbiter for its orbital maneuvering system (OMS).

Tanks for subsystems, like those used for orbital maneuvering or attitude control, operate at much lower thrust than launch vehicles, and may be designed in three basic configurations: pressurized liquid propellant, liquid propellant with a positive displacement expulsion device, and pressurized gas propellant. A purely liquid propellant system may operate in the bipropellant mode using an oxidizer and a fuel as illustrated in Figure 7.67 or in the monopropellant mode using solely MMH or hydrogen peroxide passing over a catalyst to generate the necessary propulsive gas. Under conditions of weightlessness, as in orbit, propellants will float freely within the tank unless confined by a positive expulsion device like a metallic bellows, a piston, or a collapsible bladder and thereby fed to the engine for expansion through the nozzle. Purely gas propellant systems rely

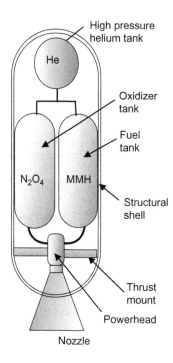

FIGURE 7.67

General layout of a propellant tank system for attitude control or orbital maneuvering using a liquid bipropellant rocket.

on storing gas at very high pressure (20 MPa) for expulsion through a nozzle after electrical heating (e.g., a resistojet) or at the ambient temperature of the tank. For such systems, the greater structural efficiency of spherical tanks yields a weight advantage over other geometries and as such should be designed to the highest working pressure consistent with the capabilities of the material and the downstream exhaust system to achieve the minimum possible diameter.

The mass of the propellant storage tanks constitutes a significant part of the structural mass of a launch vehicle. Any excess mass of the propellant tanks reduces the payload mass so fuel tank design must be carefully performed. The tanks are subject to both static and dynamic loads. The main static loads are as follows:

- Hydrostatic pressure—due to the state of the fluid contained
- Overpressure—sometimes used to avoid cavitation in the fuel pumps
- Payload weight—the tanks generally also serve as structural members.

The dynamic loads are as follows:

- Inertia loads—due to vehicle acceleration
- Control loads—due to thrust vectoring induced bending
- Aerodynamic loads—due to vehicle geometry and wind shear.

Having calculated the mass of each propellant required to accomplish the mission, including a small reserve for contingencies, one may determine the tank volume needed. The density of each liquid propellant is known as a function of the environmental temperature so that the initial volume of the propellant is determined from $v_p = m_p/\rho_p$. Some additional volume above the free surface of the liquid in the tank v_u, called ullage, is needed for the gas used to pressurize the tank and this volume must be consistent with the predicted operational environment. The initial ullage is at least 3% of the total tank volume. Extra volume v_i must also be provided to account for unusable propellant which remains trapped in feed lines and in the internal structural details of the tank. If a cryogenic propellant is used, the loss of propellant due to boil-off must be accounted for by an appropriate increment during filling and this requires a volume v_b. The volume of the tank is then given by

$$v_t = v_p + v_u + v_i + v_b \tag{7.164}$$

Once the total fluid volume required is established, the tank shape may be selected. Large tanks are typically cylinders with hemispherical or ellipsoidal end caps; other shapes are not often used because they are usually heavier than these simpler shapes. Wagner (1974) illustrates the variety of end cap shapes that have been analyzed and shows that ellipsoidal end caps with a ratio of minor to major axis of about 0.7 tend to have the minimum weight as well as a smaller overall length for a given diameter. Smaller subsystem level tanks are usually simple spheres or cylinders with hemispherical end caps.

Liquid rocket propulsion system tanks are thin wall structures of thickness much less than one-tenth the tank radius. Therefore simple membrane stress formulas are acceptable for analyzing simple geometrical shapes like spheres and cylinders. Using mechanical property data along with arbitrarily selected factors of safety permits preliminary selection of tank materials, wall thickness, and operating stress levels. In the case of stresses due to internal pressure a design safety factor $k_d > 1$ is used to define the maximum allowable working stress. The yield or ultimate stress (σ_y or σ_u) of the material is divided by the safety factor k_d to define a maximum allowable working stress which accounts for possible variations in loads, construction quality, material properties, etc. Design safety factors for vehicle tanks range from 1.0 to 1.1 for yield stress and from 1.25 to 1.5 for ultimate stress, the higher values in each category generally being used for manned flight vehicles.

We assess the basic loading on the tanks by assuming that they will be of thin shell construction. Sections of a cylindrical shell and a spherical shell are shown in Figure 7.68 along with the forces acting due to internal pressure. The shells are thin in the sense that the thickness is everywhere much smaller than the length and/or radius of the shell, i.e., $t \ll l, R$. Also shown in Figure 7.68 is the axial load on a cylindrical shell tank with spherical shell end caps. In an infinitesimal segment dl of an infinite cylindrical tank, as shown in Figure 7.68(a), equilibrium of forces produced by the internal pressure p requires the minimum thickness t_c based on the yield stress σ_y of the tank material to be

$$t_c = R_c \frac{p}{(\sigma_y/k_d)} \tag{7.165}$$

However, along an equator of a sphere, as shown in Figure 7.68(b), the equilibrium of forces due to internal pressure p requires that the minimum thickness t_s based on the yield stress σ_y of the tank material to be

$$t_s = \frac{1}{2} R_s \frac{p}{(\sigma_y/k_d)} = \frac{1}{2} t_c \tag{7.166}$$

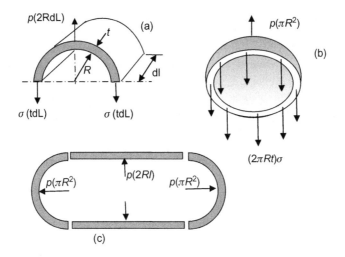

FIGURE 7.68

(a) Stress in circular cylindrical tank due to internal pressure force, (b) stress in spherical tank due to internal pressure force, and (c) forces due to internal pressure on a circular cylindrical tank with hemispherical end caps e.

Thus the wall of a circular cylindrical tank with spherical end caps must be twice as thick as that of a spherical tank of the same radius. However, by increasing the length of the circular cylindrical tank the volume enclosed can be arbitrarily increased while keeping the radius of the tank fixed. The mass of the two types of tank made of the same material leads to the following results for a spherical tank of radius R_s and a circular cylindrical tank of radius R_c with hemispherical end caps and overall length l, respectively:

$$m_s = 2\pi R_s^3 \frac{pk_d}{(\sigma_y/\rho_w)} \tag{7.167}$$

$$m_c = 2\pi R_c^3 \left(\frac{l}{R_c} - 1\right) \frac{pk_d}{(\sigma_y/\rho_w)} \tag{7.168}$$

Here ρ_w is the density of the tank material, k_d is the design factor of safety, and the ratio of yield stress to material density σ_y/ρ is called the specific yield stress. Typical behavior of the specific stress of three common metallic tank materials is shown as a function of operating temperature in Figure 7.69. Stainless steel and aluminum alloys had been favored for some time and now titanium alloy and aluminum–lithium alloy are being used more widely because of their higher specific strength. For example, at room temperature aluminum–lithium alloy 2195 has a specific strength 60% greater than that of aluminum alloy 2219. The latest version of the Space Shuttle external tank employed aluminum–lithium alloy to reduce the weight of the tanks which carry a total of 729 t of propellant. Filament wound tanks with aluminum liners provide higher specific strength than metallic materials alone because the fiber is completely in tension when containing the hoop stress due to internal pressure. Fiberglass in epoxy resin had been used in the past for such tanks up to about 70 °C and now carbon fiber has supplanted it and is used up to 150 °C.

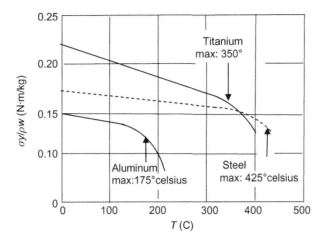

FIGURE 7.69

Approximate behavior of the specific yield stress of three common metallic tank materials is shown as a function of operating temperature. The recommended maximum working temperatures are also indicated.

Some improved resins can increase the operating temperature to as much as 250 °C. Hybrid tanks where only the cylindrical section is filament wound for support have also been used successfully according to Tam, Hersh, and Ballinger (2003).

Comparing tanks of equal volume contained, the ratio of the mass of a circular cylindrical tank with spherical end caps to a spherical tank is given by

$$\frac{m_c}{m_s} = \frac{4\left(\dfrac{l}{R_c} - 1\right)}{3\dfrac{l}{R_c} - 2} \tag{7.169}$$

Equation (7.169) is based on the supposition that the wall thickness in the end caps of the cylindrical tank is that appropriate to a sphere of equivalent radius. For $l/R_c = 2$ the cylindrical tank becomes a spherical one and $m_c/m_s = 1$ while for a long cylindrical tank with $l/R_c \gg 1$ the ratio m_c/m_s approaches 4/3. Therefore the lightest tank to contain a given volume is a spherical one; however, it will also have the largest radius because

$$\left(\frac{R_c}{R_s}\right) = \left(\frac{3}{4}\frac{l}{R_c} - \frac{1}{2}\right)^{-\frac{1}{3}} \tag{7.170}$$

For example, a cylindrical with a length to diameter ratio of 5.66 will have a radius half that of a spherical tank containing the same volume. Because the aerodynamic drag is proportional to the frontal area the cylindrical tank will experience one-quarter that of the spherical tank. On the other hand, using a small diameter tank means that the tank will be very long, with the attendant problem of buckling under bending loads. Thus the driving factor is how large a diameter tank is practical. Spherical tanks are commonly used for small propellant volume applications because of their

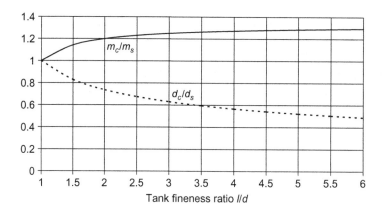

FIGURE 7.70

The mass ratio m_c/m_s and the diameter ratio d_c/d_s of cylindrical and spherical tanks containing equal volumes are shown as a function of the cylindrical tank fineness ratio l/d_c.

superior weight characteristics. An illustration of the variation of the mass ratio m_c/m_s and the tank diameter ratio d_c/d_s for tanks of equal volume is shown as a function of the cylindrical tank fineness ratio l/d_c in Figure 7.70.

Note that a 40% reduction in diameter is realized when using a cylindrical tank of fineness ratio $l/d = 4$ with a penalty of less than a 30% increase in weight. The choice of tank configuration must also account for dynamic loading in addition to the hydrostatic load. Aerodynamic drag is proportional to the frontal area so spherical tanks will suffer a disadvantage. However, the inertial load due to acceleration will increase the hydrostatic pressure loading and this increase is proportional to the length of the tank, so that here the spherical tank has an edge. Mounting and securing the tanks also provides extra weight that must be taken into account.

For a cylindrical tank with a diameter of 2.54 m and a length of 7.62 m under an internal pressure of 345 kPa and a reasonable temperature environment, the minimum thickness would be 0.32 mm for a steel tank and 1.06 mm for an aluminum tank. Practice suggests that stock thickness for tank materials should be greater than 0.25 mm and 0.5 mm for stainless steel and aluminum alloy, respectively. It is clear that thin-walled cylinders can support a substantial pressure load, but they are prone to failure by buckling. The critical (buckling) compressive axial stress for a cylinder having no stiffeners is given empirically by

$$\frac{\sigma_{crit}}{E} = 9\left(\frac{t}{R_c}\right)^{1.6} + 0.16\left(\frac{t}{L}\right)^{1.3} \tag{7.171}$$

Considering the cylinder to have a stainless steel skin with modulus of elasticity $E = 190$ GPa and yield strength $\sigma_y = 1380$ MPa the buckling stress is $\sigma_{crit} = 5.4$ MPa while for 6000 series aluminum alloy with $E = 71$ GPa and yield strength $\sigma_y = 414$ MPa the buckling stress is $\sigma_{crit} = 6.5$ MPa. Note that for either material the buckling stress is much less than the yield stress.

Thus buckling is the critical factor and as a result cylindrical tanks which must support axial loads may require both longitudinal and ring stiffeners, which add to the weight. The critical compressive axial force is

$$F_{a,crit} = (\pi dt)\sigma_{crit}$$

For the tanks being considered as examples, the critical axial loads are 18.1 kN and 50.8 kN for the stainless steel and 6000 series aluminum alloy, respectively. On the other hand, because the tanks are pressurized to a level $p' = 345$ kPa they are preloaded in tension by a force of $\pi R_c^2 p' = 1747$ kN. Therefore the pressurized tanks could support compressive loads considerably larger than the critical buckling load. Thus internal pressurization of the tanks aids in stiffening the cylinders, acting like a balloon to delay buckling to higher load levels. This stiffening approach is called pressure stabilization.

Atlas launch vehicles and Agena second stages used pressure stabilization to achieve extremely lightweight propellant tanks. The Atlas RP-1 301 stainless steel fuel tank measured 3 m in diameter by 24.4 m long with thickness varying from 0.71 to 0.97 mm and a maximum operating pressure of 515 kPa while the LOX 301 stainless steel tank measured 3 m in diameter for 12.2 m and then reduced in diameter to 1.2 m over the next 3 m with thickness varying from 0.43 to 0.71 mm and a maximum operating pressure of 280 kPa. The Agena 6061-T6 aluminum UMDH fuel tank was 1.5 m in diameter and 1.7 m long with a thickness of 1.52 mm and a maximum operating pressure of 480 kPa while the 6061-T6 aluminum RFNA oxidizer tank was 1.5 m in diameter and 2.3 m long with a thickness of 1.52 mm and a maximum operating pressure of 480 kPa. Though pressure stabilization works well it poses practical difficulties in handling the fragile tanks and requires complex controls to ensure that pressurization is maintained to avoid buckling failure. As a consequence of these practical considerations, it is generally necessary to stiffen the sidewalls of large tanks by attaching stringers and frames as in an airplane fuselage, or by an orthogonal array of ribs machined into the sidewall in a waffle pattern, in addition to any stiffening provided by the pressure in the tank. Methods for analyzing such strengthening mechanisms are beyond the scope of this book; see, for example, Weingarten, Seide, and Peterson (1968) for a review of the problem in the case of propellant tanks, and Jones (2006) for fundamental analyses.

In addition to internal pressure loads we must consider the effect of inertial loads on the tank. Consider the cylindrical tank shown in Figure 7.71 containing liquid propellant of density ρ_p with an ullage region containing a gas at pressure p'. For simplicity in the calculations we arbitrarily assume that the ullage volume is equal to the upper end cap hemispherical volume, as illustrated in Figure 7.71. The same approach may be used for other choices for the location of the liquid free surface and ullage volume.

Assuming that the tank is moving vertically (the negative y-direction in Figure 7.71) with acceleration ng, the hydrostatic pressure in the liquid propellant through the length of the tank is given by

$$p = p' + \rho_p n g_E y \tag{7.172}$$

From Eqn (7.165), the minimum thickness required for the cylindrical portion of the tank is

$$\frac{t_{cp}}{d} = \frac{p}{2(\sigma_y/k_d)} = \frac{p' + \rho_p n g y}{2(\sigma_y/k_d)} \tag{7.173}$$

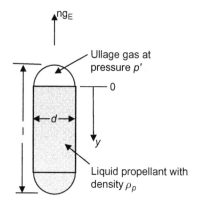

FIGURE 7.71

Schematic diagram of a cylindrical tank subjected to a load factor n due to vertical acceleration.

The minimum thickness required for the cylindrical portion of the tank clearly depends upon the length of the tank and the acceleration experienced. If the density of the wall material is ρ_w, then the mass of any segment of length dy is $dm_c = (\rho_w \pi d)t\,dy$ so that the total mass of the cylindrical portion, whose length is l_c, is given by

$$m_{cp} = \int_0^{l_c} \rho_w \pi d^2 \frac{p' + \rho_p n g y}{2(\sigma_y/k_d)} \, dy$$

Noting that $l_c = l - d$ and carrying out the integration yields

$$m_{cp} = \rho_w \frac{\pi d^3}{4(\sigma_y/k_d)} \left[2p' \left(\frac{l}{d} - 1\right) + \rho_p n g d \left(\frac{l}{d} - 1\right)^2 \right] \tag{7.174}$$

The development above assumes that the location of the free surface $y = 0$ remains fixed with respect to time. For upper stages that are not firing this is correct and the maximum value of the load factor may be used to set the maximum cylinder weight. However, for propellant tanks in an operating stage, the load factor is increasing during the flight while the free surface is moving closer to the bottom of the tank as the propellant is consumed, continually emptying the tank. On the basis of the analysis at the beginning of this chapter concerning vehicle acceleration and rate of propellant consumption, one may show that the maximum inertial pressure increase occurs early in the flight, when the acceleration is quite modest, say around $1.2g_E$, and continually decreases as the ascent progresses. Therefore for the first stage of manned vehicles the maximum thickness of the cylinder may be determined on the basis of the magnitude of the initial vehicle acceleration and full tanks of fuel and oxidizer. But for inoperative upper stages full of propellant, the thickness, and therefore the mass, of their tanks should be based on the highest acceleration to be experienced.

The minimum thickness of the spherical end caps is given by Eqn (7.166) yielding for the upper and lower caps, respectively, the following results:

$$\frac{t_{sc,u}}{d} = \frac{p'}{4(\sigma_y/k_d)} \tag{7.175}$$

$$\frac{t_{sc,l}}{d} = \frac{p' + \rho_p ng(l - d/6)}{4(\sigma_y/k_d)} \tag{7.176}$$

The same arguments concerning the inertial effects for operating and passive stages follow here and the mass of the two end caps is therefore

$$m_{sc} = \rho_w \frac{\pi d^2}{2}(t_{sc,u} + t_{sc,l}) = \rho_w \frac{\pi d^3}{4(\sigma_y/k_d)}\left[p' + \frac{1}{2}\rho_p ng\left(\frac{l}{d} - \frac{1}{6}\right)\right] \tag{7.177}$$

Then the total mass of the tank is

$$m_t = \rho_w \frac{\pi d^3}{4(\sigma_y/k_d)}\left\{2p'\left(\frac{l}{d} - \frac{1}{2}\right) + \rho_p ngd\left[\left(\frac{l}{d} - 1\right)\left(\frac{l}{d} - \frac{1}{2}\right) + \frac{5}{12}\right]\right\} \tag{7.178}$$

Note that if one had assumed that the pressure is constant through the lower end cap and equal to that just at the bottom of the cylindrical portion of the tank, then the fraction 5/12 would not appear in Eqn (7.178).

In the determination of an acceptable wall thickness, practical criteria must be employed. A material with a high strength/density ratio may provide a weight advantage as far as wall thickness is concerned, but other factors may serve to increase the weight of the complete tank. Wagner (1974) points out an example: fiberglass appeared to provide the lightest weight tank for the Apollo Service Module, but the addition of reinforcements for various loads increased the weight substantially so that 6AI-4V titanium alloy ultimately was used. Sometimes an acceptably thin tank wall may be difficult or impossible to fabricate. Again, Wagner (1974) offers as an example the shells for titanium tanks in the Apollo Service Module, which could be as thin as 0.011 in. based on pressure alone, but fabrication constraints led to the 0.023-in. thickness that was ultimately used.

The mass of propellant carried in the tank (note that the ullage volume has been assumed to be equal to that of one end cap and the gas mass therein is neglected) is

$$m_p = \rho_p\left[\frac{\pi d^3}{12} + \frac{\pi d^2}{4}(l - d)\right] = \rho_p \frac{\pi d^3}{4}\left[\left(\frac{l}{d} - \frac{2}{3}\right)\right] \tag{7.179}$$

Having selected the propellant composition previously, the propellant density ρ_p may be determined, for example, from Table 7.1 which covers the most widely used fuel-oxidizer combinations for propellants. Thus both the oxidizer density ρ_o and the fuel density ρ_f are fixed and the oxidizer to fuel ratio O/F is known for the propellant chosen. Because the diameter of the vehicle has already been estimated in Section 7.6.5 it forms an upper bound for the tank diameter d, which would be the same for both tanks for a tandem tank configuration like those shown in Figures 7.65 and 7.66. For other arrangements like the side-by-side mounting in Figure 7.67 or the Space Shuttle ET the tank diameter is dictated by other factors, such as aerodynamic or structural requirements. The total mass of propellant m_p required for the mission was determined in the trajectory analysis of Sections 7.5.2–7.5.4 and the oxidizer and fuel components are given by

$$m_p = m_o + m_f = m_f[(O/F) + 1] \tag{7.180}$$

The corresponding fineness ratio l/d is now determined by Eqn (7.179). The trajectory designed for the mission likewise poses known limits on the load factor n arising from vehicle acceleration. Thus the only parameters available in Eqn (7.178) for controlling the tank mass are the ullage pressure p' and the specific strength of the tank material σ_y/ρ_w. The usual range of ullage pressure is 200 kPa $< p' <$ 700 kPa for turbopump-fed stages and 700 kPa $< p' <$ 2.8 MPa for pressure-fed stages while fineness ratios of individual tanks are generally less than 4.

There are several tank components that are important for effective tank operation but add mass unaccounted for in the purely structural calculations. The tanks have thus far been considered to contain only liquid propellant along with gas occupying the ullage volume. The unrestrained liquid propellant may be readily set into sloshing motion upon small vehicle accelerations not aligned with the local gravitational acceleration. Such bulk movement of the propellant can adversely influence vehicle stability as well as the integrity of the tank structure. To damp out such motions baffles are installed within the tank. In cylindrical tanks, flat rings attached to the sidewalls serve the dual purpose of limiting fluid motion and stiffening the sidewall of the tank against buckling. Other fluid motions which affect tank operation are vortices which may form at propellant outlet ports. Radial vanes are generally installed within the tank to retard the formation of vortices.

So-called propellant positioning devices provided to ensure proper propellant feed when weightless during orbital flight also add to the structural mass load. These include positive expulsion devices which mechanically force propellant through the feed tubes to the engine.

Furthermore, if cryogenic propellants are used, thermal insulation is required to retard loss of propellant through boil-off as well as to exploit increased strength of tank materials at very low temperatures. For example, the yield stress for two aluminum alloys used for propellant tanks at several characteristic temperatures is shown in Table 7.13. Applying insulation to the inside of the tank keeps the tank material at or near the warmer ambient temperature so as to preserve ductility and toughness. On the other hand, applying insulation to the outside of the tank allows the tank to take advantage of increased yield strength. For example, the Saturn S-IVB (Saturn V third stage) has internal insulation while the Saturn S-II (Saturn V second stage) has external insulation. Insulation may be applied to the tank walls by bonding, bolting, or spray foaming.

In order to gain an appreciation of the complexity of flight weight propellant tanks let us consider an optimized structural design for a 10-m-diameter metallic tank for LH_2 as considered in detail by Johnson, Sleight, and Martin (2013). This design was used as a baseline against which composite tank designs which promise substantial weight savings would be compared. The basic design and dimensions are shown in Figure 7.72. A design condition is the containment of LH_2 at 20 K and a

Table 7.13 Yield Stress of Two Aluminum Alloys at Cryogenic and Room Temperatures

	Aluminum 2219	Aluminum 2195[a]
Density (kg/m³)	2840	2685
σ_y (MPa) at LH_2 temperature (20 K)	483.5	608.6
σ_y (MPa) at LOX temperature (90 K)	462.9	589.9
σ_y (MPa) at room temperature (298 K)	388.2	521.6

[a]Aluminum–lithium alloy.

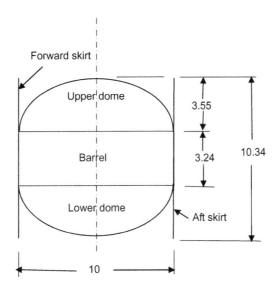

FIGURE 7.72

Liquid hydrogen tank design considered by Johnson et al. (2013). All dimensions are in meters

pressure of 318.4 kPa. The baseline metallic tank was to be constructed of aluminum lithium alloy 2195 whose properties are listed in Table 7.13. The barrel was a circular cylinder while the domes were ellipsoidal with a ratio of minor to major semiaxis equal to 0.707 (Figure 7.72).

The wall thickness of the cylindrical barrel section is simply

$$t_c = \frac{pR_c}{(\sigma_y/k_d)} = \frac{(318,400 \text{ N/m}^2)(5 \text{ m})}{\left(\dfrac{608.6 \times 10^6 \text{ N/m}^2}{1.1}\right)} = 2.88 \text{ mm}$$

The dome thickness can be estimated on the basis of a hemispherical end cap with a radius equal to the crown radius of the ellipsoid $R_e = R_c/0.707 = 7.07$ m so that we may approximate the dome thickness as follows:

$$t_d \approx \frac{1}{2}\frac{pR_e}{(\sigma_y/k_d)} = \frac{1}{2}\frac{(318,400 \text{ N/m}^2)(7.07 \text{ m})}{\left(\dfrac{608.6 \times 10^6 \text{ N/m}^2}{1.1}\right)} = 2.03 \text{ mm}$$

The volume of the tank is

$$v_t = 2\left[\frac{2}{3}\pi R_c^2(0.707R_c)\right] + \pi R_c^2 l_c = 370.1 + 254.5 = 624.6 \text{ m}^3$$

The volume of the walls of the tank is

$$v_w = 2\pi\left[R_c l_c t_c + R_c^2 t_d + \frac{1}{4}\frac{R_c^2 t_d}{e}\ln\left(\frac{1+e}{1-e}\right)\right] = 2\pi[0.0467 + 0.0508 + 0.0444] = 0.892 \text{ m}^3$$

The mass of the tank, based solely on the prescribed pressure load is then

$$m_t = \rho_w v_w = (2685 \text{ kg/m}^3)(0.892 \text{ m}^3) = 2393 \text{ kg}$$

The optimized design presented by Johnson et al. (2013), taking into account stiffening, fasteners, closures, skirts, etc., is 4349 kg which is twice as large as that given by a purely membrane analysis. Note that the mass of LH_2 carried by the tank is

$$m_p = \rho_p v_t = (69.5 \text{ kg/m}^3)(829 \text{ m}^3) = 57,616 \text{ kg}$$

The tank mass to propellant mass is

$$\frac{m_t}{m_p} = \frac{4349 \text{ kg}}{57,616 \text{ kg}} = 0.075$$

Three composite tanks designed by Boeing, Lockheed-Martin, and Northrop Grumman were less than 3000 kg in mass, thereby exceeding NASA's target of a 30% saving in mass over the state-of-the art aluminum lithium alloy metallic tank. NASA successfully tested a 2.4-m subscale composite tank built by Boeing in 2013 and a 5.5-m tank is under construction for subsequent test trials (Morring, 2014).

As mentioned previously, low-thrust propulsion applications like orbital maneuvering generally use pressurized gas propellants. The tanks are spherical, usually made of titanium, and contain inert propellant gases like helium or nitrogen at high pressure, usually around 20 MPa. Using data for 14 spherical titanium tanks listed by Wagner (1974), a simple correlation for the mass of the tank m_t in kilograms was constructed in terms of the tank diameter d in meters and the design burst pressure $p_b = p(\text{FOS})$ in kilopascals where p is the nominal operating pressure and FOS is the factor of safety. The correlation is given by

$$m_t \approx 0.004 p_b d^3 \tag{7.181}$$

Equation (7.181) yields tank masses within about +10%, as illustrated in Figure 7.73. This suggests that 250 kPa-m^3/kg is an appropriate scaling factor for spherical titanium pressure tanks based on burst pressure. Tanks made with Kevlar wrapping over metallic liners, called composite overwrapped pressure vessels (COPV), have supplanted purely metallic pressure vessels on manned spacecraft like the Space Shuttle providing significant weight reductions. Fully composite tanks using materials like graphite epoxy provide about a 20% improvement over metallic tanks developing a scaling factor of 300 kPa · m^3/kg. Theoretically, the mechanical properties of composite materials may be tailored as necessary, and therefore the tanks need not necessarily be constrained to spherical shape. However, in the final analysis the factors of cost and ease of manufacture must be given due weight.

7.7.2 SOLID PROPELLANT ROCKETS

The solid rocket motor consists of a mixture of oxidizer and fuel in the form of a solid completely encased in a tank which serves both as pressure vessel and support for the propulsive nozzle. Thus there is no need for separate tanks of oxidizer and fuel or for means to introduce them into the

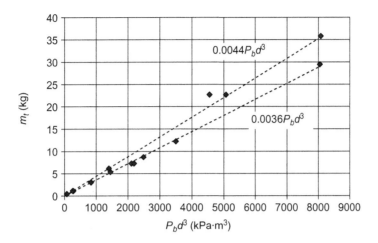

FIGURE 7.73

Correlation of the mass, in kilogram, of spherical titanium pressurized gas tanks with the product of tank burst pressure in kilopascal and diameter d in meters.

combustion chamber in the correct ratio. The solid rocket tank, or case, must contain the high pressure gases formed by the chemical reaction of the fuel and the oxidizer and therefore it is relatively heavy. On the other hand, because of its sturdiness it requires no special means for structurally stabilizing it. Solid rocket motors are discussed in more detail by Sforza (2012).

The main advantage of solid rocket motors is simplicity. They have no moving parts, no injection system, and do not have to be fueled prior to launch. Therefore the storage, handling, service, and auxiliary equipment connected with solid rockets are much simpler than those for liquid rockets. Because of the low part count and few, if any, moving parts, the reliability of solid rockets is high, around 99%. For the same reasons, the payload mass ratio is high and the overall cost low compared to liquid rockets. These attributes of simplicity of structure, easy storability of the solid propellant, and virtually instantaneous availability for launch make the solid rocket motor an attractive propulsion device.

The solid propellant contains both fuel and oxidizer and they are ignited by electric or pyrotechnic means in order to permit gasification to initiate. Thereafter the propellant burns in the gas phase with the heat of the combustion being sufficient to continue the gasification process. The rocket design is very simple, as shown below in Figure 7.74.

Such assets do not come without penalties. For example, since the solid propellant mixture must be cast into the required shape as one piece there are limitations in the size of the propellant blocks achievable. Thus, application of solid rocket motors to lower thrust levels is more attractive and clusters of smaller such motors are typically used to achieve higher thrust levels. However, there are successful large-scale solid rocket motors, like those used to aid in launching the Space Shuttle, which produce more than 11MN (2.47 million pounds of thrust). However, the performance, as measured by specific impulse, is lower relative to LOX–LH₂ rocket engines and

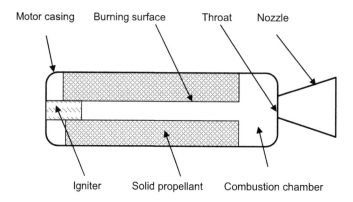

FIGURE 7.74

Schematic diagram of a solid propellant rocket motor.

in addition modulating and terminating the thrust are both difficult. Burn duration is relatively short compared to that in liquid rockets and therefore the total impulse delivered is smaller. Finally, there is no means for adequately cooling the nozzle of a solid rocket motor like there is in a liquid rocket motor where the liquid propellants may be circulated through cooling jackets built into the nozzle. The performance of solid propellants is quite sensitive to temperature and the associated manufacture of the energetic materials is involved, costly, and difficult. As a result, solid rockets have had limited application to manned launch vehicles. Only the Space Shuttle has made use of them, and then purely as boosters to provide the very large thrust level necessary.

A good illustration of a large-scale solid rocket motor is the reusable solid rocket motor (RSRM) built by ATK and which is the major component of the Space Shuttle solid rocket booster (SRB) system. The SRB is shown schematically in Figure 7.75. The RSRM is 38.43 m long and 3.708 m in diameter and it is housed in the SRB which is 45.47 m long. The configuration is made up of four motor segments with an igniter system in the forward segment and an articulated nozzle for thrust vectoring attached to the aft segment.

The solid propellant grain is composed of the following:

70% Ammonium perchlorate (NH_4ClO_4)
14% Polybutadiene acrylonitrile (PBAN)
16% Aluminum powder
0.07% Iron oxidized powder as a catalyst

Ammonium perchlorate produces hydrochloric acid (HCl) in the combustion products and forms a white cloud when exhausted into even mildly humid air. PBAN is a polymeric rubber-based binder that also serves as fuel. Aluminum is added to the grain to enhance the thrust. The RSRM operating temperature range is from 20 °F to 120 °F (-6 °C to 49 °C). The total mass of propellant is about 5×10^5 kg while the total mass of the RSRM is about 5.698×10^5 kg. The burn-out weight

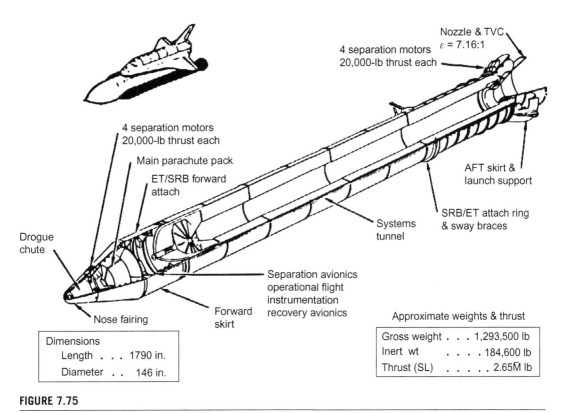

FIGURE 7.75

Schematic diagram of the SRB for the space shuttle.

is about 63,840 kg which suggests a propellant mass fraction of roughly 89%. The average vacuum level of thrust is 11.52MN (2.59×10^6 lb) and the associated vacuum-specific impulse is 268 s. This is about 30% more thrust than an F-1 engine used on the Saturn V but at about the same vacuum-specific impulse. The action time, the elapsed time from a specified pressure (here 3.88 MPa) to a specified final pressure (152 kPa) is about 124 s. Expanded versions of the SRM are planned for the Space Launch System which will combine the manned capsule technology of the of Saturn V/Apollo program and the solid and liquid rocket engine technology of the Space Shuttle to provide access to space for humans. The new larger SRMs use five solid propellant segments rather than the four segments used for the Space Shuttle launch system.

A new large solid rocket motor manufactured by Avio in Italy, the P80, is designed for future evolution of the boosters for the ESA Ariane-5 launch vehicle. It is 3 m in diameter and 10.5 m long and has a mass of 88,000 kg. The propellant is based on hydroxyl terminated polybutadiene (HTPB) fuel and has high aluminum powder content. The maximum thrust is about 3040 kN (683 klb) and the burn time is about 107 s. The specific impulse is given as 280 s. Another large solid rocket motor is the J-1 first stage motor developed by the National Space Development Agency of Japan (NASDA). This is a 21-m-long and 1.8-m-diameter rocket motor with a mass of 71 t (71,000 kg). The propellant is also based on polybutadiene fuel and delivers 159 t of thrust at a specific impulse of 273 s with a burn time of 94 s.

7.8 SUMMARY OF CONSTANTS AND PARAMETERS

Some constants and conversion factors which are useful in dealing with the subjects in this chapter are collected below along with Figure 7.76 which gives the characteristics of several uniform solids.

Gravitational acceleration at earth's equator $g_E = 9.80665$ m/s
Gravitational constant for earth $k = 398,600$ km^3/s^2
Mean radius of earth $R_E = 6371$ km
Rotational period of the earth $= 23$ h 56 m
Circular orbit velocity at earth's surface $V_E = 7.9087$ km/s
Scale height of the atmosphere $H = 7.16$ km.

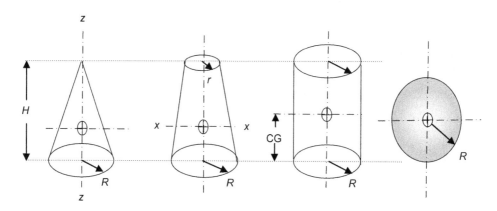

Cone	Cone frustum	Cylinder	Sphere
$v = \pi R^2 H/3$	$(\pi H/3)(R^2+Rr+r^2)$	$\pi R^2 H$	$4\pi R^3/3$
$CG = H/4$	$(H/4)(R^2+2Rr+3r^2)/(R^2+Rr+r^2)$	$H/2$	R
$I_{xx} = (3m/20)(R^2+H^2/4)$		$(m/12)(3R^2+H^2)$	$2mR^2/5$
$I_{zz} = 3mR^2/10$	$(3m/10)(R^5-r^5)/(R^3-r^3)$	$mR^2/2$	$2mR^2/5$
$S_s = \pi R(R^2+H^2)^{1/2}$	$\pi(R+r)[H^2+(R-r)^2]^{1/2}$	$2\pi RH$	$4\pi R^2$

FIGURE 7.76

Volume v, center of gravity location H measured from base, moments of inertia I_{xx} and I_{zz} about two axes passing through the CG, and area of the curved surface S_s for several uniform solids.

Density:	$1 \text{ kg/m}^3 = 16.018 \text{ lbm/ft}^3$
Length:	$1 \text{ in} = 2.54 \text{ cm} = 0.0254 \text{ m}$
	$1 \text{ mi} = 1609.3 \text{ m}$
	$1 \text{ nm} = 1.1515 \text{ mi} = 1.8531 \text{ m}$
Area:	$1 \text{ in}^2 = 6.452 \times 10^{-4} \text{ m}^2$
	$1 \text{ ft}^2 = 0.09290 \text{ m}^2$
Volume:	$1 \text{ L} = 0.001 \text{ m}^3$
	$1 \text{ US gallon} = 0.003785 \text{ m}^3$
	$1 \text{ ft}^3 = 0.02832 \text{ m}^3$
Speed:	$1 \text{ ft/s} = 0.3048 \text{ m/s}$
	$1 \text{ mi/h} = 1.467 \text{ ft/s} = 0.4471 \text{ m/s} = 1.604 \text{ km/h}$
Mass:	$1 \text{ metric ton (tonnes)} = 1000 \text{ kg}$
	$1 \text{ lbm} = 0.4536 \text{ kg}$
Force:	$1 \text{ lbf} = 4.448 \text{ N}$
Pressure:	$1 \text{ psia} = 6.894 \text{ kPa} = 6894 \text{ N/m}^2$
	$1 \text{ lbf/ft}^2 = 0.04788 \text{ kPa}$
	$1 \text{ atm} = 1.013 \text{ bar} = 101.3 \text{ kPa} = 14.7 \text{ psia} = 2116 \text{ lbf/ft}^2$
Temperature:	$1 \,°F = 1.8 \,°C + 32 = 1 \text{ R-459.7}$
	$1 \text{ R} = 1.8 \text{ K}$

7.9 NOMENCLATURE

A	a parameter in Eqn (7.95)
A_b	axial force due to the body
A_e	nozzle exit area
A_f	axial force due to the tail fins
APS	auxiliary propulsion system
A_t	nozzle throat area
a	acceleration or speed of sound
B	parameter in Eqn (7.95)
b	span of wing or fin
C	parameter in Eqn (7.125)
C_A	axial force coefficient, A/qS
C_D	drag coefficient, D/qS
$C_{D,n}$	drag coefficient of cylinder based on cross-flow velocity, D_n/qS
C_m	pitching moment coefficient, M/qSd
C_N	normal force coefficient, N/qS
CG	center of gravity
CP	center of pressure
C_L	wing lift coefficient, L/qS
$C_{L,f}$	fin lift coefficient, L_f/qS
$C_{l,a}$	wing lift curve slope, dC_L/da
C_m	pitching moment coefficient, M/qSd
C_N	normal force coefficient, N/qS
$C_{N,f}$	fin normal force coefficient, N_f/qS

c	wing or fin chord length
c_l	two-dimensional airfoil section lift coefficient
$c_{l,\alpha}$	two-dimensional airfoil lift curve slope, $dc_l/d\alpha$
c_r	wing or fin root chord length
c_t	wing or fin tip chord length
D	drag force or a parameter used in Eqn (7.95)
D_n	drag normal to body due to cross-flow at angle of attack
d	body or tank diameter
d_m	mean diameter of flared section $= (4\pi\upsilon_f/h_f)^{1/2}$
E	modulus of elasticity
E''	elliptic integral of the second kind
ET	external tank
F	thrust or a parameter used in Eqn (7.95)
$F_{a,crit}$	critical axial load
F_{vac}	thrust in a vacuum environment
G_0	parameter, see Eqn (7.31)
G_1	parameter, see Eqn (7.32)
G_2	parameter, see Eqn (7.33)
G_3	parameter, see Eqn (7.34)
g	gravitational acceleration
g_E	gravitational acceleration at the earth's surface $= 9.80665$ m/s
H	scale height of the atmosphere $= 7.16$ km
h	length of conical nose cone
h_f	length of flared section
h_i	distance between the center of gravity of vehicle segment i and its base
I_{sp}	specific impulse, see Eqn (7.8)
$I_{sp,vac}$	specific impulse in a vacuum environment
$\bar{I}_{sp}$	$I_{sp}/I_{sp,vac}$, see Eqn (7.9)
K_f	fin normal force factor, see Eqn (7.122)
k	gravitational constant for earth $= 398,600$ km^3/s^2
k_d	design safety factor
L	lift
L_a	latitude angle
LH_2	liquid hydrogen
LOX	liquid oxygen
l	length
l_f	distance from fin center of pressure to body center of gravity
l_w	height of atmospheric pressure water column
M	Mach number or pitching moment
MMH	monomethyl hydrazine fuel
m	mass
m_c	mass of cylinder
m_{cp}	mass of cylindrical barrel of tank
m_p	mass of propellant
m_s	mass of spherical tank
m_{sc}	mass of hemispherical end cap
m_t	mass of tank
N	force normal to body axis
N_f	force normal to fin surface

N_2O_4	nitrogen tetroxide oxidizer
n	load factor
O/F	oxidizer to fuel ratio by mass
OMS	orbital maneuvering system
p	pressure
p'	ullage pressure
Q	parameter, see Eqn (7.45)
q	dynamic pressure, see Eqn (7.18)
R	radius
R_c	cylinder radius
R_E	radius of earth at the equator $= 6378.14$ km
Re	Reynolds number based on diameter Vd/ν
RFNA	red fuming nitric acid oxidizer
RP1	rocket propellant kerosene fuel
R_s	sphere radius
r	radial coordinate
S	body reference area equal to $\pi d^2/4$
S_b	vehicle base area
S_f	fin planform area
S_p	body planform area
S_s	area of curved surface of a given body
S-IC	Saturn V launch vehicle first stage
S-II	Saturn V launch vehicle second stage
S-IVB	Saturn V launch vehicle third stage
T	temperature
t	time or thickness
t_c	cylinder thickness
TVC	thrust vector control
t_{cp}	thickness of cylindrical barrel
t_s	sphere thickness
$t_{sc,l}$	thickness of upper hemispherical end cap
$t_{sc,u}$	thickness of lower hemispherical end cap
UMDH	unsymmetrical dimethyl hydrazine fuel
V	velocity
V_E	velocity of circular orbit at earth's surface
V_e	velocity at the surface of the earth due to earth's rotation
V_{eff}	effective exhaust velocity, see Eqn (7.7)
V_r	velocity relative to the vehicle
V_w	wind velocity
V_ω	linear speed of rotation of the earth
W	weight
x	distance downrange or along axis of body measured from the nose tip
x_{CG}	body center of gravity location measured from the nose tip
x_{CP}	body center of pressure location measured from the nose tip
x_p	body planform area centroid location measured from the nose tip
y	distance from gimbal axis to actuator attachment point or depth of fluid in propellant tank
z	altitude
A	aspect ratio b^2/S_f
α	angle of attack

α_w	angle of attack caused by wind, see Figure 7.60
β	Prandtl–Glauert factor, see Eqn (7.129) or (7.133) or azimuth angle
Γ	function of specific heat ratio, see Eqn (7.10)
γ	flight path angle or ratio of specific heats
δ	pressure ratio p/p_{sl} or thrust vector deflection angle
ε	nozzle expansion ratio or a small angle $\ll 1$
η	cross-flow drag coefficient effectiveness
κ	ratio of actual to theoretical airfoil lift slope at a given M, $\kappa = (c_{l,\alpha})_M/(2\pi/\beta)$
$\Lambda_{c/2}$	sweepback angle of wing 50% chord line
Λ_{LE}	sweepback angle of wing leading edge
Λ_{TE}	sweepback angle of wing trailing edge
λ	taper ratio, c_t/c_r
ν	kinematic viscosity or index for chord location, see Eqn (7.132)
ρ	density
ρ_p	propellant density
ρ_w	tank wall density or density of water
σ	density ratio ρ/ρ_{sl} or tensile stress
σ_{crit}	critical buckling stress
σ_u	material ultimate stress
σ_y	material yield stress
χ	thrust inclination angle with respect to V
θ	angle between vehicle velocity and vehicle longitudinal axis
υ	volume of the launch vehicle
υ_b	boil-off volume for cryogenic propellants
υ_i	volume occupied by unusable trapped propellant
υ_f	volume of the flared section of the vehicle
υ_p	volume occupied by useful propellant
υ_u	ullage volume above free surface of a liquid propellant
ω	trajectory range angle, see Figure 7.2

7.9.1 SUBSCRIPTS

bo	propellant burn-out condition
c	circular orbit conditions
e	nozzle exit plane conditions
eng	engine
i	inertial conditions
n	nominal condition
p	propellant
pay	payload
sl	sea level condition
str	structure
t	nozzle throat conditions
vac	vacuum conditions
0	initial condition
∞	ambient atmospheric conditions

7.9.2 **SUPERSCRIPTS**

$\bar{t}$ $t/I_{sp,vac}$
$\bar{V}$ $V/g_E I_{sp,vac}$
$\bar{x}$ x/R_E
$\bar{z}$ z/R_E

REFERENCES

Abbott, I. H., & Von Doenhoff, A. E. (1959). *Theory of wing sections.* New York, NY: Dover.

Allen, H. J., & Perkins, E. W. (1951). *A study of the effects of viscosity on flow over slender inclined bodies of revolution.* NACA Report 1048.

Ashley, H. (1974). *Engineering analysis of flight vehicles.* Reading, MA: Addison-Wesley.

Bate, R. R., Mueller, D. D., & White, J. E. (1971). *Fundamentals of astrodynamics.* New York, NY: Dover.

Blair, J. C., Ryan, R. S., Schutzenhofer, L. A., & Humphries, W. R. (2001). *Launch vehicle design process: Characterization, technical integration, and lessons learned.* NASA/TP—2001—210992, May 2001.

Bursnall, W. J., & Loftin, L. K., Jr. (1951). *Experimental investigation of the pressure distribution about a yawed cylinder in the critical reynolds number range.* NACA Technical Note 2463.

Daniele, C. J., & Teren, F. (1968). *A wind profile for generating control requirements for rocket vehicles using liquid-injection control systems.* NASA TM X-1708.

Haeussermann, W. (1965). *Guidance and control of saturn launch vehicles.* AIAA Paper No. 65-304, AIAA Second Annual Meeting, July.

Hamner, R. L., & Leff, A. D. (1966). *Linear Aerodynamic Loads on Cone—Cylinders at Mach Numbers from 0.7 to 2.0.* NASA CR-413.

Heaslet, M. A., & Lomax, H. (1954). Supersonic and transonic small perturbation theory. In W. R. Sears (Ed.), *General theory of high speed aerodynamics.* Princeton, NJ: Princeton University Press.

Hemsch, M. J., & Nielsen, J. N. (July—August, 1983). Equivalent angle-of-attack method for estimating non-linear aerodynamics of missile fins. *Journal of Spacecraft and Rockets, 20*(4), 356—362.

Hoak, D. E., et al. (1978). *USAF stability and control DATCOM.* Wright-Patterson AFB: Flight Control Division, Air Force Flight Dynamics Laboratory.

Hoerner, S. F. (1958). *Fluid-dynamic drag.* Published by the author.

Huzel, D. K., & Huang, D. H. (1967). *Design of liquid propellant rocket engines.* NASA SP-125.

Huzel, D. K., & Huang, D. H. (1992). *Modern engineering for design of liquid propellant engines.* Reston, VA: American Institute of Aeronautics and Astronautics.

Isakowitz, S. J. (1995). *International reference guide to space launch systems.* Reston, VA: AIAA.

Jernell, L. S. (1968). *Aerodynamic characteristics of bodies of revolution at Mach numbers from 1.50 to 2.86 and angles of attack to 180°.* NASA TM-X-1658.

Johnson, T. F., Sleight, D. W., & Martin, R. A. (2013). *Structure and design phase I summary for the NASA composite cryotank technology demonstration project.* AIAA Paper 2013-1825.

Jones, R. M. (2006). *Buckling of bars, plates, and shells.* Blacksburg, VA: Bull Run Publishing.

Jones, G. W., Jr., Cincotta, J. J., & Walker, R. W. (1969). *Aerodynamic forces on a stationary and oscillating cylinder at high reynolds numbers.* NASA TR R-300.

Jones, R. T., & Cohen, D. (1960). *High speed wing theory.* NJ: Princeton University Press, Princeton Aeronautical Paperback No. 6.

Jorgensen, L. H. (1973a). *A method for estimating static aerodynamic characteristics for slender bodies of circular and noncircular cross section alone and with lifting surfaces at angles of attack from 0° to 90°.* NASA TN D-7228.

Jorgensen, L. H. (1973b). *Prediction of static aerodynamic characteristics for space-shuttle-like and other bodies at angles of attack from 0° to 180°*. NASA TN D-6996.

Kit, B., & Evered, D. S. (1960). *Rocket propellant handbook*. New York, NY: Macmillan.

Laitone, E. V. (1989). Lift-curve slope for finite-aspect-ratio wings. *Journal of Aircraft*, 26(8), 789–790.

Loftin, L. K., & Bursnall, W. J. (1948). *The effects of variations in Reynolds number between 3.0×10^6 and 25.0×10^6 upon the aerodynamic characteristics of a number of 6-series airfoil sections*. NACA Technical Note No. 1773.

McDevitt, J. D. (1955). *A correlation by means of transonic similarity rules of experimentally determined characteristics of a series of symmetrical and cambered wings of rectangular planform*. NACA Report 1253.

Morring, F. (July 8, 2014). Composite breakthrough. *Aviation Week & Space Technology*, 18.

Muraca, R. J. (1966). *An empirical method for determining static distributed aerodynamic loads on axisymmetric multistage launch vehicles*. NASA TN D-3283.

Myers, J. A. (1962). *Handbook of equations for mass and area properties of various geometric shapes*. U.S. Naval Ordnance Test Station Report NOTS TP 2838, NAVWEPS Report 7827.

Piland, R. O. (1949). *Summary of the lift, damping-in-roll, and center-of-pressure characteristics of various wing plan forms at supersonic speeds*. NACA Technical Note 1977.

Pinson, G. T. (1971). *Apollo/Saturn V postflight trajectory—AS-510*. NASA CR-120464.

Pitts, W. C., Nielsen, J. N., & Kaattari, G. E. (1957). *Lift and center of pressure of wing-body-tail combinations at subsonic, transonic, and supersonic speeds*. NACA Report 1307.

Polhamus, E. C. (1971). *Charts for predicting the subsonic vortex-lift characteristics of arrow, delta, and diamond wings*. NASA TN D-6243 (1971).

Sforza, P. M. (2012). *Theory of aerospace propulsion*. New York, NY: Elsevier.

Sforza, P. M. (2014). *Principles of commercial airplane design*. Oxford, UK: Elsevier.

Sherer, A. D. (1966). *Analysis of the linearized supersonic flow about pointed bodies of revolution by the method of characteristics*. NASA TN D-3578.

Smith, J. H. B., Beasley, J.A., & Stevens, A. (1961). *Calculations of the lift slope and aerodynamic center of cropped delta wings at supersonic speeds*. Aeronautical Research Council CP No. 562.

SSRM (1988). *Space shuttle reference manual*. <http://science.ksc.nasa.gov/shuttle/technology/sts-newsref/stsref-toc.html#>.

Stoney, W. E. (1961). *Collection of zero-lift drag data on bodies of revolution from free-flight investigations*. NASA Technical Report R-100.

Tam, W., Hersh, M., & Ballinger, I. (2003). *Hybrid propellant tanks for spacecraft and launch vehicles*. 39th AIAA Propulsion Conference, AIAA Paper 2003-4607.

Teren, F., Davidson, K. I., Borsody, J., & Daniele, C. J. (1968). *Thrust-vector control requirements for large launch vehicles with solid-propellant first stages*. NASA TN D-4662.

Wagner, W. A. (1974). *Liquid rocket metal tanks and tank components*. NASA SP-8088.

Walker, C. E. (1968). *Results of several experimental investigations of the static aerodynamic characteristics for the Apollo/Saturn V launch vehicle*. NASA TM X-53770.

Weingarten, V. I., Seide, P., & Peterson, J. P. (1968). *Buckling of thin-walled cylinders*. NASA SP-8007.

Wertz, J. R., & Larson, W. J. (1999). *Space mission analysis and design* (3rd ed.). New York, NY: Microcosm/Springer.

SPACECRAFT FLIGHT MECHANICS

8.1 SPACE VEHICLE FLIGHT MECHANICS AND PERFORMANCE ANALYSIS

The analyses of launch, orbit, and entry carried out so far have considered the spacecraft to be a point mass. Manned spacecraft have substantial volume requirements and therefore must be at least considered as rigid bodies. At closer levels of scrutiny, it becomes necessary to evaluate the effects of flexure of the structure. The mechanics of subsonic and supersonic flight vehicles are well understood and are described in many books (e.g., Stengel (2004) and Etkin and Reid (1995)). Critical flight segments for spacecraft are carried out at hypersonic speeds, a flight regime not understood and documented nearly as well. At hypersonic speeds, adiabatic compression through shock waves basically determines the force field, unlike subsonic speeds where isentropic expansions serve that role. Furthermore, although the temperature extremes produced by such compressions affect gas density and chemical composition, they do not strongly influence the pressure levels, and therefore the force structure is not greatly affected. Thus, the shape of the windward side of a spacecraft has an important effect on the total force structure, while the leeward side shape is much less important in that regard although it can influence vehicle stability by generating appropriate righting moments.

8.2 HYPERSONIC AERODYNAMICS

However, the extreme nature of hypersonic flight also provides us with limiting cases that are quite practical for use in preliminary design and even relatively simple to apply. We will discuss hypersonic aerodynamics from the standpoint of the simple, yet powerful, Newtonian flow theory. This approach permits relatively simple computation of the inviscid forces and moments acting on a realistic spaceplane geometry. Because the analysis depends on local conditions at a point on the spaceplane's surface, a panel-based method is easily constructed. Contributions to the forces and moments arising from boundary layer friction are also easily accommodated. The general behavior of slender bodies in hypersonic flight is examined in terms of L/D performance and methods for calculating laminar and turbulent skin friction are presented and the effects of boundary layer transition and high-altitude rarefaction phenomena on spaceplane performance are discussed. Because extreme temperatures are achieved in hypersonic flight, a sound accounting of the thermodynamic and transport properties of high-temperature air is required. Approximations for enthalpy, molecular weight, specific heat, and sound speed are developed. Real gas stagnation temperature is treated

Manned Spacecraft Design Principles. DOI: http://dx.doi.org/10.1016/B978-0-12-804425-4.00008-8

along with simple curve fits for enthalpy and the $\rho\mu$ product. Blunt bodies in hypersonic flight, like space capsules, and their L/D capability are addressed separately, and the special case of forces on spherically blunted cones is analyzed. Having treated the force and moment structure on realistic configurations, we move on to investigate longitudinal and lateral static stability of such vehicles.

8.2.1 NEWTONIAN FLOW THEORY

Newtonian theory for hypersonic flow, which is discussed in some detail in Appendix A, is based on the notion that as $M \to \infty$ and $\gamma \to 1$, the shock layer becomes coincident with the body surface. This occurs because the density ratio across the shock, $\varepsilon = \rho_1/\rho_2 = (\gamma - 1)/(\gamma + 1)$ approaches zero and the density behind the shock is so high the shock layer becomes thinner and thinner. Momentum considerations can then be used to deduce a relationship between the pressure on the surface and the flow inclination. Although shock layers in hypersonic flow can be rather thin, they are not coincident with the body. Indeed, it is really correct to consider Newtonian flow in terms of shock inclination θ, and in Appendix A, we show that in the limit as the density ratio across the shock becomes smaller and smaller, the pressure coefficient may be written as follows:

$$\lim_{\varepsilon \to 0} C_p = 2(1 - \varepsilon)\sin^2 \theta \to 2 \sin^2 \theta \tag{8.1}$$

In the limit as the shock angle θ approaches the flow inclination δ and the density ratio across the shock ε approaches zero, we find that

$$\lim_{\theta \to \delta, \varepsilon \to 0} C_p = \frac{2 \sin^2 \delta}{(1 - \varepsilon)\cos^2(\theta - \delta)} \to 2 \sin^2 \delta \tag{8.2}$$

This approach is illustrated in Figure 8.1 where the flow deflection and shock angles are essentially coincident and the pressure rise is entirely due to the conversion of the normal component of momentum. Note that for the pressure rise to occur by transformation of the normal component of the momentum, the velocity direction must be opposite to that of the outward unit normal vector, as in case 1 in Figure 8.1. When the flow deflection is zero, as in case 2 of Figure 8.1, the pressure coefficient is zero, and when the flow deflection increases further, the pressure coefficient is taken as zero.

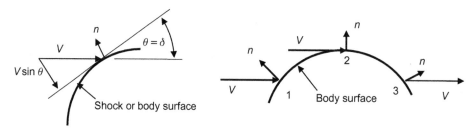

FIGURE 8.1

Schematic diagrams of the application of Newtonian flow theory.

We may generalize this expression, for a point on the surface of a shock wave, by the following vector expression:

$$C_p = 2\frac{(\vec{V}_\infty \cdot \hat{n})^2}{\vec{V}_\infty \cdot \vec{V}_\infty} \cdot (1 - \varepsilon) \quad \text{for } \vec{V}_\infty \cdot \hat{n} < 0 \tag{8.3}$$

$$C_p = 0 \quad \text{for } \vec{V}_\infty \cdot \hat{n} \geq 0 \tag{8.4}$$

Here $\hat{n}$ is the outward normal to the shock surface, that is, pointing toward the upstream side. In Appendix A, Section A.8, we show that the maximum pressure coefficient on the body is

$$C_{p_{max}} = 2(1 - \varepsilon) + \varepsilon = 2 - \varepsilon \tag{8.5}$$

Thus, for real flows past blunt bodies, the value of $C_{p,max}$ is less than two. The same result is obtained for both two-dimensional and axisymmetric flows. Furthermore, the same first terms appear in a constant density analysis. Lees (1956) introduced the idea that one should always modify the Newtonian flow rule for blunt bodies by using the actual value of $C_{p,max}$, that is,

$$C_p = (2 - \varepsilon)\frac{(\vec{V}_\infty \cdot \hat{n})^2}{\vec{V}_\infty \cdot \vec{V}_\infty} \tag{8.6}$$

This so-called modified Newtonian theory is found to give quite useful results. Of course, as the body inclination becomes parallel to the free stream direction $C_p \to 0$, the results are no longer accurate. Accuracy starts to fall off when the flow deflection reaches about 40° or 50°.

Pointed bodies are assumed to have a tip segment, which is a cone or a wedge, and when there is an attached shock wave, a similar modification is made, that is

$$C_p = \frac{C_{p_N}}{\sin^2 \delta}\frac{(\vec{V}_\infty \cdot \hat{n})^2}{\vec{V}_\infty \cdot \vec{V}_\infty} \tag{8.7}$$

Here, $C_{p,N}$ is the pressure coefficient of the pointed nose segment. For wedges of half-angle δ and cones of half-angle δ_c, we have the following nose pressure coefficients:

$$C_{p,Nw} = \frac{2\sin^2 \delta}{(1 - \varepsilon)\cos^2(\theta - \delta)}$$

$$C_{p,Nc} = \frac{2\sin^2 \delta_c}{\left(1 - \dfrac{\varepsilon}{4}\right)\cos^2(\theta - \delta_c)} \tag{8.8}$$

Here, the correct $C_{p,max}$ values are greater than two. Note that in the hypersonic limit, $\theta \to \delta$ so that $\cos^2(\theta - \delta) \to 1$. This approach gives useful results until the body contour becomes nearly parallel to V_∞. In the case of two-dimensional bodies, the $C_{p,Nw}$ for the wedge would be used, while for axisymmetric bodies, $C_{p,Nc}$ for the cone would be used. In the case of axisymmetric bodies at angle of attack, $C_{p,N}$ of the conical tips would vary with azimuthal angle ϕ as would the body slope, that is, $\delta = \delta(\phi)$. Discussion of such details, which are not directly related to the problems considered here appears in Appendix A.

The Newtonian method works reasonably well and lends itself to application to bodies of quite arbitrary shape. Of course, we can apply the Prandtl−Meyer (P−M) expansion relations at any point along the body at which the Newtonian method begins to become inaccurate. The P–M expansion may be used for two-dimensional bodies in general, while for axisymmetric (or three-dimensional) bodies, it should be used only where the transverse slope is small. That is, one should expect that streamline curvature due to the shape of the body will influence the pressure development to an extent beyond that described by the two-dimensional P–M expansion alone. At high Mach numbers, the shock-expansion method for pointed bodies works very well with the matching point for the P–M expansion fixed anywhere downstream of the shock. If the body still has substantial slope at its aft end, it usually is sufficient to use the Newtonian method alone.

8.2.2 APPLICATIONS TO PRESSURE FORCE AND MOMENT ANALYSIS OF A SPACEPLANE

The Newtonian theory, being based on the idea that the shock layer is so thin that the shock practically coincides with the body surface, is locally applicable to every surface point on a smooth body. As a result, a body surface may be subdivided into a number of individual panels and each may be treated separately to determine the pressure force acting. The total force and moment structure on the whole vehicle may then be obtained by summing the contributions of each element of the body surface. A general element of the body surface is shown in Figure 8.2.

In Figure 8.2, station i denotes a plane normal to the x-axis of the flight vehicle at which a transverse body surface contour is defined, and station $i + 1$ denotes the next axial station along the x-axis at which the next contour of body shape is defined. Rays j and $j + 1$ denote longitudinal lines along the body surface defining the shape of the vehicle. Arbitrarily selecting four points on the neighboring contours and rays to define a quadrilateral panel does not ensure a flat panel. However, it is usually convenient to lay out panels in this fashion, and if reasonable care is used in

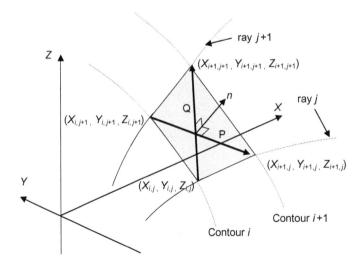

FIGURE 8.2

The shaded area is a general quadrilateral surface element having a normal vector n. The spatial locations of the four corner points are shown with the axial contours and longitudinal rays defining the body surface shape.

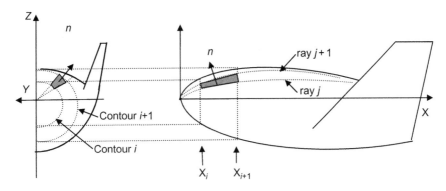

FIGURE 8.3

General illustration of a typical surface element on a flight vehicle.

choosing their size and shape, they are suitable for preliminary design purposes. If desired, one may ensure flat panels by considering triangular panels, like the two triangles on each side of either the vectors $\vec{P}$ or $\vec{Q}$ forming the shaded quadrilateral panel in Figure 8.2 although this of course doubles the number of panels employed.

The approach used here to calculate the hypersonic flow over a spaceplane depends primarily on knowing the area, location, and normal vector of each panel, and it is readily adapted to other, more accurate, paneling schemes, such as those embedded in CAD packages. In Figure 8.2, the coordinate x is measured along the axis of the vehicle from the nose tip $x = 0$. The x-z plane is taken as the symmetry plane of the vehicle with z measured positive vertically and y measured positive outboard along the right (starboard) side of the vehicle as shown. For stability and control studies, the body axes usually originate at the center of mass of the vehicle with the x-axis pointing forward, the y-axis as shown, and the z-axis pointing down, that is, the x-z plane is rotated $180°$ counterclockwise about the y-axis. The general panel illustrated in Figure 8.2 is shown as it might appear on a spaceplane drawing in Figure 8.3.

A vector $\vec{N}$ normal to the panel may be found by forming the cross-product between the vectors $\vec{P}$ and $\vec{Q}$ such that the normal points outward as shown. Thus

$$\vec{N} = \vec{P} \times \vec{Q} = [(x_{i+1,j} - x_{i,j+1})\hat{i} + (y_{i+1,j} - y_{i,j+1})\hat{j} + (z_{i+1,j} - z_{i,j+1})\hat{k}]$$
$$\times [(x_{i+1,j+1} - x_{i,j})\hat{i} + (y_{i+1,j+1} - y_{i,j})\hat{j} + (z_{i+1,j+1} - z_{i,j})\hat{k}]$$

Then, the outward normal vector becomes

$$\vec{N} = [(y_{i+1,j} - y_{i,j+1})(z_{i+1,j+1} - z_{i,j}) - (y_{i+1,j+1} - y_{i,j})(z_{i+1,j} - z_{i,j+1})]\hat{i}$$
$$+ [(x_{i+1,j+1} - x_{i,j})(z_{i+1,j} - z_{i,j+1}) - (x_{i+1,j} - x_{i,j+1})(z_{i+1,j+1} - z_{i,j})]\hat{j} \qquad (8.9)$$
$$+ [(x_{i+1,j} - x_{i,j+1})(y_{i+1,j+1} - y_{i,j}) - (x_{i+1,j+1} - x_{i,j})(y_{i+1,j} - y_{i,j+1})]\hat{k}$$

The unit vector normal to the panel may be obtained by dividing the normal vector by its magnitude as follows:

$$\hat{n} = \frac{\vec{N}}{\sqrt{\vec{N} \cdot \vec{N}}}$$

The area of the element may be found by using the simple formula

$$A = \frac{1}{2}|\vec{P} \times \vec{Q}| = \frac{1}{2}|\vec{N}|$$ (8.10)

Once having the unit normal to the surface element, one may determine the pressure force acting on it using the Newtonian theory described previously, as follows:

$$d\vec{F} = -C_p qA\hat{n} = -(2-\varepsilon)\frac{(\vec{V}_\infty \cdot \hat{n})^2}{\vec{V}_\infty \cdot \vec{V}_\infty} qA\hat{n}$$ (8.11)

Recall that the limiting value for the density ratio is $\varepsilon = (\gamma - 1)/(\gamma + 1)$ and that from Eqn (8.4), the pressure force on an element in the shadow region is zero. The resultant pressure force on the νth surface element may be expressed as a vector incorporating the elemental axial, side, and normal force as follows:

$$d\vec{F}_\nu = dF_{a,\nu}\hat{i} + dF_{s,\nu}\hat{j} + dF_{n,\nu}\hat{k}$$ (8.12)

For the special case of flight in a plane where the angle of attack α is the angle between the velocity vector and the i-direction, which defines the x-axis, the magnitude of the lift and drag increments acting on the νth surface element may be obtained from the following relations:

$$dD_\nu = d\vec{F}_\nu \cdot \frac{\vec{V}_\infty}{\sqrt{\vec{V}_\infty \cdot \vec{V}_\infty}}$$ (8.13)

$$dL_\nu = \sqrt{d\vec{F}_\nu \cdot d\vec{F}_\nu - (dD_\nu)^2}$$ (8.14)

Recall that the lift and drag forces are referred to the free stream velocity: lift acts normal to the velocity vector and drag acts in the direction of the velocity vector.

The differential moment about a reference point, for example, the nose tip (x_0, y_0, z_0), produced by the differential force on the νth surface element is given by

$$d\vec{M}_{0,\nu} = \vec{r}_\nu \times d\vec{F}_\nu = [(x_\nu - x_0)\hat{i} + (y_\nu - y_0)\hat{j} + (z_\nu - z_0)\hat{k}] \times d\vec{F}_\nu$$ (8.15)

In this expression, the radius vector $\vec{r}_\nu$ is measured from the reference point to the centroid of the surface element denoted as the point (x_ν, y_ν, z_ν). The components of the moment about the x, y, and z axes are defined as the rolling, pitching, and yawing moments of the vehicle, that is, L, M, and N, respectively; thus

$$d\vec{M}_{0,\nu} = dL_\nu \hat{i} + dM_\nu \hat{j} + dN_\nu \hat{k}$$ (8.16)

Summing up the contributions of all the panels provides the complete force and moment description of the vehicle for the flight condition chosen. One may develop information like the drag polar and the associated moment variation by rotating the vehicle through a number of angles of attack at a particular Mach number and altitude.

8.2.3 BOUNDARY LAYER CONSIDERATIONS

The preceding analysis considered forces and moments arising from normal stresses alone. However, the tangential stresses due to friction and the resultant heating must also be accounted for in the aerodynamic design of an atmospheric entry vehicle. The geometric layout of the body

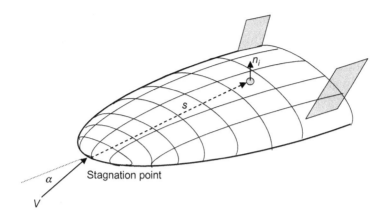

FIGURE 8.4

A schematic diagram illustrating the approximate distance s from the stagnation point to the centroid of the ith panel.

surface panels, the determination of the pressure on each element, and the determination of the velocity component normal to the body surface panels may all be used in an approximate analysis of the skin friction and heating loads experienced by the vehicle during the hypersonic portion of the entry process. The major parameter that forms the basis of the analysis is the Reynolds number based on the local tangential velocity, temperature, and distance from the stagnation point, that is

$$\mathrm{Re}_s = \frac{\rho_e V_e s}{\mu_e} \tag{8.17}$$

The variable s denotes the distance along the surface of the vehicle measured from the relevant stagnation point (for a body) or stagnation line (for a wing), while V_e denotes the magnitude of the inviscid velocity at the surface of the body. The subscript e is introduced to indicate that these are conditions that are experienced at the outer edge of the boundary layer. In order to evaluate the density and viscosity, the pressure and temperature are required. The pressure on the panel is obtained from the pressure coefficient in which the Newtonian theory assumes that all the momentum of the stream normal to a surface to be given up in a pressure force (Figure 8.4). We further assume that the direction of V_e is the direction of the tangential component of the free stream velocity on the panel, which is given by

$$\vec{V}_{\infty,t} = \vec{V}_\infty - (\vec{V}_\infty \cdot \hat{n})\hat{n} \tag{8.18}$$

In Section 8.3.4, we show that the known pressure coefficient may be used to derive the inviscid surface Mach number on the surface, or more accurately, at the outer edge of the viscous boundary layer, as follows:

$$M^2 \approx \frac{2}{\gamma - 1}\left[\left(\frac{\gamma^{3/4} M_\infty^2}{1 + \frac{\gamma}{2} M_\infty^2 C_p}\right)^{\frac{\gamma-1}{\gamma}} - 1\right] \tag{8.19}$$

This equation is as accurate as the pressure coefficient prediction, and the Newtonian theory loses accuracy as the tangent to the body becomes closer to the free stream velocity vector. It will be shown in Section 8.3.4 that using the P–M expansion described in Appendix A at stations where $\theta < 20°$ can provide accuracy superior to that achieved using Newtonian theory alone.

The temperature at the edge of the boundary layer may be found by first considering the inviscid adiabatic energy equation on the body streamline as given by

$$h_t = h_\infty + \frac{1}{2}V_\infty^2 = h_e + \frac{1}{2}V_e^2$$

The stagnation enthalpy h_t is constant along streamlines. If we use $V_e = a_e M_e$ in the energy equation, where a_e is the sound speed at the edge of the boundary layer defined by $a_e^2 = k_e Z_e R T_e$, we may solve for the temperature at a given point at the edge of the boundary layer on the body as follows:

$$T_e = \frac{2(h_t - h_e)}{k_e Z_e R M_e^2} \tag{8.20}$$

This is an implicit equation for T_e because although we know h_t from the flight conditions, M_e from Eqn (8.19), and R, which is the gas constant for air under standard conditions, the quantities k_e, Z_e, and h_e, all vary considerably with both p_e and T_e for $T_e > 2000$ K. In order to accurately determine the temperature T_e from Eqn (8.20), it is necessary to use tables or models for the thermodynamic properties of high-temperature air and iterate. These thermodynamic properties are treated in some detail subsequently in Section 8.5 based on the approximations given by Hansen (1959). Once M_e increases sufficiently along the body surface to result in a value for $T_e < 2000$ K, it would be reasonable to set $h_e = c_{p,e}T_e$, $Z_e = 1$, and $k_e = \gamma_e$ and use the following reduced form of Eqn (8.20):

$$T_e \approx \frac{h_t}{\frac{1}{2}\gamma_e R\left(M_e^2 + \frac{2}{\gamma_e - 1}\right)} \tag{8.21}$$

For 200 K $< T_e < 1500$ K, Section 8.5.2 shows that $1.4 > \gamma_e > 1.3$. At 2000 K, the lower limit drops to 1.26 and some effects of pressure begin to be noticed. Once the temperature is known, one may calculate the velocity from the Mach number and the density and transport properties may also be determined.

8.2.4 SOME SPECIAL CONSIDERATIONS IN COMPUTING SPACEPLANE FLOW FIELDS

In general, spaceplanes are characterized by a blunt nose, and therefore the first set of panels will be triangular in shape, as shown in Figure 8.5. At the nose tip, station $i = 0$, the coordinates $x_0 = y_0 = z_0 = 0$ for all values of j, and the equations described in Section 8.2.2 can be applied. However, it is important to select the location of the first station $i = 1$ close enough to the nose to ensure that $x_1/R_N < 0.4$, which results in the first panel being inclined to the x-axis by more than about $68°$. The pressure coefficient approximation for these blunt bodies is that given by Eqn (8.6) as used in Eqn (8.11). If desired, the nose section may instead be approximated by a spherical segment for fuselage-like bodies or a cylindrical segment for wing-like bodies. These nose segments are treated separately to determine the forces acting on them.

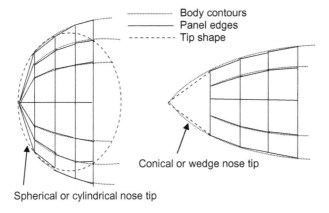

.... Body contours
———— Panel edges
--------- Tip shape

Conical or wedge nose tip

Spherical or cylindrical nose tip

FIGURE 8.5

Treatment of nose tips and leading edges as either panels or simple geometric shapes.

Pointed bodies that have attached shock waves must be treated as having a conical nose for fuselage-like bodies or a wedge nose for wing-like bodies, and the pressure coefficient to use in Eqn (8.11) is that given by Eqn (8.7).

It is important to locate the panel with the minimum value of $\sin\theta_{i,j}$, which is given by

$$\sin\theta = \frac{\vec{V}_\infty \cdot \hat{n}}{\sqrt{\vec{V}_\infty \cdot \vec{V}_\infty}} \tag{8.22}$$

This is needed because the panel with $\sin\theta_{i,j}$ closest to -1 will contain the stagnation point. The shock wave at this panel is considered to be a normal shock, and the stagnation conditions are calculated here. At the same time, it is useful to locate those panels lying in the shadow region on which the pressure force is zero according to Newtonian flow theory. These panels are defined by $0 \le \sin\theta_{i,j} \le 1$ and although they generate no pressure load within Newtonian theory, they can carry shear and heating loads due to friction. In addition, if a more detailed analysis, like the Newtonian plus P−M shock-expansion method is used, it is likely that some small additional pressure loading will be found.

Control point locations must also be selected for each panel because we assume that the elemental forces on the panel are concentrated at this point. The centroid of the panel would be a logical choice but the calculation of this location becomes cumbersome, especially for quadrilateral panels. The control point is used when calculating the distance to the panel from some specified point of interest, for example, the stagnation point or the center of gravity of the vehicle. However, the level of accuracy of the current approximation to the flow field permits any convenient point on the panel as an acceptable choice. Obviously, the smaller the panels, the less important the exact location of the control point on the panel. A simple approximation for the radius vector defining the control point of the νth panel for panels of approximately rectangular shape is

$$\vec{r}_\nu = \frac{1}{2}\left[(x_{i,j} + x_{i+1,j+1})\hat{i} + (y_{i,j} + y_{i+1,j+1})\hat{j} + (z_{i,j} + z_{i+1,j+1})\hat{k}\right] \tag{8.23}$$

8.3 BLUNT BODIES IN HYPERSONIC FLIGHT

The preponderance of manned space vehicles are capsule shaped, such as the Gemini vehicle, as shown in Figure 8.6. This shadowgraph of a Gemini capsule model in hypersonic flight in a ballistic range illustrates the shock wave and wake flow field. Such vehicles may be characterized by the idealized form as shown in Figure 8.7 for the symmetric case of zero angle of attack. The important parameters for such vehicles are the heat shield radius R_N, the maximum radius of the capsule R_b, and the associated angle $\phi_1 = \sin^{-1}(R_b/R_N)$.

In the hypersonic regime of interest, we may determine the zero-lift drag coefficient, $C_{D,0}$ using the Newtonian flow approximation. The local tangent to the spherical cap heat shield is shown in Figure 8.7 at a generic point on the heat shield (R_N, ϕ). The outward normal to the surface at any point is defined by the nose cap radius R_N. As will be seen in Chapter 9, a large nose radius is the foundation of the spacecraft's passive thermal protection system because convection heating is

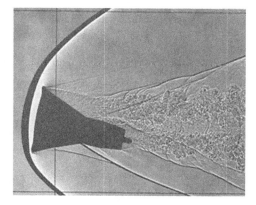

FIGURE 8.6

Shadowgraph of a model of the Gemini space capsule traveling at hypersonic speed in a ballistic range.

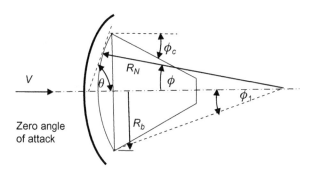

FIGURE 8.7

Schematic diagram of a blunt Apollo-like space capsule at a zero angle of attack. The angle θ defines the local tangent to the spherical nose cap and ϕ_c is the afterbody cone angle.

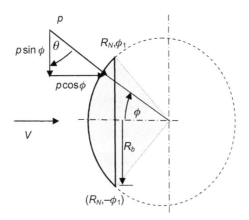

FIGURE 8.8

Axisymmetric segment of a spherically blunted nose of radius R_N.

inversely proportional to the local radius of curvature. At the same time, the large frontal area produces high drag due almost entirely to the adiabatically compressed high-pressure shock layer formed over it. Skin friction is a minor contributor to the drag unlike the case for the slender bodies treated subsequently in this chapter. Therefore, we will be primarily concerned with pressure forces in our consideration of blunt bodies.

Consider a segment of spherically blunted spacecraft forebody as shown in Figure 8.8. We see from the pressure field for the axisymmetric situation posed that the lift distribution is antisymmetric about $\phi = 0$ so the lift is zero. The drag distribution is symmetric about $\phi = 0$, so the differential drag coefficient is given by

$$dC_D = \frac{dD}{S} = \frac{C_p}{S} 2\pi r^2 \cos \phi \sin \phi d\phi$$

We have shown previously that Newtonian theory for a blunt body gives the pressure coefficient on the surface as follows:

$$C_p = (2 - \varepsilon)\sin^2 \theta = (2 - \varepsilon)\cos^2 \phi \qquad (8.24)$$

The quantity ε is the ratio of density upstream of the shock to that downstream of the shock. In Section A.2.2 of Appendix A, we show that for constant γ, the density ratio is given by

$$\frac{\rho_1}{\rho_2} = \varepsilon = \frac{\gamma - 1}{\gamma + 1}\left(1 + \frac{2}{(\gamma - 1)M^2}\right) = \varepsilon \lim_{M \to \infty}\left(1 + \frac{2}{(\gamma - 1)M^2}\right)$$

The density ratio ε is shown as a function of Mach number for different constant, but practical, values of γ in Figure 8.9. It is clear that ε becomes essentially constant once the hypersonic regime is entered. Because the Newtonian approximation is appropriate in the hypersonic limit where $M \gg 1$, the density ratio $\varepsilon = \varepsilon_{\lim} = (\gamma - 1)/(\gamma + 1)$ is a good approximation. The error incurred by using $\varepsilon_{\lim}$ rather than the actual value of ε are below 10% for $1.2 < \gamma < 1.4$ when $M > 10$. For $\gamma \sim 1.1$, a 10% error threshold is not reached until $M > 15$.

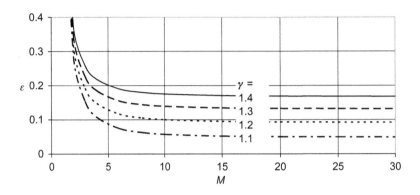

FIGURE 8.9

The density ratio $\varepsilon = \rho_1/\rho_2$ as a function of Mach number for different constant values of γ.

Integrating around the segment of the nose, the zero-lift drag coefficient is found to be

$$C_{D,0} = \int_0^{\phi_1} \frac{(2 - \varepsilon)2\pi r^2}{S} \left(-\frac{d\cos^4 \phi}{4} \right) = \frac{(2 - \varepsilon)\pi r^2}{2S} \left[1 - \cos^4 \phi_1 \right] \tag{8.25}$$

Beyond $\pm \phi_1$, the Newtonian pressure coefficient is zero, so there is no contribution from the base region. In practice, there is actually a small contribution from the capsule afterbody, and this detail will be discussed subsequently. The reference area is chosen to be $S = \pi R_b^2$, where R_b is the maximum transverse radius of the segment, as illustrated in Figure 8.8. The quantity R_b is related to R_N by $R_b/R_N = \sin \phi_1$, and therefore the zero-lift drag coefficient may be written as

$$C_{D,0} = \left(1 - \frac{\varepsilon}{2} \right)(1 + \cos^2 \phi_1) = (2 - \varepsilon) \left[1 - \frac{1}{2} \left(\frac{R_b}{R_N} \right)^2 \right] \tag{8.26}$$

When $\phi_1 \ll 1$, the drag coefficient $C_{D,0} \sim 2 - \varepsilon$, the result for a flat plate normal to the flow, and when $\phi_1 \sim \pi/2$, the drag coefficient $C_{D,0} \sim 1 - \varepsilon/2$, the result for a hemisphere.

Note that the shaded portion of Figure 8.8 resembles the shape of a typical capsule-type spacecraft. Several candidate capsule designs for the NASA Mercury program illustrating the different afterbody cone angles ϕ_c considered as shown in Figure 8.10. The afterbody cone angle basically defines the maximum angle of attack that can be safely employed. Larger angles would require cone sidewalls to be protected from the high-temperature flow. Astronaut John Glenn's Mercury spacecraft, Friendship 7, which is shown in Figure 8.11 following its return to Cape Canaveral after recovery in the Atlantic Ocean, has a configuration closest to capsule C in Figure 8.10.

The variation of $C_{D,0}$ with capsule bluntness ratio R_b/R_N according to Eqn (8.26) is shown in Figure 8.12 for two different values of the ratio of specific heats: $\gamma = 1.4$, the standard air value, and $\gamma = 1.2$, a value more representative of the hot gas around a re-entering space capsule. The range of bluntness ratios for Apollo-like capsules is also shown on the figure. Note that as R_b/R_N increases, the forward face of the capsule, that is, the heat shield, becomes relatively rounder resulting in a lower drag coefficient. In order to produce some lift to aid in the reentry process, it is necessary to put the capsule at some angle of attack, as discussed in the next section.

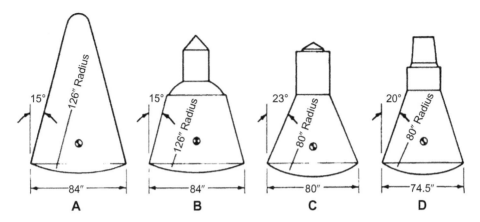

FIGURE 8.10

Early capsule designs showing the different afterbody cone angles considered. Bluntness ratio $R_b/R_N = 0.33$, 0.33, 0.5, and 0.465 for capsules A, B, C, and D, respectively.

Courtesy NASA.

FIGURE 8.11

Astronaut John Glenn's Mercury spacecraft, Friendship 7, is shown here after recovery in the Atlantic Ocean.

Courtesy NASA.

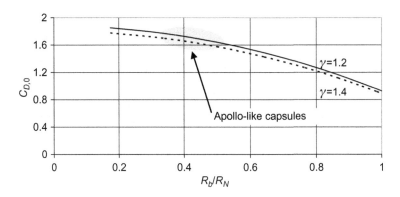

FIGURE 8.12

The zero-lift drag coefficient of Apollo-like space capsule as a function of the capsule bluntness ratio R_b/R_N; for Apollo $R_b/R_N = 0.426$.

8.3.1 BLUNT BODY L/D

Past and current manned space capsules are illustrated in Figure 8.13, and the forces acting on a typical blunt capsule are illustrated in Figure 8.14. The lift and drag components of the resultant aerodynamic force may be obtained from the components of that resultant force resolved normal to and along the axis of the capsule. If we choose to consider the angle of attack $\alpha > 0$ as shown, the following relations apply:

$$L = F_a \sin \alpha - F_n \cos \alpha \tag{8.27}$$

$$D = F_a \cos \alpha + F_n \sin \alpha \tag{8.28}$$

Note that since only pressure forces on the spherical cap are considered important and because pressure acts normal to the surface, the resultant force must pass through the origin of the spherical cap. Because no moment can act at that point, it is also the center of pressure. Note that pressures are highest where the angle between the velocity vector and the surface normal is smallest. Figure 8.15 is a shadowgraph of a model of a Mars Lander capsule at high Mach number in a ballistic range in which the degree to which the shock follows the body shape is evident. A diagram of a capsule on an entry trajectory has a similar attitude, as shown in Figure 8.16. For completeness in considering the forces to be acting at the center of gravity of the capsule, we show a pitching moment that must be also acting there. The moment must be trimmed out, that is made zero, for equilibrium to be possible. This is typically achieved by offsetting the center of gravity as shown in Figure 8.14.

The remarkably general accuracy of the simple Newtonian theory was of great benefit during the early manned spaceflight program because it put calculation of hypersonic aerodynamic parameters within relatively easy reach of researchers and designers during a period when computational capabilities were quite limited. The clearest manifestation of the breadth of application of Newtonian theory may be found in Ried and Mayo (1963). They provide equations based on Newtonian theory for the important static and dynamic stability coefficients for bodies of revolution with an offset center of gravity location flying at angles of attack, yaw, and bank. Although the final equations are often quite lengthy, solutions are reduced to carrying out a single integration

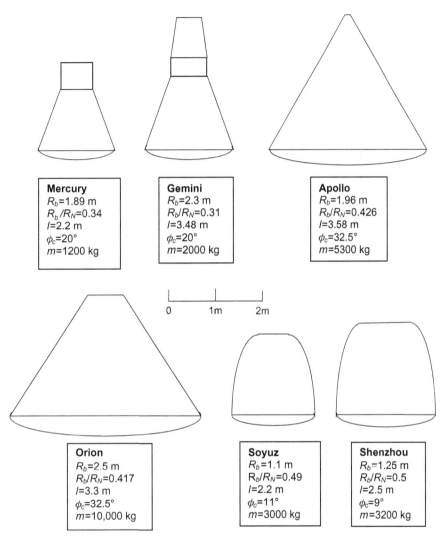

FIGURE 8.13

Scaled sketches of past and current space capsules with pertinent configuration data.

along the longitudinal axis of the vehicle for each of the aerodynamic coefficients sought. It is worthwhile to examine this work to appreciate the extent of effort on analysis that was necessary to formulate the equations in a manner which then-current computers could readily handle. Computational power is now within such easy reach that the Newtonian theory may be readily applied to an array of independent panels representing the actual vehicle, so that the primitive forces on each may be combined to generate overall forces and moments as described previously in this chapter.

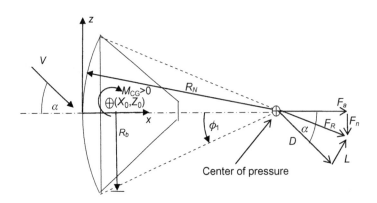

FIGURE 8.14

Blunt Apollo-like space capsule at an angle of attack showing resultant force field, direction of positive pitching moment, and center of pressure.

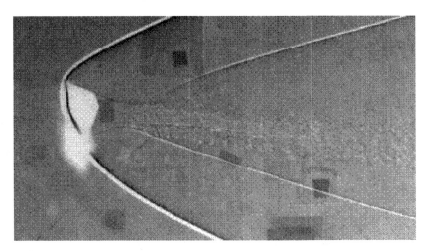

FIGURE 8.15

Shadowgraph of a model of a Mars Lander capsule at high Mach number in a ballistic range showing the shock following the body shape.

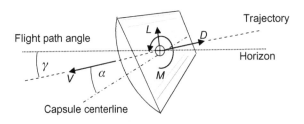

FIGURE 8.16

Orientation of capsule for positive lift coefficient at a practical angle of attack.

The value of analytically based approaches is that they often provide general information about the behavior of a solution rather than just the solution itself. The equations for the axial and normal force coefficients given by Ried and Mayo (1963), when simplified for the conditions of zero side-slip and bank angles and zero pitch, yaw, and roll rates are as follows:

$$C_a = 8\left(1 - \frac{\varepsilon}{2}\right)\left(\frac{R_N}{d}\right)^2 \int_0^{\phi_1} (2\cos^2\alpha\cos^3\phi\sin\phi + \sin^2\alpha\sin^3\phi\cos\phi)d\phi \tag{8.29}$$

$$C_n = 8\left(1 - \frac{\varepsilon}{2}\right)\left(\frac{R_N}{d}\right)^2 \sin 2\alpha \int_0^{\phi_1} \sin^3\phi\cos\phi d\phi \tag{8.30}$$

Noting that $R_N/d = (2\sin\phi_1)^{-1}$, we may integrate these equations to yield

$$C_a = \left(1 - \frac{\varepsilon}{2}\right)\left[\cos^2\alpha(1 + \cos^2\phi_1) + \frac{1}{2}\sin^2\alpha\sin^2\phi_1\right] \tag{8.31}$$

$$C_n = \frac{1}{2}\left(1 - \frac{\varepsilon}{2}\right)\sin 2\alpha\sin^2\phi_1 \tag{8.32}$$

We may also write Eqn (8.31) in terms of the zero-lift drag coefficient as follows:

$$C_a = C_{D,0}\cos^2\alpha + \frac{1}{2}\left(1 - \frac{\varepsilon}{2}\right)\sin^2\alpha\sin^2\phi_1 \tag{8.33}$$

The moment coefficient, after applying the simplifying assumptions concerning the flight attitude of the vehicle, becomes

$$C_m = 8\left(\frac{R_N}{d}\right)^2 \sin 2\alpha \int_0^{\phi_1} \left[(x - x_0) + \left(\frac{R_N}{d}\right)\cos\phi\right](-\sin^3\phi\cos\phi)d\phi - \left(\frac{z_0}{d}\right)C_a$$

Carrying out the integration yields

$$C_m = \left[\left(\frac{x_0}{d}\right) - \left(\frac{R_N}{d}\right)\right]C_n - \left(\frac{z_0}{d}\right)C_a \tag{8.34}$$

The angle of attack used in the general development presented by Ried and Mayo (1963) is defined in the opposite sense to that shown in Figure 8.14 because the latter is more convenient for presenting force and moment data for the specific case of space capsules. The only special care to be taken is that the sense of the axial and normal forces should be accounted for in the moment Eqn (8.34). That is, the contribution of the normal force coefficient to the moment about the center of gravity in Eqn (8.34) is positive because the term in square brackets is negative and, as depicted in Figure 8.14, C_n acts in the negative z-direction. Of course, Eqn (8.34) might have been written down by inspection because, as mentioned previously, only pressure forces are considered and they pass through the origin of the circular nose making that point the center of pressure. The normal and axial forces may be considered to act there with no associated moment. Then, the moment about the center of gravity is simply the sum of the moments produced by the axial and normal forces acting at the center of pressure. In practical cases where boundary layer details, afterbody wake effects, and surface protuberances are considered, the center of pressure may be somewhat removed from the location of the center of the nose radius.

Using the above equations for the axial and normal force coefficients, along with our convention for α, in Eqns (8.27) and (8.28), the lift and drag coefficients become

$$C_L = \left[C_{D,0} \cos^2 \alpha + \frac{1}{2} \left(1 - \frac{\varepsilon}{2} \right) \sin^2 \phi_1 \left(3 \sin^2 \alpha - 2 \right) \right] \sin \alpha \tag{8.35}$$

$$C_D = \left[C_{D,0} \cos^2 \alpha + \frac{3}{2} \left(1 - \frac{\varepsilon}{2} \right) \sin^2 \phi_1 \sin^2 \alpha \right] \cos \alpha \tag{8.36}$$

The variation of the lift and drag coefficients with angle of attack as given by Eqns (8.35) and (8.36) is shown in Figure 8.17 for $\gamma = 1.4$ and two bluntness ratios $R_b/R_N = 0.426$ and 0.5 which are characteristic of the Apollo and Soyuz capsules, respectively. The variation of the lift to drag ratio L/D is shown in Figure 8.18 for the same two bluntness ratios. Also shown on both figures are curves representative of experimental data for the Apollo command module (CM) presented by DeRose (1969). In general, the theoretical results for manned capsule shapes are limited to angles of attack about equal to the afterbody cone angle ϕ_c. When $\alpha > \phi_c$, the afterbody surfaces are exposed to the shock-heated flow environment rather than the relatively benign environment in the Newtonian shadow region. In addition, at such high angles, the lift begins to drop off very much like the stalling of the lift on a wing at high angle of attack. For relatively flat-backed axisymmetric bodies, like some unmanned planetary reentry vehicles, the results of Eqns (8.35) and (8.36) are applicable although the high angle of attack restriction still applies.

It is clear from Figure 8.17 that the Newtonian theory overestimates the experimentally observed drag coefficient by around 7.5%. Miller and Lawing (1966) present experimental data from their experiments and from those of several others on the pressure distribution over the blunt nose of an Apollo-like capsule. The results they present are for $\alpha = 0$ and $\alpha = 33°$ (approximately the capsule cone angle ϕ_c) over the Mach number range $6 < M < 24.5$, and all data show quite good agreement with the predictions of Newtonian theory. Interestingly, the best agreement occurs for $\alpha = 33°$, whereas at $\alpha = 0$, the pressure is somewhat overestimated over the outermost portion of the heat shield. They also compare the experimental data for surface pressure with theoretical

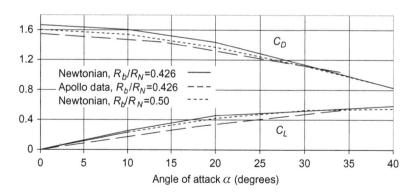

FIGURE 8.17

Variation of lift and drag coefficients of Apollo-like capsules with angle of attack for two bluntness ratios, both with $\gamma = 1.4$.

Experimental Apollo data taken from DeRose (1969).

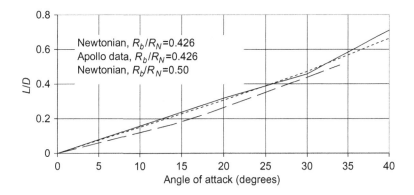

FIGURE 8.18

Variation of *L/D* for Apollo-like capsules with angle of attack for two bluntness ratios, both with $\gamma = 1.4$.

Experimental Apollo data taken from DeRose (1969).

results based on the approach presented by Kaattari (1962), which show even better agreement than do the Newtonian predictions for the zero angle of attack case.

Kaattari (1962) also provides a method for quite accurately calculating the shock shape over a blunt body rather than relying on the Newtonian assumption that the shock and body shapes are coincident. His approach, although more accurate than the simple Newtonian theory, is more cumbersome to apply, and the relatively small improvement in the results for surface pressures, particularly in the practical case of non-zero angle of attack, has limited its use. The method does however give useful results for shock shape, shock standoff distance, and stagnation point locations for capsule-type bodies.

This overestimation of the front face pressure contributes to predictions for the zero-lift drag coefficient that are around 5–10% higher than those obtained by experiment. It must also be noted that part of the higher estimate for the drag coefficient arises from the fact that the Newtonian theory predicts the afterbody (shadow) region to have $C_{p,a} = 0$. However, Miller and Lawing (1966) also provide experimental data which show that actually $C_{p,a} > 0$, as described subsequently in Section 8.3.2, and this also contributes to a slight reduction in the drag coefficient from that predicted by Newtonian theory alone.

8.3.2 CAPSULE AFTERBODY PRESSURE

The assumptions of the Newtonian theory discussed in Appendix A require that the thin shock layer follow the blunt nose and not turn to follow the capsule afterbody. Instead, the shock layer departs from the body leaving the afterbody in the so-called shadow region where $C_p = 0$. However, experimental results from more than a half dozen different investigations were presented by Miller and Lawing (1966) for hypersonic flow over an Apollo-like capsule model which show that the afterbody pressure p_a increases with Mach number from values close to free stream pressure (for which $C_p = 0$) at $M \sim 5$ to as much as an order of magnitude greater than free stream pressure at $M = 20$. The data presented could be approximated by the following simple relation:

$$\frac{p_a}{p_\infty} \approx 1 + 0.025 \, M_\infty^2 \tag{8.37}$$

Then, the pressure coefficient on the afterbody would be essentially constant with Mach number and equal to

$$C_{p,a} = \frac{p_a - p_\infty}{\frac{1}{2}\gamma p_\infty M_\infty^2} \approx \frac{0.05}{\gamma} \tag{8.38}$$

Miller and Lawing (1966) also show data for the circumferential variation of the afterbody pressure field at zero angle of attack and at the essentially maximum angle of attack $\alpha = 33°$. At $\alpha = 0$, the pressure is reasonably constant around the afterbody, and although at $\alpha = 33°$ the pressure varies circumferentially, the average value is about the same as that at $\alpha = 0$. Under these conditions, the pressure force on the afterbody is $C_{p,a}q\pi R_b^2$, and the zero-lift drag coefficient predicted by Newtonian theory could probably be corrected as follows:

$$C_{D,0} = (2 - \varepsilon)\left[1 - \frac{1}{2}\left(\frac{R_b}{R_N}\right)^2\right] - \frac{0.05}{\gamma} \tag{8.39}$$

Such a correction amounts to about a 2–3% reduction in the drag coefficient. We mentioned previously that accounting for the Newtonian theory overestimation of the front face pressure is likely to contribute another 5% or so.

8.3.3 FORCES ON SPHERICALLY BLUNTED CONES

Some reentry vehicles, particularly ballistic missile warheads, are configured as spherically blunted cones. At high Mach numbers, Newtonian flow theory can be used to estimate the aerodynamic forces on these bodies as well. A schematic of the blunted cone showing all the variables appears in Figure 8.19. The behavior of the pressure coefficient for the spherical nose portion follows directly from the results of the previous sections, while on the conical portion of the body, the pressure coefficient remains constant. For the spherical nose cap, the differential drag coefficient is

$$dC_D = (2 - \varepsilon)\frac{2\pi r^2}{S}\left[d\left(-\frac{\cos^4\phi}{4}\right)\right] \tag{8.40}$$

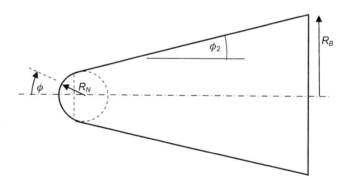

FIGURE 8.19

Schematic diagram of spherically blunted cone.

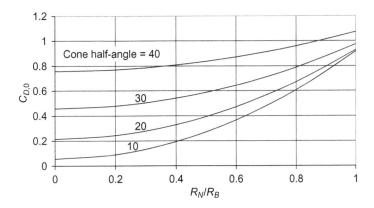

FIGURE 8.20

Zero-lift drag coefficient for spherically blunted cones as a function of the ratio of nose to base radius for various cone half angles and $\gamma = 1.4$.

On the conical portion of the body, the differential drag coefficient is

$$dC_D = (2 - \varepsilon)\sin^2 \phi_2 \tan \phi(2\pi r dx) \tag{8.41}$$

The complete zero-lift drag coefficient is given by

$$C_{D,0} = (2 - \varepsilon)\left[\frac{1}{2}\left(\frac{R_N}{R_B}\right)^2 \left(1 - \sin^4 \phi_2\right) + \sin^2 \phi_2 - \left(\frac{R_N}{R_B}\sin \phi_2 \cos \phi_2\right)^2\right] \tag{8.42}$$

Note that when $R_N = R_B$ (and therefore $\phi_2 = 0$), the body is a hemisphere-cylinder and $C_D = (1 - \varepsilon/2)$, the same as that for a sphere. When $R_N = 0$, the body is a sharp cone and the drag coefficient $C_D = (2 - \varepsilon)\sin^2\phi_2$, the same as for a cone of half-angle equal to ϕ_2. The variation of the zero-lift drag coefficient with the ratio of nose radius to base radius for various values of the cone half-angle is shown in Figure 8.20 for the case of $\gamma = 1.4$. Other values of γ may be accommodated by using $\varepsilon = (\gamma - 1)/(\gamma + 1)$.

It is clear that small nose radii can be accommodated without substantial drag penalties. Nose bluntness mitigates reentry heating effects because the heat flux at the stagnation point is inversely proportional to the square root of the nose radius. Thus, thermal protection can be enhanced without much sacrifice in drag. The effect of the thermodynamic state of the gas is contained solely in the initial term multiplying the square brackets; recall that

$$\varepsilon = \frac{\gamma - 1}{\gamma + 1} \tag{8.43}$$

Thus at higher temperatures where $\gamma < 1.4$, the drag coefficient will be larger than the perfect gas value as shown in Figure 8.20. For example, when $\gamma = 1.2$, multiply the results in Figure 8.20 by 126/111 which increases the drag coefficient by about 4%.

An example of a sphere-cone body used as a practical entry vehicle is the General Electric Mark-6 as shown in Figure 8.21. This entry vehicle is large, with a mass of 3360 kg, a length of 3.1 m, a base radius of 1.15 m, and a nose radius of 0.59 m. The ballistic coefficient $B = mg/C_{D,0}S$ for this vehicle is

FIGURE 8.21

The 3.1 m long Mark-6 reentry vehicle as shown in the Titan-2 ICBM exhibit at the National Atomic Museum Albuquerque, NM.

Photograph by Stephen Sutton.

therefore 7.86 kPa/C_D. The cone half-angle is 12.5° and $R_B/R_N = 0.5$ which, from Figure 8.20, suggests $C_{D,0} \sim 0.3$ and therefore a ballistic coefficient $B \sim 26$ kPa, about five times higher than a Project Mercury capsule. This provides the capability for very steep high-velocity entry, as discussed in Chapter 6, making the Mark-6 difficult for ABM systems to intercept. Advances in ablative thermal protection system design led to smaller vehicles with smaller nose bluntness ratio compared to the Mk-6. A more modern sphere-cone entry vehicle is the 360 kg Mark-12 with a 10° cone half-angle and a very small nose bluntness resulting in $C_{D,0} \sim 0.05$ and B = 325 kPa.

8.3.4 NEWTONIAN FLOW WITH A P−M EXPANSION

Consider the hypersonic inviscid flow over a hemisphere-cylinder blunt body as shown in Figure 8.22. Under the Newtonian flow approximation, the ratio of the local surface value of C_p to the stagnation point value $C_{p,max}$ is

$$\frac{C_p}{C_{p,\max}} = \sin^2 \theta = \cos^2 \phi \tag{8.44}$$

The variation of $C_p/C_{p,max}$ as a function of non-dimensional distance $s/R_b = \phi$ along the surface starting from the stagnation point where $\phi = 0$ is shown in Figure 8.23. The predicted

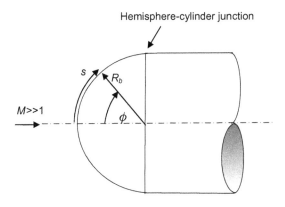

FIGURE 8.22

Schematic diagram of a hemisphere-cylinder in hypersonic flow.

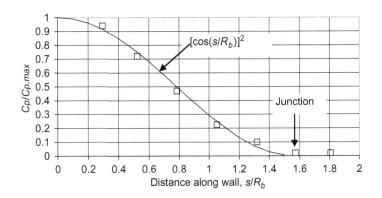

FIGURE 8.23

$C_p/C_{p,max}$ as a function of non-dimensional distance $s/R_b = \phi$ along the body surface from the stagnation point to beyond the hemisphere-cylinder junction. Symbols denote $M = 6.8$ data from Crawford and McCauley (1957).

pressure distribution of Eqn (8.44) agrees well with the measurements on a hemisphere-cylinder presented by Crawford and McCauley (1957) for $M_1 = 6.8$. They also show data from other researchers for $M_1 = 1.97$, 3.8, and 5.8. Only for the case of $M_1 = 1.97$ is there a noticeable departure from the Newtonian prediction and that only for $\phi > 45°$, which is in the vicinity of the sonic point. This overexpansion compared to Newtonian theory is only in evidence for $M_1 < 3$. Otherwise, the pressure recovers to approximately the free stream value $p \sim p_1$ as the hemisphere-cylinder junction is approached, as predicted by the Newtonian theory. However, it can be seen in Figure 8.23 that the measured pressure coefficient beyond the junction is small, but not zero as predicted.

According to the Newtonian pressure distribution, the sonic point on the surface of the body is approximately located at the angle ϕ_c where

$$\cos \phi_c = \left(\frac{\gamma+1}{2}\right)^{\frac{\gamma}{2(\gamma-1)}} = \sqrt{\frac{p^*}{p_{t2}}} \tag{8.45}$$

Using the definition of the pressure coefficient, one may solve for the static-to-stagnation pressure ratio and therefore M on the surface of the body. This approach yields

$$\frac{p}{p_{t2}} = \left(1 + \frac{\gamma-1}{2}M^2\right)^{\frac{-\gamma}{\gamma-1}} = \frac{C_p}{C_{p,\max}} + \left(1 - \frac{C_p}{C_{p,\max}}\right)\left(\frac{p_{t2}}{p_1}\right)^{-1} \tag{8.46}$$

Solving Eqn (8.46) for the surface, Mach number results in

$$M^2 = \frac{2}{\gamma-1}\left\{\left[\frac{C_p}{C_{p,\max}} + \left(1 - \frac{C_p}{C_{p,\max}}\right)\left(\frac{p_{t2}}{p_1}\right)^{-1}\right]^{\frac{-(\gamma-1)}{\gamma}} - 1\right\} \tag{8.47}$$

The ratio of the stagnation pressure behind a normal shock to the static pressure upstream of the shock p_{t2}/p_1 comes from the normal shock relations and is given by

$$\frac{p_{t2}}{p_1} = \left[\frac{\gamma+1}{2}M_1^2\right]^{\frac{\gamma}{\gamma-1}}\left[\frac{(\gamma+1)}{2\gamma M_1^2 - (\gamma-1)}\right]^{\frac{1}{\gamma-1}} \tag{8.48}$$

For $M \gg 1$, Eqn (8.48) becomes

$$\frac{p_{t2}}{p_1} \approx \left(\frac{\gamma+1}{2}\right)^{\frac{\gamma+1}{\gamma-1}}\gamma^{\frac{-1}{\gamma-1}}M_1^2 \approx \gamma^{0.75}M_1^2 \tag{8.49}$$

Equation (8.49) is accurate to within about 4% for $M_1 > 3$ and improves as M_1 increases. Rather than relying completely on the Newtonian pressure distribution as expressed in Eqn (8.44), one may apply the P–M expansion discussed in Appendix A for points on the body beyond the sonic point ϕ_c, where $M = 1$ on the body and, by definition, the P–M angle $\nu = 0$. The change in the P–M angle as one proceeds around the body is given by $\Delta\nu = \phi - \phi_c$, and for each value of ν, there is associated a Mach number and static-to-stagnation pressure ratio as given in Tables A.1 and A.2 of Appendix A.

In Figure 8.24, the Mach number calculated from Eqn (8.47) on the basis of the Newtonian pressure distribution of Eqn (8.44) is compared to that calculated using the P–M expansion beyond the sonic point as defined by Eqn (8.45). It is clear that up to $s/R_b = 1.2$ or $\phi = 69°$, the two results are approximately equal, but beyond that point, the results for the two methods are different with the coupled Newtonian and P–M approach showing better agreement with experiment.

Using Eqn (8.47) with the Newtonian pressure distribution to calculate the surface, Mach number yields $M = 3.33$ at the hemisphere-cylinder junction for a free stream Mach number of 6.8 which is about 19% high compared to experiment. Although the pressure difference between the two methods is small and the effect on pressure force is not significant, the difference in surface Mach number prediction has an effect on the boundary layer and therefore the skin friction and heat transfer.

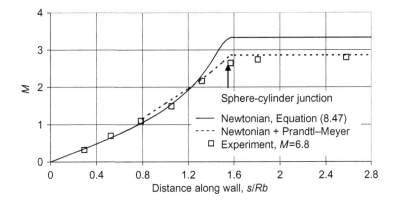

FIGURE 8.24

Mach number distributions around a hemisphere-cylinder according to two simple theories using $\gamma = 1.4$ are compared with experimental results.

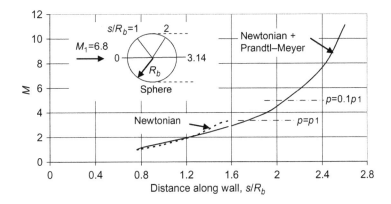

FIGURE 8.25

Mach number distributions around a sphere according to two simple theories using $\gamma = 1.4$. The dashed lines behind the sphere delineate the region within which Newtonian theory does not apply.

If we consider the flow around a sphere alone, the Newtonian theory predicts $C_p = 0$ at $\phi = 90°$ and a "dead water" region behind the body beyond that point where the theory does not apply. Instead, the P–M expansion, which takes over from the Newtonian theory when $M > 1$, can be continued around the body continually dropping the pressure and increasing the inviscid Mach number on the body. This effect on the surface Mach number is illustrated in Figure 8.25 for flow of a gas with $\gamma = 1.4$ over a sphere at $M_1 = 6.8$. The surface pressure is equal to the free stream pressure p_1 at $M = 3.33$, which occurs at $\phi = 90°$ according to Newtonian theory and about $\phi = 97.4°$ for the continued P–M expansion. Also shown on Figure 8.25 is the point $\phi = 120°$ and $M = 5.11$ for

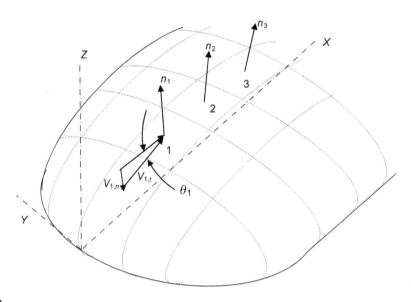

FIGURE 8.26

Schematic of a typical adjacent panels used to implement the P−M expansion for values of $\theta < 20°$.

which $p = 0.1p_1$, which is about as far as a real flow can proceed. In general, somewhere between these values of ϕ, viscous effects become important and a recirculation region is established behind the body causing the inviscid flow to separate from the body.

In the general panel method described in Section 8.2.3, we know θ as the angle between the free stream and the tangent to the surface on each panel, as shown in Figure 8.26. When $\theta < 20°$, the P−M expansion is more accurate than the Newtonian theory and should be applied. The flow deflection between adjacent panels 1 and 2 is simply $\Delta\theta = \theta_2 - \theta_1$. Knowing M_1 on panel 1 at a given x-station, we also know ν_1 on that panel and we may find ν_2 on panel 2, the adjacent panel at the next x-station, because $\nu_2 = \nu_1 + \Delta\theta$. Using this value of ν_2, we can find the corresponding value of M_2 on panel 2 either by using Tables A.1 and A.2 in Appendix A or one the equations for ν given there. We also can determine p_2/p_t and then proceed for the other variables as described previously in Section 8.2.3. This unit process may be carried out for any pair of panels and thereby determine the flow conditions on each.

An example of a blunt body like a hemisphere-cylinder is the entry vehicle for the KH-7 (Keyhole-7) high-resolution space reconnaissance system as shown in Figure 8.27. The entry vehicle was about 0.7 m in diameter and 0.8 m long and had a mass of 160 kg. Satellites with film-return capabilities were used for two basic functions: search or surveillance. CORONA satellites, first launched in 1960, were search systems that photographed wide swaths of land to identify airfields and missile sites, among other things. The KH-7 series of satellites were operational during 1963−1967 fulfilling the need for surveillance using stereo high-resolution cameras. These satellites were launched by Atlas-Agena rockets from Vandenberg AFB in California, and the film capsules were recovered in flight near Hawaii.

FIGURE 8.27

KH-7 reconnaissance satellite the National Museum of the US Air Force. The blunt body protected film containers during entry from orbit.

Courtesy US Air Force.

8.4 SLENDER BODIES IN HYPERSONIC FLIGHT

8.4.1 SLENDER BODY *L/D*

Consider a slender, relatively flat-bottomed body of planform area S, span b, maximum thickness t, sharp leading edge angle $\theta \ll 1$, and overall length l ($t \ll l$) at angle of attack α, as shown in Figure 8.28. In terms of the axial and normal forces, F_a and F_n, the lift to drag ratio is

$$\frac{L}{D} = \frac{(F_n/F_a)\cos\alpha - \sin\alpha}{(F_n/F_a)\sin\alpha + \cos\alpha} \tag{8.50}$$

The slender body has planform area S, wetted surface area S_w, and base area S_{base}. The normal force F_n is determined primarily by pressure acting on the upper and lower surfaces of the vehicle, while the axial force F_a is determined mainly by the frictional force and the pressure acting on the forebody and base of the vehicle, as suggested by Figure 8.28. These force components may be written as follows:

$$\begin{aligned} F_n &= (C_{p,l} - C_{p,u})qS \\ F_a &= C_F qS_w + C_{D,b}qS_b \end{aligned} \tag{8.51}$$

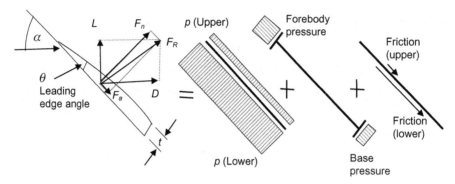

FIGURE 8.28

Schematic of forces acting on an idealized slender body in hypersonic flow as seen in an elevation view.

The pressure, skin friction, and form drag coefficients in Eqn (8.51): C_p, C_F, and $C_{D,b}$, respectively, are values integrated over the surfaces on which they act. Then the ratio of normal to axial force becomes

$$\frac{F_n}{F_a} = \frac{(C_{p,l} - C_{p,u})}{C_F \frac{S_w}{S} + C_{d,b} \frac{S_b}{S}} \tag{8.52}$$

In the Newtonian flow approximation for hypersonic flow, the lower surface pressure coefficient is constant over the relatively flat surface of planform area S and is given by $C_{p,l} = 2 \sin^2 \alpha$. On the other hand, the pressure over the upper surface of the body, which lies in the shadow region for angles $\alpha > \theta$, is constant and equal to the free stream pressure and therefore $C_{p,u} = 0$. The normal force is then merely the product $C_{p,l} q S$. Then, the ratio of normal to axial force in Eqn (8.52) becomes

$$\frac{F_n}{F_a} = \frac{2 \sin^2 \alpha}{C_F \frac{S_w}{S} + C_{d,b} \frac{S_b}{S}} \tag{8.53}$$

The form drag may be considered to be the axial component of the force due to the pressure acting over the entire front surface of the vehicle less the pressure force on the base. In Newtonian flow theory, the base pressure coefficient would be considered zero because the base is within the shadow region, in which case it would contribute no axial force, and the form drag would rise mainly from the force on the nose. In the ideal case of a thin flat plate, the form drag would be zero. The form drag of other practical spacecraft shapes will be discussed subsequently. For the current general analysis, we will consider the form drag term parametrically.

Therefore, following the usual drag build-up estimation procedure described, for example, by Sforza (2014), we write the denominator of Eqn (8.53) as

$$C_F \frac{S_w}{S} + C_{d,b} \left(\frac{S_b}{S}\right) = C_F \frac{S_w}{S} \left[1 + \frac{C_{d,b}}{C_F} \frac{S_b}{S_w}\right] = K C_F \tag{8.54}$$

Here, C_F denotes the skin friction coefficient integrated over the wetted surface of the body excluding the base region, wherein the contribution of frictional forces is negligible. The coefficient K defined in Eqn (8.54) depends not only on the body geometry but also on the Reynolds and

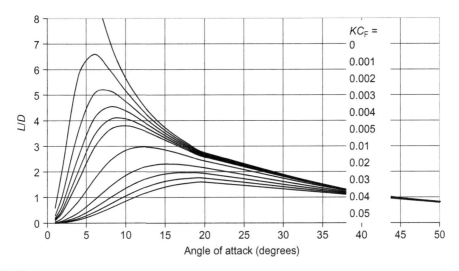

FIGURE 8.29

The lift to drag ratio according to Eqn (8.10) is shown as a function of the friction and form drag loss parameter KC_F in the order shown.

Mach numbers, because the drag and skin friction coefficients depend on them. Resolving the normal and axial forces into the standard lift and drag yields

$$L = C_L qS = F_n \cos \alpha - F_a \sin \alpha \tag{8.55}$$

$$D = C_D qS = F_n \sin \alpha + F_a \cos \alpha \tag{8.56}$$

Then the lift and drag coefficients become

$$C_L = 2 \sin^2 \alpha \cos \alpha - KC_F \sin \alpha \tag{8.57}$$

$$C_D = 2 \sin^3 \alpha + KC_F \cos \alpha \tag{8.58}$$

The lift to drag ratio follows directly as

$$\frac{L}{D} = \frac{2 \sin^2 \alpha - KC_F \sin \alpha}{2 \sin^3 \alpha + KC_F} \tag{8.59}$$

The variation of L/D with angle of attack is shown in Figure 8.29 as a function of the frictional and base drag loss parameter KC_F. The maximum lift to drag ratio decreases and the angle of attack for maximum lift to drag ratio increases as the loss parameter increases. Equation (8.59) illustrates that increasing the ratio of wetted to planform area of a spacecraft will necessarily decrease L/D. But manned spacecraft must provide sufficient internal volume for the crew and life support equipment, and this tends to increase the base drag as well. The quantity $\tau = v^{2/3}/S$ is sometimes used to characterize hypersonic vehicle shapes, and it is related to the ratio of wetted area to planform area in the factor K by the following:

$$\frac{S_w}{S} = \frac{S_w}{v^{2/3}} \frac{v^{2/3}}{S} = \frac{S_w}{v^{2/3}} \tau \tag{8.60}$$

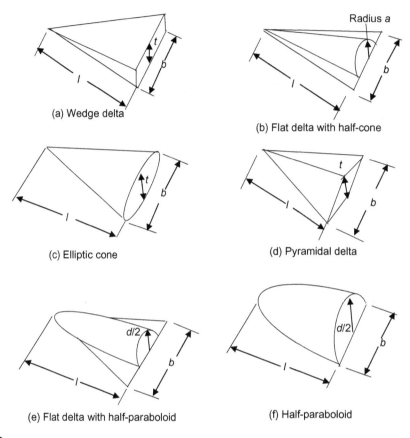

FIGURE 8.30

Generic configurations for high-speed slender bodies with equivalent ratios of base area to planform area S_b/S (drawings are not to scale).

For a variety of slender shapes like wedge deltas, flat deltas with a half-cone on the upper surface, elliptic cones, and pyramidal deltas, cones, etc., some of which are depicted in Figure 8.30, the shape factor $S_w/v^{2/3}$ remains relatively constant, lying in the range $10 < S_w/v^{2/3} < 12$. The shape factors $S_w/v^{2/3}$ for the bodies illustrated in Figure 8.30 were determined under the constraint of fixed b/l with the results shown in Table 8.1. The volume parameter based on planform area, $\tau = v^{2/3}/S$, was constant for several of these similar slender bodies. Adding volume by placing a half-paraboloid on top of a flat delta or a half-paraboloid as shown in Figure 8.30e and f, respectively, yields increased values of τ and of S_w/S. Calculations based on drawings of proposed and actual spaceplanes reveal the characteristics as shown in Table 8.2. Thus, it is unlikely that practical manned spacecraft will generate lift to drag ratios of the more slender configurations.

A simple estimate of the maximum L/D for a given configuration may be obtained by noting that the frictional and form drag contribution to the lift in the numerator of Eqn (8.59) is very

Table 8.1 Characteristics of Typical Slender Body Geometries with Equal Length and Span

Body Shape	Span, b	Thickness, t	Radius, $d/2$	$S_w/v^{2/3}$	$v^{2/3}/S = \tau$	S_w/S	S_b/S
1. Wedge delta	$0.4l$	$0.157b$		10.0	0.206	2.0	0.125
2. Flat delta/half-cone	$0.4l$		$0.316b$	11.6	0.206	2.37	0.125
3. Elliptic cone	$0.4l$	$0.200b$		11.2	0.206	2.19	0.125
4. Pyramidal delta	$0.4l$	$0.314b$		10.0	0.206	2.10	0.125
5. Flat delta/paraboloid	$0.4l$		$0.316b$	9.56	0.256	2.45	0.125
6. Half-paraboloid	$0.4l$		$0.5b$	6.94	0.375	2.60	0.472

Table 8.2 Configuration Characteristics of Typical Spaceplanes

Spaceplane	S (m^2)	S_w (m^2)	Volume (m^3)	Volume parameter $\tau = v^{2/3}/S$	Volume parameter $v^{2/3}/S_w$	S_w/S	$b/2l$
SSTS	290.2	1146	1040	0.354	11.2	3.95	0.352
Hermes	80.46	295.8	184.3	0.403	9.12	3.68	0.225
HIMES	29.46	153.6	78.91	0.625	8.35	5.22	0.385
Soviet SP	5.574	20.25	3.567	0.419	8.67	3.63	0.373
HOTOL	254.3	1073	1038	0.403	10.5	4.22	0.157
X-15	18.67	111.1	26.67	0.478	12.43	5.95	0.178
HL-10	14.86	47.10	13.22	0.376	8.42	3.17	0.321
Average value				0.437	9.60	4.27	0.284
Reference sphere				0.827	4.853	4.00	

small, that is, $2 \sin \alpha \ll KC_F$ for slender bodies. Using this assumption, the maximum lift to drag ratio is found to be

$$\left(\frac{L}{D}\right)_{max} \approx \frac{2}{3}(KC_F)^{-1/3} \tag{8.61}$$

This maximum occurs at an angle of attack

$$\alpha_{max(L/D)} \approx (KC_F)^{\frac{1}{3}} \tag{8.62}$$

The increase of the estimated $\alpha_{max(L/D)}$ with KC_F is evident in Figure 8.29, and $(L/D)_{max}$ is shown as a function of KC_F in Figure 8.31. In that figure, the range of $(L/D)_{max}$ measured in hypersonic flight for the Space Shuttle and the X-15 is outlined by the dashed curve. Because $C_F \sim 10^{-3}$ and $S_w/S \sim 4$, the form drag factor $(C_{D,b}/C_F)(S_b/S) \sim 10$. The base area ratio for both vehicles is about $S_b/S = 10^{-1}$ suggesting that $C_{D,b}/C_F \sim 10^2$ or $C_{D,b} \sim 10^{-1}$, a value that is reasonable. Results from NASA wind tunnel studies of the maximum L/D of slender bodies like those shown in Figure 8.30a–d with $0.1 < \tau < 0.3$ were reported by Becker (1964). That data may be correlated by the following relation:

$$\left(\frac{L}{D}\right)_{max} \approx 6 - 7.3\tau \tag{8.63}$$

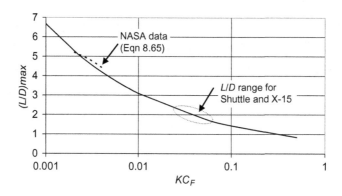

FIGURE 8.31

Approximate variation of the maximum L/D is shown as a function of KC_F. The L/D range for the Space Shuttle and the X-15 measured in hypersonic flight is outlined by the dotted curve, while the dashed curve indicates the range of L/D for some NASA hypersonic wind tunnel data using Eqn (8.65).

The NASA experiments were performed at $Re = 1.5 \times 10^6$ and $M = 6.8$, a condition for which the boundary layer is at least transitional in nature and is more likely to be turbulent because boundary layer transition in wind tunnels always occurs at lower Re than in free flight. The integrated skin friction for turbulent flow over an adiabatic flat plate in this case is $C_F \sim 1.86 \times 10^{-3}$. The quantity KC_F for the NASA wind tunnel data may be estimated as

$$KC_F \approx 10\tau(1.86 \times 10^{-3}) = 0.0186\tau \tag{8.64}$$

Solving Eqn (8.64) for τ and substituting the result into Eqn (8.61) yields

$$\left(\frac{L}{D}\right)_{max} \approx 6 - 7.3 \left(\frac{KC_F}{0.0186}\right) \tag{8.65}$$

This result is shown in Figure 8.31 as the dashed curve of the NASA data range. Wind tunnel experiments are of short duration, so that the model wall heats up to a temperature considerably less than the adiabatic wall temperature for which the skin friction coefficient is larger than in the adiabatic case. On the other hand, if in the experiment there is laminar flow to some extent over the body, a lower value for the integrated skin friction coefficient will be obtained. Therefore, the curve for the adiabatic wall case shown in Figure 8.4 should be representative of the NASA data range. Note that form drag has been considered negligible in this analysis of the slender body results reported in the NASA experiments.

It should be apparent from Figure 8.31 that as spacecraft volume becomes important, the maximum L/D achievable in the hypersonic portion of the spacecraft's flight is likely to be, at the best, $(L/D)_{max} \sim 2$. Recall that

$$KC_F = C_F \frac{S_w}{S} + C_{D,b} \frac{S_b}{S} \tag{8.66}$$

The skin friction contribution, which is the first term on the right-hand side of Eqn (8.66), will be on the same order of magnitude for typical manned spacecraft, independent of the details of its shape.

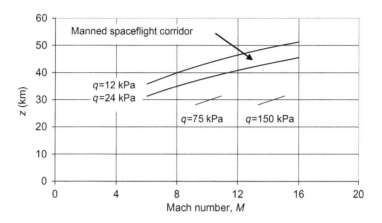

FIGURE 8.32

Dynamic pressure contours for safe manned spaceflight are shown along with values of q for flight carried out at a constant altitude of $z = 30$ km.

However, the form drag contribution, represented by the second term on the right-hand side of Eqn (8.65), tends to grow larger as the spacecraft internal volume increases. As indicated by Figure 8.31, very slender bodies are characterized by the range $10^{-3} < KC_F < 10^{-2}$, while more practical spacecraft configurations are characterized by the range $10^{-2} < KC_F < 10^{-1}$. The data in Tables 8.1 and 8.2 suggest that in terms of the volume parameter τ, one may classify as slender missiles those shapes with $0.1 < \tau < 0.3$ and as practical manned spacecraft those with $0.35 < \tau < 0.65$.

The preceding development illustrates the fact that in hypersonic flight, the lift to drag ratio is dominated by the drag-related term KC_F. In subsequent sections, we will show that the integrated skin friction coefficient $C_F = C_F(Re_l, M, T_w/T_e)$. Flight at constant altitude and increasing Mach number causes the dynamic pressure $q = \gamma p M^2/2$ to increase dramatically, as shown in Figure 8.32. For $z = 30$ km, the dynamic pressure varies from about 25 kPa to over 150 kPa for the Mach number range shown. Also shown on Figure 8.32 is the typical dynamic pressure corridor for safe human flight in the atmosphere, which was discussed in Chapter 6. Inertial, structural, and heating constraints would confine manned hypersonic cruise vehicles to flight at a nearly constant dynamic pressure between 2.4 and 24 kPa.

Flight at constant altitude or constant dynamic pressure also impacts the Reynolds number of the flow over the vehicle, and this affects the state of the boundary layer. The Reynolds number variation with Mach number for flight at constant $z = 30$ km or flight at constant $q = 24$ kPa is shown in Figure 8.33 for flat plate bodies with length $l = 10$ and 30 m. Note that at $M = 6$, the Reynolds numbers are equivalent for the two cases but the Reynolds number increases with Mach number for the constant altitude case and decreases with Mach number for the constant dynamic pressure case. At $M = 16$, there is about an order of magnitude difference between the values of Re_l for the corresponding cases. Also shown on Figure 8.6 is a nominal curve for boundary layer transition based on momentum thickness, which will be described in a subsequent section of this chapter. Basically, for $Re_l < Re_t$, the boundary layer should be laminar over the entire plate, while for $Re_l > Re_t$, the boundary layer may be considered to be laminar for $Re_x < Re_t$ and turbulent thereafter.

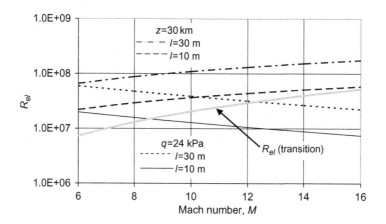

FIGURE 8.33

Variation of the Reynolds number based on plate length for flight at either constant $q = 24$ kPa or constant altitude $z = 30$ km. The gray curve indicates the nominal Reynolds number based on x for transition on the plate. For interpretation of the references to color in this figure legend, the reader is referred to the web version of this book.

Taking the simplest case of a slender plate-like body with negligible form drag $(C_{D,b} \sim 0)$ so that $K = S_w/S \sim 2$, the maximum lift to drag ratio is

$$\left(\frac{L}{D}\right)_{max} = \frac{2}{3}(KC_F)^{-\frac{1}{3}} = \frac{0.529}{C_F^{1/3}} \tag{8.67}$$

In a subsequent section, we will show that for laminar or turbulent flow, the behavior of the integrated skin friction coefficient is reasonably well represented by

$$C_F \sim Re_l^{-m} \tag{8.68}$$

The exponent $m = 1/2$ is exact for laminar flow and $m = 1/7$ is reasonable for turbulent flow. The Reynolds number based on body length l may be written as

$$Re_l = \frac{\rho_e u_e l}{\mu_e} = \left(\frac{2ql}{a_e\mu_e}\right)M_e^{-1} \tag{8.69}$$

Therefore, the maximum L/D is given by

$$\left(\frac{L}{D}\right)_{max} \sim Re_l^{m/3} = \left(\frac{2ql}{a_e\mu_e}\right)^{m/3} M_e^{-m/3} \tag{8.70}$$

The quantity $(a_e\mu_e)^{m/3} \sim$ constant over the altitude range of interest so that

$$\left(\frac{L}{D}\right)_{max} \sim \left(\frac{ql}{M_e}\right)^{m/3} \tag{8.71}$$

We see that $(L/D)_{max}$ tends to rise slowly with q and to drop slowly with M for the case of negligible form drag. In the case of the somewhat less slender configurations that are more typical of spaceplanes, the form drag will be significant and will reduce the achievable $(L/D)_{max}$. In hypersonic flight where $M \gg 1$, the form drag coefficient would not vary with M; this is one aspect of the so-called Mach number independence principle first derived for inviscid flow by Oswatitsch (1950), described by Hayes and Probstein (2004) and assessed for viscous flows by Kliche, Mundt, and Hirschel (2011). Thus, it appears that at high Mach numbers, $(L/D)_{max}$ will be essentially constant with M for flight at constant q. Remember that both L and D are proportional to q so that flight at constant q maintains constant lift and drag loading while flight at increasing q increases both.

8.4.2 LAMINAR SKIN FRICTION

Boundary layer theory for compressible flow is treated in many textbooks, such as Schlichting and Gersten (2000), Schetz (1992), and White (2006). Compressibility brings about substantial variations in the thermodynamic properties of the gas leading to complexity in analyzing such flows. Although CFD approaches, like those, for example, discussed by Cebeci and Cousteix (2005) are currently accessible, they can be a bit cumbersome for preliminary design applications. A simple, versatile, and reasonably accurate technique for laminar and turbulent compressible boundary layers is the FPRE method generally associated with Eckert (1956). The method has enjoyed wide and continued use over the years. Meador and Smart (2005) revisited the concept and made extensive use of computational solutions to explain the foundation of and evaluate the accuracy of the FPRE method. An interesting facet of the FPRE method is that although it is based on the assumption of constant pressure flow over constant temperature walls, it has often been applied to cases lying outside these constraints.

The basic idea is that the skin friction coefficient in a compressible flow is equivalent to that for an incompressible flow if the gas properties are evaluated at a "reference" enthalpy h^*. We describe this by considering a laminar flow over a flat plate where subscript e denotes conditions at the edge of the boundary layer. We postulate that the reference skin friction coefficient has the same form as the incompressible flow counterpart so that

$$C_f^* = \frac{\tau_w}{\frac{1}{2}\rho^* u_e^2} = \frac{0.664}{\sqrt{Re_x^*}} = \frac{0.664}{\sqrt{\frac{\rho^* u_e x}{\mu^*}}} \tag{8.72}$$

Then the actual skin friction coefficient may found by solving for τ_w in Eqn (8.72) and dividing that by the dynamic pressure at the edge of the boundary layer leading to

$$C_f = \frac{\tau_w}{\frac{1}{2}\rho_e u_e^2} = \frac{0.664}{\sqrt{Re_x^*}} \frac{\rho^*}{\rho_e} = \frac{0.664}{\sqrt{Re_x}} \sqrt{\frac{\rho^* \mu^*}{\rho_e \mu_e}} = C_{f,inc}\sqrt{C^*} \tag{8.73}$$

The integrated skin friction coefficient

$$C_F = \frac{1}{l}\int_0^l C_f dx = 1.328\sqrt{\frac{C^*}{Re_l}} \tag{8.74}$$

The question remains as to what is the appropriate reference enthalpy. Meador and Smart (2005) propose the following reference enthalpy model for laminar flows:

$$\frac{h^*}{h_e} = 0.45 + 0.55\frac{h_w}{h_e} + 0.16r\frac{\gamma - 1}{2}M_e^2 \tag{8.75}$$

Here, h_e and h_w are the enthalpies at the edge of the boundary layer and at the wall, respectively, and Pr is the Prandtl number $Pr = \mu c_p / \kappa$. The quantity $r = Pr^{1/2}$ is the recovery factor for laminar flow over an adiabatic flat plate. Meador and Smart (2005) report that this approach yields quite accurate skin friction coefficient results for hot and cold walls in air-like mixtures where the Prandtl number is around 0.7. The results they report are for the range $0 \leq M \leq 6$.

It is worth noting that supersonic flow over a flat plate or over a sharp cone are both constant surface pressure flows. White (2006) derives the result that for such flows with equal local Reynolds number, Mach number, and wall temperature ratio, the skin friction coefficient of the cone is

$$C_{f,cone} = \sqrt{3} C_{f,plate} \tag{8.76}$$

In the forward portion of a vehicle, the local Reynolds number Re_x is usually low enough to maintain laminar flow for some distance downstream. Therefore, it is appropriate to calculate the skin friction on a vehicle with a conical nose using the laminar cone rule given above because C_f will be 73% larger than for a flat surface at equivalent flow conditions. Although hypersonic vehicles designed for $M < 6$ may have conical noses, entry vehicles that must traverse a much broader Mach number range typically have blunted noses because of thermal constraints so the cone rule is not applicable in such cases.

8.4.3 TURBULENT SKIN FRICTION

In the case of turbulent flow, White (2006) suggests that a power law expression, like Prandtl's power law formulation $C_f = 0.027 Re^{-1/7}$, is adequate. Meador and Smart (2005) choose the form given in Schlichting and Gersten (2000):

$$C_f = \frac{0.02296}{Re_x^{0.139}} \tag{8.77}$$

Following the process delineated earlier for laminar flow, the following result may be derived for the local skin friction coefficient in turbulent flow:

$$C_f = \frac{0.02296}{Re_x^{0.139}} \left(\frac{\rho^*}{\rho_e} \right)^{0.8961} \left(\frac{\mu^*}{\mu_e} \right)^{0.139} = \frac{0.02296}{Re_x^{0.139}} C^{*0.861} \left(\frac{\mu^*}{\mu_e} \right)^{-0.722} \tag{8.78}$$

The integrated skin friction coefficient is

$$C_F = \frac{0.0266}{Re_l^{0.139}} \left(\frac{\rho^*}{\rho_e} \right)^{0.861} \left(\frac{\mu^*}{\mu_e} \right)^{0.139} = \frac{0.0266}{Re_l^{0.139}} C^{*0.861} \left(\frac{\mu^*}{\mu_e} \right)^{-0.722} \tag{8.79}$$

The reference enthalpy Meador and Smart (2005) propose for turbulent flow is

$$\frac{h^*}{h_e} = 0.5 \left(1 + \frac{h_w}{h_e} \right) + 0.16 r \frac{\gamma - 1}{2} M_e^2 \tag{8.80}$$

The recovery factor for turbulent flow is often taken as $r = Pr^{1/3}$. Meador and Smart (2005) find that for turbulent compressible flow, the results are not as accurate as for laminar flow with the FPRE method overestimating skin friction coefficients for cold walls and underestimating them for hot walls. The error was around 9% for $h_w / h_e = 0.25$, reducing to about 0.7% as the wall warmed to $h_w/h_e = 1$, and continuing to fall to -6% at $h_w/h_e = 2$, all the way down to almost -27% for $h_w/h_e = h_{aw}/h_e$, the adiabatic wall case where

$$h_{aw} = h_e + r\frac{1}{2}u_e^2 = h_e + Pr^m\frac{1}{2}u_e^2 \tag{8.81}$$

Although the FPRE method is not as accurate for turbulent flow as for laminar flow, it is easy to implement and provides reasonable results in many cases of practical interest. A more accurate and more complicated method that does not require CFD approaches to implement was presented by Van Driest (1956) and is popularly known as the Van Driest II method. White (2006) gives a succinct discussion of the method and notes that it is still widely used. Indeed, Hopkins and Inouye (1971) carried out extensive comparisons of six different methods for predicting turbulent skin friction on flat surfaces, with and without heat transfer, and determined that the Van Driest II method gave the most satisfactory agreement with experiment over a Mach number range of 1.5–9. Hopkins (1972) presents charts for rapid estimation skin friction coefficients in the following ranges: $0 < M < 10$, $10^5 < Re_x < 10^9$ and $0.2 < T_w/T_{aw} < 1$. The Van Driest II method is like the FPRE in that a transformation is sought which will relate a turbulent flat plate flow to its incompressible counterpart. Following White (2006), the skin friction coefficient is given implicitly by

$$\frac{\sin^{-1} A + \sin^{-1} B}{\sqrt{C_f\left(\frac{T_{aw}}{T_e} - 1\right)}} = 4.15 \log_{10}\left(Re_x C_f \frac{\mu_e}{\mu_w}\right) + 1.7 \tag{8.82}$$

The quantities in Eqn (8.82) are defined as follows:

$$A = \frac{2a^2 - b}{\sqrt{b^2 + 4a^2}} \tag{8.83}$$

$$B = \frac{b}{\sqrt{b^2 + 4a^2}} \tag{8.84}$$

$$a = \sqrt{r\frac{\gamma - 1}{2}M_e^2\frac{T_e}{T_w}} \tag{8.85}$$

$$b = \frac{T_{aw}}{T_w} - 1 \tag{8.86}$$

The adiabatic wall temperature T_{aw} appearing in Eqn (8.86) is the temperature appropriate to the adiabatic wall enthalpy h_{aw} given by Eqn (8.81). In incompressible flow, where $M_e = 0$, Eqn (8.82) reduces to the Karman-Schoenherr (Schlichting and Gersten, 2000) relation

$$\frac{1}{\sqrt{C_{f,inc}}} = 4.15 \log_{10}(Re_x C_{f,inc}) + 1.7 \tag{8.87}$$

There is an analog to the laminar cone rule in the case of a turbulent boundary layer over a cone with a constant pressure surface, but it turns out that the increase in skin friction on the cone over that of a flat plate is in the range of only 10–15% as opposed to the 73% predicted for laminar flow.

The FPRE method is compared to the Van Driest II method for the case of an adiabatic flat plate ($T_w = T_{aw}$) of length $l = 10$ and 30 m in Figures 8.34 and 8.35, respectively. The extent to which the FPRE underestimates the integrated skin friction coefficient for this case of a hot wall is clearly apparent on the figure.

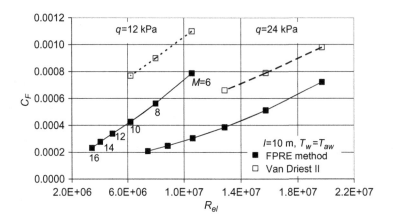

FIGURE 8.34

Comparison of the FPRE method to the Van Driest II method for an adiabatic (hot wall) flat plate of length $l = 10$ m. The Mach numbers considered are called out on one of the C_F curves.

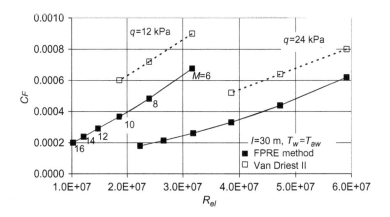

FIGURE 8.35

Comparison of the FPRE method to the Van Driest II method for an adiabatic (hot wall) flat plate of length $l = 30$ m. The Mach numbers considered are called out on one of the C_F curves.

In the case of a relatively cool wall where $T_w = T_{aw}/5$, the FPRE method compares more favorably with the Van Driest II method, as shown for $l = 10$ and 30 m in Figures 8.36 and 8.37, respectively.

8.4.4 BOUNDARY LAYER TRANSITION

At the hypersonic speed of entry, the transition of the boundary layer from laminar to turbulent flow of substantial importance not only because of the substantial drag increase in turbulent flow but also because of the attendant increase in heat transfer. Inaccuracy in predicting transition for entry

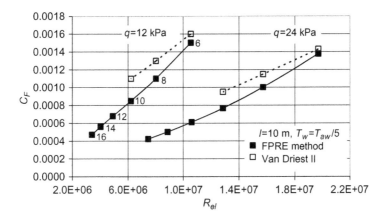

FIGURE 8.36

Comparison of the FPRE method to the Van Driest II method for a relatively cool wall ($T_w = T_{aw}/5$) flat plate of length $l = 10$ m. The Mach numbers considered are called out on one of the C_F curves.

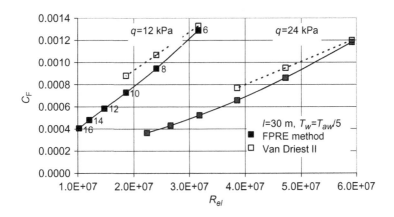

FIGURE 8.37

Comparison of the FPRE method to the Van Driest II method for a relatively cool wall ($T_w = T_{aw}/5$) flat plate of length $l = 30$ m. The Mach numbers considered are called out on one of the C_F curves.

vehicles may lead to inadequate or overly conservative designs for heat shield ablation and thermal protection system weight thereby diminishing safety margins or reduced performance and stability. Lin (2008) gives a detailed discussion of the influence of transition on entry vehicle design and shows supporting flight test and experimental data and points out those aspects which are still not well understood. An earlier review of flight tests aimed at obtaining transition data at supersonic and hypersonic speeds was presented by Schneider (1999). He describes a generic transition criterion proposed for the National Aerospace Plane (NASP) and similar hypersonic maneuvering vehicles given in terms of a momentum thickness Reynolds number for transition as follows:

$$Re_{\theta,t} = \frac{\rho_e u_e \theta}{\mu_e} = 150 M_e \tag{8.88}$$

Lin (2008) reports other engineering correlations for estimating boundary layer transition, including a variation of Eqn (8.88)

$$Re_{\theta,t} = 110 M_e + 250 \tag{8.89}$$

The correlations in Eqns (8.88) and (8.89) are based on natural transition of flow over smooth walls and yield similar results. In the case of entry vehicles with charring and ablating heat shields, which provide cooling by effectively injecting material from the wall into the boundary layer, the effects of the transverse flow due to blowing should be considered. Lin (2008) notes the following correlation for flows with blowing based on Apollo data

$$Re_{\theta,t} = 200 \left[1 - \frac{0.05}{C_h} \left(\frac{\rho_w v_w}{\rho_e u_e} \right) \right] M_e \tag{8.90}$$

Here the blowing parameter $(\rho_w v_w / \rho_e u_e)/C_h$, which was in the range of 4−6 for the Apollo spacecraft, accounts for transverse injection of ablation products into the boundary layer. Equation (8.90) shows that this effective boundary layer blowing acts to make transition occur sooner than in the smooth wall case. When the surface is rough, as characterized by the surface roughness height, still another correlation emerges:

$$Re_{\theta} = 250 \left(\frac{\theta}{k} \right) \tag{8.91}$$

This correlation suggests that transition will occur sooner for flows where the roughness height becomes larger in comparison with the momentum thickness. A variant of this correlation is the modified passive nose cone tip program correlation given by

$$Re_{\theta} = C'' \left(\frac{k}{\theta} \frac{T_e}{T_w} \right)^{-n} \tag{8.92}$$

This correlation has $200 < C'' < 750$ and $0.7 < m < 1.5$. These correlations were developed for making engineering estimates, and all have been used at one time or another in design projects. The phenomenon of boundary layer transition is complicated and arises from different causes so that a firm predictive capability is not yet available. An appreciation of the difficulties associated with the basic physics of transition is presented by Reshotko (2008) who also provides guidelines for experimental work aimed at refining transition estimation techniques.

To employ the various transition criteria, we must be able to calculate the Reynolds number based on the momentum thickness which is defined as follows:

$$\theta(x) = \int_0^{\delta(x)} \frac{\rho u}{\rho_e u_e} \left(1 - \frac{u}{u_e} \right) dy \tag{8.93}$$

The integral is taken over the boundary layer thickness $\delta(x)$ and $\theta(x)$ and represents the growing loss of momentum in the boundary layer due to friction and pressure forces. In a constant pressure two-dimensional flow, the integrated momentum equation is simply

$$\frac{d\theta}{dx} = \frac{1}{2}C_f \tag{8.94}$$

Using Eqns (8.72) and (8.73) in Eqn (8.94) yields

$$\frac{Re_\theta}{\sqrt{Re_l}} = 0.664C^* \tag{8.95}$$

Then, using the simple criterion of Eqn (8.88) in Eqn (8.95), the transition Reynolds number is found to be

$$Re_{x,t} = 5.1 \times 10^4 \left(\frac{M_e}{C^*}\right)^2 \tag{8.96}$$

In constant pressure flow over an axisymmetric cone with an attached shock, the integral equation for a right circular cone of given semivertex angle is

$$\frac{d(x\theta)}{dx} = 2xC_f \tag{8.97}$$

Following the same development as for the flat plate and incorporating the laminar flow cone rule, we find

$$\frac{Re_\theta}{\sqrt{Re_x}} = \frac{0.664C^*}{\sqrt{3}} \tag{8.98}$$

Then, using the criterion of Eqn (8.88) in Eqn (8.98), the transition Reynolds for the cone flow is found to be

$$Re_{x,t} = 1.53 \times 10^5 \left(\frac{M_e}{C^*}\right)^2 \tag{8.99}$$

The laminar skin friction would be computed starting from the leading edge of the plate using Eqn (8.73) until the transition Reynolds number $Re_{x,t}$ given, for example, Eqn (8.99), is reached, whereon the computation would be continued according to the turbulent FPRE method using Eqn (8.78) or the Van Driest II method using Eqn (8.82). This approach is necessarily inexact, but it is a consistent means of incorporating the effects of transition and should be appropriate for comparison of similar configurations.

8.4.5 HIGH-ALTITUDE RAREFACTION EFFECTS

The Space Shuttle was supposed to fly like an airplane in the upper atmosphere, and one of the major difficulties in achieving this goal, according to Woods, Arrington, and Hamilton (1983), is the complexity of modeling low-density real gas environments over practical vehicle configurations. The rarefied nature of the upper atmosphere introduces a number of different flow regimes. Wilhite, Arrington, and McCandless (1984) showed that a parameter that served to correlate low-density effects on the aerodynamics of aeroassisted orbital transfer vehicles was a modified hypersonic viscous interaction factor

$$V' = \frac{M_\infty}{\sqrt{Re_{l,\infty}}} \sqrt{\frac{\mu^* T_\infty}{\mu_\infty T^*}} = \frac{M_\infty}{\sqrt{Re_l}} \sqrt{C_\infty^*} \tag{8.100}$$

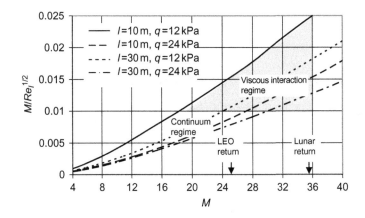

FIGURE 8.38

Variation of the hypersonic interaction parameter $M/Re_l^{1/2}$ is shown as a function of M manned spacecraft flight corridor defined by 12 kPa $< q <$ 24 kPa.

Here, the starred quantities are evaluated at the reference temperature T^* appropriate to the reference enthalpy h^* discussed previously. Wilhite et al. (1984) point out that that the following regimes based on V' appeared to apply:

- $V' < 0.007$ corresponds to continuum flow
- $0.007 < V' < 0.07$ corresponds to the viscous interaction regime
- $0.07 < V' < 5$ corresponds to the transitional flow regime
- $V' > 5$ corresponds to the free molecular flow regime

In particular, as V' rises above 0.007, L/D tends to drop and other aerodynamic performances measures are similarly compromised.

The parameter $M/Re_l^{1/2}$ is shown as a function of M in Figure 8.38 specifically for the typical manned spacecraft flight corridor defined by 12 kPa $< q <$ 24 kPa. The behavior of the μ^*/T^* ratio is described subsequently in Section 8.3, but in general, the correction factor $\sqrt{C_\infty^*}$ will be around 0.4–0.7. Therefore, the viscous interaction regime discussed in detail by Hayes and Probstein (2004) will lie in the shaded region above $M/Re_l^{1/2} = 0.01$ shown in Figure 8.38. Vehicles with $l = 10$ m will likely be subject to rarefaction effects somewhat above an altitude of $z = 50$ km, while those with $l = 30$ m will encounter such effects at somewhat higher altitude, perhaps $z = 55$ km.

8.5 THERMODYNAMIC PROPERTIES OF AIR

The compressibility of air complicates flow calculations in high subsonic and low supersonic flight Mach numbers ($0.6 < M_0 < 3$) because the density is no longer constant but changes with pressure, and according to the perfect gas law, the temperature also changes. In this flight regime, however, air behaves like a thermally perfect gas. That is, air follows the perfect gas law with the molecular weight remaining constant up to a temperature of about 500 K. Air is also approximately calorically perfect, that is, the specific heats remain approximately constant up to 500 K. Beyond 500 K, gas imperfections arise which must be taken into account to maintain a reasonable degree of accuracy. To get an idea of what flight conditions are appropriate for using perfect gas relations, consider that the maximum temperature of the air passing over a vehicle to be no greater than the free stream stagnation temperature. The ratio of free stream stagnation temperature to static temperature is a function of the flight Mach number M_0 according to the adiabatic relation for a perfect gas given by

$$\frac{T_{t,0}}{T_0} = 1 + \frac{\gamma - 1}{2} M_0^2 \tag{8.101}$$

For flight through the atmosphere at standard conditions, we choose $\gamma = 1.4$ and the temperature in the undisturbed atmosphere to be a typical value, say $T_0 = 215$ K. For a stagnation temperature $T_{t,0} = 500$ K, the flight Mach number is $M_0 = 2.57$. Therefore, it is reasonable to use perfect gas relations for flight in the atmosphere up to a flight Mach number of about 3. At higher supersonic Mach numbers and at hypersonic Mach numbers, say $M_0 > 4$, more accurate descriptions of the thermodynamic properties of air must be used.

We may examine the stagnation conditions using the hypersonic approximation of $M \gg 1$ applied along the stagnation streamline across an adiabatic shock. The stagnation enthalpy and pressure are given approximately by the following relations:

$$h_t \approx \frac{1}{2} u^2 = \frac{q}{\rho} = \frac{q}{p} RT \tag{8.102}$$

$$p_t \approx \rho u^2 = 2q \tag{8.103}$$

The dynamic pressure range for manned spaceflight is typically 12 kPa $< q <$ 24 kPa so that the stagnation pressure is in the range 1/4 atm $< p_t <$ 1/2 atm. We may determine the stagnation temperature from Eqn (8.102) providing we have an accurate state equation of the form $h = h(p,T)$.

8.5.1 ENTHALPY AND COMPRESSIBILITY

A detailed set of approximations for the thermodynamic and transport properties of air over the temperature and pressure range of 500 K $< T <$ 1500 K and 10^{-4} atm $< p <$ 100 atm, respectively, is presented by Hansen (1959). The perfect gas equation of state must be corrected for changes in the molecular weight of the air from the value at standard pressure and temperature W_{stp} as it goes through dissociation of oxygen and nitrogen molecules as the gas is heated, and this is accomplished by introducing so-called compressibility $Z = W_{stp}/W$. The equation of state is then

$$p = \rho Z \left(\frac{R_u}{W_{stp}}\right) T = \rho ZRT \tag{8.104}$$

A simple curve fit for the compressibility Z was given by Hansen and Heims (1958) in the following form:

$$Z \approx 2.5 + 0.1 \tanh\left(\frac{\theta}{500} - 7\right) + 0.4 \tanh\left(\frac{\theta}{1000} - 7\right) + \tanh\left(\frac{\theta}{2500 - 5.8}\right)$$

The reduced temperature includes the pressure and is given by

$$\theta = T\left(1 - \frac{1}{8}\log\frac{p}{p_{sl}}\right)$$

Here p_{sl} is the standard atmospheric pressure at sea level and the logarithm is to the base 10. This correlation for Z gives fairly accurate results, generally less than 10%.

The enthalpy h and compressibility Z given by Hansen (1959) appear in Table 8.3 for the temperature range $500\,\mathrm{K} < T < 7500\,\mathrm{K}$ and the pressure range $0.01\,\mathrm{atm} < p < 1\,\mathrm{atm}$; for temperatures between 200 and 500 K, the properties are taken from Kays, Crawford, and Weigand (2005). Also shown in Table 8.3 is the enthalpy calculated from $h = c_p T$ where the specific heat is taken to be constant and equal to the standard value $c_p = 1.004$ kJ/kg · K. The variation of enthalpy with temperature is contrasted with the prediction based on constant c_p in Figure 8.39. The deviation between the actual and constant c_p prediction starts to become noticeable at $T = 1000$ K, and by $T = 2000$ K, the error exceeds 10% and increases thereafter.

Table 8.3 Compressibility and Enthalpy as a Function of Temperature for Several Pressures

T (K)	p = 1 atm Z	p = 0.1 atm Z	p = 0.01 atm Z	p = 1 atm h (kJ/kg)	p = 0.1 atm h (kJ/kg)	p = 0.01 atm h (kJ/kg)	p = 1 atm $c_p^a\,T$
200	1	1	1	200.2	200.2	200.2	201
300	1	1	1	300.5	300.5	300.5	301
400	1	1	1	401.3	401.3	401.3	402
500	1	1	1	505.1	505.1	505.1	502
1000	1	1	1	1047.6	1047.6	1047.6	1004
1500	1	1	1	1635.9	1635.9	1635.9	1506
2000	1	1.001	1.002	2250.1	2250.1	2255.8	2008
2500	1.004	1.011	1.033	2934.6	3063.7	3451.2	2510
3000	1.026	1.072	1.149	3969.2	4778.6	6138.9	3012
3500	1.092	1.167	1.197	5775.9	7111.9	7654.3	3514
4000	1.165	1.198	1.208	7737.5	8357.4	8644.4	4016
4500	1.196	1.213	1.245	9014.7	9466.7	10,512.8	4518
5000	1.214	1.252	1.359	10,188.5	11,422.6	15,038.8	5020
5500	1.248	1.348	1.599	11,965.0	15,358.8	23,898.5	5522
6000	1.316	1.529	1.849	14,981.4	22,265.5	33,234.6	6024
6500	1.437	1.752	1.961	19,848.9	30,706.1	37,962.9	6526
7000	1.607	1.904	1.997	26,518.8	36,845.1	40,200.1	7028
7500	1.778	1.971	2.017	33,320.7	40,165.7	42,059.9	7530

[a]Value at standard conditions, $c_p = 1.004$ kJ/kg · K.

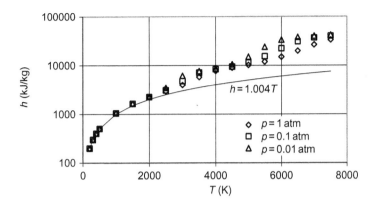

FIGURE 8.39

The variation of the enthalpy of air with temperature for different pressures. The enthalpy calculated as $c_p T$ with $c_p =$ constant $= 1.004$ kJ/kg · K is shown for comparison.

Table 8.4 Speed of Sound Parameter $\gamma = a^2 \rho/p$ and Specific Heat at Constant Pressure for Air at Several Pressures

T (K)	$p = 1$ atm γ	$p = 0.1$ atm γ	$p = 0.01$ atm γ	$p = 1$ atm c_p (kJ/kg · K)	$p = 0.1$ atm c_p (kJ/kg · K)	$p = 0.01$ atm c_p (kJ/kg · K)	γ Eqn (8.105)
200	1.397	1.397	1.397	1.006	1.006	1.006	1.40
300	1.399	1.399	1.399	1.005	1.005	1.005	1.40
400	1.395	1.395	1.395	1.013	1.013	1.013	1.40
500	1.387	1.387	1.387	1.030	1.030	1.030	1.40
1000	1.337	1.337	1.337	1.137	1.137	1.137	1.33
1500	1.312	1.312	1.312	1.205	1.205	1.208	1.29
2000	1.296	1.285	1.259	1.266	1.312	1.464	1.26
2500	1.247	1.196	1.144	1.567	2.190	3.897	1.24
3000	1.181	1.147	1.15	2.761	4.776	5.068	1.22
3500	1.166	1.187	1.265	4.276	3.533	1.868	1.21
4000	1.204	1.254	1.208	3.171	1.983	2.437	1.20
4500	1.241	1.196	1.133	2.190	2.758	5.648	1.19
5000	1.202	1.143	1.111	2.749	5.447	13.27	1.18
5500	1.161	1.124	1.113	4.549	10.69	20.86	1.17
6000	1.141	1.124	1.135	7.735	16.53	14.17	1.16
6500	1.137	1.136	1.193	11.80	15.60	5.912	1.16
7000	1.142	1.167	1.249	14.32	8.949	3.699	1.15
7500	1.156	1.216	1.231	12.25	4.971	4.015	1.145

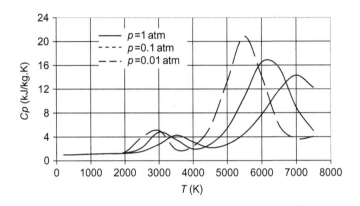

FIGURE 8.40

The variation of the specific heat at constant pressure of air with temperature for different pressures.

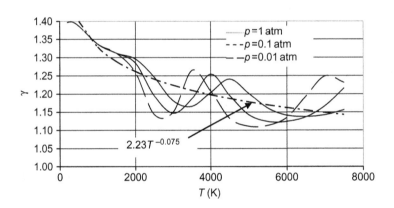

FIGURE 8.41

The variation of the ratio of specific heats of air with temperature for different pressures.

8.5.2 SPECIFIC HEAT AND SOUND SPEED

The specific heat at constant pressure c_p has substantial variation with temperature and pressure as shown in Table 8.4 and illustrated in Figure 8.40. As a result, the speed of sound parameter $\gamma = a^2 \rho/p$ also varies widely with temperature and pressure, as can be seen in Table 8.4 and Figure 8.41. A simple correlation that is accurate within $5-10\%$ is given by

$$\gamma = \frac{a^2 \rho}{p} \approx 2.23 T^{-0.075} \tag{8.105}$$

Even this small difference can be important because γ often appears in the ratio $\gamma/(\gamma - 1)$, and γ approaches unity as the temperature increases.

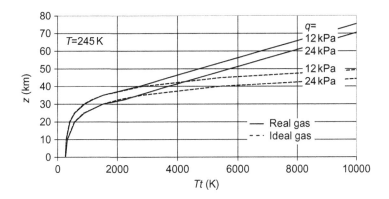

FIGURE 8.42

The variation of the stagnation temperature with altitude for flight at constant dynamic pressure q in considering air to be a real gas or an ideal gas.

8.5.3 REAL GAS STAGNATION TEMPERATURE

Using the more accurate thermodynamic properties described above, we may now proceed to calculate the likely stagnation temperature levels as a function of altitude during flight along a constant q trajectory. Results for the stagnation temperature when considering air as a real gas are contrasted with those for a perfect gas with $\gamma = k = 1.4$ as shown in Figure 8.42, assuming a constant temperature atmosphere with $T = 245$ K, which corresponds to our model exponential atmosphere. Above about 35−45 km, the ideal gas assumption yields stagnation temperatures which are too high because the energy absorption characteristics of molecular vibration, dissociation, and ionization have been neglected.

The behavior of enthalpy as a function is not easily correlated, particularly at higher temperatures and lower pressures. However, a simple approximation may be given as follows:

$$T < 1500 \ K : h \approx 1.004T$$
$$T > 1500 \ K : h \approx 550 \exp\left(\frac{T}{T_r}\right) \tag{8.106}$$

For the temperature range $1500 < T < 4000$ K, use $T_r = 1500, 1420, 1350$ K for $p = 1, 0.1,$ and 0.01 atm, respectively. For the temperature range $4000 < T < 7500$ K, use $T_r = 1800, 1620,$ and 1550 K for $p = 1, 0.1,$ and 0.01 atm, respectively. These are merely guidelines for estimation purposes.

8.5.4 SIMPLE ENTHALPY CURVE FITS

At high temperatures, air in chemical equilibrium exhibits the interesting characteristic that all the molecular oxygen is dissociated by the time nitrogen dissociation begins and then ionization begins only after all the oxygen and nitrogen is in the atomic state. Therefore, it is possible to consider that only one reaction is taking place at a given time, depending upon the thermodynamic state. Hansen and Heims (1958) exploit this observation and present the following approximate fits for the energy of the air $\theta' = T/1000$. First, only oxygen dissociation is taking place and this occurs in the compressibility range $1.0 < Z < 1.2$ for which the non-dimensional energy is approximated by

$$\frac{ZE}{RT} \approx (2 - Z)\left[\frac{5}{2} + \frac{3/\theta'}{(e^{3/\theta'} - 1)}\right] + (Z - 1)\left(3 + \frac{59}{\theta'}\right) \tag{8.107}$$

The nitrogen dissociation reaction then takes over in the range $1.2 < Z < 2.0$ for which the non-dimensional energy is approximated by

$$\frac{ZE}{RT} \approx (2 - Z)\left[\frac{5}{2} + \frac{3/\theta'}{e^{3/\theta'} - 1}\right] + 0.2\left(3 + \frac{59}{\theta'}\right) + (Z - 1.2)\left(3 + \frac{113}{\theta'}\right) \tag{8.108}$$

Finally, when only atoms are present, the ionization reaction dominates and in the compressibility range $2.0 < Z < 2.2$, the non-dimensional energy is approximated by

$$\frac{ZE}{RT} \approx (4 - Z)\left(\frac{3}{2} + \frac{51}{\theta'}\right) + (Z - 2)\left(3 + \frac{220}{\theta'}\right) \tag{8.109}$$

The non-dimensional enthalpy may then be obtained from the following relation:

$$\frac{ZH}{RT} = \frac{ZE}{RT} + Z \tag{8.110}$$

The dimensional form of the specific enthalpy h or the specific energy e in kJ/kg may be found by multiplying the ZH/RT or ZE/RT by $0.287T$, where T is in degrees Kelvin. These approximations can easily be incorporated into a computation scheme for preliminary design purposes. For more extensive and detailed calculations, the comprehensive curve fits for air in chemical equilibrium presented by Srinivasan, Tannehill, and Weilmunster (1987) may be used.

8.5.5 THE $\rho\mu$ PRODUCT

The product $\rho\mu$ generally appears as a ratio across a boundary layer over which the pressure is essentially constant so that we may write

$$\rho\mu = \frac{p}{ZRT}\mu = \left(\frac{p}{R}\right)\frac{\mu}{ZT} \sim \frac{\mu}{ZT} \tag{8.111}$$

The viscosity of high-temperature air calculated from the data of Hansen (1959) is shown in Table 8.5. The product $\rho\mu$ according to Eqn (8.111) based on the results presented by Hansen (1959) for high-temperature air is shown in Tables 8.5 and 8.6, respectively. A similar table for air at $T < 1000$ K is shown in Table 8.7 for completeness. Note that for $T < 1000$ K, μ is not a function of pressure.

The $\rho\mu$ ratio at the reference enthalpy condition is

$$C^* = \frac{\rho^*\mu^*}{\rho_e\mu_e} = \frac{p^*}{p_e}\frac{Z_e}{Z^*}\frac{T_e}{T^*}\frac{\mu^*}{\mu_e} = \frac{\mu^*}{Z^*T^*}\left(\frac{\mu_e}{Z_eT_e}\right)^{-1} \tag{8.112}$$

We note again that the pressure is assumed to be constant across the boundary layer. The variation of μ/ZT for 500 K $< T < 7500$ K is shown in Figure 8.43 along with two suggested correlation curves which depend on pressure:

$$p \approx 1 \; atm: \frac{\mu}{ZT} \approx \frac{6.4 \times 10^{-7}}{T^{0.4}} \tag{8.113}$$

$$0.01 \; atm \leq p \leq 0.1 \; atm: \frac{\mu}{ZT} \approx 1.5 \times 10^{-7}(1 - 0.105 \ln T) \tag{8.114}$$

Table 8.5 Viscosity of High-Temperature Air

T (K)	p = 1.0 atm μ (Pa · s)	p = 0.1 atm μ (Pa · s)	p = 0.01 atm μ (Pa · s)
500	2.671E − 05	2.671E − 05	2.671E − 05
1000	4.158E − 05	4.158E − 05	4.158E − 05
1500	5.269E − 05	5.269E − 05	5.269E − 05
2000	6.192E − 05	6.192E − 05	6.192E − 05
2500	7.025E − 05	6.997E − 05	6.997E − 05
3000	7.920E − 05	7.720E − 05	7.720E − 05
3500	9.152E − 05	8.431E − 05	8.465E − 05
4000	1.048E − 04	9.175E − 05	9.193E − 05
4500	1.144E − 04	9.885E − 05	9.933E − 05
5000	1.228E − 04	1.063E − 04	1.086E − 04
5500	1.326E − 04	1.154E − 04	1.218E − 04
6000	1.463E − 04	1.276E − 04	1.365E − 04
6500	1.665E − 04	1.424E − 04	1.479E − 04
7000	1.935E − 04	1.558E − 04	1.586E − 04
7500	2.218E − 04	1.662E − 04	1.668E − 04

Table 8.6 The Product $\rho\mu \sim \mu T/Z$ for Specific Pressures in High-Temperature Air

T (K)	p = 1 atm μ/TZ (Pa · s/K)	p = 0.1 atm μ/TZ (Pa · s/K)	p = 0.01 atm μ/TZ (Pa · s/K)
500	5.342E − 08	5.342E − 08	5.342E − 08
1000	4.158E − 08	4.158E − 08	4.158E − 08
1500	3.513E − 08	3.513E − 08	3.513E − 08
2000	3.096E − 08	3.093E − 08	3.090E − 08
2500	2.799E − 08	2.768E − 08	2.709E − 08
3000	2.573E − 08	2.400E − 08	2.239E − 08
3500	2.395E − 08	2.064E − 08	2.021E − 08
4000	2.249E − 08	1.915E − 08	1.902E − 08
4500	2.126E − 08	1.811E − 08	1.773E − 08
5000	2.022E − 08	1.698E − 08	1.598E − 08
5500	1.932E − 08	1.557E − 08	1.385E − 08
6000	1.853E − 08	1.391E − 08	1.231E − 08
6500	1.783E − 08	1.251E − 08	1.160E − 08
7000	1.720E − 08	1.169E − 08	1.134E − 08
7500	1.663E − 08	1.124E − 08	1.103E − 08

In the case of relatively low temperatures, 200 K $< T <$ 2000 K, Eqn (8.113) is appropriate for 0.01 atm $< p <$ 1 atm.

In Eqn (8.73), the compressibility factor evaluated at the reference temperature, $C^* = \rho^* \mu^* / \rho_e \mu_e$, appears. For purposes of illustration, we arbitrarily set $T_e = 500$ K and evaluate C^* with the results as shown in Table 8.8 and Figure 8.44. Similarly, in Eqn (8.78), another compressibility factor $C' = C^{*0.861}(\mu^*/\mu_e)^{-0.722}$ appears. Again setting $T_e = 500$ K, we calculate C' and present the results

Table 8.7 The Viscosity and the Product $\rho\mu \sim \mu/T$ in Low-Temperature Air

T (K)	μ^a (Pa·s)	μ/T (Pa·s/K)
200	1.325E − 05	6.627E − 08
300	1.843E − 05	6.146E − 08
400	2.284E − 05	5.711E − 08
500	2.670E − 05	5.342E − 08
600	3.017E − 05	5.030E − 08
700	3.334E − 05	4.764E − 08
800	3.627E − 05	4.534E − 08
900	3.900E − 05	4.334E − 08
1000	4.158E − 05	4.158E − 08

^{a} μ *independent of* p *in this temperature range.*

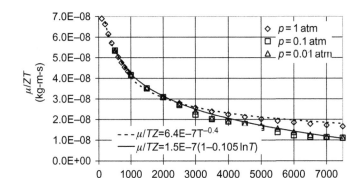

FIGURE 8.43

Variation of μ/ZT for different pressures. Data symbols are calculated values from Hansen (1959) and suggested correlation curves are indicated.

in Table 8.3 and Figure 8.45. Examining Figures 8.44 and 8.45, we see that C' is not as dependent on pressure as is C^*. Errors accrued in using Eqns (8.113) or (8.114) are within $+10\%$ for 0.01 atm $<p<1$ atm and 300 K $<T<7500$ K while for 0.1 $<p<1$ atm the same error pertains to $5000<T<7500$ K with $p=0.01$ atm case showing greater error in this temperature range.

The viscosity is independent of pressure until there is appreciable molecular dissociation whereon the viscosity becomes larger until ionization sets in and the viscosity dramatically drops. The pressure effect can be approximated by the following relation presented by Hansen and Heims (1958):

$$\frac{\mu}{\mu_o} = \left\{1 + 0.023\theta'\left[1 + \tanh\frac{\theta'\left(1 - \frac{1}{8}\log\frac{p}{p_{sl}}\right) - 6.5}{1.5 + \frac{1}{8}\log\frac{p}{p_{sl}}}\right]\right\}\left\{1 + \exp\left[\frac{\theta' - 14.5 - 1.5\log\frac{p}{p_{sl}}}{0.9 + 0.1\log\frac{p}{p_{sl}}}\right]\right\}^{-1} \qquad (8.115)$$

Table 8.8 Compressibility Factors C^* and C' for High-Temperature Air[a]

T* (K)	p = 1 atm C^*	p = 0.1 atm C^*	p = 0.01 atm C'	p = 1 atm C'	p = 0.1 atm C'	p = 0.01 atm C^*
500	1	1	1	1	1	1
1000	0.7783	0.7783	0.7783	0.5855	0.5855	0.5855
1500	0.6576	0.6576	0.6576	0.4268	0.4268	0.4268
2000	0.5795	0.5790	0.5784	0.3407	0.3404	0.3401
2500	0.5239	0.5182	0.5072	0.2851	0.2833	0.2781
3000	0.4817	0.4494	0.4192	0.2432	0.2334	0.2199
3500	0.4483	0.3864	0.3783	0.2060	0.1923	0.1883
4000	0.4210	0.3584	0.3561	0.1770	0.1696	0.1684
4500	0.3981	0.3390	0.3319	0.1582	0.1532	0.1499
5000	0.3786	0.3178	0.2992	0.1441	0.1375	0.1285
5500	0.3617	0.2914	0.2592	0.1310	0.1202	0.1046
6000	0.3469	0.2604	0.2304	0.1177	0.1015	0.0870
6500	0.3337	0.2341	0.2172	0.1037	0.0856	0.0781
7000	0.3220	0.2188	0.2123	0.0902	0.0757	0.0728
7500	0.3114	0.2104	0.2064	0.0794	0.0698	0.0685

[a]Reference condition is taken as $T_e = 500$ K.

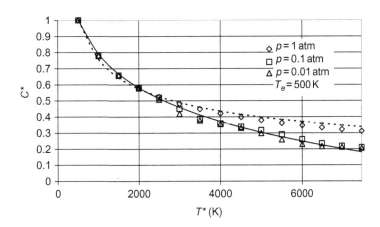

FIGURE 8.44

Compressibility factor C^* evaluated for $T_e = 500$ K and several pressures.

Here, again $\theta' = T/1000$, p_{sl} is the standard atmospheric pressure at sea level and the logarithm is to the base 10. The viscosity μ_o in units of Pa $\cdot$ s is given by the Sutherland formula

$$\mu_o = 1.46 \times 10^{-6} \frac{T^{3/2}}{T + 112} \tag{8.116}$$

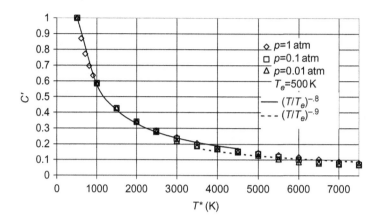

FIGURE 8.45

Compressibility factor C' evaluated for $T_e = 500$ K and several pressures.

8.6 DYNAMICS OF SPACECRAFT

8.6.1 LONGITUDINAL STATIC STABILITY

In the analysis of entry mechanics the spacecraft was treated as a point mass so that for equilibrium only the net forces acting at the mass concentrated at the vehicle CG need to be zero. However, when the spacecraft is considered to be a rigid body, both the net forces and net moments acting at the CG of the vehicle must be zero for equilibrium flight. A discussion of longitudinal static stability for aircraft is presented by Sforza (2014). The rolling moment occurs about the longitudinal x-axis of the vehicle, the pitching moment occurs about the y-axis which is normal to the longitudinal x-axis, and the yawing moment occurs about the z-axis which is normal to the x-y plane and which to the crew appears to be the vertical axis. The net moment about each axis through the CG must be zero for equilibrium, but we start by considering the pitching moment because it is most strongly affected by the lift and forces acting on the vehicle. It is conventional to deal with dimensionless coefficients, so we define the pitching moment about the center of gravity as follows:

$$M = C_m qSl = C_m \frac{1}{2}\rho V^2 Sl \qquad (8.117)$$

The quantities l and S denote a reference length and a reference area, respectively, q is the free stream dynamic pressure, and C_m is the pitching moment coefficient about the center of gravity. For symmetric flight, that is, flight in the x-y symmetry plane of the spacecraft, as depicted in Figure 8.46, the rolling and yawing moments are zero, as is any side force.

A spacecraft can be trimmed, that is, put into an equilibrium state where the net forces and moments about the center of gravity are all zero. But the question remains as to whether the equilibrium so achieved is statically stable. The spacecraft is said to be statically stable if its response

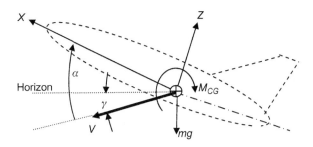

FIGURE 8.46

Elevation view in the symmetry plane of a spacecraft showing the pitching moment about the CG. Positive direction of the moment and angle of attack are shown.

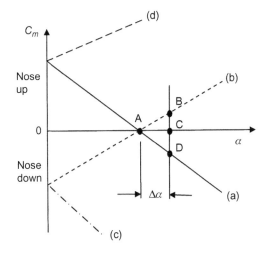

FIGURE 8.47

Moment coefficient variation with α of a spacecraft about its center of gravity illustrating static stability of response curve (a), instability of response curve (b), and response curves (c) and (d) which cannot be trimmed.

to a disturbance in angle of attack is to tend to return to the original equilibrium position. Thus, if flying in equilibrium at one angle of attack and a disturbance increases the angle of attack, the moment produced at this new angle of attack must act to reduce itself, that is, to tend back toward the original equilibrium state angle of attack. Conversely, if a disturbance decreases the angle of attack, the moment at the new lower angle of attack must serve to increase that angle. In other words, the rate of change of the moment about the center of gravity must be negative for static stability: an increase in α should reduce C_m and a decrease in α should increase C_m.

Possible behavior of the moment coefficient with angle of attack α is shown in Figure 8.47. Assuming the forces are in equilibrium, at point A, we also have moment equilibrium because $C_m = 0$, and therefore, the spacecraft is in equilibrium flight. However, if we displace the angle of attack from this trim point by a small positive angle of attack, that is, with the nose rising, the spacecraft may respond in three different ways. Along the dashed response line (b), we see that α moves from A to B producing a positive (nose-up) pitching moment which tends to drive the spacecraft further from equilibrium, and it is statically unstable. However, along response line (a), the spacecraft has a negative (nose-down) pitching moment which tends to drive the nose-back down to the trim point, thereby demonstrating statically stable behavior in pitch. Movement from A to C induces no change in pitching moment, and the spacecraft is said to be statically neutral in pitch. Spacecraft with response lines like (c) or (d) cannot even be trimmed, must less put into a statically stable state.

These results show that we wish to have a spacecraft design that can be trimmed and that the trimmed states are statically stable in pitch. In order to achieve trim at positive (lifting) angles of attack, the zero-lift moment coefficient must be positive

$$C_{m,0} > 0 \tag{8.118}$$

Furthermore, for stability, the derivative of the moment coefficient with respect to angle of attack must be negative

$$\frac{\partial C_m}{\partial \alpha} \equiv C_{m,\alpha} < 0 \tag{8.119}$$

One method for ensuring longitudinal static stability is to place the spacecraft CG far forward enough to make $C_{m,\alpha} < 0$. However, to achieve trim, we must also have $C_{m,0} > 0$, and in aircraft applications, this is usually done by adding an aft horizontal tail set to provide negative lift and positive (nose-up) moment. Because of the extreme thermal environment of hypersonic flight, the use of a separate horizontal tail on spacecraft is avoided.

8.6.2 LONGITUDINAL STATIC STABILITY OF SPACE CAPSULES

The force structure on typical entry capsules is shown in Figure 8.48 for ballistic nonlifting entry and in Figure 8.49 for lifting entry. According to the sign convention for moments in Figure 8.48, the pitching moment coefficient about the center of gravity is given by the sum of the contributions of the axial and normal forces as follows:

$$C_{m,CG} = C_a \frac{z_0}{R_N} + C_n \left(1 - \frac{x_0}{R_N}\right) \tag{8.120}$$

If the center of gravity is placed on the longitudinal axis, that is, if $z_0 = y_0 = 0$, as depicted in Figure 8.48, then in order to trim the vehicle, Eqn (8.120) requires either the normal force $F_n = 0$ or the center of gravity is located outside the body, at the origin of the nose radius $x_0 = R_N$. A symmetrical capsule geometry would satisfy the first possibility with a trim angle of attack $\alpha_t = 0$, while the second possibility is impossible to achieve. Furthermore, from Eqn (8.120), the pitching moment slope is given by

$$\frac{\partial C_{m,CG}}{\partial \alpha} = \frac{\partial C_a}{\partial \alpha} \frac{z_0}{R_N} + \frac{\partial C_n}{\partial \alpha} \left(1 - \frac{x_0}{R_N}\right) \tag{8.121}$$

Under the assumption that the center of gravity is on the symmetry axis, Eqn (8.121) shows that the slope of the pitching moment coefficient must be zero, that is, the static stability of the capsule

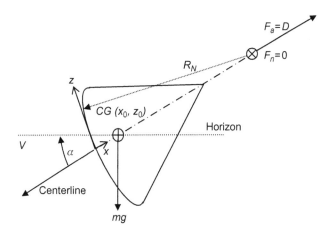

FIGURE 8.48

Force structure on blunt capsule during ballistic (nonlifting) entry.

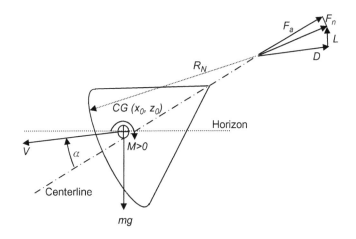

FIGURE 8.49

Force structure on blunt capsule during lifting entry.

is neutral. Therefore, a symmetric capsule configuration would be able to achieve neutrally stable flight along a ballistic trajectory. In the analysis of atmospheric entry in Chapter 6, the deceleration was given by Eqn (6.18), which is repeated below for convenience:

$$\frac{1}{g_E}\frac{dV}{dt} = -\frac{q}{B} - \sin\gamma$$

The flight path angle γ can only be modulated by the lift force normal to the trajectory, and such modulation also can alter the dynamic pressure history of the spacecraft. Without the lift force, it was shown that deceleration levels can become unacceptably high for human flight. The determination that trained and fit fighter pilots were capable of briefly withstanding decelerations of around $10g_E$ typical of ballistic entry led to the use of capsules with CG locations on the centerline of symmetry. The first manned space capsules like Mercury in the US space program and Vostok in the USSR space program took advantage of the simplicity of the ballistic approach although it posed a greater stress on the human crew.

As flight experience and associated laboratory research became broader and deeper, it was possible to move to more capable capsules which developed low, but useful, values of L/D. Acceptable levels of longitudinal static stability was achieved by simply offsetting the center of gravity from the centerline of symmetry as illustrated in Figure 8.49. From Eqn (8.120), we note that trim is achieved when

$$\frac{z_0}{R_N} = \frac{C_n}{C_a}\left(1 - \frac{x_0}{R_N}\right) \tag{8.122}$$

The question of how to choose the coordinates of the center of gravity then arises and appears to depend largely on the magnitudes of C_n and C_a. Previously in this chapter, we used Newtonian theory to determine the normal and axial force components, and the results may be expressed generally as follows:

$$C_a = 1 \quad b\alpha^2 \tag{8.123}$$

$$C_n = a\alpha \tag{8.124}$$

For typical capsule geometries, $a = O(10^{-1})$ per radian and $b = O(1)$ per radian^2. Then, the trim point and the slope of the moment coefficient curve are as follows:

$$\frac{z_0}{R_N} = \frac{a\alpha_t}{C_{n,0} - b\alpha_t^2}\left(1 - \frac{x_0}{R_N}\right) \tag{8.125}$$

$$\frac{\partial C_m}{\partial \alpha} = -a\left(1 - \frac{x_0}{R_N}\right) \tag{8.126}$$

Because $C_n/C_a = O(10^{-1})$, the L/D capability of the capsule may be approximated by

$$\frac{L}{D} = \frac{\tan\alpha - (C_n/C_a)}{1 + (C_n/C_a)\tan\alpha} \approx \tan\alpha - (C_n/C_a) \tag{8.127}$$

The greater the desired lift to drag ratio at the trimmed condition, $(L/D)_t$, the greater the trim angle of attack α_t required. Of course at some value of α, L/D begins to drop, analogous to the case of wing stall.

For a given capsule geometry, the values of a and b are fixed so Eqn (8.126) shows that the stability of the capsule increases as x_0/R_N decreases, that is, as the capsule CG moves closer to the heat shield. Of course, the center of gravity location is not completely arbitrary because the interior equipment and fittings are determined by other than aerodynamic factors. The capsule interior will tend to be used as fully as possible so that the apparent density of the capsule will be relatively uniform.

Assuming that the capsule is approximated by the sector of a uniform sphere of radius R_N, the center of gravity will be located at a height $x_0/R_N = 1 - 3(1 + \cos\phi_1)/8$. For typical capsules, we

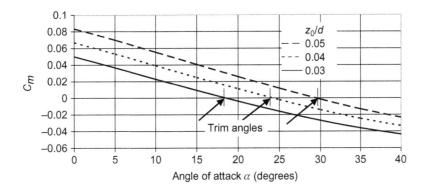

FIGURE 8.50

The moment coefficient for the Apollo capsule according to Newtonian theory as a function of angle of attack for different vertical displacements of the CG from the vehicle centerline.

may take $\phi_1 = 25°$ so that $x_0/R_N = 0.285$. Barring ballasting, which involves carrying parasitic mass, rarely a desirable choice for spacecraft, this value for the longitudinal location of the center of gravity is difficult to vary by much. As a consequence, the static stability of the capsule is essentially fixed by a, the slope of the C_a curve, as illustrated by Eqn (8.126). Equation (8.125) then shows that to achieve the high $(L/D)_t$, which means high trim angle α_t, the required value of z_0/R_N increases. In practice, just as the center of gravity location is constrained, so is the aerodynamic performance of the capsule, such that $z_0/x_0 = O(10^{-1})$.

Offsetting the center of gravity of the space capsule from the vehicle centerline provides the means for achieving static longitudinal stability. That is, the moment about the center of gravity can be nulled out at some angle of attack called the trim angle. Using Eqn (8.34) to calculate the moment coefficient about the center of gravity located at $x_0/d = 0.27$ and offset by an amount $z_0/d = 0.03$, 0.04, and 0.05 leads to the results shown in Figure 8.50 for an idealized Apollo capsule. The trim angles α_t are shown as the angles of attack at which $C_m = 0$.

Moving the center of gravity further off the centerline moves the trim angles to higher values, as shown in Figure 8.51. The shaded area in Figure 8.51 includes laboratory test data on high fidelity Apollo models as well as Apollo flight test data reported by DeRose (1969). The simple Newtonian method yields results which are in reasonably good agreement with experiment. In a similar vein, the variation of L/D as a function of trim angle for an idealized Apollo capsule according to Newtonian theory is shown in Figure 8.52. The results are compared there with laboratory and flight test data reported by DeRose (1969) shown as a shaded region. The calculated results overestimate the experimental data slightly and show the proper trend.

The longitudinal stability parameter $C_{m,\alpha} = dC_m/d\alpha$ for the Apollo capsule according to Newtonian theory is shown in Figure 8.53 as a function of angle of attack for different vertical displacements of the CG from the vehicle centerline. The variation of $C_{m,\alpha}$ with respect to both angle of attack and center of gravity offset is seen to be quite small so that longitudinal static stability is readily achievable by passive means in space capsules.

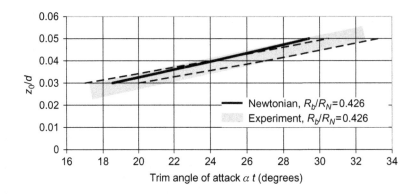

FIGURE 8.51

The variation of trim angle with vertical displacement of the CG from the vehicle centerline for an Apollo capsule according to Newtonian theory. The shaded area includes laboratory and flight test data reported by DeRose (1969).

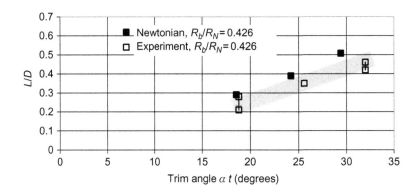

FIGURE 8.52

The variation of L/D as a function of trim angle for an Apollo capsule according to Newtonian theory. The shaded area includes laboratory and flight test data reported by DeRose (1969).

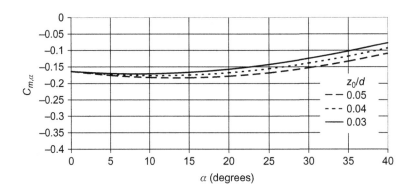

FIGURE 8.53

The longitudinal stability parameter $C_{m,\alpha} = dC_m/d\alpha$ for the Apollo capsule according to Newtonian theory as a function of angle of attack for different vertical displacements of the CG from the vehicle centerline.

8.6.3 **LONGITUDINAL STATIC STABILITY OF SPACEPLANES**

Consider the family of thin power law planform wings like those shown in Figure 8.54. The pitching moment about the center of gravity of plate-like wings is due primarily to the pressure field about the wing. The friction and form drag are essentially in the plane of the slender wing, and their contribution to the pitching moment is assumed negligible. Integrating the incremental moment about a general axial station $x = x_a$ due to the difference in pressure between the upper and lower surfaces of the wing yields

$$M_{CG} = \int_0^l (C_{p,u} - C_{p,l})q(2ydx)(x - x_a) \tag{8.128}$$

The wing is assumed to be symmetrical about the longitudinal axis and the leading edge is defined by

$$y = \frac{b}{2}\left(\frac{x}{l}\right)^n \tag{8.129}$$

In hypersonic flight, the pressure coefficient given by the basic Newtonian approximation is

$$C_{p,l} - C_{p,u} = 2\sin^2\alpha \tag{8.130}$$

The pitching moment is therefore

$$M_{CG} = 2\sin^2\alpha\left(\frac{1}{n+1}\frac{x_a}{l} - \frac{1}{n+2}\right)qbl^2 \tag{8.131}$$

The planform area and mean aerodynamic chord c_{mac} of the power law wing are given by the following:

$$S = 2\int_0^l ydx = \frac{bl}{n+1} \tag{8.132}$$

$$c_{mac} = \frac{\int_0^l [c(y)]^2 dy}{\int_0^l c(y)dy} = \frac{2l}{n+2} \tag{8.133}$$

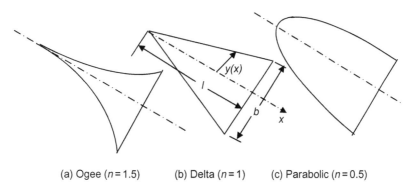

(a) Ogee ($n = 1.5$)   (b) Delta ($n = 1$)   (c) Parabolic ($n = 0.5$)

FIGURE 8.54

Several types of power law wings with leading edges with $y = ax^n$.

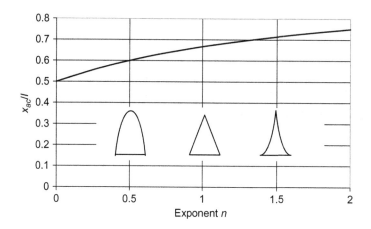

FIGURE 8.55

Variation of the aerodynamic center location with the exponent n which defines the planform shape of the wing.

We may use the previous definitions to write the pitching moment coefficient of the wing as follows:

$$C_{m,a} = \left[(n+2)\frac{x_a}{l} - (n+1)\right]\sin^2 \alpha \tag{8.134}$$

The slope of the pitching moment coefficient is

$$\frac{\partial C_{m,a}}{\partial \alpha} = 2\left[(n+2)\frac{x_a}{l} - (n+1)\right]\sin \alpha \cos \alpha \tag{8.135}$$

The aerodynamic center of the wing is that point on the wing where the pitching moment is independent of angle of attack. From Eqn (8.135), we see that the aerodynamic center for slender power law wings in hypersonic flow is defined by

$$\frac{x_{ac}}{l} = \frac{n+1}{n+2} \tag{8.136}$$

The variation of the aerodynamic center location with the exponent n which defines the planform shape of the wing is shown in Figure 8.55. Under the present assumptions, we see that the aerodynamic center coincides with the center of pressure because both the moment coefficient and its slope vanish at the same point on the body. This is not generally true as was clearly seen in the case of the force structure on blunt capsule-like bodies as discussed in Section 8.3.1. To consider stability of the wing, we look to satisfy Eqns (8.118) and (8.119) by placing the CG of the wing appropriately. Assuming, for example, the case of a delta wing with $n = 1$, we may study the behavior of C_m as a function of α for several values of x_{CG}/l. Equation (8.136) gives the aerodynamic center location for a delta as $x_{ac}/l = 2/3$. Using Eqn (8.134) with $x_a/l = x_{CG}/l = 1/2$, $2/3$, and $3/4$, we obtain the results as shown in Figure 8.56. It is clear from the figure that if the center of gravity is located ahead of the aerodynamic center, the wing will exhibit static longitudinal stability.

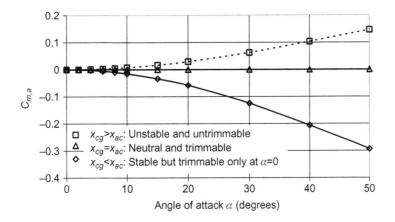

FIGURE 8.56

Variation of the moment coefficient about the CG for a delta wing ($n = 1$, $x_{ac}/l = 2/3$) with $x_{CG}/l = 1/2$, $2/3$, and $3/4$.

Aircraft longitudinal static stability is usually expressed in terms of a quantity called the static margin, which is typically written as $h - h_n$. The quantity $h = (x_0 - x_{lemac})/c_{mac}$ is the normalized distance to the aircraft center of gravity measured from the leading edge of the mean aerodynamic chord, while $h_n = (x_{ac} - x_{lemac})/c_{mac}$ is the normalized distance to the neutral point, or aerodynamic center, measured from the leading edge of the mean aerodynamic chord. Thus, the quantity $h - h_n$ expresses how far the center of gravity of the airplane is forward of the neutral point, expressed as a fraction of the mean aerodynamic chord. We may substitute Eqns (8.133) and (8.136) in Eqn (8.135) to reformulate the slope of the moment coefficient about the center of gravity as follows:

$$\frac{\partial C_{m,0}}{\partial \alpha} = 4 \left[\frac{x_0}{c_{mac}} - \frac{x_{ac}}{c_{mac}} \right] \sin \alpha \cos \alpha \tag{8.137}$$

From the numerator of Eqn (8.53) for Newtonian flow, we may show that

$$\frac{\partial C_n}{\partial \alpha} = \frac{\partial}{\partial \alpha} \left(\frac{F_n}{qS} \right) = \frac{\partial}{\partial \alpha} (2 \sin^2 \alpha) = 4 \sin \alpha \cos \alpha \tag{8.138}$$

We may also write

$$h - h_n = \left(\frac{x_0}{c_{mac}} - \frac{x_{lemac}}{c_{mac}} \right) - \left(\frac{x_{ac}}{c_{mac}} - \frac{x_{lemac}}{c_{mac}} \right) = \frac{x_0}{c_{mac}} - \frac{x_c}{c_{mac}} \tag{8.139}$$

Therefore, the slope of the moment coefficient about the center of gravity is

$$\frac{\partial C_{m,0}}{\partial \alpha} = (h - h_n) \frac{\partial C_n}{\partial \alpha} = - (h_n - h) \frac{\partial C_n}{\partial \alpha} \tag{8.140}$$

Because the normal force coefficient slope is positive, Eqn (8.140) shows that the larger $h_n - h$, the more stable the vehicle, and therefore the less maneuverable it is. Although for stability the static margin $h - h_n < 0$, it is traditionally expressed as a positive percentage figure for stability and as a negative percentage for instability, probably because stability is considered a positive attribute. For example, typical commercial aircraft have a static margin of around 5–10%. Flight data for three Space Shuttle orbiters presented by Suit and Schiess (1988) suggest a static margin in hypersonic flight of around 2–3%. The desired range of center of gravity location for the Space Shuttle was given by Surber and Olsen (1978) as $0.65 < x_0/l < 0.675$, and it is clear from Eqn (8.128) that wing planforms with $n > 1$ would have an aerodynamic center aft of that particular range of desired center of gravity locations.

Greater stability reduces the pilot workload because fewer control inputs are required to keep a particular course. The airplane is said to be stiffer as the static margin increases. When the static margin is zero the airplane is neutrally stable, while if the static margin is negative, the aircraft is statically unstable. Modern fighter aircraft employ relaxed static stability in order to achieve higher maneuverability, but this requires a flight control system that senses motions and uses redundant computers to provide stabilizing control inputs, thus relieving the pilot of a heavy and continuous workload.

It should be clear from Figure 8.56 that a simple flat plate wing may be statically stable in pitch, but it cannot be trimmed. Bodies having symmetry in the vertical plane suffer from this practical difficulty, and such a wing alone cannot satisfy Eqn (8.118), which requires the zero-lift pitching moment to be greater than zero. On conventional aircraft, a horizontal tail is added and configured to produce negative lift and, because the tail is downstream of the CG, a positive (nose-up) moment. The stress of hypersonic flight precludes such an approach, but the addition of outboard vertical fins produces a drag force, which also acts to produce a positive pitching moment at zero-lift as illustrated in Figure 8.57. Obviously, the addition of vertical fins destroys vertical plane symmetry and thereby provides means for trimming the vehicle.

Not only does the flat plate wing lack the ability to be trimmed but it also possesses little volume and is impractical in that regard. Let us assume a rectangular wing ($n = 0$) of length l and span b that has some depth so as to accommodate a half-wedge forebody, as shown in Figure 8.58.

We made the wing nonsymmetric in the vertical plane and look toward producing a positive zero-lift pitching moment so that the wing might be trimmed. Assuming two-dimensional flow over

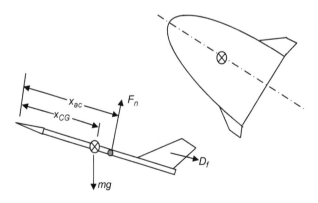

FIGURE 8.57

Outboard vertical fins produce drag to provide a positive pitching moment at the zero-lift angle of attack ensure the capability of trimming.

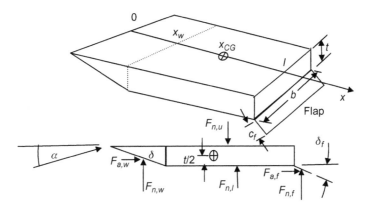

FIGURE 8.58

Sketch of a rectangular wing with a wedge forebody and aft body flap. The normal force designations are shown as is the CG location.

the wing, the normal forces (that is, perpendicular to the x-axis) on the flat wing upper surface, the lower wedge surface, the lower flat wing surface, and the aft flap surface in hypersonic flight are given respectively as follows:

$$\alpha < 0 : F_{n,u} = -2 \sin^2 \alpha (S)$$
$$\alpha \geq 0 : F_{n,u} = 0$$

$$\alpha + \delta > 0 : F_{n,w} = 2 \sin^2(\alpha + \delta)\left[qS \frac{x_w}{l} \right]$$

$$\alpha < -\delta : F_{n,w} = 0$$

$$\alpha \geq 0 : F_{n,l} = 2 \sin^2 \alpha \left[qS\left(1 - \frac{x_w}{l} \right) \right]$$

$$\alpha < 0 : F_{n,l} = 0$$

$$\alpha < 0 : F_{n,f} = 0$$

$$\alpha \geq 0 : F_{n,f} = 2 \sin^2(\alpha + \delta_f)\left[qS \frac{c_f}{l} \sin \delta_f \right]$$

Note that the aft flap is not completely in the Newtonian shadow region within the following angle of attack range:

$$-\frac{(c_f/l)\sin \delta_f}{1 - (x_w/l) + (c_f/l)\cos \delta_f} < \alpha < 0$$

For realistic values of c_f/l, x_w/l, and δ_f, this range constitutes perhaps two or three degrees, and within the assumptions of the theory, the flap contribution there may safely be neglected. Because the pressure force is always normal to the surface, the axial components of the pressure force (that is, parallel to the x-axis) on the wing are

$$F_{a,u} = 0$$

$$F_{a,w} = 2 \sin^2(\alpha + \delta) \left[qS \frac{t}{l} \right]$$

$$F_{a,s} = 0$$

$$F_{a,f} = 2 \sin^2(\alpha + \delta_f) \left[qS \frac{c_f}{l} \cos \delta_f \right]$$

The pitching moment about the CG is

$$M_{CG} = F_{n,u} \left(x_{CG} - \frac{l}{2} \right) + F_{n,w} \left(x_{CG} - \frac{x_w}{2} \right) + F_{n,l} \left[x_{CG} - \left(\frac{l + x_w}{2} \right) \right]$$

$$+ F_{n,f} \left[x_{CG} - \left(l + \frac{1}{2} c_f \cos \delta_f \right) \right] + F_a \frac{1}{2} (t + c_f \sin \delta_f)$$

Frictional forces have been neglected here because they typically make little contribution to the pitching moment compared to the pressure forces. The lift and drag components of the resultant pressure force are given by

$$L = [(F_{n,u} + F_{n,w} + F_{n,l} + F_{n,f}) \cos \alpha - (F_{a,w} + F_{a,f}) \sin \alpha]$$
$$D = [(F_{n,u} + F_{n,w} + F_{n,l} + F_{n,f}) \sin \alpha + (F_{a,w} + F_{a,f}) \cos \alpha]$$

Assume that the wing in Figure 8.58 has the following dimensions: $l = b = 1$, $t = 0.1$, $c_f = 0.1$, $\delta = 20°$, $\delta_f = 0$ or $20°$, and that the center of gravity is located centrally within the body at $x_{CG} = 0.4$. The pitching moment coefficient about the center of gravity $C_m = M_{CG}/qSl$ for this wing is shown in Figure 8.59. With the aft flap undeflected, the wing has neutral stability up to $\alpha = -5°$, then is unstable to $\alpha = 10°$, and becomes stable as α is increased further, but the wing cannot be trimmed. With the aft flap deflected to $\delta_f = 20°$, we see that as α is increased from negative values, the wing is stable, then neutral, and finally stable, with trim achieved at $\alpha = 8°$. The lift curve for the wing is shown in Figure 8.60, where we see that the zero-lift angle of attack is $\alpha_0 = -7°$ and that deflection of the aft flap boosts the lift to levels above that achieved with no flap deflection. Figure 8.61 shows the

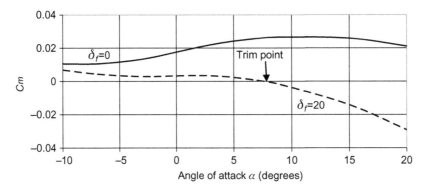

FIGURE 8.59

The moment coefficient for the wing in Figure 8.58 with the aft flap at $\delta_f = 0$ and $\delta_f = 20°$. In the latter case, the wing is trimmed and stable at about $\alpha = 8°$.

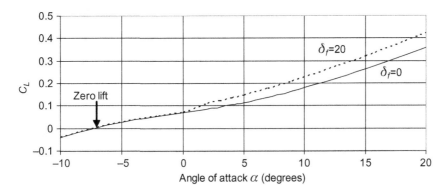

FIGURE 8.60

The lift curve is shown for the wing considered in the previous figure. The zero-lift angle of attack is $\alpha_0 = -7°$.

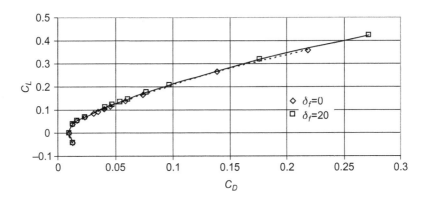

FIGURE 8.61

The drag polar is shown for the wing considered in the previous figure. The lift to drag ratio is $L/D \sim 2$ at the trim lift coefficient $C_{L,trim} \sim 0.2$.

drag polar for the wing, and it is clear that the effect of flap deflection is negligible. From Figure 8.60, we find the lift coefficient at the trim point to be $C_{L,trim} \sim 0.2$, and from Figure 8.61, we see that at trim, $L/D = C_L/C_D \sim 2.2$. However, a substantial drag penalty must be paid for trimming the vehicle, about 30% in this case, as shown in Figure 8.62. The deflection of the aft body flap does not change L/D much, as illustrated by Figure 8.61, because the lift and drag both increase as a result of the deflection.

8.6.4 LATERAL STATIC STABILITY

Lateral stability of a spacecraft refers to the stability of its motion in the x-y plane or the y-z plane and may be treated in the same manner as longitudinal static stability. Motions in the x-y plane involve yawing moments about the z-axis due to heading changes caused by the angle of sideslip β.

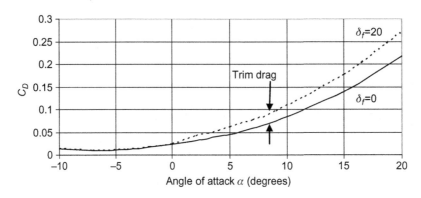

FIGURE 8.62

The drag coefficient is shown as a function of angle of attack for the wing considered in the previous figure. The trim drag ratio is indicated at the trim $\alpha = 8°$.

Motions in the y-z plane involve rolling moments about the x-axis caused by a change in the bank angle ϕ. In stability and control studies, it has been traditional to denote L, M, and N as the moment about the x, y, and z axes and C_l, C_m, and C_n, the moment coefficients about those axes. Similarly p, q, and r denote the rotational velocities about the x, y, and z axes, respectively. However, the coordinate axes in such studies, called stability axes, orient the x-axis with the velocity of the aircraft center of gravity and the y-axis outward along the right (starboard) wing. As a result, for a right-handed system, the z-axis points downward from the x-y plane. The origin of this coordinate system is typically the center of gravity of the vehicle. This is different from the axes traditionally used in aerodynamic studies where the origin of the coordinate system is usually taken at the nose of the vehicle with the x-axis aligned in the opposite sense to the vehicle velocity. The y-axis is still taken outward along the right wing and therefore the z-axis points upward from the x-y plane. It is important to keep these distinctions in mind when using the literature from the different fields.

The same basic ideas as discussed for longitudinal static stability apply to the case of lateral static stability. If the spacecraft is subjected to a small change in the sideslip or bank angle, the tendency of the spacecraft to respond by reducing that angle is considered to be indicative of static stability. Spacecraft are transatmospheric vehicles and must be capable of performing at altitudes where aerodynamic forces are negligible. As a result, the stability of such vehicles must be augmented by applying onboard external forces like rocket thrusters, and such systems are discussed in a subsequent section of "Spacecraft Attitude Control."

8.6.5 LATERAL STATIC STABILITY OF SPACE CAPSULES

The geometry of a space capsule is very close to that of a spherical sector, and because of the degree of rotational symmetry about the longitudinal axis, its static stability depends almost exclusively on the position of its center of gravity. Thus, space capsules depend upon banking about the longitudinal axis to place the center of gravity in the desired location. The placement of the center of gravity to achieve static longitudinal stability was described previously.

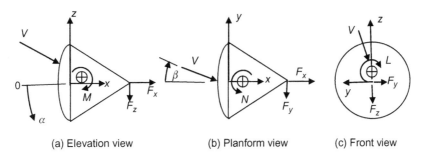

(a) Elevation view   (b) Planform view   (c) Front view

FIGURE 8.63

Projections of the velocity and force field on a space capsule in sideslip. Positive senses for angles and moments are shown.

A space capsule in static longitudinal equilibrium at hypersonic speed is depicted in an elevation view in Figure 8.63a, where its center of gravity is located such that changes in the angle of attack α (here the trim angle $\alpha_t < 0$) will be countered by moments produced by corresponding changes in the axial and vertical forces F_x and F_z. If α is increased to $\alpha = \alpha_t + \Delta\alpha$ (the velocity becoming more axial in direction), then F_z decreases in magnitude and F_x increases in magnitude making $M < 0$ rotating the capsule counterclockwise to restore trim. Conversely, an increase in α to $\alpha = \alpha_t - \Delta\alpha$ (the velocity becoming more normal in direction) increases the magnitude of F_z and decreases the magnitude of F_x making $M > 0$ and once again returning the capsule to the trim condition. This is the situation described by the stability derivative in pitch $C_{m,\alpha} < 0$.

Suppose the capsule then encounters a small disturbance that induces a sideslip angle β, as shown from above in the planform view of Figure 8.63b. The side force component F_y generated by the sideslip produces a yawing moment $N < 0$ about the z-axis, which acts to reduce the sideslip angle, thereby indicating static lateral stability in yaw. The axial force component produces no moment because it is colinear with the center of gravity location in the x-y plane.

One must keep in mind that the coordinate systems used in reporting stability derivatives may differ so that the sign of the derivative alone may be misleading. For example, in Figure 8.63, we have shown our coordinate system that has the z-axis pointing up so that the yawing moment $N < 0$ and therefore $C_{n,\beta} < 0$, whereas if stability axes were used, where the z-axis points down, the stability derivative $C_{n,\beta} > 0$. Therefore, in the stability and control literature, one will see that the requirement for static stability in yaw is $C_{n,\beta} > 0$. The difference is due to the choice of coordinate system, and the important idea should be that the induced moment serves to reduce the disturbance.

Wind tunnel measurements on Apollo capsule models reported by Moseley, Graham, and Hughes (1968) showed that for hypersonic flight speed, they are statically stable in yaw. In the coordinate system used here, the yawing stability derivative, that is, the change in the yawing moment with change in sideslip angle β, has the following magnitude

$$\frac{\partial C_n}{\partial \beta} = C_{n,\beta} \approx -0.125$$

At the same time, F_y is causing a negative N moment about the z-axis and thereby rotating the capsule so as to reduce β as shown in Figure 8.63b, it is also causing a negative L moment rolling

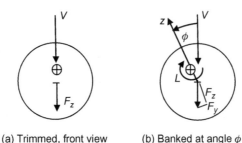

(a) Trimmed, front view   (b) Banked at angle ϕ

FIGURE 8.64

Projections of the velocity and vertical force on the front face of a space capsule (a) trimmed at a given value of α at zero bank angle and (b) for a bank angle of ϕ.

the capsule about the x-axis so as to drive the bank angle ϕ into the sideslip, as shown in Figure 8.63c. Thus, sideslip causes cross-coupling between the yawing moment and the rolling moment, which affects the bank angle, as may be seen in the front view of the capsule in Figure 8.63c. The effect of the yaw-induced roll is to change the bank angle ϕ and may be considered similar to the dihedral effect in winged vehicles. The rolling moment stability derivative, that is, the change in the rolling moment with change in sideslip angle β is generally much smaller than the yawing stability derivative mainly because of the small offset of the center of gravity from the x-axis. Moseley et al. (1968) report values typified by the following:

$$\frac{\partial C_l}{\partial \beta} = C_{l,\beta} \approx -0.02$$

An interesting aspect of banking the capsule is that it has no effect on the aerodynamic forces acting on the capsule. The only effect of banking is on the displacement of the center of gravity. As shown in Figure 8.64, the effect of banking is to reduce the F_z component from the original trimmed situation and increase the F_y component from zero while F_x remains the same. The cross-coupling discussed previously is now seen to produce a yaw effect from the roll to the bank angle ϕ and a positive rolling moment L which tends to reduce the bank angle.

8.6.6 LATERAL STATIC STABILITY OF SPACEPLANES

We may apply the same definition of longitudinal static stability to the case of lateral yawing motion about the z-axis due to a sideslip angle β, as illustrated in Figure 8.65 and discussed by Sforza (2014). The sideslip angle β displaces the motion from the original direction and generates a lift and drag force on the vertical tail supplying a positive moment N, which tends to reduce the sideslip. In this sense, the vertical tail ensures lateral static stability in sideslip. This type of stability is often called weathervane stability.

The addition of a separate vertical tail surface is not as temperature-critical as would be a horizontal tail surface, because the vertical tail can operate on the lee surface of the spacecraft where it may be partially shielded from the thermal stress affecting the windward surface. Of course, this

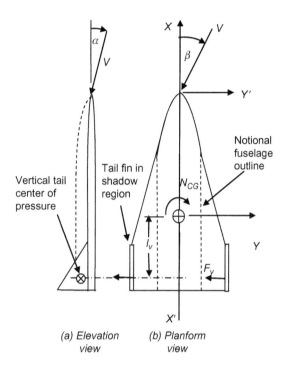

(a) Elevation
view

(b) Planform
view

FIGURE 8.65

Elevation and planform views of a vehicle at sideslip angle β showing positive direction of the yawing moment, angle of sideslip, and angle of attack. Outline of a notional fuselage is indicated by dashed lines.

type of shielding may compromise the effectiveness of the vertical tail, so dual outboard vertical fins are often employed to ensure that one is always operative.

The yawing moment on a vertical fin like that in Figure 8.65 may be calculated according to Newtonian theory which yields

$$C_N = \frac{N}{qSb} = 2 \sin^2 \beta \left(\frac{S_v l_v}{Sb} \right) = 2 \sin^2 \beta (V_v) \tag{8.141}$$

The pressure force on the active fin is normal to the surface so the moment produced by that fin is $N = F_v l_v$. The quantity S_v is the area of the active vertical fin, and $V_v = S_v l_v /Sb$ is called the vertical fin volume coefficient. The magnitude of V_v is indicative of the effectiveness of the vertical tail in providing lateral stability. This can be seen by considering the rate of change of the restoring moment with sideslip angle, which is given, for small sideslip angles, by

$$\frac{\partial C_N}{\partial \beta} = 4V_v \sin \beta \cos \beta \approx 4\beta V_v \tag{8.142}$$

For stability in sideslip, we need to have $\partial C_n/\partial \beta > 0$, so that for positive sideslip angle β, as shown in Figure 8.65, the active vertical tail will produce a positive moment swinging the vehicle into the velocity vector. For negative β, the opposite will occur. The rapidity of the restoration of proper direction depends upon the magnitude of V_y. The competing design demand here is that increasing V_y by having larger S_y must be paid for by accepting an increase in the drag of the vehicle and a reduced aerodynamic efficiency parameter L/D.

As an example, consider the Space shuttle, which has vertical tail area $S_y = 38.4$ m^2, wing area $S = 250$ m^2, and span $b = 23.8$ m, so that the vertical fin volume coefficient $V_y = 0.00645l_y$. The vertical fin moment arm is given approximately by the distance from the aft end of the fuselage to the center of gravity $l_y = l - x_0 = 0.35l = 11.5$ m so that $V_y = 0.074$. This value is in the range for commercial airliners, which is shown by Sforza (2014) to be $0.065 < V_y < 0.13$.

Using the nominal value $V_y = 0.1$, Eqn (8.142) shows that the stability in sideslip derivative $C_{N,\beta} = 0.4\beta$ when β is measured in radians and 0.007β when β is measured in degrees. It must be noted here that this result is for a slender wing alone. The presence of a fuselage, as indicated by the dashed lines in Figure 8.63, tends to be destabilizing and typically reduces $C_{N,\beta}$ substantially. Consider that the portion of the fuselage ahead of the center of gravity will be subjected to a side force that produces a negative destabilizing moment.

In addition, if the angle of attack of the vehicle is too high (greater than about 15° for the Space Shuttle), the wake from the fuselage can blanket the vertical tail and dramatically reduce its effectiveness. In the case of the Space Shuttle, $-0.002 < C_{N,\beta} < -0.0015$ in hypersonic flight where the angle of attack is typically greater than 20°. As M and α decrease and the vertical tail becomes effective, $C_{N,\beta}$ increases to positive values.

8.6.7 LONGITUDINAL DYNAMIC STABILITY

We derived Eqns (6.6)–(6.8) to describe the motion of the entering spacecraft when it is considered to be a point mass. Then, in this chapter, we treated the spacecraft as a rigid body and examined the longitudinal static stability in terms of the pitching moment about its center of gravity. A schematic diagram of an entering space capsule and the associated attitude angles is shown in Figure 8.66. For convenience, we rewrite the equations, including the equation for the spacecraft's pitch angle θ as follows:

$$\frac{dV}{dt} = -\frac{\rho V^2 S}{2m} C_D - g \sin \gamma \tag{8.143}$$

$$\frac{d\gamma}{dt} = \frac{\rho VS}{2m} C_{L,\alpha}\alpha - \left(\frac{g}{V} - \frac{V}{r}\right) \cos \gamma \tag{8.144}$$

$$\frac{d^2\theta}{dt^2} = \frac{\rho V^2 Sd}{2I} C_m \tag{8.145}$$

The pitching moment coefficient is generally assumed to be a function of the angle of attack α, the pitch rate $\theta = d\theta/dt$, the time rate of change of α, and a general control input δ. Assuming that the aerodynamic force and moment coefficients are all linear functions, we may write

$$C_m = C_{m,\alpha}\alpha + C_{m,q}\left(\frac{d\theta}{dt}\right)\frac{l}{V} + C_{m,\dot{\alpha}}\left(\frac{d\alpha}{dt}\right)\frac{l}{V} + C_{m,\delta}\delta \tag{8.146}$$

We use the convention that the flight path angle $\gamma < 0$ during descent, so that the pitch angle is $\theta = \alpha - \gamma$. Tobak and Allen (1958) showed that the equations of motion could be split into two parts,

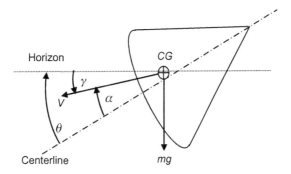

FIGURE 8.66

Capsule attitude angles during entry.

one uniformly time-dependent and the other oscillatory in time. The uniformly time-dependent part includes Eqns (8.143) and (8.144), which describe the evolution of the flight path angle of the center of gravity with time, while Eqn (8.145) specifies how the control input (δ) should be managed to keep the spacecraft trimmed during this motion. The oscillatory part describes how the spacecraft moves around the center of gravity (α and θ) and how the flight path angle (γ) oscillates about its uniformly time-dependent path. Tobak and Allen (1958) show that the oscillatory part of the motion may be combined into the following single equation for the oscillatory angle of attack α_o:

$$\frac{d^2\alpha_o}{dt^2} + f_1(t)\frac{d\alpha_o}{dt} + f_2(t)\alpha_o = 0 \tag{8.147}$$

The functions in Eqn (8.147) are given by

$$f_1(t) = \left(\frac{\rho VS}{2m}\right)\left[C_{L,\alpha} - (C_{m,q} + C_{m,\dot{\alpha}})\left(\frac{ml^2}{I}\right)\right] \tag{8.148}$$

$$f_2(t) = \frac{d}{dt}\left(C_{L,\alpha}\frac{\rho VS}{2m}\right) - \frac{C_{m,q}C_{L,\alpha}}{mI}\left(\frac{\rho VSl}{2}\right)^2 - C_{m,\alpha}\frac{\rho V^2 Sl}{2I} \tag{8.149}$$

Using Eqn (8.143) and the assumption of an exponential atmosphere $\sigma = \exp(-z/H)$, the first term on the right-hand side of Eqn (8.149) may be expanded further to

$$\frac{d}{dt}\left(C_{L,\alpha}\frac{\rho VS}{2m}\right) = \left[C_D - \left(\frac{2m}{\rho SH}\right)\sin\gamma\right]\left(\frac{\rho VS}{2m}\right)^2 C_{L,\alpha}\alpha \tag{8.150}$$

Then Eqn (8.149) may be rewritten as

$$f_2(t) = \left(\frac{\rho VS}{2m}\right)^2\left[\left(C_D - \frac{ml^2}{I}C_{m,q} - \frac{2m}{\rho SH}\sin\gamma\right)C_{L,\alpha} - \frac{2m}{\rho Sd}\frac{ml^2}{I}C_{m,\alpha}\right] \tag{8.151}$$

Assuming that the ratio $I/ml^2 = O(10^{-1})$ and that the density of the spacecraft may be defined such that $m = \rho_s Sd$ permits us to approximate the first term in the square brackets on the right-hand side of Eqn (8.151) as follows:

$$C_D - \frac{ml^2}{I}C_{m,q} - \frac{2m}{\rho SH}\sin\gamma \approx C_D - 10C_{m,q} - 2\left(\frac{\rho_s}{\rho}\right)\left(\frac{l}{H}\right)\sin\gamma \tag{8.152}$$

In the case of a practical space capsule, for example, the Apollo CM, the reference length $l = d$, the capsule maximum diameter, the magnitude of $C_{m,q} = O(10^{-1})$ and $\rho_s = O(10^2)\text{kg/m}^3$, while $l/H = O(10^{-4})$ and $\sin\gamma = O(10^{-1})$. In the case of a practical space plane, for example, the Space Shuttle Orbiter, the reference length $l = c_{mac}$, the mean aerodynamic chord of the wing, the magnitude of $C_{m,q} = O(10^{-1})$ and $\rho_s = O(10^2)\text{kg/m}^3$, while $l/H = O(10^{-3})$ and $\sin\gamma = O(10^{-1})$. Therefore, in both types of spacecraft, all the terms on the right-hand side of Eqn (8.152) are around $O(1)$ with the last term becoming dominant for altitudes above 60 km. We may now compare this last term with the one remaining in the square brackets of Eqn (8.151) to find

$$-2\left(\frac{\rho_s}{\rho}\right)\left(\frac{d}{H}\right)C_{L,\alpha}\sin\gamma - 2\left(\frac{\rho_s}{\rho}\right)C_{m,\alpha} = -2\left(\frac{\rho_s}{\rho}\right)\left(\frac{l}{H}C_{L,\alpha}\sin\gamma + C_{m,\alpha}\right)$$

Here, the last term dominates so that we may finally approximate Eqn (8.151) by

$$f_2(t) \approx \left(\frac{\rho VS}{2m}\right)^2\left(-\frac{2m^2 l}{\rho SI}\right)C_{m,\alpha} = \frac{\rho V^2 Sl}{2I}C_{m,\alpha}$$

The oscillatory angle of attack Eqn (8.147) may now be written as

$$\ddot{\alpha}_o + \left(\frac{\rho VS}{2m}\right)\left[C_{L,\alpha} - \left(\frac{ml^2}{2I}\right)(C_{m,q} + C_{m,\dot{\alpha}})\right]\dot{\alpha}_o - \left(\frac{\rho V^2 Sl}{2I}\right)C_{m,\alpha}\alpha_o = 0 \qquad (8.153)$$

The non-dimensional Eqns (6.6)–(6.8) may be used to calculate the equilibrium (nonoscillatory) motion of the spacecraft during entry. The corresponding non-dimensional angular acceleration equation becomes

$$\frac{d^2\theta}{d\tau^2} = \left(\frac{\rho_{sl}g_E R_E Sl}{2I}\right)C_m e^{-s} v^2 \qquad (8.154)$$

Equations (8.143)–(8.145) and (8.153) constitute four equations in the six unknowns V, γ, ρ, z, θ, and α. The equation set is rounded out by the auxiliary equations

$$\frac{dz}{dt} = V\sin\gamma \qquad (8.155)$$

$$\rho = \rho_{sl}\exp\left(-\frac{z}{H}\right) \qquad (8.156)$$

The longitudinal dynamic stability of the spacecraft is determined by the damping or amplification of perturbations to its angle of attack. Allen (1957) provided an approximate analysis of the problem that yielded a useful dynamic stability parameter

$$\xi = C_D - C_{L,\alpha} + \frac{ml^2}{I}(C_{m,q} + C_{m,\dot{\alpha}}) \qquad (8.157)$$

The quantity $C_{m,q} + C_{m,\dot{\alpha}}$ in Eqn (8.157) is called the pitch damping coefficient and should be negative for effective damping. When the parameter $\xi < 0$, the vehicle is considered dynamically stable, and the more negative, the more dynamically stable.

The moment of inertia may be expressed as $I = mr_g^2$, where r_g is the radius of gyration about the same axis used for the moment of inertia, here the y-axis, so that Eqn (8.157) may be written

$$\xi = C_D - C_{L,\alpha} + \left(\frac{l}{r_g}\right)^2(C_{m,q} + C_{m,\dot{\alpha}}) \qquad (8.158)$$

A space capsule like Apollo is cone-like in shape with a cone half-angle $\phi_c = 25.2°$. Considering it to be uniform in density, the radius of gyration is

$$r_g = d\sqrt{\frac{3}{80}\left(1 + \frac{1}{4\tan^2\phi_c}\right)}$$

Then, for a typical cone-shaped capsule, we obtain

$$\left(\frac{l}{r_g}\right)^2 = \left(\frac{d}{r_g}\right)^2 = \frac{320\tan^2\phi_c}{3(1+4\tan^2\phi_c)} \tag{8.159}$$

Equation (8.159) suggests that a capsule the size of the Apollo CM has $(d/r_g)^2 \sim 12.5$. Reported data for the Apollo capsule leads to $(d/r_g)^2 \sim 13$, and therefore the approximation of a uniform cone for a capsule appears acceptable. In addition, for a capsule, we have $C_D = O(1)$ and $C_{m,q} = -O(10^{-1})$, so the stability parameter ξ can become positive if d/r_g is too small. Note that for a thin conical shell of the same shape as the solid cone, we find that $(d/r_g)^2 \sim 1$, and this would likely lead to dynamic instability.

Sommer, Short, and Compton (1960) fired Mercury capsule models in a ballistic range to study their static and dynamic stability characteristics. They found that the models were statically stable throughout the angle of attack range tested for a Mach number range of $3 < M < 14$. Detailed dynamic studies at $M = 3$ and 9.5 showed that the capsules were dynamically unstable at those Mach numbers. They reported that flow conditions over the relatively long afterbody appeared to have a strong effect on dynamic stability of the models. Values of the dynamic stability parameter ξ ranged from one to six for cases where the flow was separated over the afterbody, and values as high as 14.4 were observed for cases where the flow remained attached over the afterbody although $(d/r_g)^2 = 7$.

In the case of a spaceplane like the Space Shuttle Orbiter, Suit (1989) reports that the radius of gyration $r_g = 9.75$ m while the reference length $l = c_{mac} = 12.06$ m so that $(d/r_g)^2 = 1.53$, which is considerably smaller than the corresponding values for a space capsule. However, the values of the damping coefficient $C_{m,q} + C_{m,\dot\alpha} \approx -2$ so that $\xi < 0$ and the vehicle is dynamically stable. A similar result ensues for the X-15 suborbital spacecraft, which has $(d/r_g)^2 = 0.64$ but $C_{m,q} + C_{m,\dot\alpha} \approx -5$ and again $\xi < 0$ thus preserving dynamic stability.

Detailed determination of the degree of stability of spacecraft is beyond the scope of this book although the approach outlined above is still actively used. For a recent discussion of space capsule stability, refer to the survey of blunt body dynamic stability presented by Kazemba, Braun, Clark, and Schoenberger (2012).

8.7 SPACECRAFT CONTROL SYSTEMS

To be able to point a spacecraft in a given direction, it is necessary to control its angular orientation. For roll control about any axis, the following equations apply

$$I_{xx}\ddot\phi = L = r_L F_L$$
$$I_{yy}\ddot\theta = M = r_M F_M$$
$$I_{zz}\ddot\beta = N = r_N F_N$$

The quantities I_{xx}, I_{yy}, and I_{zz} denote the moments of inertia about the indicated axes, while r_L, r_M, and r_L denote the moment arms of the thrusts F_L, F_M, and F_N, which produce torques L, M, and N about the x, y, and z axes, respectively.

Consider, as an example, control of the bank angle ϕ, which describes rotation about the x-axis. If the force that produces roll about the x-axis, F_L is constant with time, and if that force is applied over a burn time t_b, the following results apply:

$$\dot{\phi} = \frac{r_L F_L}{I_{xx}} t_b + \dot{\phi}(0)$$

$$\phi = \frac{r_L F_L t_b^2}{2 I_{xx}} + \dot{\phi}(0)t_b + \phi(0) \qquad (8.160)$$

During the burn, the spacecraft angular velocity increases and the bank angle grows. After the thrust F_L is terminated at $t = t_b$, the spacecraft spin velocity remains constant and the bank angle grows at a constant rate. If an equal and opposite thrust is initiated at a later time t' and is applied over the same burn time t_b, the spacecraft spins down and the bank angle becomes constant at some new angle. This acceleration and braking procedure is shown in Figure 8.67. Thrusters are typically used in opposed pairs so that positive and negative impulses may be applied to the spacecraft.

The parameters r_L and I_{xx} in Eqn (8.160) are fixed by the spacecraft configuration while the thrust applied, the burn time, and, implicitly, the coast time are variable. The angular velocity depends upon the total impulse delivered, $I = F_L t_b$. This may be written as $I = I_{sp}\dot{m}_p g_E \approx I_{sp}m_p$, and we see the reason for desiring high-specific impulse to obtain a given impulse with less expenditure of propellant mass m_p. However, in manned spacecraft, spin rates must be low to be acceptable to the human crew. Assuming $d\phi/dt = O(10^{-1})$, about 1 rpm and that the thruster

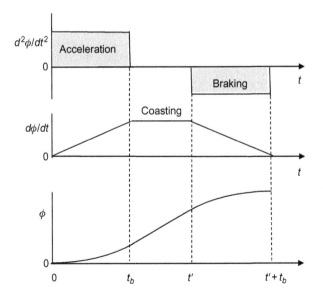

FIGURE 8.67

Rotation about the x-axis for a positive thruster torque, followed by a coast period with no thruster torque and then an equal and opposite thruster torque.

moment arm is about equal to the corresponding radius of gyration of the spacecraft, $r_L = r_g$, the impulse level required is approximately $I = mr_g/10$. For a spacecraft with $m = O(10^4)$kg and a characteristic dimension $r_g = O(1)$m, the impulse $I = O(10^3)$N-s. As an example, the Apollo CM characteristics would yield $I \sim 500$ N · s. Then for a 1-second burn, a 500 N thruster would be required.

If the specific impulse of the thruster is around 260s, then each one-second burn would use about 2 kg of propellant. It is clear that in order to determine the amount of propellant to be carried on a mission, there must be a plan of maneuvers that would be required. The mission design addresses this issue. The number and type of orbital transfers and corrections, docking and undocking maneuvers, station keeping, etc., must be planned beforehand and connected with each are ΔV requirements that entail impulses of various magnitudes. Then, impulse delivery in connection with attitude control and changes, pointing, and stabilization also require consideration. These represent planned thrusting for which propellant consumption can be readily calculated. However, there must be additional propellant allotted to account for off-nominal performance and to satisfy safety reserve requirements. After these allowances are made, an additional small amount, about 5%, should be made to offset unusable propellant that may be trapped in lines, valves, and tanks as well as uncertainty in propellant loading and, in the case of bipropellant systems, mixture asymmetries that cause one propellant to be expended prior to the other.

For spacecraft that must rendezvous and dock with other spacecraft or a space station, some degree of translational control is also required. The same approach as used for rotational control may be used for translational control by adding thrusters capable of providing forces in the x, y, and z directions. The thrusters are also operated in opposed pairs so that linear acceleration and braking may be accomplished. Control systems and reaction control systems (RCS) for unmanned spacecraft are covered in some detail by Brown (1996).

8.7.1 RCS CHARACTERISTICS

Rocket thrusters are used to provide attitude and stability control outside the sensible atmosphere where aerodynamic surface forces provide negligible control authority. General performance characteristics required of the RCS thrusters are as follows:

- Sufficient supply of easily storable propellant
- Capability for both intermittent and continuous operation
- Accurate and repeatable thrust levels
- Simple and reliable operation

Although solid propellant rockets satisfy the first and fourth requirements, the lack of start–stop operation makes them unsuitable for RCS applications. Liquid propellant rockets can be pulse-modulated through electronic control of flow rates by fast-acting valves. However, reliability concerns render the use of liquid propellants that require ignition systems to ensure combustion unattractive. Liquid hypergolic propellants are those that spontaneously react upon contact so that an ignition system is unnecessary. Characteristics of commonly used hypergolic propellants are described in Table 8.9. Both the Apollo CM and the Space Shuttle Orbiter use the monomethyl hydrazine (MMH)–N_2O_4 combination in their RCS. More common in other rockets is the use of UMDH-N_2O_4 or UMDH/N_2H_4-N_2O_4, but their performance and density are not as attractive as the MMH-N_2O_4 combination. These materials are dangerous and toxic. As a result, they must be

Table 8.9 Characteristics of Some Common Hypergolic Propellants

Name	Type[a]	Formula	Boiling Point, K	Freeze Point, K	Density (kg/m^3)	Molecular Weight
Hydrazine	F	N_2H_4	386	274	1011	32.05
Nitrogen Tetroxide	O	N_2O_4	294	262	1450	92.02
MMH	F	CH_3NHNH_2	360	221	880	46.03
Unsymmetrical-dimethylhydrazine (UDMH)	F	$(CH_3)_2NNH_2$	336	221	790	60.04
Red fuming nitric acid (RFNA)	O	$HNO_3 + 13\%$ $N_2O_4 + 3\%H_2O$	216	358	1550	63

[a]F and O denote fuel and oxidizer, respectively.

Table 8.10 General Properties of Representative RCS Rocket Propellants

Type	Propellant	Energy Source	I_{sp} (s) Vacuum	F (N) Range	Specific Gravity	Advantages	Disadvantages
Cold gas	N_2, NH_3, He, Freon	High pressure	50–75	0.05–250		Simple, reliable, low cost	Low performance, high weight
Liquid monopropellant	H_2O_2 N_2H_4 (hydrazine)	Exothermic decomposition	150 200	0.05–0.5	1.46 1	Simple, reliable, low cost	Low performance, higher weight
Hypergolic bipropellant	UMDH/ N_2O_4	Chemical	270–340	$10-10^6$	1.14	Storable, good performance	Complicated
Hypergolic propellant	N_2H_4/F_2	Chemical	425	$10-10^6$	1.1	Very high performance	Toxic, dangerous, complicated

handled under strict safety protocols and they are now very expensive due to stringent environmental regulations. Detailed discussion of rocket propulsion and propellant characteristics may be found in Sforza (2012).

Monopropellants are liquid substances which require no oxidizer but rather rapidly decompose to release their chemical energy. To be practical, such substances must be stable under normal operating conditions but decompose rapidly upon ignition. Obviously, a shortcoming of such propellants for RCS is the need for an ignition system. On the other hand, a monopropellant RCS has no need for a second propellant feed system. The most common monopropellants include hydrogen peroxide, hydrazine, and ethylene oxide. Another class of propellant that satisfies the RCS requirements is simply an inert gas, which relies on high-storage pressure rather than chemical reaction for energy release. Typical so-called "cold gas" systems use inert gases like nitrogen and helium as well as other gases like ammonia and Freon. General performance characteristics of the propellant systems that are suitable for RCS applications are shown in Table 8.10.

8.7.2 PULSED THRUST OPERATION

The thrust of an RCS nozzle of exit radius r_e exhausting into the essentially zero pressure of the upper atmosphere with exit density and velocity ρ_e and V_e, respectively, may be expressed as

$$F = \pi r_e^2 \rho_e V_e^2 \tag{8.161}$$

Similarly, the mass flow passing through the nozzle is

$$\dot{m} = \frac{1}{\lambda} \pi r_e^2 \rho_e V_e \tag{8.162}$$

The quantity λ represents the losses due to angular dispersion of the flow at the nozzle exit and is given by

$$\lambda = \frac{1}{2}(1 + \cos \theta_w) \le 1 \tag{8.163}$$

If the nozzle wall angle $\theta_w = 0$, the flow leaves the nozzle parallel to its axis while if it does not, as in the case of a conical nozzle, the mass flow is greater although the thrust produced is the same. As discussed in Chapter 7, the figure of merit for nozzles is the specific impulse, and here, because we are considering exhaust into a vacuum, it is the vacuum-specific impulse

$$I_{sp,vac} = \frac{F}{\dot{m} g_E} = \lambda \frac{V_e}{g_E} \tag{8.164}$$

This development shows that although a conical nozzle may be simpler and less expensive than a contoured nozzle, the specific impulse will be lower for the same level of thrust. Consider a nozzle operating over a time interval t_1 to time t_2 as illustrated in Figure 8.68. If we assume that the rise and decay of the nozzle momentum flux is linear with time and define the momentum flux pulse duration $\tau = t_2 - t_1$, we find the impulse to be

$$I = \int_{t_1}^{t_2} F(t) dt \approx \pi r_e^2 \rho_e V_e^2 t_{avg} = F t_{avg} \tag{8.165}$$

In this analysis, t_1 is the time at which the propellant valve is opened, t_i is the time at which steady nozzle flow is initiated, t_c is the time at which the propellant valve is closed, and t_2 is the time at which the moment flux decays to zero. The quantity t_{avg} is the average time and τ the total time of nozzle operation, respectively. Assuming that the propellant valves act instantaneously so that the transients for mass flow are negligible, the mass consumed is

$$m_p = \int_{t_1}^{t_2} \frac{1}{\lambda} \pi r_e^2 \rho_e V_e dt \approx \pi r_e^2 \rho_e V_e (t_c - t_1) \tag{8.166}$$

Here $t_c - t_1$ is the duration of the valve opening.

Then, combining Eqns (8.165) and (8.166), the vacuum-specific impulse is

$$I_{sp,vac} \approx \frac{t_{avg}}{t_c - t_1} \lambda \frac{V_e}{g_E} \tag{8.167}$$

We see that the specific impulse of the transient pulse in Figure 8.68 is equal to the steady state value of the specific impulse modulated by the ratio of average nozzle operation time to valve operation time. Of course, if the rise and decay times of the momentum flux were instantaneous,

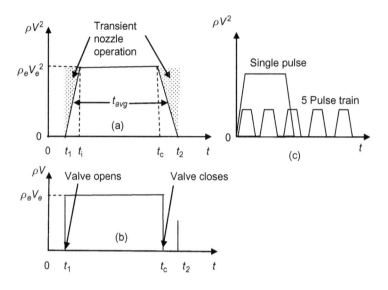

FIGURE 8.68

(a) Typical single pulse of momentum flux showing rise and decay times, (b) associated single pulse of mass flux for instantaneous valve action, and (c) single pulse and train of five identical pulses at lower momentum flux.

then the steady value of specific impulse is recovered. Notice also in Figure 8.68a, the transient nozzle operation has momentum flux increasing after the valve is opened at t_1, but it also has momentum flux decaying after the valve is closed at t_c so that t_{avg} is generally quite close to $t_c - t_1$. Indeed, practice suggests that the ratio $t_{avg}/(t_c - t_1) = 1 \pm 0.2$. Therefore, for preliminary design purposes, it is reasonable to consider ideal pulsed operation with $t_{avg} = t_c - t_1 = t_b$, the so-called burn time and this means that the ideal specific impulse is assumed to be realized.

Extending this ideal operation to a train of n identical pulses of duration t_b as shown in Figure 8.68c, the total impulse delivered is

$$I_n = \sum_{i=1}^{n} F_i t_{avg,i} = n(\pi r_{e,n}^2)\rho_e V_e^2 t_b \tag{8.168}$$

For a single pulse of duration $t_{b,1}$, the impulse delivered is

$$I_1 = (\pi r_{e,1}^2)\rho_e V_e^2 t_{b,1} \tag{8.169}$$

For similar nozzle geometries, the flow properties are, neglecting friction and heat transfer in the nozzle, independent of scale, so the thrust depends primarily on the chamber pressure p_c and is given by

$$\rho_e V_e^2 = \gamma_e p_e M_e^2 = \gamma_e p_c M_e^2 \left[1 + \frac{\gamma_e - 1}{2} M_e^2\right]^{\frac{-\gamma_e}{\gamma_e - 1}} = \gamma_e p_c f\left(\frac{r_e}{r*}\right)$$

Chamber pressures are typically in the range of $0.5\ \text{MPa} < p_c < 3\ \text{MPa}$.

If the impulse delivered is to be the same for single or multiple pulse operation, we may set Eqns (8.168) and (8.169) equal and solve for the ratio of the nozzle diameters:

$$\frac{r_{e,n}}{r_{e,1}} = \sqrt{\frac{t_{b,1}}{nt_{b,n}}} \tag{8.170}$$

By extending the pulse train duration so that $nt_{b,n} \gg t_{b,1}$, we may use a much smaller nozzle in the pulsed system and still deliver the same total impulse although the thrust of each bit is smaller than the thrust of the larger single pulse system. In addition, by delivering impulse in small bits, very fine adjustments to the attitude of the spacecraft may be achieved. Still another advantage of using small RCS thrusters is that their size and weight make it relatively easy to incorporate redundant systems for increased reliability and safety. For example, commercially available monopropellant hydrazine engines in the steady state thrust range of $1\,N < F < 500\,N$ have a mass in kilograms given approximately by $m = (F/100)^{1/3}$. Thus, engine mass is relatively small so other parts like valves, heaters, and propellant tanks make up the bulk of the RCS system mass.

The period of a pulse train is equal to the total time between sequential propellant valve openings, so that it may be defined as follows:

$$P = (t_c - t_1) + t_{off} \approx t_b + t_{off} \tag{8.171}$$

Here, t_b denotes the time during which valves are on and t_{off} is the downtime between closing the valve and reopening it. The duty cycle is then t_b/P and denotes the fraction of time the valves are open. Commercially available RCS thrusters can range in duty cycle from 0.01% to more than 99%.

The mass flow of propellant consumed during the train of n pulses is

$$\dot{m} = \frac{n\tau}{\lambda} \pi r_e^2 \rho_e V_e \tag{8.172}$$

Combining Eqns (8.168) and (8.172) shows that within our assumptions the specific impulse of the pulsed operation is

$$I_{sp,n} = \lambda \frac{n\pi r_e^2 \rho_e V_e^2 \tau}{n\pi r_e^2 \rho_e V_e g_E \tau} = \lambda \frac{V_e}{g_E}$$

Therefore, for ideal pulse operation of the thrust, the specific impulse is equal to the steady state value. But the steady state value depends upon the extent of losses in the nozzle due to heat transfer and friction, both of which consume energy otherwise used to accelerate the gas in the nozzle. For a given gas composition, the velocity at the exit of the nozzle is proportional to $T_c^{1/2}$. In the representative case of hydrazine used as a monopropellant, the exit velocity ratio for a "hot" pulse to a "cold" firing of a single pulse is

$$\frac{V_{e,n}}{V_{e,1}} \sim \sqrt{\frac{T_{c,n}}{T_{c,1}}} = \sqrt{\frac{900\,K}{300\,K}} = 1.732 \tag{8.173}$$

When the number of pulses increases the thruster and the propellant are at nominal "hot" temperatures, while the first pulses may be relatively "cold". According to Eqn (8.173) hydrazine which typically provides $I_{sp,n} = 200$ s in hot (900 K) firing may deliver only $I_{sp,1} = 115$ s for initial pulses with a relatively cold (300 K) thruster system. In the case of bipropellants such as hydrazine–nitrogen tetroxide (N_2H_4-N_2O_4) where the combustion temperature is around 3000 K and $I_{sp} = 260$ s, the cold firing specific impulse would be around $(260 \text{ s})/10^{1/2} = 82$ s.

8.7.3 SPACE CAPSULE CONTROL SYSTEMS

Capsule geometry is such that aerodynamic control surfaces are difficult to implement and when tried they prove to be, in the end, impractical, although many approaches for improving stability or effecting control have been investigated (see, for example, Moseley et al. (1968)). The space capsule instead relies on center of gravity placement to maintain stability during entry in the sensible atmosphere. RCS are relied on altitudes above those for which aerodynamic forces demonstrate sufficient control authority. The Apollo CM RCS is illustrated in Figure 8.69. The Apollo CM, like that of Gemini, was a symmetrical body with an offset center of gravity, and control of the entry flight path is accomplished by rolling the vehicle and thereby changing the location of the center of gravity.

The Apollo CM RCS uses two redundant systems each incorporating six 414 N (93 lb) thrust engines to provide adequate attitude control should one system fail. The liquid propellants are hypergolic: MMH fuel and nitrogen tetroxide (N_2O_4) oxidizer.

8.7.4 SPACEPLANE CONTROL SYSTEMS

Spaceplanes like the X-15 and the Space Shuttle achieve high enough altitudes that aerodynamic forces are insufficient and must be supplemented by thrust forces produced by rocket thrusters. The Space Shuttle's aerodynamic control surfaces and the rocket-powered RCS are illustrated in

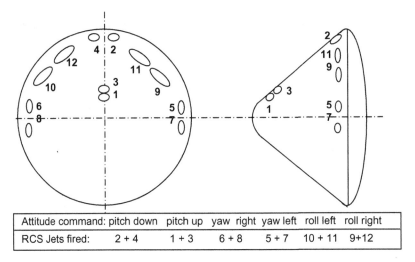

Attitude command:	pitch down	pitch up	yaw right	yaw left	roll left	roll right
RCS Jets fired:	2 + 4	1 + 3	6 + 8	5 + 7	10 + 11	9+12

FIGURE 8.69

The Apollo CM RCS has two redundant thruster systems shown in pairs. Nozzles of the roll thrusters 9–12 are canted to achieve a tangential force, thereby producing a moment about the longitudinal axis.

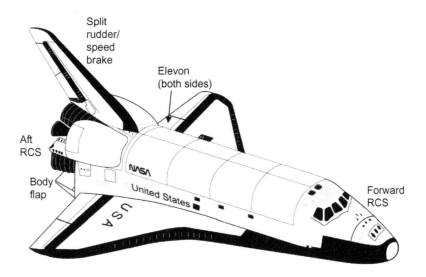

Split
rudder/
speed
brake

Elevon
(both sides)

Aft
RCS

Body
flap

Forward
RCS

NASA

United States

USA

FIGURE 8.70

Diagram of the Space Shuttle showing the aerodynamic control surfaces and the RCS.

Courtesy NASA.

Figure 8.70. The elevons are combined elevators and ailerons, that is, they can be deflected up or down in unison or differentially. When both are deflected in the same sense, they act like an airplane's elevator to provide a pitching moment which moves the nose up or down. When they are deflected differentially, they act like an airplane's ailerons and produce a rolling moment which banks the vehicle. The rudder can be deflected from one side to another as a unit to generate a yawing moment which moves the nose from one side to the other. The rudder is split so that the two halves may be deflected in opposite directions to increase the frontal area of the vertical tail producing an aerodynamic braking effect. The body flap located on the bottom of the aft end of the fuselage may be deflected downward to provide a nose-up pitching moment during high angle of attack flight of entry at the greatest speeds and thermal loads.

The RCS is divided into two parts as shown in Figure 8.70. The forward RCS is located in the nose of the fuselage. The aft RCS is located with the orbital maneuvering system (OMS) in the left and right OMS/RCS pods at the top of the aft end of the fuselage. The more detailed view of the forward RCS given in Figure 8.71 shows 14 primary thrusters:

- 3 pointing in the negative x-direction to generate axial deceleration
- 3 pointing in the positive z-direction to provide pitch-down of the nose
- 2 on the left side of the nose pointing in the negative y-direction to provide a yawing motion to the right
- 2 on the right side of the nose pointing in the positive y-direction to provide a yawing motion to the left

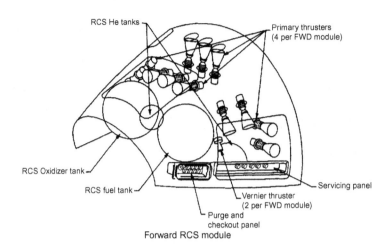

FIGURE 8.71

A schematic diagram of the left side and top of the Space Shuttle nose showing the layout of the forward RCS.

Courtesy NASA.

- 2 on the left side of the nose slanted in a downward (negative z) and outboard (negative y) direction to produce a rolling moment that raises the left wing and the nose
- 2 on the right side of the nose slanted in a downward (negative z) and outboard (positive y) direction to produce a rolling moment that raises the right wing and the nose

Note that the slanted thrusters which are used individually to produce a rolling moment may also be fired simultaneously to pitch the nose up without a rolling moment. The forward RCS also has a small vernier thruster on each side of the nose for fine attitude adjustments when in orbit. Figure 8.72 is a photograph showing the general arrangement of the thruster nozzles of the forward RCS as viewed from the left and above.

The aft RCS has 12 primary and two vernier engines in each of the OMS/RCS pods that are mounted either side of the upper aft fuselage. The more detailed view of the left aft pod RCS given in Figure 8.73 shows the following primary thrusters:

- 2 pointing in the positive x-direction to generate axial acceleration
- 3 pointing in the positive z-direction to provide pitch-up of the nose
- 4 on the left side of the left pod pointing in the negative y-direction to provide a yawing motion to the left.
- 3 on the bottom of the pod pointing in the negative z-direction to provide a pitch-down of the nose

By virtue of the distance of the thrusters from the Shuttle's longitudinal axis, the side and vertical force thrusters can be fired so as to produce a rolling moment as well. The aft RCS also has two small vernier thrusters on each pod for fine attitude adjustments when in orbit. Figure 8.74 is a

FIGURE 8.72

Photograph of the top and left side of the nose of the Space Shuttle showing the layout of the forward RCS.

Courtesy NASA.

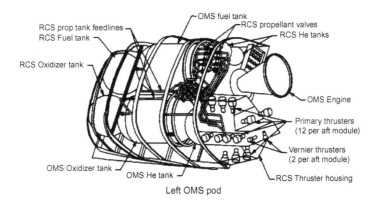

FIGURE 8.73

A schematic diagram of the left side OMS/RCS pod of the Space Shuttle aft fuselage showing the layout of the aft RCS.

Courtesy NASA.

FIGURE 8.74

Photograph of the aft RCS of the Space showing both the left and right OMS pods. The three large nozzles are for the SSME, the two next largest are the OMS nozzles, and the smallest are 5 of the 12 RCS thrusters.

Courtesy NASA.

photograph showing the general arrangement of the thruster nozzles of the forward RCS as viewed from the left and above.

The primary RCS engines provide 3.87 kN (870 lb) of vacuum thrust each, and the vernier RCS engines provide 106 N (24 lb) of vacuum thrust each. The primary engines operate at a nominal chamber pressure of about 1 MPa (152 psia), while the vernier engines operate at 0.76 MPa (110 psia). The propellants are nitrogen tetroxide (N_2O_4) oxidizer and MMH fuel at an oxidizer to fuel ratio O/F = 1.6. Helium tanks pressure feed the liquid oxidizer and fuel to the RCS engines. The propellants are hypergolic and ignite upon contact when injected in the combustion chamber. The hot gas passes through the nozzles of the primary thrusters, which have expansion ratios in the range of $22 < \varepsilon < 30$. The primary thrusters can fire continuously for a period of 1 s up to 150 s or in a pulsed mode with minimum pulse duration of less than 0.1 s, and a multiplicity of engines is used to ensure redundancy. The vernier engines have expansion ratios in the range of $20 < \varepsilon < 50$ and can fire in about the same fashion as the primary engines but there are considerably fewer such engines indicating that redundancy is not a major concern.

8.8 SUMMARY OF CONSTANTS AND CONVERSION FACTORS

Length	1 m = 3.2804 ft
Area	1 m^2 = 10.7639 ft^2
Volume	1 m^3 = 35.3147 ft^3
Speed	1 ft/s = 0.3048 m/s = 0.6815 mph
Force	1 lbf = 4.448 N
Pressure	1 kPa = 10^3 N/m^2 = 0.145 psia

Universal gas constant: $R_u = 8.31451$ kJ/kmol $\cdot$ K Molecular weight of air at standard T and p: $W_{stp} = 28.96$ kg/kg $\cdot$ mol

Atmospheric pressure at standard sea level conditions: $p_{sl} = 101.3$ kPa

Atmospheric temperature at standard sea level conditions: $T_{sl} = 288.15$ K

Atmospheric density at standard sea level conditions: $\rho_{sl} = 1.225$ kg/m^3

Atmospheric viscosity at standard sea level conditions: $\mu_{sl} = 1.7894 \times 10^{-5}$ Pa $\cdot$ s

8.9 NOMENCLATURE

A	aspect ratio or general coefficient
a	sound speed or general coefficient
B	ballistic coefficient $mg_E/C_D S$ or general coefficient
b	wingspan or general coefficient
C	Chapman-Rubesin factor $\rho\mu/\rho_e\mu_e$
C_a	axial force coefficient F_a/qS
C_D	drag coefficient D/qS
$C_{D,b}$	base drag coefficient D_b/qS_b
$C_{D,0}$	zero-lift drag coefficient, $D_0/q\pi R_b^2$
C_f	skin friction coefficient τ_w/q
$C_{f,inc}$	skin friction coefficient in incompressible flow ($M = 0$)
C_h	heat transfer coefficient $\dot{q}_w/\rho_e u_e(h_e - h_w)$
C_L	lift coefficient L/qS
$C_{l,\beta}$	stability derivative in roll
C_m	moment coefficient about the center of gravity $M/q\pi R_b^2 d$
$C_{m,\alpha}$	stability derivative in pitch
C_n	normal force coefficient F_n/qS
$C_{n,\beta}$	stability derivative in sideslip
C^*	compressibility factor $\rho^*\mu^*/\rho_e\mu_e$
C'	compressibility factor $(\rho^*\mu^*/\rho_e\mu_e)^{0.861}(\mu^*/\mu_e)^{-0.722}$
C_p	pressure coefficient
C''	coefficient in Eqn (8.92)
c_p	specific heat at constant pressure
D	drag
d	maximum diameter of the vehicle normal to its longitudinal axis, $d = 2R_b$
E	energy per mol
e	energy per unit mass
F	force
g	gravitational acceleration
H	enthalpy per mol or atmospheric scale length
h	enthalpy per unit mass
h^*	reference enthalpy
I	impulse or moment of inertia
I_{sp}	specific impulse $F/(dm_p/dt)$
K	coefficient, see Eqn (8.54)
k	surface roughness height or isentropic exponent
L	lift or rolling moment about the vehicle center of gravity

l	length
M	Mach number or pitching moment about the vehicle center of gravity
m	mass or exponent equal to 1/2 for laminar flow and 1/3 for turbulent flow
N	yawing moment about the vehicle center of gravity
n	exponent in Eqn (8.129) or number of thrust pulses
$\mathbf{Pr}$	Prandtl number $\mu c_p / \kappa$
p	pressure or rotational velocity in roll
q	dynamic pressure ($\rho u^2 / 2$ or $\gamma M^2 / 2$) or rotational velocity in pitch
$\dot{q}_w$	wall heat transfer rate
$\mathbf{Re}_x$	Reynolds number based on distance x
R	gas constant
R_b	maximum radius of the vehicle normal to its longitudinal axis
R_N	nose radius
R_u	universal gas constant
r	recovery factor Pr^m or rotational velocity in yaw or radial distance
r_g	radius of gyration $(I/m)^2$
S	reference area used in forming force and moment coefficients
S_w	wetted area
T	temperature
t	time or thickness
t_b	thruster burn time
u	velocity in the axial direction
V	free stream velocity
v	velocity in the normal direction
W	molecular weight of gas mixture
W_0	molecular weight of gas at standard conditions
x	axial coordinate measured from the vehicle nose tip
x_0	axial distance to vehicle center of gravity
y	spanwise coordinate measured from longitudinal axis
y_0	spanwise distance to vehicle center of gravity, usually $y_0 = 0$
z	vertical coordinate measured from the vehicle x_0, y_0 plane
z_0	vertical distance to vehicle center of gravity
Z	compressibility W_0 / W
α	angle of attack
β	sideslip angle
γ	ratio of specific heats or flight path angle
ϕ	angle subtended by the longitudinal axis and R_N or bank angle
ϕ_1	maximum angle subtended by the longitudinal axis and R_N, $\phi_1 = \sin^{-1}(R_b/R_N)$
κ	speed of sound parameter $a^2 \rho / p$
μ	viscosity
μ_o	viscosity given by Sutherland's law, Eqn (8.116)
ρ	density
θ	leading edge angle or reduced temperature in Section 8.5.1
θ'	$T/1000$
τ	volume parameter in Eqn (8.60)
τ_w	wall shear stress
ξ	dynamic stability parameter
υ	volume

8.9.1 SUBSCRIPTS

a   axial direction
aw   adiabatic wall conditions
b   conditions at base of vehicle
CG   center of gravity
E   conditions at earth's surface
e   boundary layer edge conditions
L   conditions pertaining to moments about the *x*-axis
M   conditions pertaining to moments about the *y*-axis
max   maximum
N   conditions pertaining to moments about the *z*-axis
n   direction normal to axis
p   propellant
R   resultant
s   spacecraft
sl   standard sea level conditions
w   wall conditions

8.9.2 SUPERSCRIPTS

( )*   reference temperature conditions

REFERENCES

Allen, H. J. (1957). *Motion of a ballistic missile angularly misaligned with the flight path upon entering the atmosphere and its effect upon aerodynamic heating, aerodynamic loads, and miss distance.* NACA TN 4048.

Becker, J. V. (1964). *"Studies of high L/D ratio hypersonic configurations", Fourth International Congress of the Aeronautical Sciences.* London: Spartan and MacMillan Company.

Brown, C. D. (1996). *Spacecraft propulsion.* Washington, DC: American Institute of Aeronautics and Astronautics.

Cebeci, T., & Cousteix, J. (2005). *Modeling and computation of boundary layer flows.* Long Beach, CA/NY: Horizon Publishing/Springer.

Crawford, D. H., & McCauley, W. D. (1957). *Investigation of the laminar aerodynamic heat-transfer characteristics of a hemisphere-cylinder in the langley 11-inch hypersonic tunnel at a mach number of 6.8.* NACA Report 1323.

DeRose, C. B. (1969). *Trim attitude, lift and drag of the apollo command module with offset center of gravity at mach numbers to 29.* NASA TN D-5276.

Eckert, E. R. G. (1956). Engineering relations for heat transfer and friction in high-velocity laminar and turbulent flows over surfaces with constant pressure and temperature. *Trans ASME*, 78(6), 1273−1283.

Etkin, B., & Reid, L. D. (1995). *Dynamics of flight: Stability and control.* New York: Wiley.

Hansen, C. F. (1959). *Approximations for the thermodynamic and transport properties of high-temperature air.* NASA TR R-50.

Hansen, C. F., & Heims, S. P. (1958). *A review of the thermodynamic, transport, and chemical reaction rate properties of high temperature air.* NACA TN 4359.

Hayes, W. D., & Probstein, R. F. (2004). *Hypersonic inviscid flow.* New York: Dover.

Hopkins, E. J. (1972). *Charts for predicting turbulent skin friction from the Van Driest method II.* NASA TN D-6945.

Hopkins, E. J., & Inouye, M. (1971). An evaluation of theories for predicting turbulent skin friction and heat transfer on flat plates at supersonic and hypersonic mach numbers. *AIAA J*, 9(6), 993−1003.

Kaattari, G. (1962). *Predicted gas properties in the shock layer ahead of capsule-type vehicles at angles of attack.* NASA TN D-1423.

Kays, W., Crawford, M., & Weigand, B. (2005). *Convective heat and mass transfer* (4th ed.). New York: McGraw-Hill.

Kazemba, C. D., Braun, R. D., Clark, I. G., & Schoenberger, M. (2012). *Survey of blunt body dynamic stability in supersonic flow.* AIAA Paper 2012-4509.

Kliche, D., Mundt, Ch., & Hirschel, E. H. (2011). The hypersonic Mach number independence principle in the case of viscous flow. *Shock Waves, 21,* 307−314.

Lees, L. (1956). Laminar Heat Transfer over Blunt-Nosed Bodies at Hypersonic Speeds. *Jet Propulsion, 26,* 259−269 and 274.

Lin, T. C. (2008). Influence of laminar boundary-layer transition on entry vehicle designs. *J Spacecraft Rockets, 45*(2), 165−175.

Meador, W. E., & Smart, M. K. (2005). Reference enthalpy method developed from solutions of the boundary-layer equations. *AIAA J, 43*(1), 135−139.

Miller & Lawing (1966). *Experimental investigation of flow characteristics of the apollo reentry configuration at a mach number of 20 in nitrogen.* NASA TM X-1258.

Moseley, W. C., Graham, R. E., & Hughes, J. E. (1968). *Aerodynamic stability characteristics of the apollo command module.* NASA TN D-4688.

Oswatitsch, K. (1950). Ahnlichkeitsgesetze fur Hyperschallstromung. *Zeitschrift fur Angewandte Mathematik und Physik (ZAMP), 2*(4), 249−264.

Park, C. (1990). *Nonequilibrium hypersonic aerothermodynamics.* New York: Wiley.

Rasmussen, M. (1994). *Hypersonic flow.* New York: John Wiley & Sons, 1994.

Reshotko, E. (2008). Transition issues for atmospheric entry. *J Spacecraft Rockets, 45*(2), 161−164.

Ried, R. C., & Mayo, E. E. (1963). *Equations for the newtonian static and dynamic aerodynamic coefficients for a body of revolution with an offset center of gravity location.* NASA TN D-1085.

Schetz, J. A. (1992). *) Boundary layer analysis.* Upper Saddle River, New Jersey: Prentice-Hall.

Schlichting, H., & Gersten, K. (2000). *Boundary layer theory* .New York: Springer.

Sforza, P. M. (2012). *Theory of aerospace propulsion.* New York: Elsevier.

Sforza, P. M. (2014). *Commercial airplane design principles.* New York: Elsevier.

Sommer, S., Short, B. J., & Compton, D. L. (1960). *Free-flight measurements of static and dynamic stability of models of the mercury reentry capsule at mach numbers 3 and 9.5.* NASA TM X-373.

Srinivasan, S., Tannehill, J. C., & Weilmunster, K. J. (1987). *Simplified curve fits for the thermodynamic properties of equailibrium air.* NASA Reference Publication 1181.

Stengel, R. (2004). *) Flight dynamics.* Princeton, New Jersey: Princeton University Press.

Suit, W. T. (1989). *Summary of longitudinal stability and control parameters as determined from space shuttle challenger flight test data.* NASA Technical Memorandum 101605.

Suit, W. T., & Schiess, J. R. (1988). *Lateral and longitudinal stability and control parameters for the space shuttle discovery as determined from flight test data.* NASA Technical Memorandum 100555.

Surber, T. E., & Olsen, D. C. (1978). Space Shuttle Aerodynamic Development. *J Spacecr Rockets, 15*(1), 40−47.

Tobak, M., & Allen, H. J. (1958) *Dynamic stability of vehicles traversing ascending or descending paths through the atmosphere.* NACA TN 4275.

Van Driest, E. R. (1956). The problem of aerodynamic heating. *Journal of the Aerospace Sciences, 23,* 1007−1011.

White, F. M. (2006). *Viscous fluid flow* (3rd ed.). New York: McGraw-Hill.

Wilhite, A.W., Arrington, J.P., and McCandless, R.S. (1984) Performance aerodynamics of aero-assisted orbital transfer vehicles, In: Thermal design of aero-assisted orbital transfer vehicles, Progress in astronautics and aeronautics, H.C. Nelson (Editor), vol. 96, American Institute of Aeronautics and Astronautics, New York, pp. 165−185.

Woods, W. C., Arrington, J. P., & Hamilton, H. H. (1983). *A review of preflight estimates of real-gas effects on space shuttle aerodynamic characteristics.* NASA CP-2283.

THERMAL PROTECTION SYSTEMS

9.1 BASIC STAGNATION POINT HEAT TRANSFER CORRELATIONS

The general nature of the heating in the vicinity of a blunt nose in hypersonic flight is shown in Figure 9.1.

Convective heat transfer rate $\dot{q}$ (W/cm^2) has been correlated in Tauber, Menees, and Adelman (1987) by the following relation:

$$\dot{q}_c = C\rho^N V^M \qquad (9.1)$$

This equation, in SI units, applies to the flight regime where boundary layer analysis is valid and provides a good approximation for both laminar and turbulent flows. For stagnation point flow, the coefficients in Eqn (9.1) are given by

$$C = 1.83 \times 10^{-8} \frac{1 - g_w}{\sqrt{R_N}}$$

$$M = 3 \qquad (9.2)$$

$$N = \frac{1}{2}$$

The quantity R_N is the nose radius (in meters), and $g_w = h_w/h_{t,e}$ is the ratio of the wall enthalpy to the stagnation enthalpy at the edge of the boundary layer. Note that for an adiabatic shock wave ahead of the stagnation point, the total enthalpy is conserved so that g_w may be written as

$$g_w = \frac{h_w}{h_\infty} \frac{1}{1 + \frac{\gamma_{eff} - 1}{2} M_\infty^2} \qquad (9.3)$$

For high Mach number flight, it is common to consider $g_w \ll 1$, the so-called cold wall approximation, which is always a conservative estimate. As the amount of time spent at hypersonic flight speeds increases, the wall temperature will rise thus decreasing the heating rate. The form of the stagnation point heat transfer rate as given by Eqns (9.1) and (9.2) is a generalization of the so-called Sutton and Graves (1971) convective heating relation for which the coefficient C in Eqn (9.2) is slightly different; Dec and Braun (2006) present it as shown below where the subscript $s - g$ denotes the Sutton and Graves (1971) value:

$$C_{s-g} = 1.74153 \times 10^{-8} \frac{1}{\sqrt{R_N}} \qquad (9.4)$$

Manned Spacecraft Design Principles. DOI: http://dx.doi.org/10.1016/B978-0-12-804425-4.00009-X

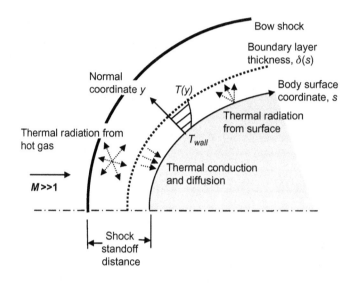

FIGURE 9.1

Heating effects in hypersonic flow over a blunt body.

Table 9.1 Coefficients for Eqn (9.5)		
Function	**V < 7.62 km/s**	**7.62 km/s < V < 9 km/s**
k_1	372.6	25.34
k_2	8.5	12.5
k_3	1.6	1.78

It is pointed out by Dec and Braun (2006) that the coefficient shown in Eqn (9.4) applies to hypersonic flight in the Earth's atmosphere and that for flight in Mars' atmosphere, the numerical coefficient changes to 1.9027×10^{-8}. The numerical difference in the coefficients vanishes if it is assumed that $g_w = 0.05$ in Eqn (9.2), that is, assuming the cold wall approximation holds. Bodies entering the atmosphere from lunar or interplanetary orbits where entry speeds are around 10 km/s or more will experience heating due to thermal radiation from the shock-compressed atmosphere in addition to the convective heating described previously. The heat transfer due to thermal radiation is substantially more difficult to predict, but Tauber and Sutton (1991) provide a method useful for design purposes. The stagnation point radiative heat transfer, as extended by Johnson, Starkey, and Lewis (2007) is given as follows:

$$\dot{q}_r = R_N k_1 (3.28084 \times 10^{-4} V)^{k_2} \left(\frac{\rho}{\rho_{sl}} \right)^{k_3} \tag{9.5}$$

The coefficients k_1, k_2, and k_3 depend upon the velocity regime considered and are listed in Table 9.1. For speeds higher than 9,000 m/s, the empirical relations are more complex and are not included here. Those higher speeds correspond to atmospheric entry from planetary missions, while the emphasis here is on return from the Earth orbit.

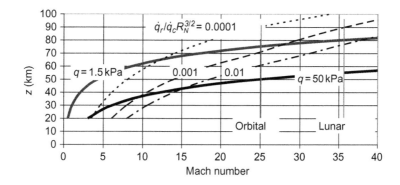

FIGURE 9.2

Contours of constant ratios of radiative to convective heat transfer in relation to the typical reentry vehicle flight corridor, $1.5 \text{ kPa} < \dot{q} < 50 \text{ kPa}$. Beyond $M = 30$, the contours are notional.

The relative importance of radiative to convective heat transfer at the stagnation point may be cast in the form of the ratio $\dot{q}_r / \dot{q}_c R_N^{3/2}$, which illustrates the importance of the nose radius R_N. This ratio is shown in Figure 9.2; beyond $M = 30$, the curves are notional since the more accurate approach for $V > 9$ km/s was not applied.

In Figure 9.2, the contours of $\dot{q}_r / \dot{q}_c R_N^{3/2}$ are plotted as an overlay to the typical reentry vehicle flight corridor to illustrate where thermal radiation effects become important. It is clear that for reentry vehicles of reasonable size, thermal radiation will be of little consequence for entry at orbital speeds. However, for reentry from lunar or planetary missions, the thermal radiation will be comparable to convective heating and must be considered. For such cases, one may turn to the methods in Tauber and Sutton (1991) and Johnson et al. (2007).

9.2 APPROXIMATE AIR CHEMISTRY

As described in the thermodynamics section of the previous chapter, the adiabatic compression of the bow shock increases the temperature and pressure to levels where chemical reactions and ionization of the air must be considered in evaluating heat transfer to the spacecraft. The chemical behavior of air under atmospheric entry conditions can be very complex and for a detailed treatment, see, for example, Park (1989) and Anderson (1989). For preliminary design purposes, we may use the approximations for the equilibrium species concentrations for high-temperature air presented by Hansen (1959). These approximations are based on the results of detailed calculations, which show that for all pressures of interest, (i) dissociation of oxygen molecules is essentially complete before dissociation of nitrogen molecules begins and (ii) that ionization of nitrogen and oxygen atoms occurs at about the same temperature with about the same energy changes. Observation (i) suggests that the dissociation reactions are approximately independent and given by

$$O_2 \rightarrow 2O \tag{9.6}$$

$$N_2 \rightarrow 2N \tag{9.7}$$

The ionization reactions are

$$O \rightarrow O^+ + e^- \tag{9.8}$$

$$N \rightarrow N^+ + e^- \tag{9.9}$$

Observation (ii) permits these reactions to be considered as a single reaction in which all atoms, N and O, are considered as a single species with appropriately weighted average properties. Under these assumptions, Hansen (1959) shows the compressibility to be given by

$$Z = 1 + \varepsilon_O + \varepsilon_N + \varepsilon_A \tag{9.10}$$

Here, the quantities ε_O, ε_N, and ε_A represent the fractions of dissociated oxygen molecules, dissociated nitrogen molecules, and ionized atoms, respectively. Carrying out an equilibrium chemical analysis as described, for example, by Sforza (2012), the following mole fractions are found:

$$X(O_2) = \frac{0.2 - \varepsilon_O}{Z} \tag{9.11}$$

$$X(N_2) = \frac{0.8 - \varepsilon_N}{Z} \tag{9.12}$$

$$X(O) = \frac{2\varepsilon_O - 0.4\varepsilon_A}{Z} \tag{9.13}$$

$$X(N) = \frac{2\varepsilon_N - 1.6\varepsilon_A}{Z} \tag{9.14}$$

$$X(e^-) = \frac{2\varepsilon_A}{Z} \tag{9.15}$$

It should be noted here that Hansen (1959) chose to set the standard temperature and pressure composition of air as $X(N_2)/X(O_2) = 4$ rather than the actual value of 3.76 in light of the overall level of approximation being employed. He also has neglected NO because its effects on the thermodynamic properties were found to be small, and there are few cases of interest for which the concentration of NO rises above 1%. Hansen's analytical results, which involve p measured in atmospheres and the equilibrium constants $K_{p,O}$, $K_{p,N}$, and $K_{p,A}$ (which are independent of p) listed in Table 9.2 are as follows:

$$\varepsilon_O = \frac{-0.8 + \sqrt{0.64 + 0.8\left(1 + \frac{4p}{K_{p,O}}\right)}}{2\left(1 + \frac{4p}{K_{p,O}}\right)} \tag{9.16}$$

$$\varepsilon_N = \frac{-0.4 + \sqrt{0.16 + 3.84\left(1 + \frac{4p}{K_{p,N}}\right)}}{2\left(1 + \frac{4p}{K_{p,N}}\right)} \tag{9.17}$$

$$\varepsilon_A = \frac{1}{\sqrt{1 + \frac{p}{K_{p,A}}}} \tag{9.18}$$

Knowing p and T at any point in the flow, we may find ε_O, ε_N, and ε_A from Eqns (9.16)–(9.18). Using these values in Eqn (9.10), we may determine Z. Substituting into Eqns (9.11)–(9.15)

Table 9.2 Equilibrium Constants for Oxygen and Nitrogen Dissociation and Ionization

T (K)	$\ln K_{p,O}$	$\ln K_{p,N}$	$\ln K_{p,A}$
500	− 104.89	− 213.29	− 331.58
1000	− 44.77	− 99.08	− 163.16
1500	− 24.57	− 60.81	− 106.59
2000	− 14.41	− 41.61	− 78.09
2500	− 8.29	− 30.005	− 60.87
3000	− 4.19	− 22.32	− 49.3
3500	− 1.26	− 16.79	− 40.98
4000	0.95	− 12.62	− 34.69
4500	2.67	− 9.37	− 29.77
5000	4.05	− 6.76	− 25.8
5500	5.18	− 4.61	− 22.53
6000	6.13	− 2.81	− 19.8
6500	6.69	− 1.28	− 17.46
7000	7.62	0.05	− 15.45
7500	8.23	1.2	− 13.7
8000	8.75	2.23	− 12.16
9000	9.64	3.96	− 9.57
10,000	10.36	5.37	− 7.48
11,000	10.96	6.56	− 5.75
12,000	11.46	7.56	− 4.29
13,000	11.89	8.44	− 3.05
14,000	12.27	9.29	− 1.97
15,000	12.6	9.87	− 1.02

yields the mole fractions of all the species considered. The natural logarithm of the equilibrium function for dissociation and ionization presented by Hansen (1959) is plotted in Figure 9.3 for temperatures up to 15,000 K. Note that the zero-crossings (temperatures for which K_p is on the order of unity) for the three equilibrium constants shown are widely separated in temperature and correspond to the separation between the dissociation and ionization temperatures. For example, when $K_{p,O} \ll 1$, then $\varepsilon_O \sim 0$ and only molecular oxygen (O_2) is present. However, when $K_{p,O} \gg 1$, then $\varepsilon_O \sim 0.2$ and no molecular oxygen is present, that is, $X(O_2) = 0$.

In addition to the effect of chemical reactions on the thermodynamic properties of air as described in the previous chapter, it is also important in the analysis of heat transfer to the spacecraft, a topic which is covered in this chapter. The brief analysis presented here is based entirely on equilibrium or frozen chemistry in which chemical reactions take place infinitely rapidly or not at all, respectively. In a hypersonic flow field, the temperature experienced by a fluid particle may vary considerably as it passes over the spacecraft. The characteristic residence time of a fluid particle is $\tau = l/u$, which, for $l = 10$ m and $u = 5000$ m/s, is only 2 ms during which time it may pass through different pressure and temperature regions. Reaction rates are not infinitely fast so that for

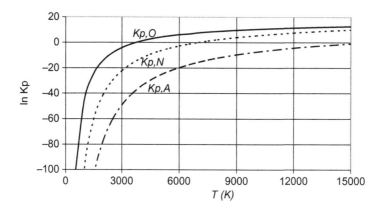

FIGURE 9.3

Equilibrium constants for dissociation and ionization for oxygen and nitrogen.

detailed investigations, it is often necessary to deal with finite-rate chemistry, which can be very complex as described by, for example, Park (1989).

9.3 STAGNATION POINT HEAT TRANSFER

The boundary layer equations in hypersonic flows are discussed in some detail by Anderson (1989). He points out that Fay and Riddell (1958) first carried out a rigorous study of stagnation point convective heat transfer $\dot{q}_w$ at hypersonic speeds and provided results in the following correlated form:

$$\dot{q}_w = 0.57 \left(\frac{4}{3}\right)^j \mathrm{Pr}^{-0.6} (\rho_e \mu_e)^{0.4} (\rho_w \mu_w)^{0.1} \sqrt{\left(\frac{du_e}{dx}\right)_t} (h_{t,e} - h_w) \left[1 + (Le^k - 1)\left(\frac{h_D}{h_{t,e}}\right)\right] \tag{9.19}$$

The index j refers to geometry with $j = 0$ for two-dimensional flow and $j = 1$ for axisymmetric flow, while the index k refers to chemistry, with $k = 0.52$ for equilibrium chemistry and $k = 0.63$ for frozen chemistry. A schematic diagram of the flow field is shown in Figure 9.4. Here, the subscripts w, e, and t denote conditions at the wall, the external flow, and the stagnation point, respectively. For simplicity, we will suppress the subscript c, which denotes convective heat transfer because that mode of heat transfer will be the primary concern. We will continue to use the subscript r in those instances where we are dealing radiative heat transfer.

The term in square brackets of Eqn (9.19) represents the effect of equilibrium chemical reactions occurring in the stagnation region and

$$Pr = \frac{\mu c_p}{k}$$

$$Le = \frac{\rho D_{12} c_p}{k} \tag{9.20}$$

$$h_D = \sum_{i=1}^{n} (Y_{i,e} - Y_{i,w}) \Delta h_{f,i}.$$

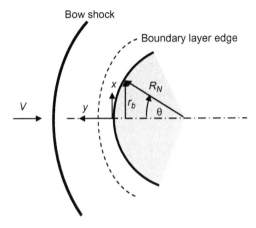

FIGURE 9.4

Schematic diagram of stagnation region and coordinate system (not to scale). The x coordinate is along the body while the y-coordinate is normal to it.

The Prandtl number Pr and the Lewis number Le are similarity parameters like the Reynolds number and measure the relative importance of friction to heat conduction and of species diffusion (mixing) to conduction, respectively. The gas considered is air which, for the purposes of mixing, is considered to be a binary mixture, that is, a mixture made up of two species: atoms (O or N) and molecules (O_2 and N_2) as discussed in detail in the previous section. The quantity D_{12} is the binary diffusion coefficient that measures the ability of species 1 to mix with species 2. The quantities Y_i and $\Delta h_{f,i}$ are the mass fractions of the individual species present (O, O_2, N, and N_2) and the chemical heat of formation of each of the species, respectively. The Lewis number for air-like mixtures is close to unity, $Le \sim 1.4Z^{-1.2}$, and for the temperatures and pressures typical of manned spacecraft trajectories, the quantity $(Le^k - 1) \ll 1$, and therefore the contribution of the chemical reaction term is often neglected in preliminary studies. The effect of chemical reactions, particularly with regard to surface catalysis, is treated later in this chapter.

The velocity gradient along the x-axis, that is, along the body surface, at the stagnation point may be found to be given by

$$\left(\frac{du_e}{dx}\right)_t = \frac{1}{R_N} \sqrt{\frac{2(p_e - p_\infty)}{\rho_e}} \tag{9.21}$$

In the Newtonian approximation, this gradient, for both cylinders and spheres, is given by

$$\left(\frac{du_e}{dx}\right)_t \approx \sqrt{2\varepsilon} \frac{V_\infty}{R_N} \tag{9.22}$$

Then, the convective heat transfer to the wall at the stagnation point given in Eqn (9.19) reduces to

$$\dot{q}_w = \frac{0.678}{\varepsilon^{1/4}} \left(\frac{4}{3}\right)^j \left(\frac{C_w}{Pr}\right)^{0.1} \sqrt{\frac{\rho_\infty V_\infty \mu_e}{R_N \ Pr}} (h_{t,e} - h_w) \tag{9.23}$$

In this equation, the variable $C_w = \frac{\rho_w \mu_w}{\rho_e \mu_e}$ is called the $\rho\mu$-ratio or the Chapman-Rubesin factor, and the Prandtl number is calculated based on conditions at the edge of the boundary layer. The flight density and altitude, and therefore the stagnation enthalpy behind the shock, as well as the body radius are all generally specified, and the density ratio across the shock can be determined from the Rankine–Hugoniot conditions, as discussed in Appendix A, or appropriate shock tables for equilibrium air chemistry. Under the hypersonic flow approximation, the stagnation conditions in the inviscid flow behind the normal shock are as follows:

$$p_{t,e} \approx p_2 \approx \rho_1 u_1^2 = \gamma_\infty p_\infty M_\infty^2$$
$$h_{t,e} \approx h_2 \approx \frac{1}{2} u_1^2 = \frac{1}{2} \frac{\gamma_\infty p_\infty}{\rho_\infty} M_\infty^2 \tag{9.24}$$

The major unknowns here are the thermodynamic properties of the gas in the stagnation region and the wall enthalpy. The former may be found from tabulations developed for this purpose, as done by Hansen (1959) and discussed in Chapter 8. A more convenient method that makes the calculation of the necessary thermodynamic properties much easier was formulated by Yoshikawa (1969) who gives the heat flux at the stagnation point in the following form:

$$\dot{q}_w = \frac{(1.34)^j}{\varepsilon^{1/4}} \sqrt{\frac{\rho_\infty V_\infty}{R_N} \left(\frac{k}{c_p}\right)_{avg}} (h_{t,e} - h_w) \tag{9.25}$$

Note that $j = 0$ for the two-dimensional stagnation point and $j = 1$ for the axisymmetric stagnation point. The quantity $(k/c_p)_{avg}$ is an integrated value (over the range zero to $T_{t,e}$), and Yoshikawa's results can be used to show that it can be fairly well approximated by the linear form

$$\left(\frac{k}{c_p}\right)_{avg} = \left[0.43 + 0.26\left(\frac{T_{t,e}}{1000}\right)\right]\left(\frac{k}{c_p}\right)_{ref} \tag{9.26}$$

The reference value $(k/c_p)_{ref} = 5.5 \times 10^{-5}$ kg/m·s. The stagnation temperature must be determined from the stagnation enthalpy, which is given in Eqn (9.24), because the specific heat at constant pressure varies substantially with temperature as discussed in detail in the previous chapter. In Figure 9.5, the results obtained using Eqn (9.25) are seen to agree favorably with those of Eqn (9.19), which are generally considered the standard for convective heat flux in the stagnation region at hypersonic speeds. Note that we have assumed in using Eqn (9.25) that the contribution of the chemical reactions is negligible. Although this is appropriate at lower hypersonic Mach numbers, that is, lower stagnation temperatures, and for preliminary design purposes, it is preferable to use the complete Eqn (9.19) for more accurate studies.

Now the only problematic variable is the wall enthalpy or, equally, the wall temperature. After a long period in space, the orbiting vehicle's surface is cold. Hughes (1990) reports that data from NASA's Long Duration Exposure Facility indicates that the surface temperature will be about that in the stratosphere, and perhaps higher, if the vehicle is slowly rotated to expose all surface equally to the sun and the shadow of space.

As entry progresses, the heat flux will increase the temperature of the surface, in other words, $h_w = h_w(t)$, and therefore the heat flux will decrease. Thus, this is a conjugate heat transfer problem where the heat transfer in the gas and the body are coupled. The heat transfer is directly proportional to the enthalpy difference $h_{se} - h_w$, but the thermodynamic properties are also dependent on

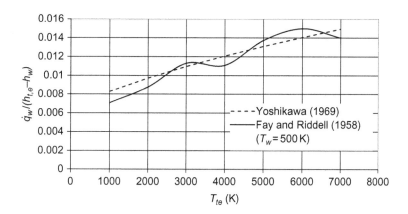

FIGURE 9.5

Comparison between Fay and Riddell (1958) results for stagnation point heat transfer and those of Yoshikawa (1969) for the cold wall case.

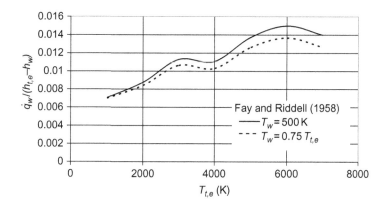

FIGURE 9.6

Comparison of the effects of wall temperature on the ratio of heat transfer rate to enthalpy difference according to Fay and Riddell's method.

the wall enthalpy h_w, which increases through the reentry. An assessment of the effect of wall enthalpy on the ratio of heat transfer rate to the enthalpy difference is illustrated in Figure 9.6, where a modest effect is noted, but it should be accounted for wherever possible.

The normalized wall heat flux $\dot{q}_w/(h_{t,e} - h_w)$ for hot and cold walls are virtually the same, as shown by Fay and Riddell's (1958) results in Figure 9.6. Yoshikawa (1969) recognized this and the averaged thermodynamic properties he uses gives results close to that of Fay and Riddell (1958) and are more convenient to use. Yoshikawa (1969) also uses this result to formulate the problem when coolant is injected through the wall, and in that case, the hot wall result must be corrected.

9.4 HEAT TRANSFER AROUND A HEMISPHERICAL NOSE

Lees (1956) presented a derivation for the variation of the heat transfer around a hemispherical nose under hypersonic flight conditions. White (2006) presents it as the ratio of the convective heat flux at any point on the nose to that at the stagnation point $\dot{q}_w(0)$ as follows:

$$\frac{\dot{q}_w(x)}{\dot{q}_w(0)} = \frac{F(x)}{\sqrt{2^{j+1}\dfrac{du_e}{dx}}}$$

$$F(x) = \frac{\rho_e \mu_e}{\rho_{t,e}\mu_{t,e}} u_e r_b^j \left[\int_0^x \frac{\rho_e \mu_e}{\rho_{t,e}\mu_{t,e}} u_e r_b^{2j} dx\right]^{-1/2}$$

(9.27)

Lees (1956) showed that the velocity gradient along the body and external to the boundary layer over a hemispherical nose in hypersonic flow varies remarkably linearly with polar angle so that

$$u_e \approx \theta \left(\frac{du_e}{d\theta}\right)_{\theta=0}$$

(9.28)

From Figure 9.4, $x = \theta r_b$ and therefore the velocity gradient external to the boundary layer may be readily determined. An expression for the heat transfer distribution is given by Lees (1956) in the form

$$\frac{\dot{q}_w(x)}{\dot{q}_w(0)} = f(\gamma_\infty M_\infty^2, \theta)$$

(9.29)

The expression for f is lengthy and is not reproduced here. Although the result given is said to be accurate down to flight Mach numbers of 2, at Mach numbers of 5 and above there is very little dependence on Mach number. An approximation for f at high Mach number may be deduced from the full expression and leads to the following relatively simple result for the heat transfer distribution as a function of the polar angle θ:

$$\frac{\dot{q}_w(\theta)}{\dot{q}_w(0)} = \frac{2\theta \sin\theta \cos^2\theta}{\sqrt{\theta^2 - \frac{1}{2}\theta \sin 4\theta + \frac{1}{8}(1 - \cos 4\theta)}}$$

(9.30)

9.5 HEAT TRANSFER AROUND A SPHERICALLY CAPPED CONE

The sharp cone at zero angle of attack is a constant pressure surface at Mach numbers where the shock is attached. As the Mach number increases, very high heat transfer rates are experienced at the sharp tip and it is difficult to maintain the sharpness due to melting or structural failure. To alleviate this effect, conical bodies for high-speed flight applications are generally capped with a spherical segment. It has already been shown that the heat transfer rate at a blunt stagnation point $\dot{q}_w(0) \sim R_N^{-1/2}$, so that the deleterious effects of heating at the nose tip will be alleviated. Of course, the flow over a spherically blunted cone will not recover to a constant pressure flow until stations fairly far from the nose are reached, and this modifies the heat transfer distribution from that of the

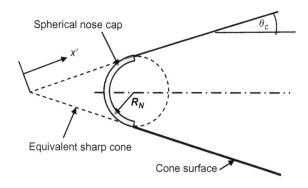

Spherical nose cap

x'

θ_c

R_N

Equivalent sharp cone

Cone surface

FIGURE 9.7

Schematic diagram of spherically capped cone showing an additional coordinate measuring distance along the equivalent sharp-nosed cone.

equivalent sharp cone. Lees (1956) has also considered this problem, a schematic of which is given in Figure 9.7. The distance x' measured along the surface of the equivalent sharp-nosed cone is related to the distance along the surface of the spherically capped cone by the relation

$$\frac{x'}{R_b} = \cot \theta_c + \left[\frac{x}{R_N} - \left(\frac{\pi}{2} - \theta_c\right)\right] \tag{9.31}$$

The ratio of heat transfer at any point on the surface to the stagnation value is given by

$$\frac{\dot{q}_w(x')}{\dot{q}_w(0)} = A(\theta_c)\frac{x'}{R_N}\left[B(\theta_c) + \left(\frac{x'}{R_N}\right)^3\right]^{-1/2}$$

This is valid for $x'/R_b \geq \cot \theta_c$, and detailed expressions are given by Lees (1956) for the functions $A(\theta)$ and $B(\theta)$. Simplified versions for high flight Mach number are

$$A(\theta_c) \approx \frac{\sqrt{3}}{2}\sin \theta_c \sqrt{\frac{\pi}{2} - \theta_c}$$

$$B(\theta_c) \approx \frac{3}{16}\frac{1}{\sin^4 \theta_c}\left[\frac{D(\theta)}{\theta}\right]_{\theta=\frac{\pi}{2}-\theta_c} - \cot^3 \theta_c \tag{9.32}$$

$$D(\theta) \approx \theta^2 - \frac{1}{2}\theta \sin 4\theta + \frac{1}{8}(1 - \cos 4\theta)$$

9.6 HEAT SHIELDS FOR REENTRY VEHICLES

The stagnation point convective heat transfer rate may be characterized as follows:

$$\dot{q}_w = \dot{q}_w(\rho_\infty, V_\infty, R_N, \varepsilon) \tag{9.33}$$

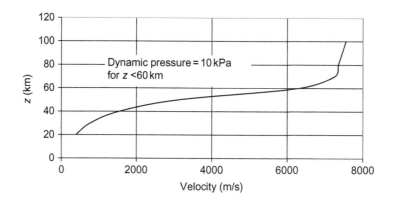

FIGURE 9.8

Variation of velocity with altitude for a typical manned reentry.

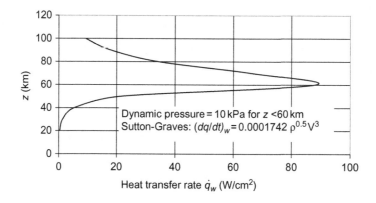

FIGURE 9.9

Variation of the stagnation point heat transfer as a function of altitude as given by Eqn (9.34) for the trajectory as shown in Figure 9.8. Note that the result has been presented in W/cm^2 rather than W/m^2.

Simplified correlations for the cold-wall convective heat transfer rate during reentry, such as the Sutton and Graves (1971) relation, are of the form

$$\dot{q}_w = K\sqrt{\frac{\rho_\infty}{R_N}}V_\infty^3 \tag{9.34}$$

The quantity K is a constant that depends upon the units used and the atmosphere being traversed. For Earth's atmosphere and SI units, $K = C_{s\text{-}g} = 1.742 \times 10^{-8}$. The velocity as a function of altitude for a reentry trajectory similar to that of the Space Shuttle Orbiter is shown in Figure 9.8, and the associated stagnation point heat transfer rate as given by Eqn (9.34) for that trajectory is shown in Figure 9.9. These results give an idea of the shape and magnitude of the stagnation point heat transfer for typical reentries from LEO. Flight test data for heat transfer rate along the

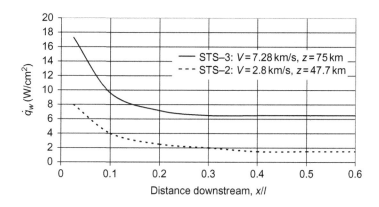

FIGURE 9.10

Flight test data for heat transfer rate along the windward centerline of the Space Shuttle for the indicated flight conditions and flight number.

windward centerline of the Space Shuttle at specific points on the trajectory of flights STS-2 and STS-3 are shown in Figure 9.10.

The reentry trajectory provides the instantaneous information on speed V and altitude z and also the density, when the atmosphere is specified. The trajectory also provides this information as a function of time and space because

$$\frac{dz}{dt} = V \sin \gamma \qquad (9.35)$$

Therefore, one knows the applied convective heat transfer at the stagnation point as a function of time, or $\dot{q}_w = \dot{q}_w(t)$. For entry from lunar or interplanetary flights, radiative heat transfer will also become important, as mentioned at the outset of this chapter. The conduction of heat into the heat shield is described by the heat conduction equation as given in Eqn (9.36). However, we will restrict our attention to the one-dimensional case of heat conduction where heat is conducted into the heat shield material in the direction normal to its surface.

$$\rho c \frac{\partial T}{\partial t} - \frac{\partial}{\partial} \left(k \frac{\partial T}{\partial x} \right) - \frac{\partial}{\partial y} \left(k \frac{\partial T}{\partial y} \right) - \frac{\partial}{\partial z} \left(k \frac{\partial T}{\partial z} \right) = 0 \qquad (9.36)$$

The material properties shown are the heat shield material density, ρ, the heat shield specific heat, c, and the heat shield material thermal conductivity, k. In Figure 9.11, the y-direction is taken as positive as shown, into the heat shield, and the material surface is the line $y = 0$.

9.6.1 HEAT SINK HEAT SHIELDS WITH CONSTANT HEAT TRANSFER

The simplest case is that for which the material properties may be considered constant so that the one-dimensional heat conduction equation becomes

$$\frac{\partial T}{\partial t} - \alpha \frac{\partial^2 T}{\partial y^2} = 0 \qquad (9.37)$$

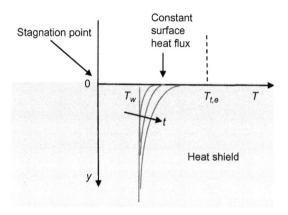

FIGURE 9.11

Schematic diagram of semi-infinite heat shield subjected to constant surface heat flux.

The quantity $\alpha = k/\rho c$ is the thermal diffusivity of the heat shield material. Again in the interests of simplicity, assume that the heat shield is a semi-infinite solid that is at a uniform temperature $T_{w,i}$ at the start of atmospheric entry as shown in Figure 9.11. Continuing with this simplified case and further considering that the convective heat transfer rate at the surface of the heat shield $\dot{q}_w$ is constant and that the heat shield material will neither change phase nor reradiate during the heating, we may determine the temperature distribution through the heat shield as a function of time. The situation as described implies the following boundary conditions:

$$T(y, 0) = T_{w,i}$$
$$\left(\frac{\partial T}{\partial y}\right)_{y=0} = -\frac{\dot{q}_w}{k} = constant \tag{9.38}$$

The assumption of an infinite slab implies that $T(\infty, t) = T_{w,i}$. Then, the temperature distribution is given by the following (where $erf(x)$ is the error function):

$$T(y, t) = T_{w,i}(y, 0) + \frac{\dot{q}_w}{k}\left[\sqrt{\frac{4\alpha t}{\pi}}\exp\left(-\frac{y^2}{4\alpha t}\right) - y\left(1 - erf\frac{y}{\sqrt{4\alpha t}}\right)\right] \tag{9.39}$$

Note that at the surface ($y = 0$), the wall temperature variation is

$$T(0, t) = T_{w,i}(y, 0) + \dot{q}_w\sqrt{\frac{4}{\pi}\frac{t}{\rho c k}} \tag{9.40}$$

If we assign a maximum temperature T_m to the material, the maximum constant heat flux that can be borne is

$$\dot{q}_{w,max} = 0.886\frac{\sqrt{\rho c k}(T_m - T_{w,i})}{\sqrt{\Delta t}} \tag{9.41}$$

Because the maximum temperature is essentially arbitrary, for example, the melting temperature, the temperature at which strength diminishes appreciably, or some other appropriate design value,

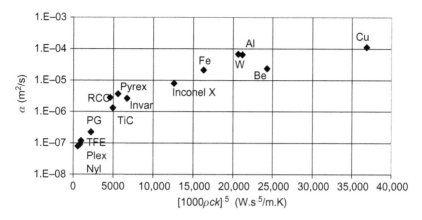

FIGURE 9.12

The heat sink parameter $(1000\rho ck)^{1/2}$ is shown here as a function of the thermal diffusivity. Notations for the materials are given in Table 9.3.

the important material parameter for heat sinks is $\sqrt{\rho ck}$. This parameter is shown in Figure 9.12 as a function of the thermal diffusivity α for a variety of materials described in Table 9.3.

The rise in the surface temperature of the heat shield depends upon the quantity ρck and the incident heat transfer rate. Relevant material properties for a number of materials useful in reentry applications are given in Table 9.3; they have been taken from various sources, including the compilation presented by Williams and Curry (1992). Material properties typically vary with temperature, and wherever possible, they have been quoted at or near their maximum operating temperatures. Furthermore, the thermal conductivity of composite materials is generally anisotropic with considerably lower values normal to the layers. The thermal conductivity of insulating materials is generally dependent on pressure and temperature, as can be seen in data presented by Williams and Curry (1992).

Equation (9.40) shows that the temperature of the surface will rise as $t^{1/2}$, and that for a given heat transfer rate, the magnitude of the temperature rise will be set by the value of ρck. To keep the surface temperature low, the value of ρck should be large and the exposure time small. The material properties presented in Table 9.3 suggest that metals, particularly copper and aluminum, might serve as good heat sink materials.

As discussed in Section 6.6.3, the total heat load on an Apollo space capsule heat shield during entry is around 30 kJ/cm^2 over a flight time of about 600 s making an average heat transfer rate on the order of 50 W/cm^2 (500 kW/m^2). Under this heat load, the final surface temperature as given by Eqn (9.40) is approximately as shown in Figure 9.13 for 10 representative heat shield materials. Several of the metals show relatively low surface temperature compared to their melting temperature, while others exceed it. Also shown is the area density ρl in kg/m^3 for a heat shield slab, which ends the $t = 600$ s flight with a back face temperature arbitrarily fixed at 200 °C above the initial slab temperature $T_{w,i} = 20$ °C. As a convenient measure, consider that the smallest diameter of a manned spacecraft, the Mercury capsule, is 1.89 m, so the heat shield area is 2.8 m^2. The smallest area density for a material that does not exceed its melting temperature is around 300 kg/m^2 for titanium carbide leading to a heat shield mass of about 840 kg, which is more than half as large as the total mass of

Table 9.3 Thermal Properties of Various Aerospace Materials

Material	Symbol	ρ (kg/m^3)	c (kJ/kg · K)	k (W/m · K)	$(10^3 \rho c k)^{1/2}$ W · s$^{1/2}$/m^2
Beryllium	Be	1840	2.72	118	24,302
Aluminum	Al	2800	0.94	171	21,181
Iron	Fe	7860	0.45	75.0	16,287
Copper	Cu	8930	0.39	390	36,854
Tungsten	W	19,300	0.13	170	20,653
Inconel X	I-X	8303	0.54	35.3	12,581
Titanium carbide	TiC	4930	0.88	5.64	4947
Invar	Inv	8000	0.52	10.9	6737
Corning Pyrex	CP	2096	1.40	10.7	5582
Pyrolitic graphite	PG	2200	2.13	1.05	37,494
Reinforced carbon–carbon composite	RCC	1655	1.67	7.64	4595
Teflon	TFE	2213	1.24	0.247	822
Lucite	Plex	1161	1.77	0.164	581
Nylon	Nyl	1133	2.51	0.336	978
Alumina-enhanced thermal barrier rigid tile	AETB	192	0.63	0.064	88
Cork	Cork	150	2.01	0.043	114

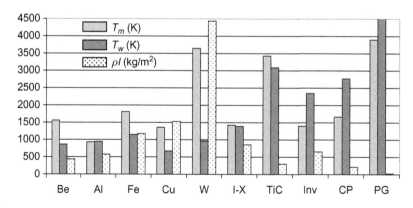

FIGURE 9.13

Melting and surface temperatures for several materials for an entry time of 600 s and a constant heat transfer of 50 W/cm^2. The areal density ρl is based on the length $y = l$ at which $T = T_{w,i} + 200\,°C$.

the capsule. This material is a "cermet," a ceramic–metallic composite and is used not only as a heat shield material but also as a replacement for tungsten in metal cutting applications.

This idealization of a semi-infinite slab as a heat shield has been carried out to illustrate the basic nature of the one-dimensional heat transfer process. Obviously, to realize this idealization, the actual mass of the heat shield would be considerably larger than even that described here where we arbitrarily truncated the heat shield at the point where the temperature had reached a prescribed condition. However, we can draw conclusions based on the relative performance of different materials subjected to the same environment. For example, we see that Invar, Corning Pyrex, and tungsten are unacceptable because the first two will melt during the flight while the last is too heavy for an efficient flight article. Invar is an iron–nickel alloy that has exceptional dimensional stability under large temperature changes and is attractive for other aerospace applications. Aluminum and Inconel X are marginal, but aluminum loses a significant amount of strength when $T > 450$ K and is therefore not suitable. Inconel X, a nickel–chromium super alloy, used extensively on the North American X-15 experimental spaceplane retains its strength at high temperature and provides almost as low heat shield mass as aluminum in this particular case. On the other hand, the Inconel X surface temperature rises to approximately its melting temperature, so it too is unacceptable. Iron and copper both satisfy the basic criterion that the surface temperature is safely less than the melting temperature, but the mass of metallic heat shield material in either case is too large. For example, if either was used for the Mercury capsule heat shield, the mass would exceed the 1060–1360 kg mass of the Mercury capsule itself. Only beryllium heat shields would come close to providing an effective solution to the heat protection problem but they would still weigh half the capsule's total weight. Pyrolitic graphite has low weight but the surface temperature is well beyond its sublimation temperature.

Safely minimizing mass is the overriding concern in spacecraft design because that is the driver in launcher design. Heat sink materials cannot perform adequately for the relatively long heating periods of manned spacecraft reentry because the mass required would be excessive. On the other hand, materials with poor thermal conductivity and low heat capacity, like AETB-12 (alumina-enhanced thermal barrier) rigid tile, like that used on the Space Shuttle Orbiter fuselage, and cork, a thermal barrier commonly used on missiles, would, according to Eqn (9.39), reach very high surface temperatures that they could not withstand. These two materials are typically used as insulating materials for spacecraft but not as high heat transfer heat shield materials. The other materials listed in Table 9.3, like Teflon, Lucite, nylon, and reinforced carbon-carbon composites (RCC) would, like pyrolitic graphite, also reach very high temperatures under our assumptions of purely thermal conduction in the material. These materials make use of other heat transfer mechanisms, particularly change of phase or ablation, for the plastics and thermal radiation emission for RCC, a subject that will be discussed subsequently.

9.6.2 HEAT SINK HEAT SHIELDS WITH TIME-VARYING HEAT TRANSFER

The actual heating rate in a reentry trajectory is not constant, as considered in the previous section, but varies with time, as described previously and illustrated in Figure 9.9. The heat transfer rate rises to a maximum somewhere in the middle of the reentry phase and then decays to low levels as the spacecraft slows in its descent. This general behavior of the typical reentry stagnation point convective heat flux is shown in Figure 9.14. Also shown on Figure 9.14 is a linear approximation to the actual heating curve. Using this approximation, the stagnation point convective heat transfer may be expressed as

$$0 \leq t \leq t_1 : \dot{q}_w = \dot{q}_{w,max} \frac{t}{t_1}$$

$$t_1 \leq t \leq t_2 : \dot{q}_w = \dot{q}_{w,max} \frac{t_2 - t}{t_2 - t_1}$$

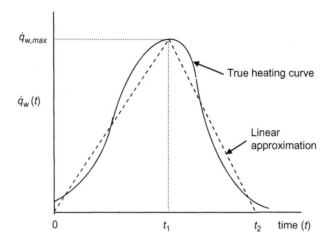

FIGURE 9.14

A linear approximation to the true heating curve.

The heat conduction Eqn (9.37) still applies under the assumption of one-dimensional unsteady conduction and constant material properties, but the appropriate boundary conditions for this case are now given as follows:

$$T(y, 0) = T_{w,i} \tag{9.42}$$

$$-k \left(\frac{\partial T}{\partial y} \right)_{y=0} = \dot{q}_w(t) \tag{9.43}$$

The solution to this general problem is given by Carslaw and Jaeger (1948) as follows:

$$T(y, t) = T_{w,i} + \frac{1}{k} \sqrt{\frac{\alpha}{\pi}} \int_0^t \frac{q_{c,w}(t - \tau)}{\sqrt{\tau}} \exp \left(-\frac{y^2}{4\alpha\tau} \right) d\tau \tag{9.44}$$

We may examine the temperature history of the heat shield at the stagnation point since this is the site at which melting would commence, and this analysis is for heat sink heat shields and is applicable only up to the point of melting. The surface temperature at the stagnation point is

$$T(0, t) = T_{w,i} + \frac{1}{k} \sqrt{\frac{\alpha}{\pi}} \int_0^t \frac{q_{c,w}(t - \tau)}{\sqrt{\tau}} d\tau \tag{9.45}$$

For our linear approximation, the quantity $\dot{q}_w(t - \tau)$ for the two intervals is given by

$$\dot{q}_w(t - \tau) = \dot{q}_{w,max} \frac{t - \tau}{t_1}$$

$$\dot{q}_w(t - \tau) = \dot{q}_{w,max} \frac{t_2 - t + \tau}{t_2 - t_1} \tag{9.46}$$

The maximum heat has been transferred to the surface when the final time $t = t_2$ has been reached, at which point, integration of Eqn (9.45) using the definitions in Eqn (9.46) yields

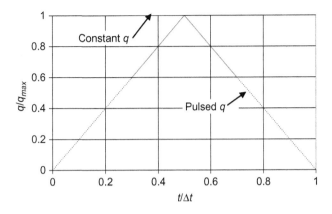

FIGURE 9.15

Two different heat flux distributions, constant and triangular pulse, are shown for the same time duration Δt. Each heat flux distribution is normalized by its own peak value.

$$\sqrt{\frac{\pi}{\alpha}}\frac{k}{\dot{q}_{w,max}}\left[T(0,t_2) - T_{w,i}\right] = \frac{4}{3}\sqrt{t_1} + \frac{2}{3}\frac{t_2^{3/2} - t_1^{3/2}}{t_2 - t_1} \tag{9.47}$$

This result defines the surface temperature at the stagnation point in terms of a flight duration of t_2 seconds and a maximum heat transfer rate $\dot{q}_{w,max}$. As an example, consider the case of a symmetrical heat pulse with $t_2 = \Delta t$ and $t_1 = \Delta t/2$. We may find the maximum value of $\dot{q}_{w,max}$ that would bring the surface to incipient melting from Eqn (9.47) as

$$\dot{q}_{w,max} = 0.98\sqrt{\frac{\rho c k}{\Delta t}}(T_m - T_{w,i}) \tag{9.48}$$

Note that the maximum convective heat flux that can be allowed, that is, that which brings the stagnation point to the melting or some other maximum temperature, may be found from Eqn (9.40) for a constant heat flux and by Eqn (9.48) for a triangular pulse heat flux. These equations, taken for the same time duration of the heat flux, as depicted in Figure 9.15, show the ratio of the maximum allowable heat flux for the two cases to be

$$\frac{\dot{q}_{w,max,const}}{\dot{q}_{w,max,pulsed}} = \frac{0.89}{0.98} = 0.91 \tag{9.49}$$

Thus, although the constant heat flux is applied for the whole time interval, it is only about 10% lower than the maximum heat flux in the pulsed case. This shows that the maximum temperature at the stagnation point is more strongly affected by the magnitude of the maximum value of the applied heat flux than by the shape of the heat flux pulse. The variation of the nondimensional stagnation point temperature difference with time for both cases is shown in Figure 9.16 with

$$\Delta T^* = \frac{T(0,t) - T_{w,i}}{\dot{q}_{w,max}\sqrt{\frac{4\Delta t}{\pi \rho c k}}} \tag{9.50}$$

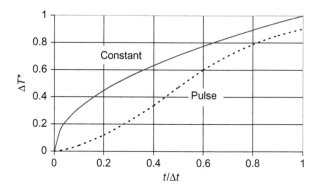

FIGURE 9.16

Variation of the nondimensional temperature difference of Eqn (9.50) with nondimensional time for the two different stagnation point heat flux distributions as shown in Figure 9.15.

The nondimensional form of Eqn (9.50) for constant heat flux case is

$$\Delta T^* = \sqrt{\frac{t}{\Delta t}}$$

For the pulsed heat flux case in the time interval $0 \le t \le t_1$, we have

$$\Delta T^* = \frac{4}{3}\left(\frac{t}{\Delta t}\right)^{\frac{3}{2}}$$

For the pulsed heat flux case in the time interval $t_1 \le t \le t_2 = \Delta t$, we have

$$\Delta T^* = \frac{2}{3}\left\{2^{-\frac{1}{2}} + 3\left(1 - \frac{t}{\Delta t}\right)\left[\left(\frac{t}{\Delta t}\right)^{\frac{1}{2}} - 2^{-\frac{1}{2}}\right] + \left[\left(\frac{t}{\Delta t}\right)^{\frac{3}{2}} - 2^{-\frac{3}{2}}\right]\right\}$$

Note that if we consider the value of $\Delta T^* = \Delta T^*_m$ in Eqn (9.50) to be that appropriate to the melting temperature $T_m(0,t)$, then the value $\Delta T^*_{m,pulsed} = 0.90\Delta T^*_{m,constant}$ because the pulsed case can withstand a higher peak heat flux than can the constant flux case (which is constantly at its peak value), as shown by Eqn (9.49).

On the other hand, we may examine the total heat absorbed by the heat shield in the two cases by integrating the heat flux over the whole heating period

$$Q = \int_0^{\Delta t} \dot{q}_w(\tau)d\tau \tag{9.51}$$

Then, the ratio

$$\frac{Q_{const}}{Q_{pulsed}} = \frac{q_{c,w,max,const}\,\Delta t}{q_{c,w,max,pulsed}\,\frac{\Delta t}{2}} = 1.82 \tag{9.52}$$

This shows that in the case of a constant heat flux about 80% more heat is absorbed than in the case of a pulsed heat flux. Thus, the total heat load is more strongly affected by the shape of the pulse than by the maximum heat flux applied.

9.6.3 CONVECTIVE HEAT TRANSFER

Instead of constant heat transfer to the surface arising from the adiabatic shock-heated gas in the stagnation region of a spaceplane, we may consider convective heat transfer due to frictional heating over broader surface areas. The heat transfer to the wall in this case may be characterized in terms of a heat transfer coefficient as follows:

$$-k\left(\frac{\partial T}{\partial y}\right)_w = C_h(T_{aw} - T_w) \tag{9.53}$$

Here it is assumed that the stagnation temperature at the edge of the boundary layer is constant and the surface temperature $T_w = T_w(0,t)$ is varying. The solution in this case may be written as

$$\frac{T(y,t) - T_{w,i}}{T_{aw} - T_{w,i}} = 1 - erf\frac{y}{2\sqrt{\alpha t}} - \left[\exp\left(\frac{C_h y}{k} + \frac{C_h^2 t}{\rho c k}\right)\right]\left[1 - erf\left(\frac{y}{2\sqrt{\alpha t}} + \sqrt{\frac{C_h^2 t}{\rho c k}}\right)\right]$$

The surface temperature variation is

$$\frac{T_w - T_{w,i}}{T_{aw} - T_{w,i}} = 1 - \left[\exp\left(\frac{C_h^2 t}{\rho c k}\right)\right]\left[1 - erf\sqrt{\frac{C_h^2 t}{\rho c k}}\right] \tag{9.54}$$

The behavior of the solution for surface temperature is illustrated in Figure 9.17. Once again, we see the parameter $\rho c k$ appears in the solution and the larger it is, the less rapidly the surface temperature increases.

We may consider a characteristic time for surface heating to be

$$\tau = \frac{\rho c k}{C_h^2}$$

For typical manned entry, heat transfer rates on the order of $10 \text{ W/cm}^2 = 10^5 \text{ W/m}^2$ with temperature differences of $O(10^3)$ K, the heat transfer coefficient C_h must be on the order of $10^2 \text{ W/m}^2 \cdot \text{K}$. The materials in Table 9.3 have values of $(\rho c k)^{1/2}$ ranging from 10^2 to 4×10^4 $(\text{W} \cdot \text{s}^{1/2}/\text{m}^2 \cdot \text{K})$, so the corresponding range for the characteristic time $C_h^2/(\rho c k)$ is 1 s for lightweight low-conductivity refractory materials like AETB tile to 1.6×10^5 s for heavy high-conductivity metals like copper. Because the model here is a semi-infinite slab, the time constants are notional but demonstrate the importance of material properties (Figure 9.17).

We see from this development that lightweight refractory materials that can withstand high temperatures and do not ablate therefore remain reusable and would be appropriate for convectively heated surfaces, like the windward fuselage and wings of the Space Shuttle Orbiter as shown in Figure 9.10, which do use AETB tiles made primarily of silica.

The heat transfer coefficient shown here is related to the Stanton number (discussed in Section 9.7) as follows

$$C_h(T_{aw} - T_{w,i}) = \rho_e u_e St(h_{aw} - h_{w,i}) = \rho_e u_e \bar{c}_p St(T_{aw} - T_{w,i})$$

At the high temperatures often experienced in hypersonic flight, the specific heat of the air in the boundary layer is not constant and an appropriately weighted value, shown here as $\bar{c}_p$, should be used. We may get an idea of the magnitude of C_h by invoking Reynolds analogy (also discussed in Section 9.7) as follows:

$$C_h = \rho_e u_e \bar{c}_p St = \rho_e u_e \bar{c}_p \frac{c_f}{2Pr_e^m} \approx \frac{1}{2}c_f Re_x\left(Pr_e^{1-m}\frac{\bar{c}_p}{c_{p,e}}\right)\frac{k_e}{x}$$

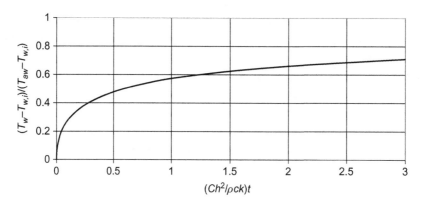

FIGURE 9.17

Increase in surface temperature as a function of time due to convection with a fixed heat transfer coefficient C_h.

In Chapter 8, we discussed boundary layer and thermodynamic characteristics that suggest that the quantity in parentheses above is likely to be around one or two while the product $c_f Re_x = O(10^2)$ for laminar flow to $O(10^4)$ for turbulent flow over flat plates. Note that this shows the convective heat transfer to be given approximately by

$$C_h(T_{aw} - T_{w,i}) \approx k_e \frac{(T_{aw} - T_{w,i})}{x}(c_f\,Re_x)$$

The thermal conductivity of air is $O(10^{-1})$ W/m $\cdot$ K at high temperature. For streamwise distances $x = O(1)$ m typical of laminar flow, the quantity $k_e/x \sim O(10^{-1})$ W/m$^2 \cdot$ K resulting in $C_h(T_{aw} - T_{w,i}) = O(10^4)$ W/m^2. For distances $x = O(10)$ m, for which the boundary layer is likely to be turbulent $C_h(T_{aw} - T_{w,i}) = O(10^5)$ W/m^2. These numbers are consistent with the measured values for the Space Shuttle Orbiter as shown in Figure 9.10.

9.6.4 FINITE SLAB HEAT SINK HEAT SHIELDS WITH CONSTANT HEAT TRANSFER

The previous discussions presumed that the heat shield was a semi-infinite slab of heat shield material. We may now assess the more practical case of a slab of material of length l supported by an adiabatic back plate. The initial condition and front face boundary condition of Eqn (9.38) still applies, but now the boundary condition at the back face is that there is no heat transfer or $\dot{q}(l, t) = -k(\partial T/\partial y) = 0$. The solution to this problem is

$$T = T_{w,i} + \dot{q}_w \frac{l}{k}\left[\frac{\alpha t}{l^2} + \frac{1}{2}\left(\frac{y}{l}\right)^2 - \left(\frac{y}{l}\right) + \frac{1}{3} - \frac{2}{\pi^2}\sum_{n=1}^{\infty}\frac{\cos(n\pi x/l)}{n^2}\exp(-n^2\pi^2\alpha t/l^2)\right] \qquad (9.55)$$

For $\alpha t/l^2 > O(1)$, the temperature rise at the front face may be reasonably approximated by

$$\frac{T - T_{w,i}}{\dot{q}_c} \approx \left(\frac{t}{\rho c l} + \frac{l}{3k} + \cdots\right)$$

Here, we see that a good heat sink material should have the heat capacity per unit volume ρc and the thermal conductivity k each independently large, whereas when we considered the semi-infinite

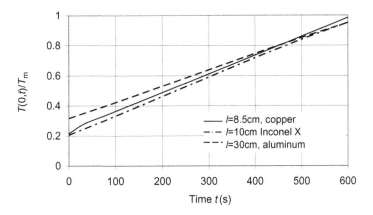

FIGURE 9.18

Ratio of front face temperature to melting temperature for finite slabs of different materials and thicknesses for 600 s flight at 50 W/cm^2.

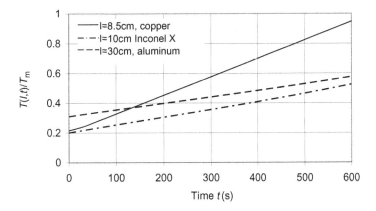

FIGURE 9.19

Ratio of back face temperature to melting temperature for finite slabs of different materials and thicknesses for a 600 s flight at 50 W/cm^2.

slab, it was only the product $\rho c k$ that needed to be large. However, if ρc and k are both large, the product $\rho c k$ will also be large and thus Figure 9.12 still gives a good idea about those materials that will function well as heat sinks. Considering the same constant heat flux of 50 W/cm^2 and initial material temperature as in the infinite slab case, we arrive at the results shown in Figure 9.18.

Here we have selected slab thicknesses that provide for a front face final temperature close to the melting temperature, so that the materials may be compared. Interestingly, the three metal slabs, although different in thickness, have mass per unit area of about 800 kg/m^2. The temperature history of the back face, which is assumed to be adiabatic, is shown in Figure 9.19. Here, we see that only the copper slab is quite close to the melting temperature at the back face and the front

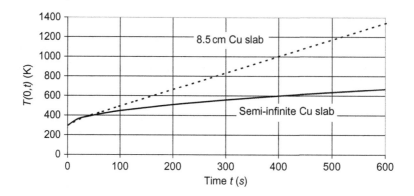

FIGURE 9.20

Comparison of the front face temperature of an 8.5-cm slab with an adiabatic back plate and a semi-infinite slab of copper, both initially at a temperature $T_{w,i} = 20\,°C$ and then exposed to a constant heat flux of 50 W/cm².

face unlike both Inconel X, which has a much smaller thermal diffusivity, and aluminum, which has a much greater thickness.

In Figure 9.20, we show the history of the front face temperature of an 8.5 cm slab of copper with an adiabatic back face compared to that previously found for a semi-infinite slab of copper. Both are initially at a temperature $T_{w,i} = 20\,°C$ and then exposed to a constant heat flux of 50 W/cm². During the first 50 seconds of the flight, the behavior is virtually identical, and thereafter, the presence of the adiabatic back face causes heat to accumulate in the finite slab heat shield, thereby driving up temperatures throughout the slab. In the semi-infinite slab, the thermal diffusivity permits heat to flow through the slab, and temperatures are therefore rising much more slowly. The manner in which the heat flux at the back face is treated very much determines the performance of the entire heat shield.

Heat sink heat shields have the virtue of simplicity and reliability and material properties are usually well known, so a high degree of confidence in designs can be expected. Because no material is lost during a flight, a heat sink heat shield can, in general, be reusable. The major drawback of such systems is their weight, and this keeps their application confined to relatively low heat-load areas of a spacecraft.

9.6.5 ABLATIVE HEAT SHIELDS WITH CONSTANT HEAT TRANSFER

If the heat transfer rate is too high for a heat sink thermal protection system (TPS) resulting in melting of the heat shield material, recourse may be made to an ablative TPS. The simplest such case is the melting or subliming ablator, a material whose latent heat of fusion may be used to expand the performance of the TPS. A schematic diagram of an ablative TPS for a leading edge or nose is shown in Figure 9.21. The stagnation region heats up under the influence of the applied convective heating until the melting or sublimation temperature is reached. At that time, a phase change takes place and the latent heat of fusion is absorbed in melting or subliming the heat shield material which reduces the heat flux entering the heat shield thereby reducing the heat load on it. The friction of the boundary layer sweeps away the sublimed vapor and can strip away some or all the liquid melt layer.

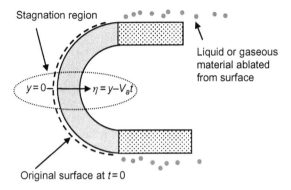

FIGURE 9.21

An ablative TPS with a moving coordinate system fixed to the receding surface of the high heat flux stagnation region.

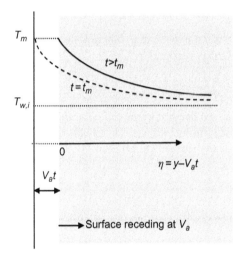

FIGURE 9.22

Diagram of semi-infinite slab with front face receding at speed V_a. The temperature profiles in the slab at and after the start of ablation are shown.

In order to focus on the major effects, we continue with the stagnation region idealized as a semi-infinite slab whose front face recedes at a speed V_a because of phase change at the melting or subliming temperature, which we denote by T_m. The governing equation for heat transfer into the material is still Eqn (9.37), assuming one-dimensional heat transfer and constant material properties, but a transformation of coordinates proves helpful. We may fix our y-coordinate to the melting surface, which recedes as melted or sublimed material is continually swept away by boundary layer friction, and new solid material is presented to the air stream, as shown in Figure 9.22.

The ablation process starts when the surface reaches the melting or subliming temperature, which, according to Eqn (9.40), would occur at time

$$t_m = \frac{\pi}{4} \rho c k \left(\frac{T_m - T_{w,i}}{\dot{q}_w} \right)$$

The new coordinate system in the stagnation region is described by the relation

$$\eta = y - V_a t \qquad (9.56)$$

Applying this coordinate transformation to Eqn (9.37) results in the following equation:

$$\frac{d^2 T}{d\eta^2} + \frac{V_a}{\alpha} \frac{dT}{d\eta} = 0 \qquad (9.57)$$

Here, the boundary conditions are that the heating maintains the receding surface at the melting or subliming temperature T_m, while the interior of the material far from the surface remains at the initial wall temperature $T_{w,i}$, or

$$T(0) = T_m$$
$$T(\infty) = T_{w,i} \qquad (9.58)$$

The solution to Eqn (9.57) is an exponential function, and when the boundary conditions are inserted, the temperature distribution becomes

$$\frac{T - T_{w,i}}{T_m - T_{w,i}} = \exp\left(-\frac{V_a}{\alpha} \eta \right) = \exp\left(-\frac{V_a}{\alpha} y + \frac{V_a^2}{\alpha} t \right) \qquad (9.59)$$

In order to use this equation, we need to know the velocity of recession of the surface. Because in this analysis the incident heat transfer rate is assumed constant, the energy balance at the surface requires that the heat transfer to the surface ($\eta = 0$ or $y = V_a t$) must equal the sum of the heat conducted into the material and the heat absorbed in melting or subliming the material, and this may be written as follows:

$$\dot{q}_w = \left(-k \frac{dT}{d\eta} \right)_{\eta=0} + \rho V_a L_m \qquad (9.60)$$

In Eqn (9.60), ρ and L_m are the density and latent heat of melting or subliming of the TPS material, respectively. The conductive heat transfer term may be determined from Eqn (9.59), and then the surface recession velocity V_a may be determined from Eqn (9.60), resulting in

$$V_a = \frac{\dot{q}_w}{\rho[L_m + c(T_m - T_{w,i})]} \qquad (9.61)$$

The recession rate of the surface is V_a and the mass flux of material lost by ablation is ρV_a, and both may be calculated using Eqn (9.61). One may also use Eqn (9.61) to determine what is commonly called the effective heat of ablation, which is defined as follows:

$$H_{eff} = \frac{\dot{q}_w}{\dot{m}_a} = \frac{\dot{q}_w}{\rho V_a} = \left[L_m + c(T_m - T_{w,i}) \right] \qquad (9.62)$$

Early in the study of TPS for atmospheric entry, the effective heat of ablation was found to be useful as a means of comparing the heat protection offered by various materials. It was relatively easy to

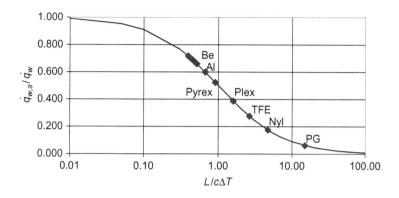

FIGURE 9.23

The ratio of heat transfer into the surface with ablation to that with no ablation $\dot{q}_{w,a}/\dot{q}_w$ is shown as a function of $L_m/c(T_m - T_{w,i})$ with $T_{w,i} = 20\,°C$. The closely grouped data points without a tag are the other metals.

measure by subjecting samples to a known heating environment and measuring the rate of material consumption, thereby immediately providing a value for the effective heat of ablation of the material.

Yet, a third way we may look at Eqn (9.61) is by solving for $\dot{q}_{w,s}$, the heat actually conducted into the solid slab:

$$\dot{q}_{w,s} = -k\left(\frac{\partial T}{\partial y}\right)_{w,s} = -k\left(\frac{dT}{d\eta}\right)_{\eta=0} = \dot{q}_w\left[1 - \frac{1}{1 + c(T_m - T_{w,i})/L_m}\right]$$

This equation shows that the reduction in heat flux to which the heat shield is subjected depends upon the ratio of the heat content of the material at the melting or subliming temperature to its latent heat of fusion. The ratio of heat transfer into the surface with ablation to that with no ablation is shown for various materials in Figure 9.23.

We see that for materials with high values of density, specific heat, melting point, and heat capacity, the recession velocity is small and the penetration of heat into the body is small. Thus, these characteristics serve to make a good ablative heat shield TPS. Recall that this analysis assumes that a constant heat flux is supplied at the surface and that the material is thick enough to permit the assumption that the heat shield acts like a semi-infinite slab, that is, the thickness $l \gg V_a t$. We have also implicitly assumed that all the ablated material is removed from the surface with no interaction with the external flow.

9.6.6 MASS TRANSFER FOR HEAT SHIELD THERMAL PROTECTION

The thermal protection afforded by taking advantage of the heat of fusion of the material of the heat shield suggests that further advantage may accrue from exploiting the heat of vaporization of a heat shield material. It is obvious that the injection of a coolant through the surface into the boundary layer reduces both skin friction and heat transfer merely by thickening the boundary layer and thereby reducing velocity and thermal gradients. In addition, the injected gas may be at lower temperature, thereby further cooling the boundary layer fluid or it may chemically react with the

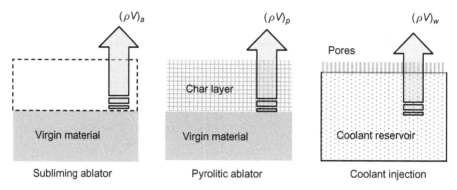

FIGURE 9.24

Surface thermal protection methods employing gaseous mass transfer to enhance heat removal from the surface.

boundary layer gas. Aside from injecting a foreign gas into the boundary layer through the surface, one may merely let the melt material heat to the vaporization temperature, thereby carry away heat from the surface. Some mass transfer techniques for thermal protection of surfaces, with schematic illustrations in Figure 9.24, are as follows:

- Sublimation or melting of the solid surface material. The case of sublimation or melting with complete melt removal is the simplest and was treated in the previous section.
- Vaporization of melted surface material. Here there will be a layer of melted material between the solid material and the vaporized material entering the boundary layer gas flow.
- Pyrolysis of surface material where the filler ablates leaving the matrix to be oxidized by the boundary layer air flow leaving a porous char layer, which provides additional insulation for the remainder of the solid material.
- Direct injection of a coolant gas or liquid through a porous surface, usually called transpiration and film cooling, respectively.

A heat balance at the surface for the melting/sublimation, pyrolysis, and transpiration mass transfer schemes yields the following result:

$$(\rho V)_a H_{eff,a} = (\rho V)_p H_{eff,p} = (\rho V)_w (h_i - h_w) = \dot{q}_w \qquad (9.63)$$

As mentioned in the previous section, the quantity H_{eff} is simply defined as the ratio of the incident heat flux to the mass flux injected into the boundary layer. In Eqn (9.63), we differentiate between the differing possibilities offered by melting or subliming ablation (subscript a), pyrolysis (subscript p), and direct injection of a gas through a porous wall (subscript w) to illustrate the fact that the heat protection processes may differ in detail but all have a common foundation.

Ablation involves mass transfer into the boundary layer in addition to phase change heat absorption and in order to get an appreciation of the effect of injecting gas into the boundary layer, we consider the results presented by Kays, Crawford, and Weigand (2005), which have been found to have a wide degree of application. For simplicity, we discuss the results applicable to constant pressure flows. The effect on heat flux and skin friction coefficient due to gas injection through a porous surface, both at the same Reynolds number based on distance x, are described as follows:

$$\frac{\dot{q}_w}{\dot{q}_{w,0}} = \frac{b_h}{\exp b_h - 1} \tag{9.64}$$

$$\frac{c_f}{c_{f,0}} = \frac{b_f}{\exp b_f - 1} \tag{9.65}$$

Here, the so-called blowing parameters b_h and b_f are given by

$$b_h = \frac{\rho_w V_w / \rho_e u_e}{St_0} \tag{9.66}$$

$$b_f = \frac{\rho_w V_w / \rho_e u_e}{c_f / 2} \tag{9.67}$$

In the above equations, ρ_w is the density of injected gas and V_w is the normal velocity component at the surface. The subscript zero refers to the case with zero injection, that is, $b_h = 0$ and $b_f = 0$. If comparisons are made at the same Reynolds number based on streamwise distance x for the case of constant free-stream velocity, the above equations fit experimental data remarkably well. Equation (9.64) is to be evaluated at the same k_e, T_w, and T_{aw} for injected and noninjected cases. A limiting case occurs for large values of blowing when the friction coefficient tends to be zero and boundary layer is literally blown off the surface, an occurrence similar to the separation of a boundary layer in an adverse pressure gradient. Two commonly used rules of thumb for "blowoff" are as follows: if $\rho_w V_w / \rho_e u_e = 0.01$ and/or $b_f = 4.0$, then it is safe to assume that blowoff has occurred.

It should be noted that the effective heat of ablation is not a property of the material, but an index of TPS performance of a material in an entry heating environment. Thus, the experimental values include heat capacity, melting, and vaporization, as well as other possible effects such as thermal radiation emission, chemical reactions, and mass transfer as discussed above. Common TPS materials are pyrolytic ablators, sometimes called charring ablators. They are primarily organic materials that decompose into charred material and gases when heated in an oxygen-poor environment. They are generally held in a supporting matrix as shown in Figure 9.24 and are gasified by the heating leaving behind charred material, which typically has about half or less of the density of the virgin material and can serve as an insulator for the remaining virgin material. Tran et al. (1997) presented experimental results showing the effective heat of ablation H_{eff} for Avcoat-5026 H/CG to be in the range of 45,000–70,000 kJ/kg. This is a phenol–formaldehyde resin with various additives loaded into a honeycomb fiberglass matrix used in the Apollo heat shield and being used now for the Orion Crew Exploration Vehicle (CEV) heat shield. The heat flux range covered included 100 W/cm^2 typical of the maximum for an 11 km/s Apollo lunar return mission as well as 1000 W/cm^2, which is typical of the 13 km/s Stardust capsule entry described subsequently in this chapter. They also tested phenolic impregnated carbon ablator (PICA) over the same wide range of heat flux (up to 3400 W/cm^2) and found H_{eff} to be comparable to that of Avcoat for the 100 W/cm^2 heat flux case but two to three times higher for the 1000 W/cm^2 heat flux case.

A general expression for the effective heat of ablation relation including the effect of the mass injection into the boundary layer due to ablation is given by

$$H_{eff} \frac{b_h}{\exp b_h - 1} \approx \left[L_m + c_s(T_m - T_{w,i}) + 0.65 c_l(T_w - T_m) \right] + f \left[L_v + \beta(h_{t,e} - h_w)_{0a} \right] \tag{9.68}$$

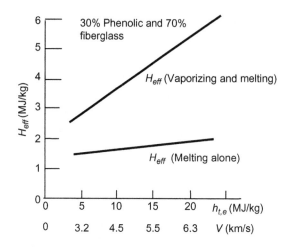

FIGURE 9.25

Effective heat of ablation for a heat shield material comprised 30% phenolic resin and 70% fiberglass matrix.

Here, L_m denotes the latent heat of fusion (melting or sublimation), while c_s and c_l denote the specific heat of the solid and liquid surface material, respectively. The temperature T_w is the temperature of the liquid at the interface with the boundary layer flow. Thus, the third term in the first set of square brackets in Eqn (9.68) represents the effect of a liquid melt layer that has not been stripped away by boundary layer friction. If the material was subliming, that third term would not appear. In the second set of square brackets in Eqn (9.68), the quantity L_v denotes the latent heat of vaporization of the surface material while f denotes the fraction of material vaporized. For direct injection of a vaporizing liquid, $f = 1$, while for charring ablators, f may be taken as the ratio of the density of the char to that of the virgin material. According to Dorrance (1962), the quantity β has been found from laminar boundary calculations and turbulent boundary layer experiments to be a constant in the range $1/6 < \beta < 2/3$. The subscript $0a$ on the difference between stagnation and wall enthalpy refers to conditions in the absence of ablation. It is found that the second term in square brackets in Eqn (9.68) is the dominant term for $V > 5$ km/s.

An example of the magnitude of the effective heat of ablation for a heat shield material comprised 30% phenolic resin and 70% fiberglass matrix carried out by Adams (1959) and presented by Dorrance (1962) as shown in Figure 9.25. Experimental results for this material agreed quite well with the theoretical results shown. The phenolic resin melts and vaporizes leaving the fiberglass matrix so that $f = 0.3$. The heat of vaporization of the resin was taken as $L_v = 1745$ kJ/kg, the factor $\beta = 0.49$, and the cold wall hypothesis, $h_{t,e} \gg h_w$. At high Mach number, we may set $h_t \sim V^2/2$ and we note that 1 kJ/kg $= 10^3$ m^2/s^2. The results demonstrate that the vaporization term in Eqn (9.65) is indeed dominant at entry speeds and because the heat flux is proportional to h_t and h_t/H_{eff} is approximately constant, the ablation rate is likewise relatively constant. A charred ablative heat shield from the first KH series of reconnaissance satellites mission is shown in Figure 9.26. Compare the appearance of the entered heat shield in Figure 9.26 to that of its original state shown in Figure 8.27.

FIGURE 9.26

Photograph of the charred ablative heat shield from the first KH-4 CORONA mission. (NASA).

Having experimental information for the effective heat of ablation permits one to estimate the surface recession for a given heat flux experienced by a surface. The data shown in Figure 9.25 suggest $H_{eff} = 1.65h_{t,e} + 2$, while experiments on Teflon ablators indicate $H_{eff} = 0.48h_{t,e} + 0.24$ where in both cases, h is measured in MJ/kg. From Eqn (9.63), the surface recession velocity may be calculated from the heat flux, the material density, and effective heat of ablation. The surface recession can be then obtained by integrating the surface recession rate.

9.6.7 THERMAL RADIATION OF SPACECRAFT SURFACES

The discussions in the previous sections show that the surface of the TPS may reach such high temperatures during entry that they are subject to ablating with the attendant material loss. The surface will of course be radiating thermal energy back into the boundary and shock layers because of this high surface temperature. Assuming a wall heat flux $\dot{q}_w$ is applied and only radiative heat transfer is operative at the wall, an energy balance at the surface requires that the steady-state wall temperature be

$$T_w = \left(\frac{\dot{q}_w}{\varepsilon \sigma} \right)^{\frac{1}{4}}$$

Here, ε is the emissivity of the surface and $\sigma = 5.6704 \times 10^{-8}$ W/m$^2 \cdot$ K^4 is the Stefan−Boltzmann constant. The only limitation on the surface is that the material can withstand the temperature appropriate to the heat transfer rate imposed. Thus, a radiative heat shield can operate indefinitely at its permissible wall temperature in an essentially passive manner with no mass loss making it attractive for reusable entry systems. The high operating temperature does put a constraint on the inner load-bearing structure unless insulating schemes are employed with the support system. Because radiative shields are limited by operating temperature and not total heat load,

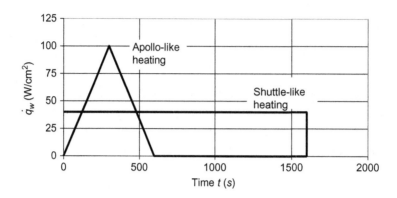

FIGURE 9.27

Generic histories of the heat flux applied to one face of the finite slab are illustrated here. They are described as typical of Apollo or Space Shuttle entry trajectories.

they are particularly suited to moderate L/D cruise-type applications. The Space Shuttle Orbiter relies heavily on radiative contributions for its RCC nose and leading edge panels as well as its AETB tiles.

To get an idea of the effect of thermal radiation emission on the surface temperature of a TPS material subjected to a particular heat flux, we consider a finite thickness slab of RCC with properties from the Space Shuttle Program Thermodynamic Design Data Book as reported by Williams and Curry (1992): $k = 7.75$ W/m · K, $\rho = 1655$ kg/m^3, and $c = 1.47$ kJ/kg · K. The maximum operating temperature is given as 1371 °C. The slab is considered to have an adiabatic back face and to be exposed to either of two heating histories as shown in Figure 9.27: a long constant heat flux like a shuttle entry and a relatively short Apollo-type triangular heat flux pulse heat flux distribution.

Assume that slab is 1 cm thick and initially at 22 °C (295 K) so that $T(y,0) = 295$ K. At $t = 0$, a triangular or rectangular distribution of heat flux is applied and maintained for a given period of time as illustrated in Figure 9.27. The front face temperature history is sought, assuming that the back face of the slab is adiabatic. Using y as distance into the slab and assuming that the gas is transparent to the emitted radiation, the boundary conditions are

$$-k\frac{\partial T}{\partial y}(0,t) = \dot{q} - \varepsilon\sigma(T_w^4 - T_\infty^4) \tag{9.69}$$

$$-k\frac{\partial T}{\partial y}(l,t) = 0 \tag{9.70}$$

To deal with the nonlinearity of the boundary condition, resort is made to a computational approach. The main point to be illustrated is that the reradiation of thermal energy at the wall reduces the maximum surface temperature experienced.

The Figure 9.28 shows the rise of the wall temperature as a function of time for the case of constant heat flux of 40 W/cm^2 applied over 1600 s. This heat flux shape and duration is similar to that for a Space Shuttle Orbiter entry, and solutions were carried out for $\varepsilon = 0$, 0.7, and 0.9. Within the first 100 s, the wall temperature attains an equilibrium value of about 1500 °C for $\varepsilon = 0.7$ and about 1400 °C for $\varepsilon = 0.9$. The maximum operating temperature for this particular specimen of RCC is

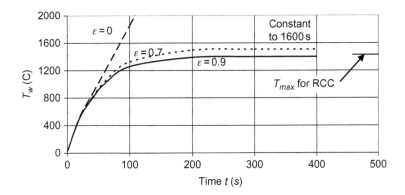

FIGURE 9.28

History of the front face temperature of a 1-cm thick slab of RCC exposed to the Shuttle-type heat constant heat pulse as shown in Figure 9.27. The back face of the slab is adiabatic, and the maximum operating temperature of the RCC is around 1400 °C.

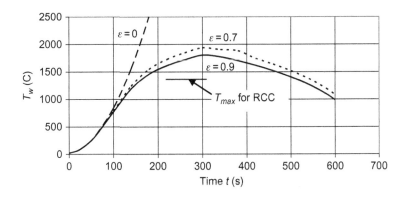

FIGURE 9.29

History of the front face temperature of a 1-cm thick slab of RCC exposed to the Apollo-type heat triangular heat pulse as shown in Figure 9.27. The back face of the slab is adiabatic, and the maximum operating temperature of the RCC is around 1400 °C.

around 1400 °C, and this is just about the equilibrium temperature reached for the case of $\varepsilon = 0.9$. Thus, the RCC is a material that can be part of the TPS of a reusable spaceplane and indeed was used on the Space Shuttle orbiter nose and wing leading edges. Surface coatings that display maximum emissivity, that is, those with values of $\varepsilon \sim 1$ are a subject of much research interest for reusable spacecraft.

Figure 9.29 shows the rise of the wall temperature as a function of time for the linear variation of heat flux from zero at $t = 0$ to a maximum of 100 W/cm^2 at $t = 300$ s and a linear drop back to zero at $t = 600$ s. This profile is similar to that experienced by an Apollo command module entry. The computations were again carried out for cases where the emissivity $\varepsilon = 0$, 0.7, and 0.9. Within the first 100 s, the wall temperature for all cases is quite close but shortly thereafter, the surface temperature of the nonradiating surface ($\varepsilon = 0$) grows rapidly to extremes well beyond the capability of the RCC material. For the cases where the emissivity of the surface of the slab $\varepsilon = 0.7$ and 0.9, the maximum temperature is reached at about the time of the maximum heat flux with values of about 1940 °C for the former and 1800 °C for the latter. These temperatures are several hundred degrees higher than the maximum operating temperature of the RCC material so that a radiative TPS is not the best choice for a capsule type entry. The peak heating is high enough to warrant the use of an ablative heat shield in these cases. For this reason, the material selected for the Orion CEV heat shield is Avcoat, a charring ablator.

9.6.8 CHEMICAL CATALYSIS OF SPACECRAFT SURFACES

The analysis of convective heating as presented by, for example, Fay and Riddell (1958) and discussed in Section 9.3 assumes that the body surface is fully catalytic to recombination. That is, the oxygen and nitrogen atoms formed by dissociation may recombine at any location wherever the thermodynamic conditions permit. As a consequence, whenever recombination occurs, chemical energy may be released keeping the heat transfer at a high level. Catalysis at the wall suggests that some surface materials may support recombination, while others, like many glassy materials, suppress it, possibly leading to reduced heat transfer. Dorrance (1962) discusses the fundamental mechanisms involved in some detail.

Flight at high altitudes is characterized by relatively low pressures and densities in the shock layer, and this suggests that chemical reactions in the boundary layer will proceed very slowly, if at all. As a consequence, the composition of the gas in the boundary layer will be diffusion-controlled and the chemistry is said to be "frozen". The only place significant chemical reactions are likely to occur is at the wall itself. The latent chemical energy of atom recombination will not be experienced by surface materials that are noncatalytic, while highly catalytic walls will experience heat fluxes similar to that occurring under equilibrium chemistry conditions. Thus, a TPS material that is noncatalytic can reduce the heat transfer substantially from the equilibrium chemistry value in regimes in which the boundary layer chemistry is frozen.

To illustrate the effect of wall catalysis, we consider the extreme case of a frozen boundary layer where reaction rates are essentially zero and the composition of the gas in the boundary layer is constant. If we permit catalytic recombination of atoms into molecules at the surface, there will be a deficit of atoms near the surface. Fick's law states that there will be a mass flux of atoms in the direction from regions of high concentration to those of low concentration as expressed by

$$\dot{m}_a = -D_{12}\frac{d\rho_a}{dy} \tag{9.71}$$

The proportionality factor D_{12}, called the binary diffusion coefficient, was introduced in Section 9.2. Fick's law of mass transfer is analogous to Fourier's law of heat transfer where heat flows in the direction of high temperature to low temperature as given by

$$\dot{q} = -k\frac{dT}{dy} \tag{9.72}$$

Thus, in a steady state, the rate of transport of atoms must be balanced by the rate of recombination of atoms into molecules at the wall. We may express the balance between diffusion of atoms toward the wall and disappearance of atoms at the wall due to recombination as follows:

$$D_{12}\left(\frac{\partial \rho_a}{\partial y}\right)_w = k_r \rho_{a,w} \tag{9.73}$$

Here the mass flux of atoms at the wall is assumed to be proportional to the mass of atoms present, and k_r is the chemical recombination rate at the wall due to catalytic reaction. The factor k_r is dependent upon material properties as well as the local thermodynamic state and has the units of velocity. From experiments, it is found that at room temperature k_r for metallic oxides is on the order of 3 m/s, while for pyrex glass, it is around 0.03 m/s, and an increase in temperature (increase in flight speed) tends to increase k_r, while an increase in altitude (decrease in pressure) tends to decrease k_r.

Knowing the pressure and temperature at a point in the inviscid flow, we may use the approximate air chemistry as described in Section 9.2 to find the mole fractions of oxygen and nitrogen atoms and molecules and the compressibility Z. The mass fraction $Y(i)$ of species k is related to its mole fraction $X(i)$ by the relation

$$Y(i) = \frac{W(i)}{W_o} Z X(i) \tag{9.74}$$

Here, $W(k)$ and W_o denote the molecular weight of species k and of the air mixture under standard conditions, respectively. As pointed out in Section 9.2, Hansen (1959) chose the ratio of nitrogen to oxygen in air to be 4 1 making the molecular weight $W_o = 28.8$, rather than the standard sea level value of 28.96. Then, the mass fraction of atoms is

$$Y_a = Y(O) + Y(N) = \frac{Z}{W_o}[X(O) + X(N)] \tag{9.75}$$

In terms of the parameters introduced in Section 9.2, this may also be written as

$$Y_a = \frac{1}{28.8}[16(2\varepsilon_1 - 0.4\varepsilon_3) + 14(2\varepsilon_2 - 1.6\varepsilon_3)] \tag{9.76}$$

We may write the stagnation point heat transfer relation of Eqn (9.19) as follows:

$$\dot{q}_w = \dot{q}_{c,w}\left[1 + (Le^k - 1)\frac{h_D}{h_{t,e}}\right] \tag{9.77}$$

Here, the term outside the square bracket is the convective heat transfer when $Le = 1$. The definition of h_D appears in Eqn (9.20) and involves the standard heat of formation of the individual species $\Delta h_{f,i}$ which are listed in Table 9.4.

The heat of formation is based on the formation of the species from the elements at standard temperature and pressure. Because molecular oxygen and nitrogen are normally found at standard temperature and pressure, there is zero heat of formation for them. The dissociation energy term

$$h_D = [Y_e(O) - Y_w(O)]\left(3980\frac{kJ}{kg}\right) + [Y_e(N) - Y_w(N)]\left(6610\frac{kJ}{kg}\right) \tag{9.78}$$

If the boundary layer chemistry is in equilibrium, the mass fractions depend upon the local temperature and pressure; note that in a boundary layer, $p_e = p_w$ to order $Re_x^{-1/2}$. If the boundary layer

Table 9.4 Standard Heats of Formation for Species in a Binary Air Mixture

Species	Δh_f (kJ/kg)
O	3980
O_2	0
N	6610
N_2	0

chemistry is frozen, the composition is constant across the layer so that $h_D = 0$ in Eqn (9.78) and the heat flux at the wall as given by Eqn (9.77) is less than the equilibrium value. If, however, the boundary layer chemistry is frozen but the wall is catalytic to recombination, the mass fraction of O and N atoms at the wall will be depleted and the dissociation enthalpy $h_D > 0$ making the heat flux higher and possibly approaching the equilibrium chemistry value.

The design intent for manned spacecraft is to keep surfaces at temperatures well below 2000 K so that the choice of $Le = 1.4$ is a reasonable one. Surface coatings that display minimal catalytic effects, that is, those with relatively small values of k_r are a subject of much research interest.

9.7 HEAT TRANSFER SIMILARITY PARAMETERS

There are several important similarity parameters in boundary layer heat transfer. The Prandtl and Lewis numbers, relating to gas properties, were already introduced in Eqn (9.20). There are two other similarity parameters that relate to the heat transfer process itself. The first is the Stanton number, which normalizes the convective heat transfer into the wall by an enthalpy flux based on the wall conditions and is given by

$$St = \frac{\dot{q}_w}{\rho_e u_e (h_{aw} - h_w)} \tag{9.79}$$

The second is the Nusselt number, which normalizes the wall heat transfer by an effective heat transfer based on a distance x along the body surface in the streamwise direction and is given by

$$Nu = \frac{\dot{q}_w}{\frac{k_e (h_{aw} - h_w)}{x} \frac{1}{c_{p,e}}} \tag{9.80}$$

The two nondimensional similarity parameters are related by the following expression

$$St = \frac{Pr_e}{Re_{x,e}} Nu \tag{9.81}$$

The wall enthalpy h_w is unambiguous, but it remains to explain the adiabatic wall temperature, h_{aw}. If there were no heat transfer to the wall, that is, if it were adiabatic, the wall would reach a certain equilibrium temperature and the gas at the wall would take on the enthalpy appropriate to that temperature. The no-slip condition requires that the velocity at the surface be zero, but since the flow is not brought to rest adiabatically through the boundary layer because of viscous dissipation, the enthalpy at the surface is not the stagnation enthalpy but a lower value, the so-called adiabatic wall enthalpy, h_{aw}. The ratio of the possible wall enthalpy increase under adiabatic conditions and the ideal isentropic enthalpy increase is given by

$$\frac{h_{aw} - h_w}{h_{t,e} - h_w} = \frac{(h_e + r\frac{1}{2}u_e^2) - h_w}{(h_e + \frac{1}{2}u_e^2) - h_w} = s \tag{9.82}$$

The recovery factor r in the adiabatic wall enthalpy represents the fraction of ordered kinetic energy that is lost to heat by friction. Experiments have long shown that the recovery factor r is well correlated by a function of the Prandtl number of the gas, as shown below:

$$h_{aw} = h_e + Pr_e^m \frac{u_e^2}{2} \tag{9.83}$$

The quantity $m = 1/2$ is for laminar flow and $m = 1/3$ is for turbulent flow. Here, h_e and h_w are the enthalpies at the edge of the boundary layer and at the wall, respectively, and Pr_e is the Prandtl number evaluated at the edge of the boundary layer.

9.7.1 REYNOLDS ANALOGY

The practical value of the Stanton number in heat transfer calculations arises from Reynolds analogy between heat transfer and skin friction, which suggests that one is proportional to the other. A detailed development of the theoretical foundation for the analogy is given by Van Driest (1959). A simplified version of the derivation is given here. The basis for the approach is the Crocco relation $h_t = a + bu$, that is, the total or stagnation enthalpy is linearly proportional to the velocity in a constant pressure boundary layer when the Prandtl number $Pr = 1$ and the wall temperature is constant. This relation is expressed as follows:

$$h_t = h_{t,w} + (h_{t,e} - h_{t,w}) \frac{u}{u_e}$$

The total enthalpy is $h_t = h + u^2/2$ so that the total enthalpy at the wall, where $u = 0$, is $h_{t,w} = h_w$ and therefore

$$h_t - h_w = (h_{t,e} - h_w) \frac{u}{u_e} \tag{9.84}$$

Using this relation, one may show that the convective heat transfer to the wall may be written as

$$\dot{q}_w = - \left(k \frac{\partial T}{\partial y} \right)_w = - \left[\frac{\mu}{Pr} \left(\frac{\partial h_t}{\partial y} \right) \right]_w \tag{9.85}$$

Then, using Eqn (9.84), for which $Pr = 1$, in Eqn (9.85) yields

$$\dot{q}_w = - \left[(h_{t,e} - h_w) \frac{\mu}{u_e} \frac{\partial u}{\partial y} \right]_w = - \frac{(h_{t,e} - h_w)}{u_e} \tau_w \tag{9.86}$$

Using Eqn (9.82) permits Eqn (9.86) to be rearranged to read

$$\dot{q}_w = - \frac{(h_{t,e} - h_w)}{u_e} \frac{1}{2} \rho_e u_e^2 c_f = - \frac{1}{s} \rho_e u_e (h_{aw} - h_w) \frac{c_f}{2} \tag{9.87}$$

Table 9.5 Coefficients for the Constant Pressure Flow Correlation in Eqn (9.90)

Type of Flow	A	a	b	c	j
Laminar	$0.332Pr^{1/3}$	0.5	0.5	0.5	–
Turbulent	$0.02296Pr^{1/3}$	0.861	0.139	0.861	–
Flat plate	–	–	–	–	0
Cone	–	–	–	–	1

Equation (9.87) for the heat flux at the wall is essentially the same as that arrived at in the detailed development due to Van Driest (1959), which eliminated the requirement of $Pr = 1$. The term shown as s in Eqn (9.87) is called the Reynolds analogy factor and in Van Driest's derivation, it is an integral involving both Prandtl number and the boundary layer enthalpy profile. Here, we simply find it to be the enthalpy fraction s in Eqn (9.82). In any event in practice, this factor is found to be well correlated by $s = Pr^{2/3}$, so we may simply write Reynolds analogy as follows:

$$\frac{-\dot{q}_w}{\rho_e u_e (h_{aw} - h_w)} = St = \frac{c_f}{2Pr^m} \tag{9.88}$$

The Nusselt number for constant pressure flow is correlated by

$$Nu = A \left(\frac{\rho^*}{\rho_e}\right)^a \left(\frac{\mu^*}{\mu_e}\right)^b Re_x^c (\sqrt{3})^j \tag{9.89}$$

The quantities ρ^* and μ^* are to be evaluated at the reference temperature T^*, which is associated with the associated reference enthalpy h^*. This relationship has been discussed in Section 8.4 of the previous chapter, and a model for laminar flow was given in Eqn (8.75) and a slightly different one for turbulent flow in Eqn (8.80). The slender planar bodies considered here will have approximately constant pressure over most of the skin surface, which should permit this approach. Under the assumption of Reynolds' analogy, the local skin friction coefficient is given by

$$C_f(x) = 2St = \frac{2Nu}{Pr^{1/3}Re_x} = \frac{2A}{Pr^{1/3}} \left(\frac{\rho^*}{\rho_e}\right)^a \left(\frac{\mu^*}{\mu_e}\right)^b Re_x^{c-1} \sqrt{3^j} \tag{9.90}$$

The coefficients used in Eqn (9.90) are defined in Table 9.5.

Extensive discussion of Reynolds analogy for chemically reacting laminar and turbulent flat plate boundary layers is presented by Dorrance (1962). Although the flat plate reference enthalpy (FPRE) method for skin friction and Reynolds analogy for heat transfer are based on flat plate (constant pressure) flows and isothermal walls ($T_w = $ constant), they are often used outside their area of applicability. It appears that for moderate positive or small negative pressure gradients, the technique works, but outside that limited range, their use is suspect. On the other hand, for preliminary design purposes, which often require comparative rather than absolute decisions, the simplicity makes the techniques attractive and commonly used.

9.7.2 CALCULATING SURFACE HEATING

Two basic methods have been introduced to calculate the local heat transfer: the FPRE method and Lees blunt body solution method. Near stagnation points, the FPRE method yields very high values of the heat transfer because it involves the distance from the stagnation point raised to a negative power. The Yoshikawa (1969) simplification of the Fay and Riddell (1958) approach for stagnation point heating is recommended. The blunt body solution due to Lees (1956) should be applied for body panels in the vicinity of a stagnation point or line. Far from a stagnation point or line, the FPRE method should give good results because body curvature of spaceplanes tends to be slight.

The FPRE method and the blunt body solution method will yield different values of heat transfer as one proceeds away from the stagnation point or line, but at least both solutions have similar behavior. Therefore, around the stagnation point or line, the blunt body heat transfer method is to be used, and the heat transfer calculated by both methods should be analyzed for each body panel as one proceeds away from the stagnation point or line. There will be a panel at which the two methods provide the same or very close heat transfer, and there, we must switch from the blunt body method to the FPRE method for panels further downstream.

In general, the two basic calculation methods are sufficient for a preliminary design assessment. However, there are certain aspects of the heating of space capsules and spaceplanes that deserve special consideration, and they will be described in the subsequent sections.

9.7.3 HEATING CHARACTERISTICS OF SPACE CAPSULES

The flow over a space capsule at an angle of attack is three dimensional in character, and a schematic diagram of a capsule at the angle of attack is illustrated in Figure 9.30. It is assumed that flight is in the symmetry plane of the capsule. The heat transfer at practical angles of attack is found to be considerably higher than that at the stagnation point with $\alpha = 0$. Wind tunnel experiments on Apollo command module capsule models have been compiled by Lee, Bertin, and Goodrich (1970). Typical results for the ratio of local heat transfer to the stagnation point value at $\alpha = 0$ are shown in Figures 9.31−9.33, and tabulations of results based on the reported data are shown in Tables 9.6−9.8. As the capsule is rotated in the pitch plane from $\alpha = 0$, we note that q/q_0 is approximately constant with s/R_b along the $\theta = 90°$ meridian. However, on the $\theta = 180°$ meridian, the heat flux diminishes with s/R_b, while along the $\theta = 0$ meridian, the heat flux increases with s/R_b reaching a maximum 40−60% greater than the $\alpha = 0$ stagnation point value as the rim is

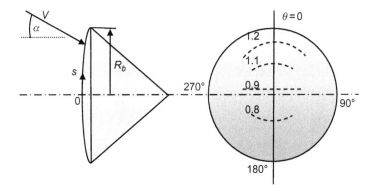

FIGURE 9.30

Schematic of contours of constant heat flux for an Apollo-type capsule at an angle of attack $\alpha = 18°$. Numbers on front face contours denote ratio of heat transfer to the stagnation value at $\alpha = 0$.

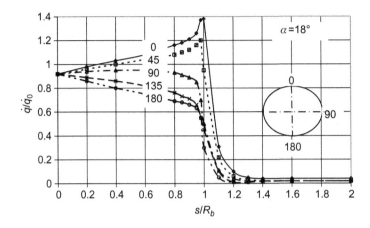

FIGURE 9.31

Wind tunnel measurements on an Apollo model at $\alpha = 18°$ showing the ratio of local heat transfer rate to that measured at the stagnation point when $\alpha = 0$.

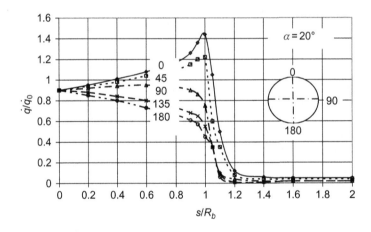

FIGURE 9.32

Wind tunnel measurements on an Apollo model at $\alpha = 20°$ showing the ratio of local heat transfer rate to that measured at the stagnation point when $\alpha = 0$.

approached. Of course, the stagnation point moves up on the $\theta = 0°$ meridian from $s/R_b = 0$ to about $s/R_b = 0.6$ for $\alpha = 18°$ to about $s/R_b = 0.72$ for $\alpha = 25°$. Note that the maximum heat flux is not at the stagnation point but closer to the shoulder on the more windward portion of the blunt face of the capsule.

Low L/D capsules fly at an angle of attack in the range of 15–30°, and this of course eliminates the axial symmetry found at $\alpha = 0$. Surface pressure measurements for such cases have also been reported by Lee et al. (1970), and a typical result is shown in Figure 9.34. As pointed out in

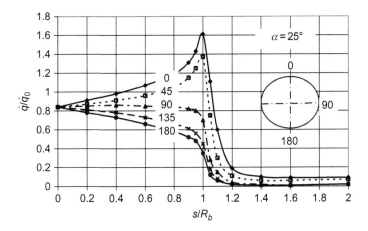

FIGURE 9.33

Wind tunnel measurements on an Apollo model at $\alpha = 25°$ showing the ratio of local heat transfer rate to that measured at the stagnation point when $\alpha = 0$.

Table 9.6 Ratio of Heat Flux at a Point on the Apollo Capsule at $\alpha = 18°$ to the Stagnation Point Value at $\alpha = 0$

α	18°	18°	18°	18°	18°
θ	0°	45°	90°	135°	180°
s/R_b	$\dot{q}/\dot{q}_0$	$\dot{q}/\dot{q}_0$	$\dot{q}/\dot{q}_0$	$\dot{q}/\dot{q}_0$	$\dot{q}/\dot{q}_0$
0	0.92	0.92	0.92	0.92	0.92
0.20	0.98	0.96	0.94	0.89	0.86
0.40	1.03	1.00	0.95	0.86	0.80
0.60	1.08	1.04	0.95	0.82	0.75
0.80	1.16	1.08	0.93	0.76	0.69
0.85	1.18	1.10	0.91	0.73	0.68
0.90	1.21	1.12	0.89	0.71	0.66
0.95	1.26	1.16	0.85	0.66	0.63
0.98	1.37	1.20	0.70	0.55	0.55
1.00	1.38	0.95	0.50	0.45	0.30
1.10	0.31	0.22	0.11	0.10	0.05
1.20	0.10	0.06	0.04	0.01	0.01
1.30	0.05	0.035	0.02	0.01	0.01
1.40	0.04	0.02	0.02	0.02	0.02
1.60	0.04	0.02	0.02	0.02	0.02
2.00	0.04	0.02	0.02	0.02	0.02
2.40	0.035	0.025	0.03	0.03	0.03

Table 9.7 Ratio of Heat Flux at a Point on the Apollo Capsule at $\alpha = 20°$ to the Stagnation Point Value at $\alpha = 0$

α	20°	20°	20°	20°	20°
θ	0°	45°	90°	135°	180°
s/R_b	$\dot{q}/\dot{q}_0$	$\dot{q}/\dot{q}_0$	$\dot{q}/\dot{q}_0$	$\dot{q}/\dot{q}_0$	$\dot{q}/\dot{q}_0$
0	0.90	0.90	0.90	0.90	0.90
0.20	0.95	0.94	0.92	0.88	0.85
0.40	1.01	0.98	0.94	0.84	0.80
0.60	1.08	1.04	0.95	0.80	0.73
0.80	1.17	1.10	0.93	0.73	0.66
0.90	1.26	1.15	0.90	0.68	0.61
0.95	1.36	1.20	0.86	0.65	0.57
1.00	1.44	1.22	0.75	0.55	0.45
1.05	1.05	0.6	0.35	0.35	0.35
1.10	0.50	0.35	0.10	0.09	0.06
1.20	0.12	0.08	0.04	0.01	0.01
1.40	0.06	0.04	0.02	0.01	0.01
1.60	0.05	0.04	0.02	0.02	0.02
2.00	0.05	0.04	0.02	0.02	0.02
2.40	0.05	0.05	0.03	0.03	0.03
2.80	0.05	0.05	0.05	0.05	0.05

Table 9.8 Ratio of Heat Flux at a Point on the Apollo Capsule at $\alpha = 25°$ to the Stagnation Point Value at $\alpha = 0$

α	25°	25°	25°	25°	25°
θ	0°	45°	90°	135°	180°
s/R_b	$\dot{q}/\dot{q}_0$	$\dot{q}/\dot{q}_0$	$\dot{q}/\dot{q}_0$	$\dot{q}/\dot{q}_0$	$\dot{q}/\dot{q}_0$
0	0.84	0.84	0.84	0.84	0.84
0.20	0.91	0.87	0.85	0.81	0.78
0.40	0.98	0.91	0.86	0.78	0.73
0.60	1.07	0.96	0.85	0.73	0.66
0.80	1.19	1.05	0.84	0.66	0.58
0.90	1.31	1.15	0.82	0.61	0.52
0.95	1.43	1.25	0.80	0.56	0.48
1.00	1.61	1.37	0.70	0.45	0.35
1.05	1.11	0.75	0.28	0.18	0.12
1.10	0.60	0.30	0.15	0.08	0.06
1.20	0.19	0.11	0.04	0.02	0.02
1.40	0.10	0.06	0.02	0.01	0.01
1.60	0.09	0.06	0.01	0.01	0.01
2.00	0.09	0.07	0.02	0.02	0.02
2.40	0.09	0.07	0.03	0.03	0.03

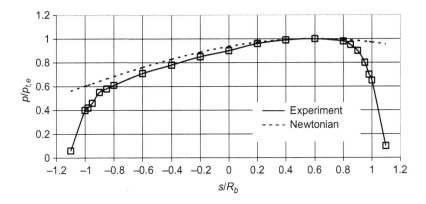

FIGURE 9.34

Experimental pressure distribution for Apollo command module at $\alpha = 18°$ taken from Lee et al. (1970) and corresponding Newtonian approximation.

Chapter 8, the Newtonian theory provides a good approximation to the surface pressure distribution except in regions very near the shoulder of the capsule, and that this is adequate for calculating integrated characteristics like lift and drag forces and associated moments. However, the actual pressure gradient rapidly accelerates the boundary layer as the flow proceeds to the shoulder of the capsule, and this thins the boundary layer and intensifies both skin friction and heat transfer. The skin friction on a blunt capsule is a negligible contributor to lift and drag, but the heat transfer is crucial to the design of a safe heat shield. As a consequence, we turn our attention to this aspect of the boundary layer on the front face of the capsule.

Dewey and Gross (1967) carried out a number of exact similar solutions to the compressible boundary layer equations for surface variables like skin friction and heat transfer using the transformed coordinates given by

$$\eta = \frac{\rho_e u_e r^j}{\sqrt{2\xi}} \int_0^y \frac{\rho}{\rho_e} dy \tag{9.91}$$

$$\xi = \int_0^x C\rho_e u_e \mu_e r^{2j} dx \tag{9.92}$$

This coordinate transformation incorporates a number of classical intermediate transformations and is discussed in some detail by Dorrance (1962), for example. Dewey and Gross (1967) chose to set the Chapman-Rubesin parameter $C = \rho_w \mu_w / \rho_e \mu_e$, so that we may write Eqn (9.92) as follows:

$$\xi = \rho_w \mu_w a_t \int_0^x \left(\frac{u_e}{a_t}\right) r^{2j} dx \tag{9.93}$$

The velocity ratio in the integrand is a function of the Mach number in the inviscid shock layer and, for constant γ, is given by

$$\frac{u_e}{a_{t,e}} = \left(\frac{M_e^2}{1 + \frac{\gamma - 1}{2} M_e^2}\right)^{\frac{1}{2}} \tag{9.94}$$

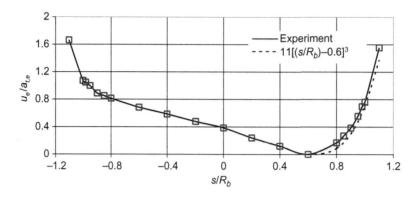

FIGURE 9.35

Nondimensional velocity distribution along the 0–180° meridian at $\alpha = 18°$ and an approximate curve fit for the 0° meridian from the local stagnation point.

The Mach number in the external inviscid flow may be found from the pressure distribution by using the following isentropic relation in the inviscid flow:

$$M_e^2 = \frac{2}{\gamma_e - 1}\left[\left(\frac{p_e}{p_{t,e}}\right)^{\frac{1-\gamma_e}{\gamma_e}} - 1\right] \qquad (9.95)$$

As mentioned in the discussion concerning Figure 9.34, the Newtonian pressure distribution is insufficiently accurate in this rapidly accelerating flow field, so that use must be made of the experimental values. An appreciation for the acceleration of the flow is provided in Figure 9.35 where the nondimensional external velocity $u_e/a_{t,e}$ is shown as a function of distance along the 0–180° meridian of the capsule. The approximation for the velocity distribution along the 0° meridian is a power law with $u_e/a_{t,e} = 11[(s/R_b) - 0.6]^3 = 11(s_1/R_b)^3$ where s measures the distance along the body from the stagnation point at zero angle of attack, and s_1 measures the distance along the body from the stagnation point at the given angle of attack as shown in Figure 9.36. The pressure distribution in Figure 9.35 shows that the stagnation point is located at $s/R_b = 0.6$ on the 0° meridian when $\alpha = 18°$. Note that the Newtonian pressure distribution does identify the stagnation point quite well.

The axisymmetric stagnation point solution ($\alpha = 0$) is readily carried out and is discussed in detail in many textbooks (see, e.g., White (2006)). The important characteristic for boundary layers is the velocity gradient in the inviscid outer flow. For the axisymmetric stagnation point, there is substantial information on the inviscid velocity, which is found to be linear with distance from the stagnation point, $u_e = Ks$. For the case where the angle of attack is not zero, the nature of the velocity gradient depends upon various factors including bow shock wave shape and standoff distance. Furthermore, the velocity curve fit shown in Figure 9.35 shows insufficient detail in the stagnation region itself, and we are interested in the higher heat transfer rate that appears from Figures 9.31–9.33 to occur closer to the capsule shoulder than at the stagnation point. Note that $u_e/a_t = M_e(T_e/T_t)^{1/2} \sim M_e$ so that the flow is subsonic until around $s/R_b = 1$. In many respects then, the velocity distribution on the 0° meridian in Figure 9.35 looks more like the flow through a nozzle starting from the low-speed plenum chamber and reaching a low supersonic Mach number at its exit and we will treat it as such.

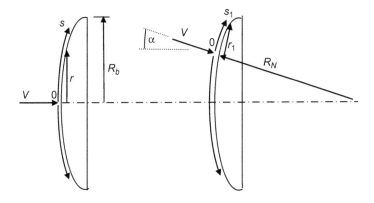

FIGURE 9.36

A schematic diagram of stagnation regions of a capsule heat shield at zero and non-zero angles of attack illustrating the coordinates considered.

Substituting the curve fit for the velocity in the angle of attack case into Eqn (9.93), we obtain

$$\xi = \rho_w \mu_w a_t R_b^3 \int_0^{s_1/R_b} 11 \left(\frac{s_1}{R_b}\right)^3 \left(\frac{r_1}{R_b}\right)^2 d\left(\frac{s_1}{R_b}\right) \tag{9.96}$$

Furthermore, because the curvature of the heat shield is small, we may approximate the local radius r_1 by s_1. The transformed coordinate then becomes

$$\xi = \frac{11}{6} \rho_w \mu_w a_t R_b^3 \left[\left(\frac{x}{R_b}\right)\right]^6 \tag{9.97}$$

We see from Eqn (9.97) that $u_e \sim \xi^m$ so that the pressure gradient parameter β that appears in the boundary layer equations is, in this case, given by

$$\beta = \frac{2\xi}{u_e} \frac{\partial u_e}{\partial \xi} = \frac{2\xi}{n\xi^m} mn\xi^{m-1} = 2m \tag{9.98}$$

Here, from Eqn (9.97) and the fit to the velocity curve in Figure 9.35, we may deduce that $m = 1/2$ and therefore $\beta = 1$. It may be recalled that the velocity at an axisymmetric stagnation point is linear where $u_e = Kx$, so in that case, $\beta = 1/2$. Much interest in the past has been on decelerating flows ($\beta < 0$) because of the deleterious effects of separation, while heat transfer appeared to be relatively unaffected by accelerating flow as described, for example, by Cohen and Reshotko (1956) who treated values of $\beta \leq 2$. However, strongly accelerating flows show singular behavior as pointed out by Dewey and Gross (1967) who provided an exhaustive compilation of similar solutions to the laminar boundary layer equations including cases up to $\beta = 5$. In their analysis, the heat transfer to the wall is given by

$$\dot{q}_w = \frac{\rho_w k_w h_{t,e}(1 - T_w/T_{t,\infty})u_e r^j}{c_{p,w}\sqrt{2\xi}} \theta'(0) \tag{9.99}$$

Here the quantity $\theta'(0)$ is the nondimensional enthalpy gradient evaluated at the surface and is one of the tabulated results presented by Dewey and Gross (1967). The heat transfer for the capsule at the angle of attack is then given by

$$\dot{q}_w = \left[\frac{\sqrt{\rho_w \mu_w}}{\mathrm{Pr}_w} h_{t,e} \left(1 - \frac{T_w}{T_{t,\infty}} \right) \right] \theta'(0) \sqrt{\frac{33 a_t}{R_b}} \left(\frac{s_1}{R_b} \right) \tag{9.100}$$

Carrying out the same procedure for the conventional axisymmetric stagnation point where $u_e = Ks$ leads to

$$\dot{q}_{w,0} = \left[\frac{\sqrt{\rho_w \mu_w}}{\mathrm{Pr}_w} h_{t,e} \left(1 - \frac{T}{T_{t,\infty}} \right) \right] \theta'_0(0) \sqrt{2K} \tag{9.101}$$

If we denote the $\alpha = 0$ case by the subscript 0, then the ratio of the heat transfer on the $\theta = 0°$ meridian to the stagnation point heat transfer at $\alpha = 0$ under identical flow conditions is

$$\frac{\dot{q}_w}{\dot{q}_{w,0}} = \frac{\theta'(0)}{\theta'_0(0)} \sqrt{\frac{33 a_t}{2 R_b K}} \left(\frac{s_1}{R_b} \right) \tag{9.102}$$

The parameter K in the axisymmetric stagnation point case is given by White (2006), for example, as

$$K \approx \frac{V_\infty}{2 R_b} \sqrt{8 \frac{\rho_\infty}{\rho_t}} = \frac{V_\infty}{R_b} \sqrt{2\varepsilon} \tag{9.103}$$

In hypersonic flow, as described in Appendix A, the stagnation enthalpy across the shock wave is constant so that

$$h_{t,\infty} = h_\infty + \frac{1}{2} V_\infty^2 = h_{t,e} = c_{p,t} T_t \approx \frac{1}{2} V_\infty^2$$

The stagnation sound speed in the shock layer is

$$a_t = \sqrt{\gamma_t R_t T_t} \approx \sqrt{\gamma_t R_t \frac{V_\infty^2}{2 c_{p,t}}} = V_\infty \sqrt{\frac{\gamma_t - 1}{2}}$$

In addition, across the shock the density ratio

$$\varepsilon \approx \frac{\gamma_t - 1}{\gamma_t + 1}$$

Using these results in Eqn (9.102) yields

$$\frac{\dot{q}_w}{\dot{q}_{w,0}} = \frac{\theta'(0)}{\theta'_0(0)} \left(\frac{33}{4} \right)^{\frac{1}{2}} (\gamma_t + 1)^{\frac{1}{4}} \left(\frac{s_1}{R_b} \right) \tag{9.104}$$

The tabulated results show that $\theta'(0)/\theta'_0(0) \sim 1$ (1.03 for a cold wall and 1.05 for a hot wall) so the ratio is considered to be unity for the current level of approximation. The results given by Eqn (9.104) are compared to experimental data for $\alpha = 18°$ in Figure 9.37. We see that the heat flux for the $\alpha = 18°$ case rises linearly with distance from the stagnation point of the $\alpha = 18°$ case and agrees well with the maximum heat flux measured. Although calculation of the boundary layer characteristics for an axisymmetric stagnation point at $\alpha = 0$ is straightforward, as described above and in earlier sections of this chapter, the same is not the case for practical, low L/D, angles of

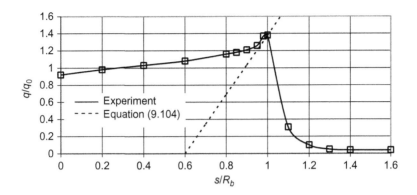

FIGURE 9.37

The variation with distance along the surface of the ratio of the local heat flux at $\alpha = 18°$ to the stagnation heat flux at $\alpha = 0$ according to experiment (solid line) is compared to the approximation of Eqn (9.104).

attack, like those considered here. The use of computational fluid dynamic (CFD) codes is necessary for detailed analyses. However, for preliminary design purposes, it is reasonable to use the extensive Apollo data to provide estimates of the maximum heat flux due to the angle of attack. From Eqns (9.101) and (9.103), we see that $\dot{q}_{w,0} \sim 1/\sqrt{R_N}$ so we may consider the maximum heat transfer to be associated with an effective nose radius R_{eff} such that

$$\frac{\dot{q}_{w,max}}{\dot{q}_{w,0}} = \sqrt{\frac{R_N}{R_{eff}}} \approx (1-0.025\alpha)^{-\frac{1}{2}} \tag{9.105}$$

The heat transfer results in Figures 9.31–9.33 are used to obtain the correlation of Eqn (9.105) which is shown in Figure 9.38.

In addition to the increased heat transfer on the $0°$ meridian due to the acceleration of the flow, there is an increase in the Reynolds number Re_s on the $180°$ meridian. We write the Reynolds number as follows:

$$\frac{Re_s}{Re_t} = \frac{\rho_e u_e s}{\mu_e} \frac{\mu_t}{\rho_t a_t R_b} = \left(\frac{\rho_e}{\rho_t}\right)\left(\frac{u_e}{a_t}\right)\left(\frac{s}{R_b}\right)\left(\frac{\mu_t}{\mu_e}\right)$$

The variation of Re_s/Re_t with nondimensional distance along the 0–$180°$ meridian for the $\alpha = 18°$ case is shown in Figure 9.39. It is interesting to see that the nondimensional Reynolds number at the bottom of the heat shield ($\theta = 180°$) grows to about five times the value at the top of the heat shield ($\theta = 0$). This effect has ramifications for larger heat shields where there is sufficient fetch for the Reynolds number to grow to the point where turbulent flow is initiated. Edquist, Liechty, Hollis, Alter, and Loomis (2006) used NASA-developed codes called LAURA and GASP to predict the three-dimensional heat transfer distributions occurring on planetary entry capsules at an angle of attack. They show that the boundary layer on the $180°$ meridian of the heat shield can become turbulent leading to high heating rates. Jones (1968) reported experiments that indicated transition on the $180°$ meridian of Apollo models at practical angles of attack. An exhaustive and thoughtful review of transition on space capsules is presented by Schneider (2006).

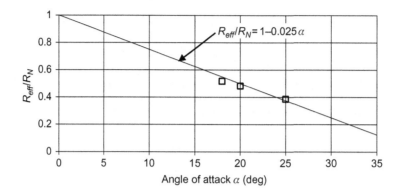

FIGURE 9.38

Correlation curve for maximum heat transfer to Apollo-like capsules as a function of the angle of attack. Symbols are data points for the 0° meridian.

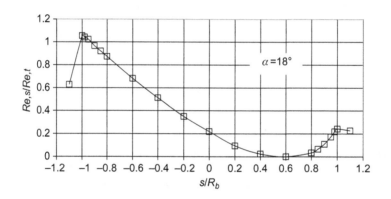

FIGURE 9.39

The variation of Re_s/Re_t with nondimensional distance along the $0-180°$ meridian for the $\alpha = 18°$ case.

Concern is not limited to the forward facing heat shield, which takes the brunt of entry heating but also the afterbodies of space capsules, the flow field over which is complicated. A Gemini capsule model fired at an angle of attack in a ballistic range is shown in Figure 9.40, and the important afterbody flow features are illustrated. The three-dimensional flow field is complex, with separated flow over the leeside of the afterbody and attached flow with shock waves over the windward side. The cabin section had relatively low heating until the aft portion containing the reaction control system (RCS) was approached. Then, the heating on the windward side rose much higher than on the leeward side until the RCS section was reached, whereon the heating was up to 75% of the stagnation value at $\alpha = 0°$. It is clear then that extended afterbodies are subject to enhanced heating due to attachment of the windward side flow and shock wave impingement while the separated leeside flow may become turbulent. Other capsules, like Soyuz and Shenzhou illustrated previously in

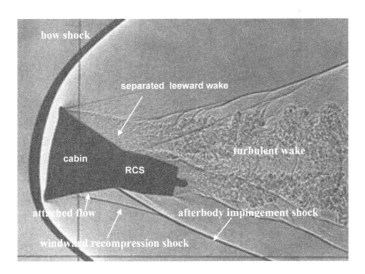

FIGURE 9.40

The important capsule afterbody flow features are illustrated on a shadowgraph of a Gemini capsule model fired in a ballistic range.

Figure 8.13 have very small afterbody cone angles, perhaps 7° or so, and therefore they are more prone to entry heat damage and require more robust TPS than capsules with larger cone angles. As mentioned, afterbody flow fields are highly three-dimensional, and their treatment is beyond the scope of this book.

9.7.4 HEATING CHARACTERISTICS OF SPACEPLANES

The relatively slender configuration of spaceplanes makes them particularly amenable to calculation on the basis of blunt body regions (axisymmetric nose or cylindrical leading edge) followed by gentle turns leading to an essentially streamwise direction. The spaceplane design is aimed at achieving relatively high L/D, and this naturally requires small displacements of the surface from the mainstream direction. The increased performance expected of spaceplanes comes at a price in the form of trim drag when aerodynamic surfaces are moved to obtain the required performance. Consider the infrared photograph of the windward side of a Space Shuttle Orbiter during entry shown in Figure 9.41. The deflection of the trailing edge elevons and the rear body flap and the associated shock waves and their interaction with the boundary layer cause increased local heating. Indeed, the mere existence of hinge lines underscores the fact that there are necessary gaps in the fuselage and wing structure that contribute to enhanced local heating even when control surfaces are not deflected.

The simple panel method for calculating flow properties provides a means for evaluating heating of the large surface acreage typical of winged spaceplanes subjected to varying intensities of heating as is obvious in Figure 9.41. Like the determination of the pressure field described in the previous chapter, the approach taken for the calculation of the heating involves some planning of the number and distribution of panels to use to properly account for the highly curved areas typical

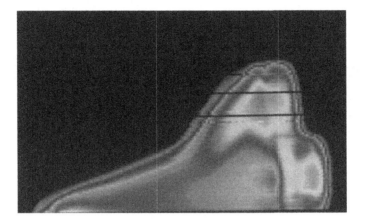

FIGURE 9.41

Heating pattern of the windward side of the Space Shuttle Orbiter during entry as measured by infrared instruments aboard the NASA Kuiper Airborne Observatory.

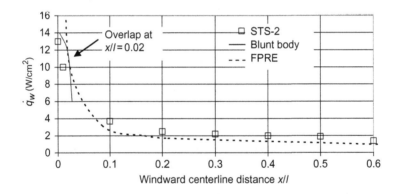

FIGURE 9.42

Windward centerline heat transfer on STS-2 ($V = 2.78$ km/s, $z = 47.7$ km, $\alpha = 34.8°$) is compared to calculations by the indicated methods applied in the regions shown. Dotted line is a fairing of the different results.

of noses and wing leading edges. Paneling and contour coordinates for several spaceplanes are presented in Appendix B.

The methods described earlier in this chapter permit relatively simple determination of the heat transfer at a stagnation point and a spherical or circular region beyond the stagnation point, while the FPRE method is applied to all the relatively planar downstream regions. The blunt body calculations for a spherical nose or a circular leading edge lead to reduced heating as distance from the stagnation point increases, while the FPRE calculations predict increasing heat transfer as the stagnation point is approached. The point of overlap is considered to delineate the regions of applicability of the two solutions. This is demonstrated in Figure 9.42 where a typical result for heating on

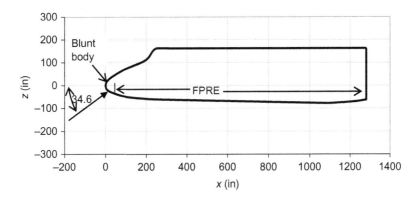

FIGURE 9.43

Space Shuttle Orbiter fuselage showing regions of application of the different estimation procedures. The case is for STS-2 with $V = 2.78$ km/s, $z = 47.7$ km, and $\alpha = 34.6°$.

the windward centerline of the Space Shuttle Orbiter fuselage during the STS-2 mission is shown along with the calculated results. The point on the trajectory shown has $V = 2.78$ km/s, $z = 47.7$ km, and $\alpha = 34.6°$. The agreement of the calculations is quite reasonable considering the simplicity of the applied techniques. The regions in which the two calculation procedures were applied are shown in Figure 9.43.

The critical importance of reducing spacecraft mass must be balanced against the more critical issue of crew safety, and more accurate design procedures serve both needs. This emphasizes the use of detailed CFD complemented by specific wind tunnel tests in the detailed design phase. A review of basic and advanced techniques for computation of high-speed convective heat transfer is given by Tauber (1989), and detailed treatment of CFD applications is presented by Anderson (1989).

9.8 HEAT SHIELD DEVELOPMENT AND PRACTICAL APPLICATIONS

Atmospheric entry conditions are so severe that the vehicle's TPS constitutes a single-point failure system. This was recognized early in the development of ICBMs when warheads could not survive the extreme heating environment of steep entry. Even more stringent were the requirements for safe manned atmospheric entry. Two innovations, one geometrical and the other chemical, were instrumental in advancing the art of safe atmospheric entry. The former is the idea of using a blunt rather than a pointed body to reduce the convective heat load and the latter is the idea of permitting some vehicle material to melt and vaporize to reduce the heat load on the remaining material. The Apollo and ICBM programs during the period 1960–1975 promoted extensive research and development of ablative TPS because of the high heat loads of lunar return and steep ballistic reentry. The first practical ablative materials were composite materials with a base matrix of phenolic resins containing silica-based materials (like fiberglass) for lower temperature applications or carbon-based materials (like graphite) for higher temperatures. An interesting account of the historical and

Table 9.9 Families of Candidate Ablative TPS Materials

Nylon Phenolics (10)[a]	Silica Phenolics (12)
Teflon Teflons (1)	Silica Silicones (14)
Phenolic Nylons (1)	Polybenzimidazoles (3)
Phenolic Phenolics (6)	Inorganics Silicone binders (6)
Carbon Phenolics (23)	Organics Silicone binders (1)
Carbon materials (2)	Other(4)
Silica Teflon (1)	

[a]*Numbers in parenthesis indicates number of materials in indicated class.*

Table 9.10 Families of Candidate Reusable TPS Materials

Carbon-based materials (10)[a]	Adhesives (1)
Silicone-based materials (17)	Coatings (2)
Polyamides (1)	Metallics (9)

[a]*Numbers in parenthesis indicates number of materials in indicated class.*

technical development of ablative heat shields is presented by Sutton (2006), an important pioneer contributor to the field.

An extensive thermophysical property literature review on materials that could be used on spacecraft TPS was compiled by Williams and Curry (1992). They divided the materials into two basic groups: ablative and reusable materials. The ablative materials are arranged into 12 categories that are descriptive of their material composition as listed in Table 9.9. The reusable material families are listed in Table 9.10. The thermophysical properties presented for both ablative and reusable TPS materials are density, specific heat, and thermal conductivity. A comprehensive table of contents is provided by Williams and Curry (1992) to aid researchers in finding the appropriate data.

After the end of the Apollo program, economic pressure on space activities shifted attention toward the reusable Space Shuttle System, which faced the less stressful environment of entry from low-earth orbit. Higher performance ablators for TPS were now needed only for some NASA science programs involving entry of probes into the atmosphere of Venus, Mars, and Jupiter. Some materials, like FM5055 carbon phenolic developed for use on ICBMs, were adapted by NASA for application and further research. By the middle of the 1990s, NASA had been engaged in the development of two new ablative materials: PICA and Silicone Impregnated Reusable Ceramic Ablator (SIRCA). In addition, the success of reinforced carbon–carbon (RCC) composite material used on the nose and leading edges of the Space Shuttle Orbiter prompted the development of advanced carbon–carbon (ACC) composite material.

The Stardust Sample Return Mission collected samples from the solar wind and returned them to Earth. The Stardust vehicle, shown after recovery in Figure 9.44, used a PICA heat shield to protect it from the 12.8 km/s entry into the Earth's atmosphere in 2006. The Genesis Sample Return

FIGURE 9.44

Stardust probe with PICA heat shield after it returned with dust from comet P/Wild 2 on January 15, 2006. The entry conditions were mass = 45.8 kg, diameter = 0.811 m, speed = 12.8 km/s (at 135 km altitude), and entry angle = $-8°$ (NASA).

Mission was to collect samples of dust from the comet P/Wild 2 and return them to earth. The Genesis vehicle, shown schematically in Figure 9.45, used an ACC heat shield to protect it from the 750 W/cm^2 heating during an 11 km/s entry into the Earth's atmosphere in 2004. Although the final stage of parachute recovery failed and the vehicle crashed on the desert (Figure 9.46), some samples returned by the vehicle still proved viable for analysis. SIRCA was used on the Mars Pathfinder and Mars Exploration Rover missions. NASA's Mars Science Laboratory (MSL) rover mission was to use a super lightweight ablator (SLA-561V), like that on the space shuttle's external tank but the SLA samples failed testing, and a switch was made to PICA tiles. The 4 m diameter heat shield, the largest ever made up to that time, surfaced with PICA tiles successfully landed the Curiosity rover on August 12, 2012.

Current research on the Orion CEV focuses again on high heat-load ablators because the mission involves a lunar return. Five different ablative systems were evaluated for the 5 m diameter heat shield, shown as a prototype in Figure 9.47. The 4.5 m diameter heat shield of the MSL heat shield described previously is similar in appearance. The PICA material used on the MSL heat shield was among the leading choices to protect the CEV from reentry heating but the final choice was Avcoat, the same material used in the Apollo program. In a concise report to the National Research Council on TPSs for future space missions, Venkatapathy et al. (2010) emphasized that "even with a detailed specification in place, getting an 'off-the-shelf' material, such as the Apollo Avcoat, back to its prior 'heritage' level requires years of intense and expensive effort due to the mothballed industrial capability and the lack of key personnel".

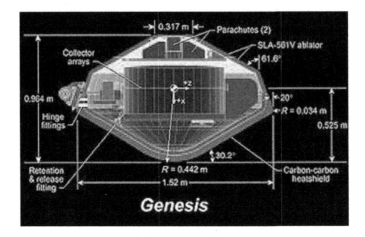

FIGURE 9.45

Genesis probe with ACC heat shield collected pristine material from the solar wind and returned the samples to Earth on September 8, 2004. The entry conditions were mass = 225 kg, diameter = 1.52 m, speed: 11.0 km/s, entry angle = $-8.0°$, peak heating = 750 W/cm², and ballistic parameter $B = W/C_D A \sim 1.2$ kPa (NASA).

FIGURE 9.46

The Genesis vehicle after it crashed in the desert after a failure of the final parachute recovery system. Some samples returned in the vehicle still proved viable for analysis.

Courtesy NASA.

FIGURE 9.47

Prototype 5 m diameter heat shield for Orion Crew Exploration Vehicle.

Courtesy NASA.

The concept of thermochemically smart materials involves the development of composite materials to exploit the performance of the constituent materials in the very specific range of environmental conditions peculiar to hypersonic vehicles. These conditions have been described in some detail in Chapter 1. Fundamentally, it would be valuable to have a material whose thermal, chemical, and structural properties vary over its surface area, through its depth, and over time, as required by the mission. The ability to specify these detailed requirements and to make progress toward constructing an amalgamated material that comes close to meeting those specifications would be a valuable investment. Some conceptual examples are as follows:

- Functionally graded composite material in which $\rho ck = \rho ck(T, y)$ where y is depth from the surface
- Variable surface mass flux by ablation, $\rho V_a = \rho V_a (T)$
- Biomimetic "sweating" materials
- Atomic oxygen protective coatings
- Memory materials for gap sealing

The idea here is that the thermal environment on a hypersonic vehicle will vary over the surface acreage as a function of time, depending upon the trajectory of the flight. Thus, the thermal protection should be tailored to the conjugate heat and mass transfer problem to achieve optimal performance. As pointed out previously, the cooling process will also provide the possibility of simultaneous skin friction reduction, which can pay for any extra weight that might be required to achieve the requisite cooling. The ideas suggested above run the gamut from simple conductivity control, to variable ablation, to two forms of passive transpiration. Here, the need for gap sealing and atomic oxygen corrosion protection is addressed.

A perennial major candidate for advanced TPS is active boundary layer control through surface transpiration. The major effects of transpiration are to reduce friction drag by forcing fluid coolant into the boundary layer through a porous surface of the vehicle. Enhancement of L/D can reduce the fuel capacity required for the mission while providing thermal protection by cooling the surface. Since friction and heating may have different distributions over a given vehicle, there is room for development of smart systems to provide just what is needed to satisfy a set of optimal requirements. Finally, the effects of transpiration on the boundary layer may be tailored to provide some degree of force alteration on a vehicle, and this aspect should be explored to determine its extent and value for aerodynamic control.

9.9 CONSTANTS, CONVERSIONS, AND TPS ACRONYMS

Some useful constants are listed below:

- Stefan–Boltzmann constant $\sigma = 5.6704 \times 10^{-8}$ W/m$^2 \cdot$K^4
- Universal gas constant $R_u = 8.3145$ J/mol $\cdot$ K
- Avogadro's number $N_A = 6.022 \times 10^{23}$/mol

Some unit conversions useful for this chapter are provided in Table 9.11
The error function is defined as follows:

$$erf(x) = \frac{2}{\sqrt{\pi}} \int_0^x \exp(-\xi^2)d\xi$$

A brief table of the error function appears in Table 9.12
The various acronyms and definitions used in TPS analyses are defined and explained as follows:

- ACC = Advanced carbon–carbon composite.
- AETB = Alumina-enhanced thermal barrier: Small amounts of alumina added to FRCI to enhance its thermal stability and conductivity.

Table 9.11 Unit Conversions

Property	Symbol	SI Units	Other	Equivalent
Density	ρ	kg/m^3	10^{-3} g/cm^3	0.0625 lb/ft^3
Heat flux	$\dot{q}$	W/m^2	10^{-4} W/cm^2	8.806×10^{-5} Btu/s $\cdot$ ft^2
Specific enthalpy	h	kJ/kg	10^3 m^2/s^2	0.4299 Btu/lb
Specific heat	c or c_p	kJ/kg $\cdot$ K		0.2389 Btu/lb $\cdot$ R
Temperature	T	K	$C + 273$	$R/1.8$
Thermal conductivity	k	W/m $\cdot$ K		0.5778 Btu/h $\cdot$ ft $\cdot$ R
Thermal diffusivity	α	m^2/s		10.76 ft^2/s
Velocity	V	m/s		3.281 ft/s
Viscosity	μ	Pa $\cdot$ s	kg/m $\cdot$ s	0.02089 lb $\cdot$ s/ft^2

Table 9.12 Error Function of x

x	$erf(x)$
0	0
0.1	0.113
0.2	0.223
0.3	0.329
0.4	0.428
0.5	0.521
0.6	0.604
0.8	0.742
1.0	0.843
1.2	0.910
1.4	0.952
2.0	0.995

- AFRSI = Advanced flexible reusable surface insulation: Silicon-based reusable material made in the form of a quilt with fibrous silica batting sewn between a woven silica outer covering and a fiberglass fabric inner covering. This material is bonded directly to the Space Shuttle Orbiter surface with a silicon adhesive and replaces LSRI tiles.
- Avcoat-5026 H/CG = A type of phenol−formaldehyde resin with various additives loaded into a honeycomb fiberglass matrix used on the Apollo heat shield. It was also selected as the heat shield material for the Orion CEV. It is interesting to note that one of the earliest plastics, Bakelite, is phenolic resin mixed with wood flour.
- BPA = Boeing phenolic ablator: Boeing's proprietary phenolic ablator contains organic and inorganic fibers and microspheres. The ablator is packed into a honeycomb matrix like Avcoat and is considered to be an alternate material for PICA.
- BRI = Boeing rigid insulation: Boeing's proprietary silica-based tiles.
- CEV = Crew exploration vehicle: the space capsule for the Orion project.
- CMCP = Chop molded carbon phenolic: So-called "heritage" carbon phenolic material made from a rayon precursor which, according to Venkatapathy et al. (2010), is in very limited supply and the US industrial base to produce it no longer exists. This is the only material with proven capability to withstand heat fluxes 10,000−30,000 W/cm^2 and could be used for super-orbital entry speeds greater than 13 km/s. The only drawback to the material is its relatively high density.
- FRCI-12 = Fibrous refractory composite insulation: Silicon-based material (192 kg/m^3) tiles used on the Space Shuttle Orbiter to replace heavier LI-2200 tiles. FRCI-12 uses aluminum−borosilicate fiber added to the basic silica tile resulting in improved thermal properties. These black tiles are used when $T < 1533$ K (1260 °C).
- FRSI = Felt reusable surface insulation.
- HETC = High efficiency tantalum-based ceramics composites: NASA patented these tantalum-based ceramics which contain tantalum disilicide, borosilicate glass, and, optionally,

molybdenum disilicide. These components are milled, along with a processing aid to facilitate sintering, and then applied to a surface of porous substrate, such as fibrous or open-pored silica, carbon, or other carbon/silicon combinations. After application, the coating is sintered resulting in a coating that can withstand temperatures above 1920 K.

- HRSI = High temperature reusable surface insulation: Tiles made of silica fibers derived from common sand and made rigid by ceramic bonding. The tiles are 90% void space, and therefore they are quite light in weight. They are coated with borosilicate glass (like pyrex) and baked to form a glossy black waterproof coating with emissivity of 85% and a low catalytic efficiency. The first HRSI were the LI-900 and LI-2200 tiles, and later the FRCI-12 tiles replaced the LI-2200 tiles on the Space Shuttle orbiter. HRSI tiles are typically 15 cm square and 5−12 cm thick.
- LI-900 = Silicon-based tiles having minimal thermal conductivity and weight used on the Space Shuttle Orbiter. Designation refers to a weight density of 144 kg/m^3 (9 lbs/ft^3).
- LI-2200 = Higher strength silicon-based tiles having minimal thermal conductivity used on the Space Shuttle Orbiter. Designation refers to a weight density of 352 kg/m^3 (22 lbs/ft^3).
- LRSI = Low Temperature Reusable Surface Insulation: Essentially the same as HRSI tiles except for the thickness, which is from 0.5 to 3.5 cm thick and the surface coating, which is white. They are used where $T < 923$ K (650 °C).
- MLI = Multilayer insulation.
- PICA = Phenolic impregnated carbon ablator: A member of the family of lightweight ceramic ablators (LCA), which consist of a fibrous ceramic substrate impregnated with an organic resin. PICA comprised a fibrous carbon matrix infused with phenolic resin (Tran et al., 1997).
- RCC = Reinforced carbon−carbon composite: A composite made of layers of rayon cloth impregnated with a phenolic resin. Repeated pyrolysis and treatment with furfural alcohol is carried out until the desired properties are obtained. Used on the nose cap and wing leading edges of the Space Shuttle Orbiter. The outer layers are converted to silicon carbide for strength and various coatings for sealing purposes. RCC is used where $T > 1500$ K, and the operating range is 120 K $< T <$ 1900 K.
- RCG = Reaction cured glass: Glass powders blended with thickeners and pigments to use as shuttle tile coatings.
- ROCCI = Refractory oxidation-resistant ceramic carbon insulation: Low-density carbon−carbon composites impregnated with a siloxane gel and pyrolyzed to produce an insulating material of carbon, silicon, and oxygen, which also provides oxidation protection as described by Skokova and Leiser (2002).
- SIRCA = Silica impregnated reusable ceramic ablator.
- SLA-561 = Super lightweight ablator: A low-density low-thermal conductivity ablator/insulator. It is a highly filled elastomeric silicone composition containing cork as one of the filler ingredients. The material forms a robust crack-free char during ablation and had been used on the Space Shuttle External Tank and the mars Viking Lander heat shield. The material can be sprayed or bonded and can be machined by sawing, sanding, and grinding.
- SPAM = SpaceX proprietary ablative material: Proprietary ablator used on the afterbody of the SpaceX Dragon capsule.
- Syntactic foams = Composite materials consisting of a polymer, metal or ceramic matrix filled with microballoons, small, generally spherical, hollow particles.

- TUFI = Toughened unipiece fibrous insulation: A coating to impart greater strength and impact resistance to AETB.
- TUFROC = Toughened unipiece fibrous reinforced oxidation-resistant composite: Developed by Stewart and Leiser (2007). It is a two-piece system consisting of a toughened high-temperature surface cap joined mechanically and adhesively to a low thermal conductivity base and can withstand temperatures up to 1970 K. The carbonaceous cap is composed of ROCCI to provide surface shape stability, while the fibrous base section provides thermal insulation for the mounting structure. The materials can be tailored to meet the specific mission requirements. TPS made of TUFROC have a density of about 400 kg/m^3, while RCC materials are about four times as dense and TUFROC is two orders of magnitude less expensive than RCC.

9.10 NOMENCLATURE

A	aspect ratio or coefficient in Eqn (9.90)
a	sound speed or coefficient in Eqn (9.90)
B	ballistic coefficient $mg_E/C_D S$
b	wingspan or coefficient in Eqn (9.90)
b_f	blowing parameter for friction, Eqn (9.67)
b_h	blowing parameter for enthalpy, Eqn (9.66)
C	Chapman-Rubesin factor $\rho\mu/\rho_e\mu_e$ or constant in Eqn (9.2)
C_a	axial force coefficient F_a/qS
C_D	drag coefficient D/qS
c_f	local skin friction coefficient τ_w/q
C_h	heat transfer coefficient $\dot{q}_w/\rho_e u_e(h_e - h_w)$
$C_{s\text{-}g}$	Sutton−Graves factor, Eqn (9.4)
c	specific heat of solid or liquid
c_p	specific heat at constant pressure of a gas
D_{12}	binary diffusion coefficient
d	maximum diameter of the vehicle normal to its longitudinal axis, $d = 2R_b$
E	energy per mol
e	energy per unit mass
$erf(x)$	error function with argument x
F	force or function in Eqn (9.27)
g	h/h_{te} or gravitational acceleration
H	enthalpy per mol or atmospheric scale height
h	enthalpy per unit mass
h^*	reference enthalpy
h_D	enthalpy of dissociation, Eqn (9.20)
j	index, zero for two-dimensional flow and one for axisymmetric flow
K	constant in Eqn (9.34) or in Eqn (9.103)
K_p	equilibrium constant
k	thermal conductivity
k_r	chemical recombination rate
$k_{1,2,3}$	coefficients in Eqn (9.5)
Le	Lewis number, Eqn (9.20)

L_m	latent heat of fusion
L_v	latent heat of vaporization
l	length
M	Mach number or coefficient in Eqn (9.1)
m	mass or exponent equal to 1/2 for laminar flow and 1/3 for turbulent flow
$\dot{m}_a$	mass flux of atoms
N	coefficient in Eqn (9.1)
Nu	Nusselt number, Eqn (9.80)
Pr	Prandtl number $\mu c_p / \kappa$
p	pressure
Q	total heat energy
q	dynamic pressure ($\rho u^2/2$ or $\gamma p M^2/2$)
$\dot{q}_w$	wall heat transfer rate
Re_x	Reynolds number based on distance x
R	gas constant
R_b	maximum radius of the vehicle normal to its longitudinal axis
R_N	nose radius
R_u	universal gas constant
r	recovery factor Pr^m or radial distance
S	reference area used in forming force and moment coefficients
St	Stanton number, Eqn (9.79)
S_w	wetted area
s	distance along a surface or quotient in Eqn (9.82)
T	temperature
t	time or thickness
u	velocity in the axial direction
V	velocity
V_a	rate of surface recession due to ablation, Eqn (9.56)
W	molecular weight of gas mixture
W_0	molecular weight of gas at standard conditions
X	mole fraction of gaseous species
x	axial coordinate measured from the vehicle nose tip or ground distance
x'	coordinate measured along cone surface from nose tip
x_0	axial distance to vehicle center of gravity
Y	mass fraction of gaseous species
y	depth coordinate in a solid
y_0	spanwise distance to vehicle center of gravity, usually $y_0 = 0$
z	altitude or vertical coordinate measured from the vehicle x_0, y_0 plane
z_0	vertical distance to vehicle center of gravity
Z	compressibility W_0/W
α	angle of attack or thermal diffusivity
β	factor in Eqn (9.68) or pressure gradient parameter in Eqn (9.98)
$\Delta h_{f,i}$	standard heat of formation of species i
ΔT^*	Eqn (9.50)
ε	density ratio or emissivity or see Eqns (9.16)–(9.18)
γ	ratio of specific heats or flight path angle
η	transformed normal coordinate, Eqn (9.56)
μ	viscosity

ρ density
θ polar angle
θ' $d(h/h_t)/d\eta$
σ Stefan−Boltzmann constant
τ dummy variable, Eqn (9.44)
τ_w wall shear stress
υ volume
ξ transformed coordinate, Eqn (9.93)

9.10.1 SUBSCRIPTS

a   axial direction
aw   adiabatic wall conditions
b   body conditions
CG   center of gravity
c   convection
E   conditions at earth's surface
e   boundary layer edge conditions
eff   effective value
i   initial condition
L   conditions pertaining to moments about the *x*-axis
M   conditions pertaining to moments about the *y*-axis
m   melting conditions
max   maximum
N   conditions pertaining to moments about the *z*-axis
n   direction normal to axis
p   propellant
R   resultant
r   radiation
ref   reference value
s   spacecraft
sl   standard sea level conditions
t   stagnation conditions
w   wall conditions
∞ free-stream conditions

REFERENCES

Adams, M. C. (1959). Recent advances in ablation. *ARS J, 29*, 625−632.

Anderson, J. D. (1989). *Hypersonic and high temperature gas dynamics.* New York: McGraw-Hill.

Carslaw, & Jaeger (1948). *Conduction of heat in solids* (2nd ed.). New York: Oxford University Press.

Cohen, C. B., & Reshotko, E. (1956) Similarity solutions for the compressible laminar boundary layer with heat transfer and arbitrary pressure gradient, NACA Report 1293.

Dec, J. A., & Braun, R. D. (2006). *An approximate ablative thermal protection system sizing tool for entry system design.* AIAA-2006-0780, 44th AIAA Aerospace Sciences Meeting and Exhibit, Reno, NV, January 2006.

Dewey, C. F., & Gross, J.F. (1967) Exact similar solutions of the laminar boundary layer equations, RAND Corporation memorandum RM-5089-ARPA.

Dorrance, W. (1962). *Viscous hypersonic flow.* New York: McGraw-Hill.

Edquist, K. T., Liechty, D. S., Hollis, B. R., Alter, S. J., & Loomis, M. P. (2006). Aeroheating environments for a mars smart lander. *Journal of Spacecraft and Rockets, 43*(2), 330−339.

Fay, J. A., & Riddell, F. R. (1958). Theory of stagnation point heat transfer in dissociated air. *Journal of the Aeronautical Sciences, 25*(2), 73−85.

Hansen, C. F. (1959). *Approximations for the thermodynamic and transport properties of high-temperature air.* NASA Technical Report R-50.

Hughes, P. C. (1990). *LDEF temperature histories.* University of Toronto Institute for Aerospace Studies UTIAS Report No. 340 (CN ISSN 0082-5255).

Johnson, J. E., Starkey, R. P., & Lewis, M. J. (2007). Aerothermodynamic optimization of reentry heat shield shapes for a crew exploration vehicle. *Journal of Spacecraft and Rockets, 44*(4), 849−859.

Jones, R. A. (1968). Wind-tunnel measurements of transition on the face of a blunt entry capsule at angle of attack. *AIAA Journal, 6*(3), 545−546.

Kays, W., Crawford, M., & Weigand, B. (2005). *Convective heat and mass transfer* (4th ed.). New York: McGraw-Hill.

Lee, D. B., Bertin, J. J., & Goodrich, W. D. (1970). *Heat-transfer rate and pressure measurements obtained during apollo orbital entries.* NASA TN D-6028.

Lees, L. (1956). Laminar heat transfer over blunt-nosed bodies at hypersonic flight speeds. *Jet Propulsion, 26,* 259−269, 274

Park, C. (1989). *Nonequilibrium hypersonic aerothermodynamics.* New York: Wiley.

Schneider, S. P. (2006). Laminar-turbulent transition on reentry capsules and planetary probes. *AIAA Journal, 43*(6), 1153−1173.

Sforza, P. M. (2012). *Theory of aerospace propulsion* NY: Elsevier.

Skokova, K. A., & Leiser, D. R. (2002). Characterization of a new refractory oxidative-resistant ceramic carbon insulation material. *Fuel Chemistry Division Preprints, 47*(2), 434−435.

Stewart, D., & Leiser, D. (2007). *Lightweight thermal protection system for atmospheric entry.* NASA Tech Briefs, pp. 20−21.

Sutton, G. W. (2006). The initial development of ablation heat protection—An historical perspective. *Space Chronicle, Journal of the British Interplanetary Society, 59*(1), 16−28.

Sutton, K., & Graves, R. A. (1971). *A general stagnation-point convective-heating equation for arbitrary gas mixtures.* NASA TR R-376.

Tauber, M. E., Menees, G. P., & Adelman, H. G. (1987). Aerothermodynamics of transatmospheric vehicles. *Journal of Aircraft, 24*(9), 594−602.

Tauber, M. (1989). *A review of high-speed convective heat transfer computation methods.* NASA TP-2914.

Tauber, M. E., & Sutton, K. (1991). Stagnation-point radiative heating relations for earth and mars entries. *Journal of Spacecraft, 28*(1), 40−42.

Tran, H. K., et al. (1997). *Phenolic Impregnated Carbon Ablators (PICA) as thermal protection systems for discovery missions.* NASA Technical Memorandum 110440.

Van Driest, E. R. (1959). Convective heat transfer in gases. In C. C. Lin (Ed.), *Turbulent flows and heat transfer.* Princeton, NJ: Princeton University Press.

White, F. M. (2006). *Viscous fluid flow* (3rd ed.). New York: McGraw-Hill.

Williams, S. D., & Curry, D. M. (1992). *Thermal protection materials.* NASA Reference Publication 1289.

Venkatapathy, E., Szalai, C. E., Laub, B., Hwang, H. H., Conley, J. L., Arnold, J., et al. (2010). *Thermal protection system technologies for enabling future sample return missions.* White Paper to the NRC Planetary Science Decadal Survey, Primitive Bodies Sub-Panel. Washington, DC: National Research Council.

Yoshikawa, K. K. (1969) *Linearized theory of stagnation point heat and mass transfer at hypersonic speeds.* NASA TN D-5246, August 1969.

SPACECRAFT CONFIGURATION DESIGN

10

10.1 THE SPACECRAFT ENVIRONMENT AND ITS EFFECT ON DESIGN

The spacecraft designer must provide a habitable environment for the crew taking into account the different demands of launch, on-orbit operation, entry, and recovery. In addition, the spacecraft designer must be conscious of the mass constraints posed by the need for cost-efficient operation.

10.1.1 PASSENGER VOLUME ALLOWANCE IN AEROSPACE VEHICLES

The pressurized cabin of a commercial airliner includes the flight deck in the nose cone, the passenger cabin, and cargo areas, and terminates at the aft pressure bulkhead. The cross-sectional shape of the cabin is generally circular, or close to circular, because of the structural and manufacturing benefits of such a shape. The fuselage diameter d_c is set by the number of passenger seats abreast N_a, while the cabin length l_c is set by the number of rows N_r, where the number of passengers $N_p = N_a N_r$. Following the approach presented by Sforza (2014), we may assess the internal space of the passenger cabin in terms of two volume measures, the pressurized volume v_p and the free volume v_f. The pressurized volume is defined as the gross volume contained within the pressure shell which is essentially $\pi l_c d_c^2 / 4$. The free volume is defined as volume that readily permits passenger mobility within the cabin. The free volume, by definition, should therefore not include the volume occupied by seats, partitions, overhead bins, equipment, etc. It is essentially the free space within which passenger activity may readily take place. As the cylindrical pressurized cabin diameter d_c grows larger, one may define an average passenger headroom h, so that the free volume is roughly $(l_c d_c h - v_o)$, where v_o is the volume of all the obstacles to crew movement within the cabin.

Reviewing the characteristics of common commercial aircraft like Boeing's 737, 767, and 777 families and the regional turboprop ATR-72, Sforza (2014) developed equations for estimating free and pressurized volume per passenger for commercial aircraft, and results for specific volumes, that is, pressurized volume and free volume per passenger, are shown in Figure 10.1 as a function of the number of seats abreast N_a.

The interesting feature in Figure 10.1 is that the free volume per passenger is essentially constant for commercial aircraft at about 1.4 m^3 per passenger. The approach used here for the free volume does not fully account for seats and other obstacles, so it may be as much as 10% higher than actual. The pressurized volume, as would be expected, grows linearly with the cabin diameter.

Business jets may have twice the value of specific volume as commercial airliners because they typically are optimized for comfort. For example, the Gulfstream G200, which carries a maximum of 10 passengers, has a specific free volume of about 2.5 m^3 per passenger and a specific

Manned Spacecraft Design Principles. DOI: http://dx.doi.org/10.1016/B978-0-12-804425-4.00010-6

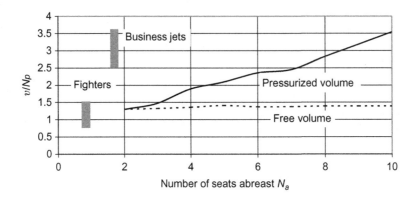

FIGURE 10.1

The variation of the nominal free and pressurized volume (m^3) per passenger for commercial aircraft is illustrated as a function of N_a. Also shown for comparison are typical ranges for business jets and military fighter aircraft.

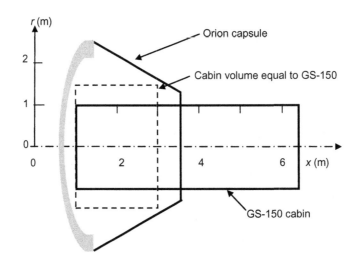

FIGURE 10.2

Comparison of dimensions of Orion capsule and a Gulfstream business jet.

pressurized volume of about 3.1 m^3 per passenger. The larger G500, which can carry 18 passengers, provides a specific free volume of 2.7 m^3 per passenger and a specific pressurized volume of 3.6 m^3 per passenger. The range of volume for these types of aircraft is also indicated in Figure 10.1 for comparison.

A comparison of the volume of a Gulfstream GS-150 business jet cabin and the cabin of NASA's Orion capsule is shown in Figure 10.2. The GS-150 can accommodate six to eight passengers while the Orion is likely to have four to six. Raymer (1989) notes that military fighter aircraft have typical cockpit dimensions that suggest a specific free volume of about 1.1 m^3 per passenger, about 20% smaller than that of commercial aircraft.

10.1.2 APPLICATION TO CREW VOLUME IN SPACECRAFT CABINS

Conclusion may be drawn from the results of the previous section that when a spacecraft cabin is to accommodate a crew requiring relatively little space for mobility, the specific free volume may be in the range of $1-1.5$ m^3 per person. Of course, using the aircraft analogy tacitly assumes that the spacecraft mission duration is brief on the order of a day or so. It seems likely that trained spacecraft crew members can be accommodated by this level of free volume for somewhat longer periods perhaps up to several days. When the mission duration is longer and the crew members have duties that require some sustained mobility, it seems that a larger specific free volume must be made available to them. For example, the specific free volume in nuclear submarines, which have mission durations measured in months, tends to be an order of greater magnitude, around 11.3 m^3 per person.

There seems some controversy on what constitutes an acceptable amount of free volume for the crew although it is generally agreed that that amount depends on the duration of the mission and the degree of activity expected. Free volume is used here to mean that portion of the pressurized volume occupied solely by the cabin atmosphere. Therefore, the free volume is calculable even if it is sometimes tedious to do. The pressurized volume is defined simply as volume defined by the pressure vessel dimensions. As an example, consider two circular cylindrical cabins with length $l_c = 4$ m and diameter $d_c = 2$ m, each of which therefore has a pressurized volume $v_p = 4\pi$ m^3 as shown in Figure 10.3.

Assume further that one cabin has a 1-m diameter cylinder (cross-sectional area $= \pi/4$ m^2) containing needed equipment and supplies mounted concentrically with the outer cabin wall and the other cabin contains the needed equipment and supplies in an annular area also equal to $\pi/4$ m^2. Now, both cabins have the same free volume given by

$$v_f = v_p - \frac{\pi}{4}(1 \text{ m})^2(4 \text{ m}) = v_p - \frac{\pi}{4}\left[(2 \text{ m})^2 - (\sqrt{3} \text{ m})^2\right](4 \text{ m}) = 3\pi \text{ m}^3$$

It seems obvious that although the free volume is the same for both cabin designs, the free volume of the cabin with the annular addition has greater utility for human occupation than the other. This approach, as considered by NASA, is represented by the notional crew compartment as shown in Figure 10.4. The difference between these calculable volumes illustrates the current

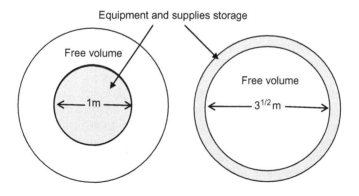

FIGURE 10.3

Cross-section of two 2-m diameter cabins with equal, but differently distributed, free volume.

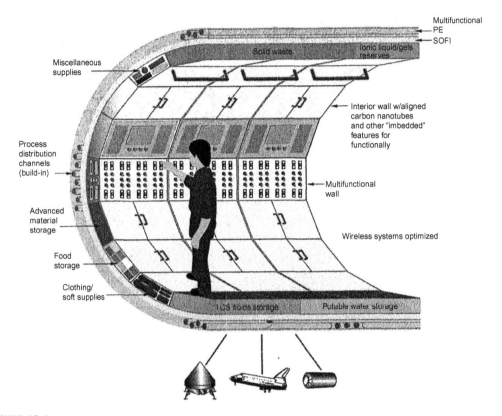

FIGURE 10.4

Multifunctional crew compartment wall system (NASA).

confusion regarding what constitutes a so-called "habitable" volume. This is a subjective term that is of concern to human factors expert when trying to fashion a suitable space in which astronauts can effectively and comfortably perform their functions. An assessment of some of these issues is presented by Hauplik-Meusberger (2011).

Because of the ambiguous nature of the definition of free or habitable volume, it is difficult to construct analytical models to aid in the preliminary design process. The basic external dimensions of spacecraft are considerably more consistently reported, and they can be successfully used to calculate a distinct volume. In general, the mass constraints on spacecraft ensure that the volume defined by gross external dimensions should provide a good approximation to the pressure vessel volume. Using this definition for the pressurized volume v_p, we show its relationship to the more general volume parameter ld^2 in Figure 10.5. The quantities l and d denote the overall length and maximum diameter of the capsule or crew cabin, respectively.

From Figure 10.5, we see that the pressurized volume is reasonably well correlated by an equation of the form $v_p \sim ld^2$ with the constant of proportionality in the range of $0.25-0.6$ for dimensions given in meters. As can be deduced from the spacecraft identified in Figure 10.5, the lower

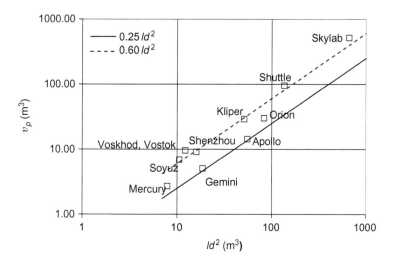

FIGURE 10.5

Nominal pressurized volumes of various spacecraft as a function of the volume parameter ld^2.

value of 0.25 corresponds to basically conical shapes while the higher value of 0.6 corresponds to spherical or cylindrical shapes. It is rather interesting to note that although the free or habitable volume figures quoted in the literature are not narrowly defined, it is possible to draw the conclusion that the free volume of a spacecraft is about 40% ± 10% of the pressurized volume. The small space station Skylab included here has a free or habitable volume equal to almost 70% of the pressurized volume, but it must be kept in mind that Skylab did not have the responsibility of manned entry into the atmosphere like the other spacecraft considered.

The pressurized volume may of course be as large as is consistent with the objectives of the mission, being constrained only by limitations of the launch system in terms of mass, size, and economic considerations. We shall show that the mass of a crew cabin is well correlated by its pressurized volume, and therefore design studies can proceed on that foundation alone. The magnitude of the free the volume is determined by the crew size, the volume and disposition of fixed and stowed equipment, life support (LS) systems, thermal management equipment, etc. The habitable volume, that is, the volume of utility to astronaut activity, is likely to be somewhat less than what we call the free volume, depending upon the final layout of the crew cabin.

We may also draw some conclusions regarding the relation between the pressurized volume and the number of crew members confined within it by examining Figure 10.6. We see that the pressurized volume of operational and proposed spacecraft falls into a fairly narrow corridor defined by two correlation curves. A the low end of the pressurized volume cases, we have

$$v_p = 2 + \frac{1}{2}N_p^2$$

At the high end of the pressurized volume cases, the correlation is

$$v_p = 10 + \frac{1}{2}N_p^2$$

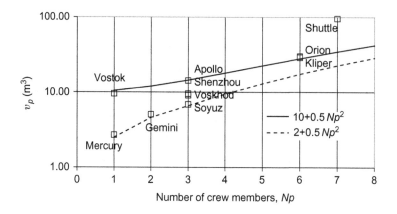

FIGURE 10.6

Nominal pressurized volumes of various spacecraft as a function of the number of crew members.

One interesting feature is that the developmental Orion capsule and the discontinued Kliper spaceplane have very similar dimensions but rather different geometries. Another is that the Space Shuttle Orbiter has substantially more pressurized volume than even the high correlation would suggest. The pressurized volume of 95 m^3 used here is a gross value calculated on the basis of outer mold line dimensions from NASA drawings, while NASA suggests a value of 74 m^3. However, even at that level, we see that it is considerably above the highest correlation line. This is in keeping with the previous comment that the crew cabin pressurized volume is not necessarily dependent on constraints posed by human factors engineering and volumes larger than those considered necessary are easily accommodated.

10.1.3 CABIN VOLUME AND FLIGHT DURATION

Although mission duration should affect cabin volume sizing, it does not necessarily follow that there must be a continuous functional relationship between specific free volume and mission duration. Perhaps, a stepped relationship may be sufficient for estimating vehicle sizing requirements, and a notional illustration of such a relationship is shown in Figure 10.7. It suggests, for example, that for intermediate duration missions, it is recommended to have more free volume than that typical of orbital spacecraft but less than that appropriate for longer duration nuclear submarines. The overlap in free volume requirements, for example, might be like that shown schematically in Figure 10.7. The Skylab space station is shown because it illustrates the notion that the spacecraft volumes may be and often are much larger than some notional acceptable value, if desired. In any event, it seems likely that a "ten-day" spacecraft would not constitute a point design but would need a broad margin of operational flexibility so that 5-day to perhaps 20-day missions could be effectively carried out with the same spacecraft design. This quantum approach also arises naturally from considering that the crew complement is figured in integer quantities. Does the increase in specific free volume arising from a reduction in crew strength significantly improve conditions for the conduct of the mission? Interestingly, the stepwise free volume function in Figure 10.7 may be

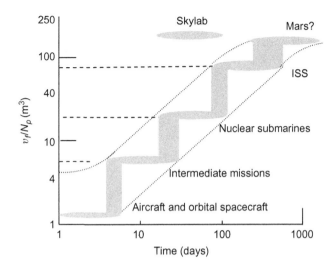

FIGURE 10.7

Notional requirements for specific free volume as a function of mission duration. Dotted lines suggest an acceptable band.

thought of as moving through a space contained within some pair of bounding curves that express the range of acceptable to desirable free volume.

10.1.4 CREW VOLUME ALLOWANCE

The number of crew members and the mission duration are basic specifications that influence the vehicle configuration to an extent not found in unmanned spacecraft. The determination of the free or habitable volume required for each crew member has been a bone of contention in the spacecraft architecture community since the start of manned space activities. One of the earliest embodiments of a prescription of spacecraft volume requirements was a set of three curves of volume as a function of mission duration presented by Celentano, Amorelli, and Freeman (1963). The curves were said to represent three habitability levels: tolerable limit, performance limit, and optimal, and they became a standard benchmark known usually as the Celentano curves or criteria (see, e.g., Woodcock, 1986). The curves were ultimately incorporated into NASA's Man-Systems Integration Standards (MSIS, 1995).

Cohen (2008) presents an exhaustive analysis of the application of the Celentano criterion and its variants as it has been applied over the intervening years as a means for selecting an appropriate volume in which astronauts can function effectively (see also, e.g., Larson & Pranke, 1999; Marton, Rudek, Miller, & Norman, 1971; Woolford & Bond 1999). Cohen (2008) collected a large body of data on existing spacecraft and their mission durations and sought simple relationships between pressurized volume and mission duration. A power law and a log law were constructed to best fit a wide array of data and are shown in Figure 10.8. As mentioned previously, the habitable volumes quoted in the literature for various spacecraft were, on the average, 43% of

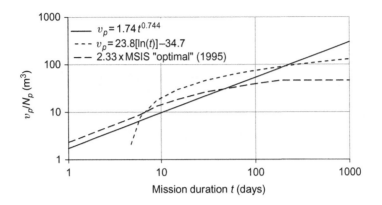

FIGURE 10.8

Power law and logarithmic data approximation curves presented by Cohen (2008) compared to Celentano "optimal" curve from MSIS (1995).

the corresponding calculated pressurized volume. Therefore, the Celentano curve for "optimal" habitable volume was scaled up by a factor of 2.33 to represent pressurized volume, and this result is also shown in Figure 10.8.

The operational data presented by Cohen (2008) was largely contained by the curves presented in Figure 10.8. The International Space Station (ISS) and the Mir space station had maximum values of pressurized volume per crew member of 200 and 180 m^3, respectively, about four times the scaled-up Celentano optimal criterion, but not far from the power law correlation. Therefore, using the greater of the three predictions for any given mission duration seems acceptable as a reasonable upper limit for the pressurized volume in a preliminary design sizing procedure. For comparison, note that the total interior volume of typical SUVs and minivans is about 3 and 4 m^3, respectively, while the interior volume of a typical hotel room is about 70 m^3.

10.1.5 SPACECRAFT MASS CHARACTERISTICS

The specified mission duration and crew complement permits us to compute the likely volume of the crew compartment. In the preliminary design phase, we are interested in the mass of the crew compartment or capsule. To this end, we consider eight manned entry capsules, the crew compartments of two spaceplanes, and the habitat provided by a small space station. The overall mass and major dimensions of these 11 different manned spacecraft are presented in Table 10.1, along with some other relevant data. The mass of the space vehicle should scale with the two-thirds power of the volume since spacecraft, like aircraft, are essentially pressurized shell structures and therefore the mass should be proportional to the shell area. The volume likewise should scale with the volume parameter ld^2, where l is the compartment length and d is the maximum diameter Consideration of the published characteristics of the 11 space vehicles yields the following correlation:

$$m = 470(ld^2)^{2/3}$$

(10.1)

Table 10.1 Characteristics of 8 Manned Capsules, 2 Spaceplane Crew Cabins, and a Small Space Station

Name	Crew	length, l (m)	diameter, d (m)	mass, m (kg)	mg/A (kPa)	B (kPa)	v_p (m³)
Mercury	1	2.21	1.89	1300	4.55	2.789	2.70
Gemini	2	3.68	2.24	3200	7.97	4.605	5.06
Vostok	1	2.30	2.30	2460	5.81	6.315	9.55
Voskhod	2 or 3	2.30	2.30	2900	6.85	7.444	9.55
Apollo	3	3.58	3.91	5900	4.82	2.887	14.30
Soyuz	1–3	2.24	2.17	2950	7.83	4.277	6.89
Shenzhou	1–3	2.50	2.52	3240	6.37	3.483	9.08
Shuttle[a]	6–8	5.00	5.20	6614[b]	3.06	NA	95.00
Kliper[a]	6	4.65	3.30	6000[b]	6.88	NA	29.00
Orion	4–6	3.30	5.00	8913	4.45	2.667	30.00
Skylab	3	14.66	6.70	35,380	9.85	NA	516.00

[a]*Crew cabin only.*
[b]*Estimated.*

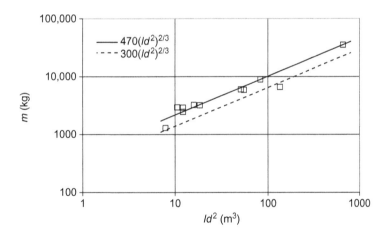

FIGURE 10.9

Spacecraft mass as a function of the volume parameter ld^2.

The mass is given in kilograms and the overall length and maximum diameter in meters. Comparison between this correlation and the tabulated data for the space vehicles is shown in Figure 10.9. Note that a second curve has been added which is closer to the two low data points corresponding to the Mercury capsule and the Space Shuttle Orbiter crew compartment.

A further correlation was made between the number of crew members and the reduced mass, $m^{2/3}$, and this is shown in Figure 10.10. There appear two similar correlation curves depending upon the geometric shape of the spacecraft. For basically conical vehicles, the correlation is

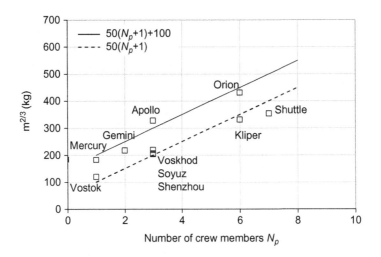

FIGURE 10.10

Correlation between the number of crew members and the reduced mass of a space vehicle.

$$m_{cone} = [50(N_p + 1) + 100]^{3/2}$$

For essentially spherical or cylindrical spacecraft, the correlation is

$$m_{cyl} = [50(N_p + 1)]^{3/2}$$

Skylab is not included in this figure because in a space station, as opposed to a transportation vehicle capable of atmospheric entry, the crew complement is not a driver of the mass. In the case of Skylab, only three crew members are on board and the reduced mass is $m^{2/3} = 1081$ kg$^{2/3}$, which is off the scale of Figure 10.10.

Therefore, with the knowledge of the mission duration and crew complement, the required pressurized volume is known. The pressurized volume is $v_p \sim ld^2$, but neither l nor d is known at this point. Figure 10.11 shows that a reasonable first estimate for the volume parameter is $ld^2 = 0.5 v_p$. Using the assumed value of the volume parameter in Figure 10.9 yields a first estimate for the mass of the capsule or crew compartment. Entering the number of crew members in Figure 10.10 provides another measure of the mass of the capsule or crew compartment with some latitude provided in terms of the general geometry desired. With this information and any external size constraints, it should be straightforward to arrive at an approximate shape, size, and mass of the crew compartment or capsule accurate enough to continue the preliminary design process.

10.1.6 BALLISTIC COEFFICIENT

Also shown in Table 10.1 is the ballistic coefficient for the eight capsule-type vehicles calculated on the basis of the given mass, dimensions, and average $C_{D,0}$ estimated from Eqn (8.26) or Figure 8.12 based on the ratio of maximum transverse radius to nose radius R_b/R_N. Then the

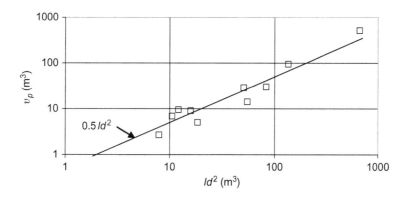

FIGURE 10.11

Correlation between the pressurized volume and the volume parameter of a space vehicle.

Data from Table 10.1.

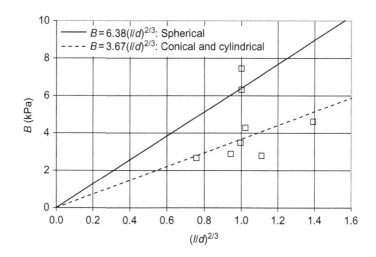

FIGURE 10.12

Correlation between the ballistic coefficient for 10 spacecraft and the fineness ratio l/d of the vehicle, l/d.

Data from Table 10.1.

ballistic coefficient may be determined using the analysis of Section 6.2.1, which gave an estimate of the ballistic coefficient as

$$B = \frac{5.87}{C_D}\left(\frac{l}{d}\right)^{2/3}$$

(10.2)

Using Eqn (10.2), the calculated value for the ballistic coefficient is shown as a function of $(l/d)^{2/3}$ in Figure 10.12. The data indicate that Eqn (10.2) works reasonably well for spacecraft with spherical, cylindrical, or conical configurations. The early Soviet capsules, Voskhod and Voskhod, are spherical while the others are conical or cylindrical, and the higher ballistic coefficient of the spherical capsule makes for a more stressful deceleration upon entry. The drag coefficient of spaceplanes like the Space Shuttle Orbiter is based on wing planform area rather than frontal area so the correlation of Eqn (10.2) is not applicable. Furthermore, spaceplane performance is more dependent upon L/D capability than ballistic coefficient.

10.1.7 SPACECRAFT CONFIGURATIONS

Generic manned spacecraft configurations are shown in Figure 10.13. One of the configurations is a traditional space capsule like the Orion capsule mentioned in other parts of this book as well as new entrants like the SpaceX Dragon capsule as shown in Figure 10.14 and the Boeing CST-100 capsule, the pressure vessel of which is shown in Figure 10.15. The maximum diameters for these two capsules fall between the Apollo (3.91 m) and Orion (5 m) capsules in terms of size. The Lockheed Martin Orion capsule pressure vessel is an aluminum—lithium alloy, but a version of the Orion capsule made of composite material has been pursued by Orbital ATK. A heavily instrumented unmanned Orion capsule was launched on a Delta IV Heavy launcher for flight testing under entry conditions close to those expected in a lunar return mission, a more stressful condition than entry from low earth orbit (LEO). The test, carried out in December 2014, was successful and returned a wealth of data pertinent to refinements of the Orion design. The manned Dragon capsule is a version of the unmanned cargo capsule that is launched on the SpaceX Falcon 9 vehicle. The cargo capsule has flown to and docked with the ISS.

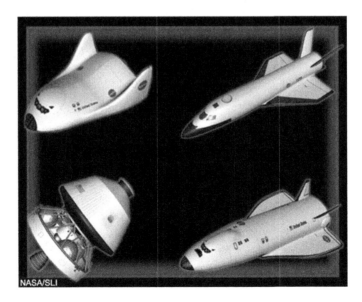

FIGURE 10.13

Generic manned spacecraft configurations. Clockwise from bottom left: space capsule, lifting body, conventional tube and wing, and winged lifting body (NASA).

FIGURE 10.14

The Dragon space capsule leaving SpaceX headquarters in Hawthorne, California, February 23, 2015 (NASA).

FIGURE 10.15

The pressure vessel for Boeing's CST-100 space capsule is shown at NASA's Kennedy Space Center in Florida (NASA).

FIGURE 10.16

The Dream Chaser, a lifting body spaceplane built by Sierra Nevada is shown during taxi and captive-carry flight testing at NASA Armstrong Flight Test Facility in California (NASA).

Another generic spacecraft configuration in Figure 10.13 is the lifting body shape, and this approach is embodied in Sierra Nevada's spaceplane Dream Chaser as shown in Figure 10.16. In October 2013, the full-scale vehicle shown in Figure 10.16 successfully completed a low-speed descent and landing test after being released from a helicopter at an altitude of 3800 m. Although the flight portion of the test was successful, the left landing gear strut did not deploy properly and upon touching down the craft skidded and ultimately ran off the landing strip. Although unmanned subscale models of the X-24A (see Figure 6.67) lifting body were successfully flown in space and recovered during the period 1966–1967, and single passenger lifting bodies were extensively tested in low-speed drop tests in the 1960s, as described in Section 6.8.1, the Dream Chaser is capable of carrying a seven-member crew to space and back. Another lifting body, a biconic shape, is being pursued by Blue Origin, which intends to boost this capsule on a proprietary launcher with a new engine design of its own. The launcher is designed to be reusable, returning to base after launch and landing vertically.

The third basic configuration shown in Figure 10.13 might be called a space airplane because it most resembles an advanced, but conventional, airplane. The only example of a manned spacecraft of this type is the Space Shuttle Orbiter, which has been discussed in some detail throughout this book. However, there is an operational unmanned spaceplane that resembles the space airplane as shown in Figure 10.13. The 5000 kg Boeing X-37B, discussed in Section 12.6.3, is 10 m long and has a wingspan of 4.5 m. It has been launched four times on an Atlas V into LEO, remained on-orbit for as much as 15 months, and then made successful automated entries and landings.

The final entry in Figure 10.13 is the winged lifting body or blended-wing body spaceplane. There are no examples of such a craft that has proceeded beyond a concept to a prototype.

The two larger spaceplanes discussed above face a major hurdle in terms of launch capability because of their high initial mass accruing from their ability to land and be reused. This difficulty was addressed by considering the transatmospheric vehicle, one which not only could take off and land like a conventional airplane but could also fly directly into orbit. This was the idea behind the X-30 National Aerospace Plane, an air breathing scramjet vehicle, and the Lockheed Martin/NASA X-33 Venture Star.

Manned suborbital flights are considered a space tourism business opportunity by several companies. Blue Origin is building the New Shepherd crew capsule for suborbital flight to 100 km with a parachute return for the crew capsule and a powered return of the reusable launch stage to the launch site and a vertical landing. XCOR Aerospace is pursuing their design of a two-place winged spaceplane, the Lynx, designed for suborbital flights. Virgin Atlantic has been testing SpaceShip Two, another spaceplane for suborbital flights, and has already been flight tested, reaching over 31 km in altitude.

10.1.8 SPACECRAFT CABINS

The relatively spacious two-level Orbiter cabin is shown in a cutaway view in Figure 10.17. The essentially cylindrical pressure shell is clearly seen in the figure. A schematic view of the layout of the two levels is shown in Figure 10.18.

A view of the interior of the Apollo capsule is shown in Figure 10.19. The figure is crowded and therefore gives a sense of the degree to which interior space was at a premium. An interesting use of the stowage space was conceived and implemented during the operation of the Skylab mini space station mission. An Apollo capsule was converted to a rescue command module (CM) by freeing the stowage space and installing two more crew seats as illustrated in Figure 10.20. The rescue CM was standing by ready to be launched by a Saturn IB launch system in the event that the problems that had developed with the Skylab thrusters became unmanageable. The plan was to

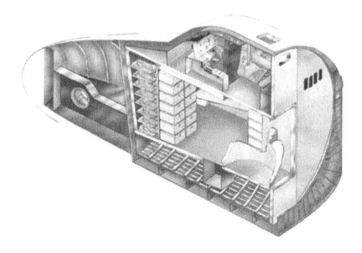

FIGURE 10.17

Cutaway view of the two-level Space Shuttle Orbiter crew cabin (NASA).

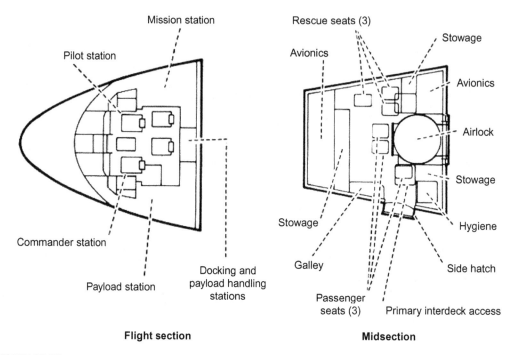

FIGURE 10.18

Schematic diagram of the layout of the Orbiter crew cabin with the upper level on the left and the lower level on the right (NASA).

launch the rescue CM with two crew members, dock with the Skylab spacecraft, and retrieve the three astronauts on board. Luckily, it was not necessary to use this rescue craft.

A similar cutaway view of the Orion space capsule is shown in Figure 10.21, where now four astronauts are accommodated. It is clear that the astronauts are seated such that during entry their backs are facing the heat shield and they are in the optimum position for withstanding deceleration effects. The area between the heat shield and the seats is occupied by stowage volume for the type of equipment described for the Apollo capsule in Figure 10.19. The docking hatch is positioned at the apex of the capsule and the docking mechanism used meets the international standards that have been developed for assuring commonality with spacecraft from different nations.

10.1.9 SERVICE MODULES

The space capsules used in expendable launch systems are designed to be sufficiently robust to survive the rigors of atmospheric entry. This additional strengthening entails mass penalties, so much of the equipment necessary to perform the space mission is carried in a lightweight, generally unpressurized, service module (SM) which is jettisoned prior to entry. In reusable spaceplanes, like

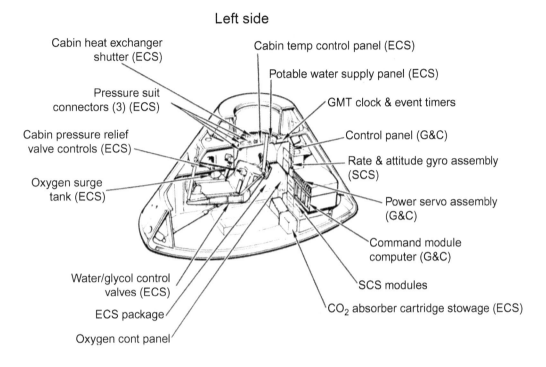

Left side

Cabin heat exchanger shutter (ECS)

Pressure suit connectors (3) (ECS)

Cabin pressure relief valve controls (ECS)

Oxygen surge tank (ECS)

Cabin temp control panel (ECS)

Potable water supply panel (ECS)

GMT clock & event timers

Control panel (G&C)

Rate & attitude gyro assembly (SCS)

Power servo assembly (G&C)

Command module computer (G&C)

Water/glycol control valves (ECS)

ECS package

Oxygen cont panel

SCS modules

CO_2 absorber cartridge stowage (ECS)

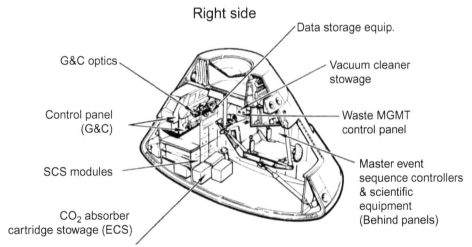

Right side

Data storage equip.

G&C optics

Vacuum cleaner stowage

Control panel (G&C)

Waste MGMT control panel

SCS modules

CO$_2$ absorber cartridge stowage (ECS)

Master event sequence controllers & scientific equipment (Behind panels)

FIGURE 10.19

A cutaway view of the interior of the Apollo capsule (NASA).

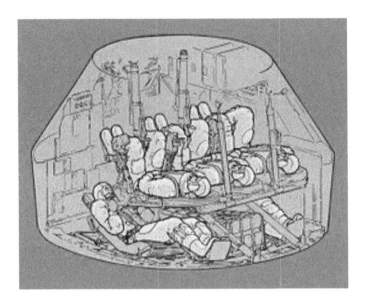

FIGURE 10.20

Artist's illustration of the Skylab rescue command module (NASA).

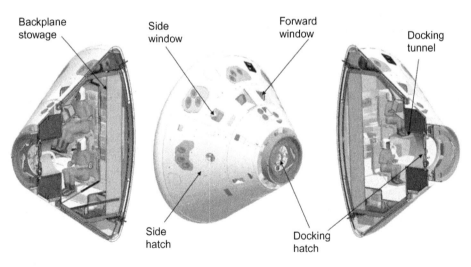

FIGURE 10.21

A cutaway view of the interior of the Orion capsule (NASA).

Table 10.2 Geometry and Mass Characteristics of Several Service Modules

Spacecraft	d (m)	l (m)	m (dry) (kg)	m_p (kg)	ld^2 (m³)	m_t (kg)
Orion-old	5.03	2.80	3700	8300	70.8	12,000
Apollo	3.91	4.46	6100	18,410	67.8	24,510
Orion-esm	4.50	2.72	3800	9200	55.1	13,000
Soyuz	2.72	2.26	2100	800	16.7	2900
Shenzhou	2.52	2.94	2000	1000	18.4	3000

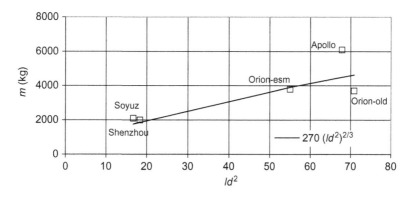

FIGURE 10.22

Geometric and dry mass characteristics of several service modules of expendable spacecraft systems.

the Space Shuttle Orbiter, the roles of service and CMs are combined. SMs serve the following general purposes:

- Carry the electrical power generation system for the spacecraft such as fuel cells, solar photovoltaic cell arrays, and batteries. The CM that carries the crew back to earth is powered by batteries.
- House storage tanks for consumable gases and liquids such as propellants, potable water, breathable atmosphere gases, and the pressurized helium or nitrogen used to force the other consumables to their point of use.
- Contain the thermal control systems for management of the environment of the spacecraft including heat exchangers, space radiators, coolant and associated pumping systems, and the like.
- Carry the components of the environmental control and life support (EC/LS) systems like CO_2 scrubbers, water monitors, waste management, fire suppression equipment, etc.
- House the propulsion system including the main engine used for achieving orbit and deorbit as well as the reaction control system (RCS) thrusters, which provide maneuvering and attitude control functions.

The geometric and dry mass characteristics of several SMs of expendable spacecraft systems are given in Table 10.2, and a plot of the dry mass as a function of the volume parameter ld^2 of the SM is presented in Figure 10.22. The dry mass m is the mass of the SM excluding liquids, and d and l denote the diameter and length of the SM structure (excluding the length of the main engine nozzle), respectively. The designations "Orion-old" and "Orion-esm" refer to two

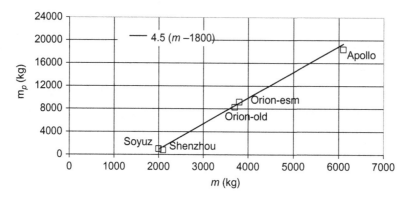

FIGURE 10.23

Variation of service module (SM) propellant mass with SM dry mass.

variants of the SM for the Orion CM. The former is the original version of the SM, which has now been superseded by a new version developed by the European Space Agency (ESA) based on their successful automated transfer vehicle (ATV), which has made many successful unmanned resupply missions to the ISS.

There is some degree of correlation between the mass of the SM and ld^2, its volume parameter, as shown in Figure 10.22, according to the following relation:

$$m = 270(ld^2)^{2/3} \tag{10.3}$$

The correlation curve given by Eqn (10.3) is of the same form as the correlation of Eqn (10.1), which was developed for the CM. Equation (10.3) yields a weaker correlation for the mass of the SM than does Eqn (10.1) for the CM. This is likely due to the greater apparent density of the CM because it must be more robust than the SM. The original version of the Orion SM has about the same mass as the current ESA/ATV-based design but a greater volume. The Apollo SM has a volume parameter about the same as that of the older Orion SM but 50% more mass, largely due to a larger and heavier propulsion system with a main engine developing more than three times the thrust of the Orion-old SM. A more satisfying correlation exists between the dry mass of the SM and the mass of propellant it carries, m_p, as shown in Figure 10.23. An equation that fits the data reasonably well is

$$m_p = 4.5(m - 1800) \tag{10.4}$$

The estimation of SM mass is best begun by estimating the propellant load required to carry out the mission as was done in the analysis of launch staging in Chapter 7. The mission plan will call for a number of relatively large ΔV maneuvers using the main engine such as orbit insertion burn, transfer to a rendezvous orbit burn, and deorbit burn. Using the methods of Chapter 7, the mass of propellant consumed in a maneuver achieving a velocity change ΔV was found to be given by

$$m_p = m_0 \exp\left(\frac{\Delta V}{g_E I_{sp}}\right) \tag{10.5}$$

Here, m_0 is the mass of the spacecraft at the start of the maneuver and I_{sp} is the specific impulse of the engine. For a given ΔV, the larger the I_{sp} of the engine the less mass of propellant consumed. The thrust F determines the amount of time required to complete the maneuver.

$$\Delta t = \frac{I_{sp}}{\left(\frac{F}{m_0 g_E}\right)}\left[1 - \exp\left(\frac{\Delta V}{g_E I_{sp}}\right)\right] \tag{10.6}$$

As will be seen subsequently in Section 10.3.2, the main SM engines of the spacecraft considered in the present discussion are in the thrust range $3\,kN < F < 90\,kN$. The liquid propellants used yield specific impulse values in the range $150\,s < I_{sp} < 340\,s$. For a discussion of such engines and propellants, see Sections 7.5, 8.7, and Sforza (2012).

Having estimated the propellant consumption for each planned mission maneuver, the total mass of propellant required for use by the main engine is their sum. The RCS uses a number of smaller thrusters to provide thrust for attitude control maneuvers that require much smaller ΔV. Assuming that these velocity changes are no more than 10% of those achieved with the main engine suggests that the the propellant load for the RCS is no more than about 10% of the main engine load. Therefore, an additional factor of perhaps 10%–20% may be added to the m_p calculated for the main engine to account for both the RCS requirement and contingencies. Careful planning is important to keep margins of safety from getting out of hand and overwhelming the mass budget of the spacecraft. Then, using Eqn (10.4), a reasonable estimate for the SM mass may be made. Entering Eqn (10.3) or Figure 10.22 with the estimated SM mass yields an estimate for the SM volume parameter ld^2. Setting the SM diameter equal to the previously determined CM diameter provides a preliminary estimate for the length of the SM. As the design progresses to more detailed stages and more information on the components is gathered, it is likely that these preliminary geometry of the SM will be somewhat altered.

10.2 EC AND LS SYSTEMS

Supporting manned systems in space requires an elaborate array of EC/LS systems. Design aspects of the variety of functions and systems used on previous manned missions, both by the United States and by Russia/USSR may be found in Wieland (1994, 2005). The various spacecraft systems and a description of their operation are presented in the following table. The basic systems that define the environmental and LS systems of a manned spacecraft are listed and described in Table 10.3.

10.2.1 THERMAL MANAGEMENT AND CONTROL

Only during entry are there periods of intense heating but even the much lower heat inputs during orbital operations must be managed to maintain the optimum environment for the crew and all the equipment aboard. Flight-proven techniques for managing and controlling heat inputs are described in Table 10.4. The major sources of heat input during on-orbit operation are solar and earth-reflected thermal radiation outside the spacecraft and heat generated by the crew, electrical, and

Table 10.3 Environmental and Life Support Systems

System, EC/LS	Description
HVAC	Thermal management and control
Water	Potable water for crew
Waste	Air purification and personal waste handling
Fire control	Fire suppression system
Emergency systems	Food, medical, and auxiliary power supplies

HVAC, heating, ventilation, and air conditioning.

Table 10.4 Spacecraft Thermal Management

	Mercury Gemini Apollo	Skylab	Space Shuttle Orbiter	Spacelab in Orbiter payload bay	ISS
Thermal control and heat rejection	Condensate boiler on Mercury, space radiators on Gemini, both on Apollo	Space radiators	Freon coolant loop to radiators, ammonia boiler, and flash evaporators	Water loop interfaced with Orbiter Freon loop	Circulating water over cold plates and heat exchangers (HEXs), external space radiators
Equipment cooling	Cold plates and cabin air cooling by passing through condensing heat exchanger	Like Mercury	Like Mercury plus dedicated liquid/air HEXs and additional systems	Like Mercury plus dedicated liquid/liquid HEXs and phase change thermal capacitors	Like Orbiter

electronic devices inside the spacecraft. The major difficulty in orbit is that heat rejection is not very effective because convection is absent when there is no atmosphere. Reliance must then be placed on space radiators but it is difficult to radiate thermal energy to space when the temperatures are relatively low. This drives up the area of the radiator which obviously will make it heavier. Spacecraft radiators reject between 100 and 350 W/m^2 of waste heat. Existing structural surface panels can serve as heat radiators, and this involves no additional weight penalty (note the thermal radiator panels for electrical equipment and for EC/LS systems on the surface of the Apollo SM shown subsequently in Figure 10.25). For optimum heat rejection, a radiator should be on the shadow side of the spacecraft facing toward deep space. This is not possible when using fixed surface radiator panels because the spacecraft is maintained in a slow spin so as to equalize the absorption of external thermal radiation over the entire surface. On the other hand, using a deployable directional radiator involves a weight penalty because they weigh around 10 kg/m^2 when the support mechanism is included. A discussion of various thermal control techniques for spacecraft is presented by De Parolis and Pinter-Krainer (1996).

A spacecraft will absorb a substantial amount of heat during entry, and this heat can be transmitted within and without the structure. In the case of capsules landing by parachute on land or water, it is possible that recovery crews may need up to several hours to get to and extract the crew. During that interval, the cabin temperature must be properly controlled. The Orion capsule uses an evaporative ammonia coolant circulation system to maintain crew and equipment at a comfortable temperature. In addition, the crew will be wearing spacesuits with cool water circulating through undergarments like those used by spacewalking astronauts on the ISS.

10.2.2 SPACECRAFT ATMOSPHERE

The maintenance of a suitable atmosphere depends upon the systems as shown in Table 10.5. Removal of carbon dioxide depends upon lithium hydroxide cartridges that scrub out the carbon dioxide according to the chemical process given below:

$$LiOH + CO_2 \rightarrow LiHCO_3$$
$$2LiHCO_3 + CO_2 \rightarrow Li_2(CO_3) + H_2O$$

Control of the cabin atmosphere in terms of composition is described in Table 10.6. Early launches of Mercury and Gemini relied upon 100% oxygen cabin atmospheres, but the experience of the Apollo 1 fire drove designers to opt for an earth-like mix of nitrogen and oxygen.

10.2.3 WATER RECOVERY AND MANAGEMENT

Water recycling and production systems for LEO operations described in Table 10.7 are suitable for these purposes, but future plans for long-duration flights to asteroids or to Mars will require advanced systems. An example of such a system is the UTC Aerospace Systems Sabatier Reactor System. It has operated aboard the ISS combining waste hydrogen from the oxygen

Table 10.5 Spacecraft Atmosphere Revitalization

Atmosphere revitalization	Mercury Gemini Apollo	Skylab	Space Shuttle Orbiter	Spacelab in Orbiter payload bay	ISS
CO_2 removal	LiOH canisters	Regenerable molecular sieve with Zeolite 5A	Similar to Mercury, etc.	Similar to Mercury, etc.	Similar to Skylab. CO_2 vacuum desorbed to space
O_2 generation	None	None	None	None	Solid polymer electrolyte device
Trace contaminant control	Activated charcoal in LiOH canisters	Activated charcoal in molecular sieve, filters for particulates	Activated charcoal, CO to CO_2 by catalysis, NH_3 absorbed by HEX condensate, filters for particulates	Similar to Orbiter	Activated charcoal with a high temperature catalytic oxidizer, filters for particulates

Table 10.6 Spacecraft Atmosphere Control and Supply

	Mercury Gemini Apollo	Skylab	Space Shuttle Orbiter	Spacelab in Orbiter payload bay	ISS
Atmosphere composition	100% O_2 at 34.5 kPa (60/40 O_2/N_2 at launch for Apollo)	72/28 O_2/N_2 at 34.5 kPa	22/78 O_2/N_2 at 101 kPa	Same as Orbiter	Same as Orbiter
Atmospheric monitoring	CO sensor (Mercury only)	CO and trace contaminant sensors	None	None	N_2, O_2, H_2, CH_4, H_2O, and CO_2 monitors in lab module
Gas storage	O_2 as 52 MPa gas in Mercury, O_2 as 6 MPa cryogenic liquid in Gemini and Apollo	O_2 and N_2 as 21 MPa gases	Same as Skylab, metabolic O_2 from cryogenic source for power system	N_2 as 21 MPa gas, O_2 from Orbiter supply at 690 kPa	O_2 as 16.5 MPa gas and N_2 as 21 MPa gas
Temperature and humidity control	Mercury Gemini Apollo	Skylab	Orbiter	Spacelab	ISS
Atmosphere temperature and humidity control	Condensing heat exchangers (CHEX) for suit and cabin. Pilot regulates T by water flow rate, condensate removed by sponge system or wicks	CHEX with Coolanol 15 coolant. Combination of air duct and wall heaters for T control	Centralized water/air CHEX. Air bypass ratio for T control, condensate removed by wiper and centrifugal separator	Like Orbiter	Like Orbiter, condensate stored in tanks
Cabin ventilation	Cabin fans	Ventilation ducts with fans and portable fans	Cabin fan with ventilation ducts	Cabin fan	Like Orbiter

generating system with CO_2 from the carbon dioxide removal system producing up to 3 L of potable water. The reactor started operation on ISS in 2010 and has proven to be reliable. It serves as a supplement for the main ISS water recovery system described in Table 10.7.

10.2.4 SPACECRAFT FIRE DETECTION AND SUPPRESSION

Spacecraft fire protection is based strongly on the danger posed by smoke inhalation. The movement away from oxygen-rich cabin atmospheres in the wake of the fatal Apollo 1 fire discussed in Chapter 11 has eliminated some of the danger posed by high temperatures and the attendant

Table 10.7 Spacecraft Water Recovery and Management

	Mercury Gemini Apollo	Skylab	Space Shuttle Orbiter	Spacelab in Orbiter payload bay	ISS
Water supply quality	Potable only	Potable only	Potable only	Not applicable	Potable only
Water monitoring	None	Iodine sampler	None	Not applicable	Online conductivity and offline microbial count
Water processing	None: vent waste water or store and send excess to CHEX (Apollo)	Store waste water until tanks full, then vent	Like Skylab	Not applicable	1. Urine: vapor compression distillation. 2. Potable and hygiene: multifiltration, ion-exchange, and catalytic oxidation

Table 10.8 Spacecraft Fire Detection and Suppression

Fire detection and suppression	Mercury Gemini Apollo	Skylab	Space Shuttle Orbiter	Spacelab in Orbiter payload bay	ISS
Suppressant	1. Water from food rehydrator 2. Manual depressurization 3. Apollo had an aqueous gel extinguisher	Portable aqueous gel (foam) extinguisher	Three remote and three portable Halon 1301 bottles. Cabin depressurization	Same as Orbiter	1. CO_2 extinguishers 2. Cabin depressurization
Detection	Crew senses	UV detectors	Ionization smoke detectors	Same as Orbiter	Photoelectric smoke sensors

rapid increase in cabin pressure that might compromise the structural integrity of the crew pressure vessel. Typical suppression and detection systems that have been proven in practice are described in Table 10.8.

10.3 STRUCTURE, PROPULSION, POWER, AND CONTROL SYSTEMS

The basic systems that define the structure, propulsion, power, and control systems of a manned spacecraft are listed and described in Table 10.9.

Table 10.9 Structure, Propulsion, Power, and Control Systems

System	Description
Structure: loads	Maintain integrity of vehicle
Structure: thermal protection system	Protect vehicle from reentry heating
Propulsion: main	Orbital insertion, transfer, and deorbit
Propulsion: RCS	Maneuvering and attitude control
Propulsion: tanks	Contain fuel for all propulsion and power
Power: generation	Fuel cells and solar cells
Power: distribution	Wiring harnesses
Power: storage	Batteries
Control: GNC	Guidance, navigation, and control system
Control: data management	Systems monitoring and control

10.3.1 STRUCTURAL LOADS AND DYNAMICS

NASA STD (1996) describes the accepted practices and requirements for the conduct of load analyses for payloads and spacecraft structures. The following excerpt illustrates the recommended analysis approaches: "Analysis methods fall into three general categories (static, transient, and random vibration analyses). Static analysis is used to predict distribution of loads and displacements in a structure due to slowly varying applied forces. This type of analysis is also used for thermal loads, which arise from temperature changes in the structure. Transient analysis is used to predict loads resulting from applied forces that are rapidly varying and deterministic functions of time. Random vibration analysis deals with applied forces that are not deterministic but are known only in terms of statistical average properties. This type of analysis predicts statistical averages of loads in the structure resulting from applied random forces."

NASA STD (1996) offers the following examples of (i) static events: maximum acceleration during ascent, descent maneuvers, steady spin, installation misalignment, and temperature variations; (ii) transient events: engine ignition and shutdown, launch pad release, staging, control system operation, and landing impact; and (iii) random vibration events: high-frequency engine thrust oscillation, aerodynamic buffeting of the payload fairing, and sound pressure on payload surfaces.

This book deals with developing the overall configuration and estimated mass of manned spacecraft and associated launch systems and the major loads they experience in operation. Developing the structural plan to effectively and efficiently contain those loads is beyond the scope of this book, and for details on structural analysis, the reader is directed toward, for example, Sarafin and Larson (1995) and Wijker (2008).

The thermal protection system (TPS), particularly the heat shield that bears the brunt of heating upon entry, has been discussed in some detail in Chapter 9. An exploded view of an early version of the Orion capsule is shown in Figure 10.24. Avcoat, rather than the originally considered phenolic impregnated carbon ablator (PICA), is used as the heat shield material in the Orion capsule that successfully completed a flight test in December 2014 under thermally stressful entry conditions lying between a LEO return and a lunar return. Details concerning the choice of the

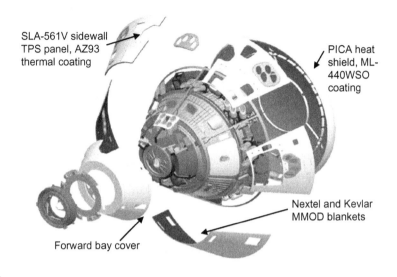

SLA-561V sidewall TPS panel, AZ93 thermal coating

PICA heat shield, ML-440WSO coating

Nextel and Kevlar MMOD blankets

Forward bay cover

FIGURE 10.24

An exploded view of an early version of the Orion capsule. Later versions use Avcoat ablator in the heat shield rather than PICA.

Apollo-era Avcoat ablator as opposed to the newer NASA-developed PICA are discussed in Chapter 9. For the afterbody panels, SLA-561V ablator with an applied thermal coating is used and Nextel and Kevlar blankets are used as insulating material underneath. The objective of the TPS is to maintain the pressure vessel temperature at a level that does not compromise the wall material strength, while the thermal management system is aimed at maintaining acceptable temperatures within the crew cabin.

One important function of the spacecraft structural wall is to contain the internal pressure against external breaches, particularly from micrometeoroid and space debris impacts, which occur with some frequency in orbit. A small puncture of the pressure shell can lead to depressurization posing a lethal threat to the crew. The TPS, which serves to protect the structure from heat, is also the first line of protection from these small high-velocity objects and it serves a sacrificial purpose, absorbing the majority of the energy so that the particle and the debris it produces is slowed considerably in the space between the TPS and the pressure shell. The impact debris still strikes the pressure shell, an aluminum—lithium alloy in the Orion capsule, but with greatly reduced force. This is the basic principle of the Whipple meteorite bumper, named after its inventor, the astronomer Fred Whipple, and used in spacecraft for many years. Such bumpers can readily handle micrometeoroids and similar sized space debris (0.1 cm or smaller), and they form by far the largest population of such bodies so are the most likely to impact a spacecraft. Large pieces of space debris (10 cm or larger) are few in number and are continually tracked by various agencies from earth-based stations. When the trajectory of such a body appears threatening, orbit of the spacecraft is altered slightly to avert a collision. The most dangerous space debris particles are those whose characteristic dimension is on the order of 1 cm because their size makes them difficult to spot.

However, their mass of around 1 g makes them high kinetic energy particles. Closing speeds can be on the order of 10 km/s making the kinetic energy upon impact as much as 50 kJ. This is the equivalent of an impact of a 1-kg rock impacting at 316 m/s. Although the population of space debris particles in this size range is small, attention must be paid to ensure that the pressure vessel of the cabin is sufficiently robust to withstand high-energy impacts. A detailed analysis of meteoroid and space debris protection is given by Christiansen (2008).

10.3.2 PROPULSION SYSTEMS

Maneuvering and positioning of the spacecraft in orbital operations is provided by an orbital maneuvering system (OMS) as discussed in Section 8.7. The propulsion systems for the command and SMs of several spacecraft are described in Table 10.10. When the mission is completed a rocket must be fired to deorbit and begin entry. Associated with the rockets are the propellant tanks for the different propulsion systems and any tanks for power generation by means of fuel cells, for example. The design and construction features of spacecraft tanks were described in some detail in Section 7.7.1. An illustration of the use of a SM to contain auxiliary equipment like propellant tanks for on-orbit maneuvering appears in the schematic diagram of the Apollo CM sitting atop the SM in Figure 10.25. The propellant tanks for the RCS of the CM are stowed in the aft compartment of the CM because the SM is jettisoned before entry.

Table 10.10 Propulsion and Power Systems for Several Spacecraft

Spacecraft	Power	Propulsion[a]
Orion CM	Six 30-Ah lithium ion batteries	12 RCS thrusters (hydrazine monopropellant)
Orion SM	11-kW solar array	One main ($F = 27$ kN) and 8 auxiliary thrusters; 24 RCS thrusters (MMH/MON-3)
Apollo CM	Three 1.4-kW H_2O_2 fuel cells	Twelve RCS thrusters (MMH/NTO)
Apollo SM	Three 40-Ah silver−zinc batteries	One main ($F = 90$ kN, Aerozine 50/NTO) 16 RCS (MMH/NTO)
Orbiter	Three 7-kW H_2O_2 fuel cells	Two main (OMS) and 14 forward RCS thrusters 14 aft RCS thrusters (all MMH/NTO)
Soyuz CM	Batteries	24 RCS thrusters (hydrogen peroxide monopropellant)
Soyuz SM	One kilowatt solar array	One main ($F = 2.94$ kN) and 28 RCS thrusters (UDMH/NTO)
Shenzhou CM	Lead-acid batteries	Eight RCS thrusters (hydrazine monopropellant)
Shenzhou SM	One kilowatt solar array	Four main ($F = 2.5$ kN each) and 24 RCS thrusters (MMH/NTO)

[a]*MMH = monomethyl hydrazine, NTO = nitrogen tetroxide, MON-3 = mixed oxides of nitrogen (NTO + 3%nitrous oxide), UDMH = unsymmetrical dimethyl hydrazine, Aerozine 50 = 50%UDMH + 50%hydrazine, RCS = reaction control system, OMS = orbital maneuvering system.*

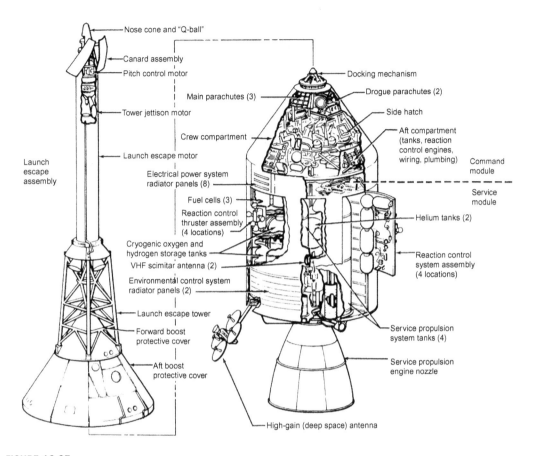

FIGURE 10.25

Schematic diagram of the Apollo CM sitting atop the SM and the launch escape system which rides above the two (NASA).

10.3.3 **POWER SYSTEMS**

Manned spacecraft generally use photovoltaic cell arrays or fuel cells for electric power production, and these are carried in the SM as illustrated in Figure 10.25. Orion was intended to use fuel cells for power generation, but that has been switched to solar arrays with the change to the ESA/ATV SM. It is generally considered that fuel cells tend to be somewhat more problematic for spacecraft use compared to solar arrays although the large solar panel areas required for sufficient power production mean that the array must be stowed folded and then be unfurled in orbit. For example, the solar array of the Orion SM is comprised of four individual wing-like structures in an X-shape, which measures about 19 m across when unfurled. Solar panels may gather about $100-300$ W/m^2. A discussion of solar cells for spacecraft may be found in Jha (2009). Battery power alone is used in the CM because the SM is jettisoned before entry. As pointed out previously, reusable space-planes carry all systems with them throughout the flight.

10.4 CREW SUPPORT SYSTEMS

The basic crew support systems of a manned spacecraft are listed and described in Table 10.11. Communication systems do not have a large impact on the preliminary design phase of manned spacecraft and for information on technical details of such systems, see, for example, Wertz, Everett, and Puschell (2011) and Pisacane (2005).

Crew escape systems are described in some detail in Chapter 11, but for convenience, a typical launch escape system for a capsule is shown in Figure 10.25. It uses a solid rocket launch escape motor to rapidly carry the crew capsule alone to a safe altitude and distance in the event of an emergency during the launch.

Landing systems for manned spacecraft are covered in Chapter 6. The parachutes for a capsule system are shown stowed at the top of the CM in Figure 10.25. This system would be used to safely land the crew either from a launch abort scenario or from a normal entry and landing process.

The ability of a spacecraft to dock with another, that is, to physically join the two spacecraft with an airtight seal to permit crew members to move from one to the other is an important aspect of manned spaceflight. There is an obvious safety aspect to this ability in the sense that crew members in a disabled spacecraft may be saved by transfer to a rescue craft. But, there is also a basic need to move people and equipment between spacecraft. The most familiar example is the transfer of crew members, supplies, and equipment between a spacecraft and the ISS. With the continued expansion of the number of spacefaring nations, there was a perceived need for a standardized docking mechanism and this has been brought about through international cooperation as evidenced by IDSS (2013). The details of the mechanism and its operation are beyond the scope of this book, and for a detailed technical foundation of spacecraft docking, the reader is directed to, for example, Fehse (2003).

Because the radiation hazard from cosmic rays, excited particles, and solar flares are variable in nature, as discussed in Chapter 3, it is imprudent to design the entire crew space to protect the crew from the highest likely radiation event. Instead, the cabin has within it the equivalent of tornado shelters, small enclosures into which the crew can withdraw when a high-energy event is expected. Orion has aluminum equipment lockers measuring about 1 m by 0.64 m by 0.71 m which, upon an emergency alert, are emptied of the contents, which are then suspended on various attachments around the cabin. When the cabinets are empty, the crew members enter the cabinets for the duration of the event. Previous spacecraft like Apollo, Shuttle Orbiter, and ISS have lockers. The Shuttle Orbiter and the ISS operate in LEO, an altitude at which the earth's magnetic field provides

Table 10.11 Crew Support Systems	
System	**Description**
Control: communications	Communication with ISS and ground
Crew systems: radiation	Protection from cosmic rays and solar flares
Crew systems: docking	Crew rescue, ingress/egress other spacecraft
Crew systems: seats	Ergonomic restraint system
Crew systems: escape	Safe abort system
Landing systems: parachutes	Stowage and release capability

substantial shielding so that the lockers are not as substantial, and Apollo's lockers were not designed for very large solar flares. In addition, the Apollo mission duration was 7 days, whereas the Orion design is for three times that and therefore requires more latitude in protective design.

An important detail that Orion capsule designers considered is an impact-absorbing seat to protect the crew's spines from compression during splashdown. The Russian Soyuz and Chinese Shenzhou capsules, which land on the ground, use a brief retrorocket firing before touchdown to cushion the impact.

10.5 NOMENCLATURE

ATV	automated transfer vehicle
B	ballistic coefficient, Eqn (10.2)
C_D	drag coefficient
CHEX	condensing heat exchanger
CM	command module
d	diameter
EC/LS	environmental control and life support
ESA	European Space Agency
F	thrust
g_E	gravitational acceleration at the surface of the earth
HEX	heat exchanger
h	average passenger headroom
ISS	International Space Station
I_{sp}	specific impulse
LEO	low earth orbit
l	length
m	mass
m_p	propellant mass
N_a	number of passenger seats abreast
N_p	number of passengers
N_r	number of rows of seats
PICA	phenolic impregnated carbon ablator
r	radius
SLA	super lightweight ablator
SM	service module
STS	space transportation system
TPS	thermal protection system
t	time
x	longitudinal distance
Δt	time required to achieve a specific ΔV
ΔV	velocity change for an orbital maneuver
v	volume
v_f	free or unobstructed volume
v_o	volume of obstructions
v_p	pressurized volume

10.5.1 SUBSCRIPTS

c cabin
0 initial state

REFERENCES

Celentano, J. T., Amorelli, D., & Freeman, G. G. (1963). *Establishing a habitability index for space stations and planetary bases*. AIAA 1963-139, AIAA/ASMA Manned Space Laboratory Conference, Los Angeles, CA.

Christiansen, E. L. (2008). *Meteoroid/Debris shielding*. NASA TP-2003-210788.

Cohen, M. (2008). *Testing the celentano curve: An empirical survey of predictions for human spacecraft pressurized volume*. SAE Technical Paper Series, 2008-01-2127.

De Parolis, M. N., & Pinter-Krainer, W. (1996). *Current and future techniques for spacecraft thermal control: 1. Design drivers and current technologies*. ESA Bulletin Nr. 87, August.

Fehse, W. (2003). *Automated rendezvous and docking of spacecraft*. New York: Cambridge.

Hauplik-Meusberger, S. (2011). *Architecture for astronauts: An activity based approach*. New York: Springer.

IDSS (2013). *International docking system standard interface definition document*. Revision C. <www.InternationalDockingStandard.com>.

Jha, A. R. (2009). *Solar cell technology and applications*. Boca Raton, FL: Taylor & Francis.

Larson, W. J, & Pranke, L. K. (Eds.), (1999). *Human spaceflight: Mission analysis and design*. New York: McGraw-Hill.

Marton, T., Rudek, F. P., Miller, R. A., & Norman, D. G. (1971). *Handbook of human engineering design data for reduced gravity conditions*. Washington, DC: NASA, NASA CR-1726.

MSIS (1995). *Man-systems integration standards*. NASA STD-3000.

NASA STD (1996). *Load analyses of spacecraft and payloads*. NASA Technical Standard STD-5002.

Pisacane, V. L. (2005). *Fundamentals of space systems* (2nd ed.). New York: Oxford.

Raymer, D. P. (1989). *Aircraft design—a conceptual approach*. Reston, VA: AIAA.

Sarafin, T. P., & Larson, W. (1995). *Spacecraft structures and mechanisms from concept to launch*, The Netherlands: Springer.

Sforza, P. M. (2012). *Theory of aerospace propulsion*. Waltham, MA: Elsevier.

Sforza, P. M. (2014). *Commercial airplane design principles*. Waltham, MA: Butterworth-Heinemann.

Wertz, J. R., Everett, D. F., & Puschell (2011). *Space mission analysis and design—The new SMAD*. Torrance CA: Microcosm Press.

Wieland P. O. (1994). *Designing for human presence in space: An introduction to Environmental Control and Life Support Systems (ECLSS)*. NASA Reference Publication 1324.

Wieland, P. O. (2005). *Designing for human presence in space: An introduction to Environmental Control and Life Support Systems (ECLSS), Appendix I, Update—Historical ECLSS for U.S. and U.S.S.R./Russian Space Habitats*. NASA/TM-2005-214007.

Wijker, J. J. (2008). *Spacecraft structures* Berlin: Springer-Verlag.

Woodcock, G. R. (1986). *Space stations and platforms*. Malabar, FL: Orbit Book Company.

Woolford, B., & Bond, R. L. (1999). Human factors of crewed spaceflight. In W. J. Larson, & L. K. Pranke (Eds.), *Human spaceflight: Mission analysis and design* (pp. 133−153). New York: McGraw-Hill.

SAFETY, RELIABILITY, AND RISK ASSESSMENT

11

11.1 SYSTEM SAFETY AND RELIABILITY

Reliability and safety have always been major goals of space missions and achieving them represents a substantial design challenge. We start with an introductory analysis of the reliability and safety of a manned space vehicle making a round trip to the International Space Station (ISS). The material follows closely the approach of Sforza, Shooman, and Pelaccio (1993) which examines a round trip to Mars using nuclear thermal propulsion for outbound and inbound transit between low Earth orbit (LEO) and Mars. An encyclopedic study of all phases of spacecraft safety is presented by Musgrave, Larsen, and Sgobba (2009).

In general we speak of reliability as the probability of mission success during a given mission time period, t, and denote it by the symbol $R(t)$, or sometimes $P(t)$. In conjunction with space missions, it is common to speak of two aspects of mission success. First, and most important, is the probability of crew safety, that is, the probability that no serious injury to or death of a crew member occurs during the mission. Second is the probability of successfully accomplishing a specific set of objectives for the mission and this is termed mission success. To generalize the notation to include both mission success and crew safety, we introduce the following:

X_i = mission success during phase i of the mission
x_i = mission is safe during phase i of the mission
R_{ms} = probability of mission success
R_{cs} = probability of crew safety.

If we make the simplifying assumption that mission success and crew safety depend on serial success of all the mission phases $1 \leq i \leq n$, then the probabilities become

$$R_{ms} = \Pr(X_1 X_2 \ldots X_n) \tag{11.1}$$

$$R_{cs} = \Pr(x_1 x_2 \ldots x_n) \tag{11.2}$$

A further simplifying assumption that is commonly made is that each of the mission phases are independent, in which case Eqns (11.1) and (11.2) become

$$R_{ms} = \Pr(X_1)\Pr(X_2) \ldots \Pr(X_n) \tag{11.3}$$

$$R_{cs} = \Pr(x_1)\Pr(x_2) \ldots \Pr(x_n) \tag{11.4}$$

Manned Spacecraft Design Principles. DOI: http://dx.doi.org/10.1016/B978-0-12-804425-4.00011-8

Table 11.1 Representative Mission Phases for Two-Stage Rocket Round Trip to ISS

Phase No.	Description	Propulsion Status	Duration
1	Ignition of ME to first stage separation	Main engines and boosters active	125 s
2	Ignition of second stage to MECO	Second stage engines active	400 s
3	MECO to ISS docking	OMS engines active	700 s
4	Orbital operations	Inactive	7 days
5	De-orbit burn to reentry interface	OMS engines active	20 min
6	Reentry	Inactive	10 min
7	Terminal operations	Inactive	15 min
8	Final approach and landing	Inactive	10 min

Table 11.1 shows the major phases that are representative of a round trip mission to the ISS using a two-stage rocket for the launch. A similar table would be constructed for a specific space transportation system and associated mission details.

The top-level functional failures that could lead to a loss-of-vehicle (LOV) event will be discussed in detail in a subsequent section and, in brief, are as follows:

- Propulsion failure—Propulsive thrust must be maintained above certain levels to assure ability to safely abort the mission or to complete it
- Vehicle configuration failure—Control features, such as thrust vectoring, must operate within certain limits to maintain vehicle control
- Containment failure—Escape of high energy gases or solid particles from their proper paths may damage other systems
- Vehicle environment failure—Life support systems must operate within certain limits to assure crew safety
- Externally initiated failure—Lightning strikes, wind shear, space debris, etc. are external events that may precipitate system failures and lead to LOV.

One of the major risk factors for space transportation systems is that of propulsion failure. In the generic round trip to orbit mission considered here there are several phases where rocket engines are firing (phases 1, 2, 3, and 5) and several where they are inactive (phases 4, 6, 7, and 8). In this chapter, we will consider the critical thrusting phases of the mission, focusing on the reliability of the propulsion system which operates during these phases. The success of these phases depends on the successful operation of a number of systems, such as those listed in Table 11.2.

One of the important uses of a top-down analysis, as outlined above, is to help select overall goals and apportion the mission success among the various mission phases and the systems functions operative within them. To reiterate, we assume a simplified model of a phase in which all systems are independent and must operate successfully or safely. The following equations,

Table 11.2 Major Systems Functioning During Thrusting Phases

Symbol	System
Y_1	Propulsion
Y_2	Communications
Y_3	Life support
Y_4	Power and thermal management
Y_5	Navigation, guidance, and control

which are analogous to Eqns (11.3) and (11.4), are applicable for a powered phase, such as phase 1, the first stage ascent:

$$\Pr(X_1) = \Pr(Y_1)\Pr(Y_2)\Pr(Y_3)\Pr(Y_4)\Pr(Y_5) \tag{11.5}$$

$$\Pr(Y_1) = \Pr(y_1)\Pr(y_2)\Pr(y_3)\Pr(y_4)\Pr(y_5) \tag{11.6}$$

The symbols are defined as follows:

Y_i = success of system i
y_i = safe performance of system i
$\Pr(X_1)$ = probability of mission success for the first stage ascent phase
$\Pr(x_1)$ = probability of safe performance of the first stage ascent phase.

The complete system reliability and safety models involve equations like Eqns (11.5) and (11.6) for all the other phases, as well as Eqns (11.3) and (11.4), or Eqns (11.1) and (11.2) if it is not possible to make the assumption that all the phases are independent.

11.2 APPORTIONING MISSION RELIABILITY

The task of choosing appropriate goals for R_{ms} and R_{cs} for the entire mission and apportioning these goals among the various mission phases is difficult. In general, such a task becomes a top-down iterative process of apportionment based on evolving system details alternating with bottom-up reliability prediction of the latest models of the system. Of course, it is expected that these estimates will tend to converge as the project progresses. We begin the top-down process by considering Eqn (11.3). For the present purposes, it is convenient to separate the equation into two parts, one in which propulsion is inactive and one in which it is active. Thus Eqn (11.3) becomes

$$R_{ms} = [\Pr(X_4)\Pr(X_6)\Pr(X_7)\Pr(X_8)][\Pr(X_1)\Pr(X_2)\Pr(X_3)\Pr(X_5)] \tag{11.7}$$

We may designate the unpowered part of the mission by the subscript 1 and the powered part of the mission by subscript 2, so that Eqn (11.7) may be rewritten as

$$R_{ms} = [R_1][R_2] \tag{11.8}$$

Table 11.3 Relationship between Mission Reliability and Phase Reliability

R_{ms}	$1-R_{ms}$	Pr_{low}	$1-Pr_{low}$	Pr_{same}	$1-Pr_{same}$
1.000	0.000	1.000	0.000	1.00000	0.00000
0.999	0.001	0.9997	0.0003	0.99987	0.00013
0.995	0.005	0.9987	0.0013	0.99937	0.00063
0.990	0.010	0.9974	0.0026	0.99874	0.00126
0.950	0.050	0.9872	0.0128	0.99361	0.00639
0.900	0.100	0.9740	0.0260	0.98692	0.01308

The apportionment process should be based on historical data for similar projects, as well as on models for the variation of phase reliabilities with system costs, weights, and other factors. Such information is seldom available during the initial system design process and therefore carefully chosen simplifying assumptions will often be necessary.

One way to proceed is to attempt to bracket the reliability estimates for the propulsion system by considering two extremes. In one case, we assume the propulsion system has low reliability compared to the other systems and is therefore the limiting system factor. In the case where the propulsion reliability is relatively low we may set $R_1 = 1$ and also set the probabilities for all the powered phases equal: $Pr(X_1) = Pr(X_1) = Pr(X_1) = Pr(X_1) = Pr_{low}$. Then Eqn (11.8) yields

$$R_{ms} = (Pr_{low})^4 \tag{11.9}$$

$$Pr_{low} = (R_{ms})^{\frac{1}{4}} \tag{11.10}$$

In the other extreme, where we assume the propulsion reliability is about the same as the other systems and all phases of the mission are about equally reliable, the following result is found:

$$R_{ms} = (Pr_{same})^8 \tag{11.11}$$

$$Pr_{same} = (R_{ms})^{\frac{1}{8}} \tag{11.12}$$

It is helpful to compare the relationship between mission success and powered phase reliabilities as defined by the apportionment rules in Eqns (11.10) and (11.12) and this is shown in Table 11.3. From these results, we can see that for a mission success goal of $R_{ms} = 0.95$, the phase reliability for the propulsion system should be between 0.99361 and 0.9872. The corresponding phase unreliability ranges between 0.00639 and 0.0128, a ratio of about 2.

We must choose the goal for R_{ms} as the start of the planning process. The best way to set such a goal is to study the probability of crew safety R_{cs} and the probability of mission success R_{ms} achieved by past projects as shown for some manned missions in Table 11.4. Note that the Apollo and Space Shuttle have demonstrated a probability of crew safety of 0.95 to almost 0.99. Also included in Table 11.4 is the mission reliability of some manned and unmanned launch systems as well as that of corresponding rocket engines.

Table 11.4 Reliabilities of Some Past Space Missions and Hardware

Project	Year	Success Ratio	R_{cs}	R_{ms}
Manned missions				
Atlas Mercury	1961–1963	7/7	0.88889[a]	0.88889[a]
Titan 2 Gemini	1964–1966	11/11	0.92308[a]	0.92308[a]
Apollo	1965–1975	17/18[b]	0.95000[a]	0.94444
Space Shuttle	1981–2011	133/135	0.98519	0.98519
Vostok (USSR)	1961–1963	6/6	0.87500[a]	0.87500[a]
Voshkod (USSR)	1964–1965	2/2	0.75000[a]	0.75000[a]
Soyuz (USSR)	1967–1991	50/66[c]	0.96970	0.75758
Soyuz (Russia)	1991–2015	59/59	0.98361[a]	0.98361[a]
Shenzhou (China)				
Launch stages				
Delta Booster	1960–1990	–		0.9403[d]
Atlas Booster	1957–1990	–		0.8063[d]
Saturn V	1967–1973	13/13		0.93333[a]
Saturn IB	1966–1975	9/9		0.83333[a]
Unmanned launchers				
Titan IVB	1997–2005	15/17		0.88235
Delta 4	2002–2015	28/29		0.96562
Proton M (Russia)	2001–2014	77/87		0.88535
Ariane V (Europe)	1996–2015	74/78		0.94872
Engines				
1–5 J-2 (LOX/LH$_2$)	1966–1975	93/96		0.96875[e]
2–6 RL-10 (LOX/LH$_2$)	1961–1986	198/198		0.99500[a,e]
3 SSME engines	1981–2011	121/121		0.99187[a,e]
SRBM pairs	1981–2011	134/135		0.99269
RS-68	2002–2015	29/29		0.96674[a]

[a]*Bayesian estimate Eqn (11.23).*
[b]*Apollo 13 mission failed but crew saved.*
[c]*Soyuz 1 and 11 suffered crew losses, 8 mission failures included 6 docking failures, 1 failure to rendezvous with Salyut 4, and 1 was rocket explosion at launch with safe evacuation of crew.*
[d]*Isakowitz (1991).*
[e]*SAIC (1989).*

11.3 THE RELIABILITY FUNCTION

In order to complete the apportionment process and relate system reliability goals to system failure rates, we must introduce and summarize a few additional aspects of reliability theory (Shooman, 1990). The reliability function $R(t)$ is defined as the probability of success for a time period t. The random variable of interest, t, is the time to failure. Using standard random variable mathematics, we can relate $R(t)$ to the cumulative density function $F(t)$, the probability of failure in the time interval from 0 to t, to the probability distribution function $f(t)$ as follows:

$$R(t) = 1 - F(t) = 1 - \int_0^t f(x)\,dx \tag{11.13}$$

Although Eqn (11.13) is mathematically complete, reliability engineers find it better to deal with the hazard, or failure rate, function $z(t)$, which is the fraction of survivors at time t which fail per unit time. We then define

$$z(t) = \mathrm{Lim}_{\Delta t \to 0} \frac{\text{failures}}{\text{survivors} \times \Delta t} \tag{11.14}$$

Then, for N items placed on test we can write

$$z(t) = \mathrm{Lim}_{\Delta t \to 0} \frac{N[f(t) - f(t + \Delta t)]}{NR(t)\Delta t} = \frac{1}{R(t)} \frac{df(t)}{dt} = -\frac{1}{R(t)} \frac{dR(t)}{dt} \tag{11.15}$$

Solution of the differential equation in Eqn (11.15), subject to the condition $R(0) = 1$, that is, the item under consideration is initially good, yields

$$R(t) - \exp\left[-\int_0^t z(x)\,dx\right] \tag{11.16}$$

Another variable often used in reliability studies is the mean time to failure (MTTF) which, using the definition of the mean of a random variable, is given by

$$MTTF = \int_0^\infty t f(t)\,dt = \int_0^\infty R(t)\,dt \tag{11.17}$$

11.4 FAILURE RATE MODELS AND RELIABILITY ESTIMATION

Many failure rate models have been used for components and systems, but the simple constant failure rate $z(t) = $ constant is suitable for a first-order approximate analysis, and is often reasonably accurate. With this assumption we have the following relations:

$$z(t) = \lambda \tag{11.18}$$

$$R(t) = \exp(-\lambda t) \tag{11.19}$$

$$MTTF = \frac{1}{\lambda} \tag{11.20}$$

Combining Eqns (11.5) and (11.18)–(11.20) yields

$$R_{ms} = e^{-\lambda_1 t'_1} e^{-\lambda_2 t'_2} \ldots e^{-\lambda_4 t'_4} = \exp\left[\sum_{i=1}^4 -\lambda_i t'_i\right] \tag{11.21}$$

In this equation, λ_i is the constant failure rate for phase i and t'_i is the time of operation from the start of phase i.

Reliability estimates are best prepared from experimental data. If it is known from prior tests and analysis that the reliability of a component or system is governed by a constant failure rate, as in Eqns (11.16)−(11.18), statistical theory suggests estimating the parameter λ by the ratio of failures to total operating hours. This maximum likelihood estimate (*MLE*) is essentially the same as that on the right-hand side of Eqn (11.12). In the event the database shows no failures then $MLE = 0$ which will lead to results that are difficult to justify. A better procedure in this case is to assume a value of 0.33 for the number of failures (Welker and Lipow, 1974).

In some cases the number of components tested, n, and the number of successes, r, are reported, but not the number of test hours. In that case one may assume that the binomial distribution is applicable and that the success parameter, p, is given by the *MLE* formula

$$p = \frac{r}{n} \tag{11.22}$$

In the event that there are zero failures such that $p = 1$, Eqn (11.22) is modified using Bayesian estimation principles to yield

$$p = \frac{r+1}{n+2} \tag{11.23}$$

The results quoted in Table 11.4 were obtained using Eqn (11.22) or (11.23). In cases where the operating time is also specified, Eqn (11.14) would be used.

11.5 APPORTIONMENT GOALS

The apportionment carried out in Section 11.2 resulted in a bracketing of the phase reliability P by P_{low} and P_{same} given by Eqns (11.10) and (11.12) and Table 11.3. We may now continue the apportionment process to a lower level so we can focus on, say, propulsion system reliability. Averaging the demonstrated reliabilities of Apollo and the Space Shuttle as given in Table 11.4 yields a value of $R = 0.963$. For the current mission we assume a mission goal of $R = 0.96$, which results in a range of powered propulsion phase reliability between 0.9898 and 0.9949. The major systems of such a phase, say X_1, are shown in Table 11.2 and again we may consider two extremes.

If the propulsion system is the limiting factor we assume that the phase reliability is equal to the reliability of the propulsion systems alone, that is, $P(X_1) = P(Y_1) = P_{low} = 0.9898$ and that the other reliability of all the other systems are all equal to unity, that is, $\Pr(Y_2) = \Pr(Y_3) = P(Y_4) = \Pr(X_5) = 1$. At the other extreme, where all the systems are considered equally reliable, $P(X_1) = [P(Y_1)]^5 = 0.9898$, so that $P(Y_1) = 0.9980$. Now the reliability for the propulsion system must lie in the range between 0.9898 and 0.9980.

We now turn to estimating the difficulty in achieving the bracketing reliability goals just calculated. We may make use of the database in Table 11.4 in which we see that five engine systems have an average demonstrated reliability of 0.98301, and the associated failure probability is 0.01699. Based on the previous data we see that it falls slightly below our required reliability level. This suggests that a more detailed and sophisticated analysis is necessary for estimating reliability.

11.6 OVERVIEW OF PROBABILISTIC RISK ASSESSMENT

Probabilistic risk assessment (PRA) in human spaceflight aims to identify failure modes that can result in the loss of the crew (LOC) or LOV and then uses existing reliability information to estimate their likelihood of occurrence. A shortcoming of PRA is that involves only the known failure modes and cannot incorporate those which are unknown or underappreciated. Therefore PRAs will always underestimate risk and a substantial factor of safety should be applied to account for those unforeseen failure modes.

The Aerospace Safety Advisory Panel (ASAP), established by Congress in 1968 to provide advice and make recommendations to the NASA Administrator on safety matter reports (ASAP, 2014) that the gap between historically based PRA and the actual risk led the NASA System Safety Handbook, SSH (2014), to call for programs to allow a "management reserve" between the PRA risk (LOC) and the maximum acceptable risk for the program set by the decision-making authority because NASA's statisticians said that actual risk may be 50% greater than the calculated risk.

ASAP (2014) also reported that NASA Exploration Systems Development (ESD) Division has now established LOC probability thresholds for manned exploration missions as shown in Table 11.5.

This appears to be not much different from the actual performance of the Space Shuttle whose actual performance over 135 flights was $R_{cs} = 0.98519$ (about 1 in 67), but this likely reflects the higher actual risk early in the then new manned exploration program because of new unknown failure modes and design weaknesses.

Carrying out a PRA requires imagining scenarios, that is, sequences of events, which lead to undesired consequences. There is an initiating, or triggering, event which sets into motion the subsequent series of events ultimately resulting in an end state. Triggering events are usually malfunctions or failures of any type that cause a deviation from the intended operation. End states are defined by the decision makers and illustrate the nature of the undesired consequence. For example, in manned spaceflight a particularly important end state is LOV. The space shuttle launch trajectory and the abort possibilities are illustrated in Figure 11.1.

Scenarios are generally presented in diagrammatic form, the development of which is not unique and often represents, to some extent, a degree of creativity that tends to come with experience. There is a hierarchy of diagrams that can be used effectively to form a framework for a rigorous review of the risk exposure in complicated systems as illustrated schematically in Figure 11.2.

Table 11.5 NASA's LOC Probability Thresholds for Manned Missions	
Flight Stage	**Maximum Probability of LOC**
Ascent	1 in 300, $R_{cs} = 0.99667$
Cislunar Mission	1 in 150, $R_{cs} = 0.99333$
Entry	1 in 300, $R_{cs} = 0.99667$
Total Mission	1 in 75, $R_{cs} = 0.98667$

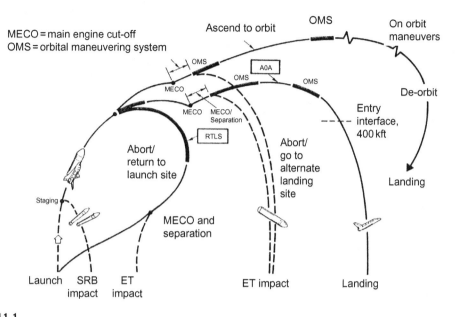

FIGURE 11.1

Nominal Space Shuttle mission and possible abort scenarios (NASA).

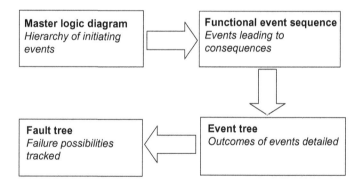

FIGURE 11.2

The flow of information used in carrying out a PRA is illustrated here as a hierarchy of diagrams.

Details of the high-level flow of information illustrated in Figure 11.2 are described as follows:

- Master Logic Diagram (MLD): This is a hierarchy of initiating events illustrating the various manners in which a damage result might occur. There is a compilation of functional categories of perturbations to the system under consideration and spreading from these are characterizations of components for each of these functions. The MLD starts with a damage event of major concern and then the subsequent events that are necessary, but not sufficient, to cause the major event are taken in greater and greater detail. A portion of such an MLD is shown in Figures 11.3–11.5.

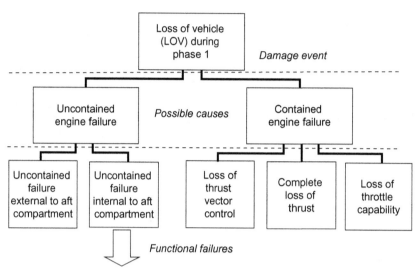

FIGURE 11.3

A portion of an MLD is shown. The top level is the damage event and the next level is the possible cause for the event. The next level is the functional failures proceeding from the possible causes. The arrow points toward the next level of the MLD continued in Figure 11.4.

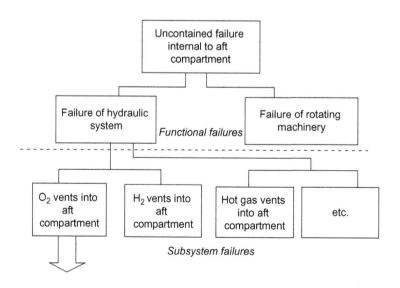

FIGURE 11.4

The continuation of the MLD shows the flow from functional failures down to subsystem failures. The arrow points to the next level of the MLD shown in Figure 11.5.

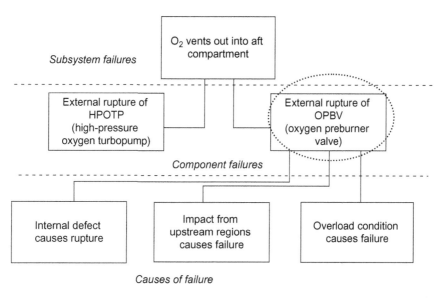

FIGURE 11.5

Concluding section of the MLD showing flow from subsystem failures to component failures to causes of failure. The circled component failure is tracked in subsequent PRA diagrams.

- Functional Event Sequence Diagram (FESD): The initiators in the MLD are screened to determine which of them are so highly unlikely that they may be eliminated from deeper analysis. Then the remaining trigger events are evaluated in the sense of establishing the sequence of events that would follow. These events are formulated in a manner that provides a binary outcome, such as yes or no, or failure or success. The FESD provides a useful interface for reliability and systems engineers to pool expert knowledge. An illustrative FESD for the failure of a main engine oxygen preburner valve (OPBV) is shown in Figure 11.6.
- Event Tree: This is a sort of detailed flow chart of binary decisions based on the FESD that may be easily developed for computer modeling. All the different event paths may be traced down to where they lead to successes or failures. An event tree for the rupture of the OPBV is shown in Figure 11.7. The branch studied is that in which the loss of oxygen flow is detected and the engine is shut down or not. If there is no hydrogen in the aft compartment the vehicle will be safe even if the engine is not shut down. However, if hydrogen is present a combustible mixture may exist and the fire suppression system should be activated and if it works the vehicle is safe. If the fire suppression system is not activated or the combustible mixture is not detected the vehicle is subject to the possibility of either a fire resulting in damage to the vehicle or a detonation and the LOV.
- Fault Tree: This is similar to an event tree except that here only the failure sequences are considered, and the components or systems that fail are examined in detail. Once again the possible modes of failure of the component or system are treated in a binary manner and every possible source of failure is traced. A fault tree for the failure to detect the loss of oxygen flow in Figure 11.6 is illustrated in Figure 11.8. The fault is considered to be either a sensor failure or a false sensor reading and these are traced down to a number of possibilities.

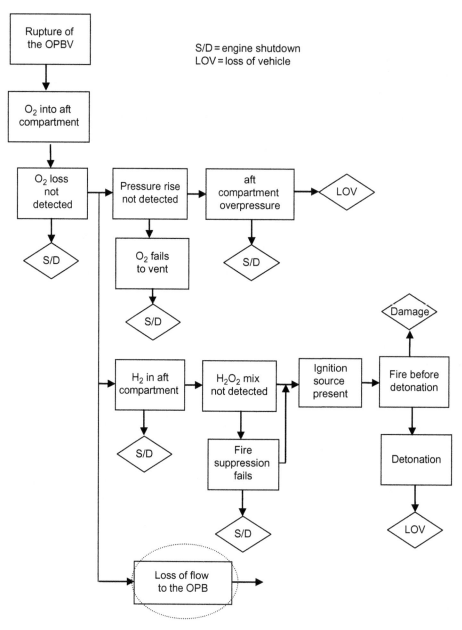

FIGURE 11.6

FESD for the rupture of the OPBV in a main engine. The circled event is treated subsequently in the event tree diagram.

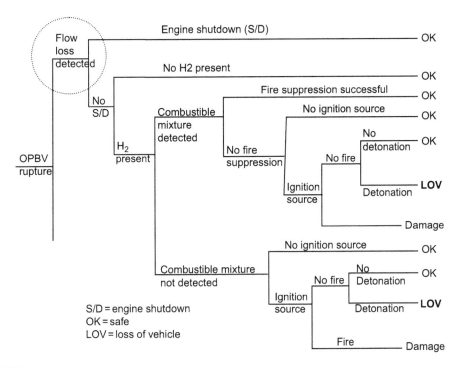

FIGURE 11.7

Functional event tree describing the possible events following the rupture of the OPBV and the lack of detection of the loss of flow. The outcomes are OK (safe), damage to the vehicle, or LOV.

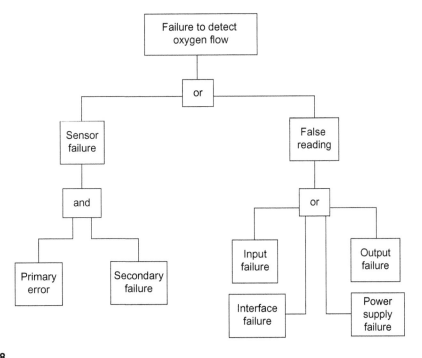

FIGURE 11.8

A fault tree for the failure to sense loss of oxygen flow in Figure 11.5 is illustrated.

FIGURE 11.9

Propulsion failure: X-15 crash at NASA Dryden Flight Research Center after rocket engine failure (NASA).

11.7 TOP FUNCTIONAL FAILURES OF SPACECRAFT

MLDs are based on functional failures and in the case of the Space Shuttle five top-level functional failures that could result in LOV were identified. These failures are discussed below using actual failures of the Space Shuttle and other manned space vehicles as illustrations.

11.7.1 PROPULSION FAILURE

Without proper levels of propulsion during powered flight phases, mission or abort success cannot be achieved. An example of this type of failure is shown in Figure 11.9 where rocket engine failure in the North American X-15 apparently resulted in the inability to dump fuel so landing was made at a heavy weight. Landing speed was high around 250 kts, and the landing skid on one side collapsed. The pilot was injured, but not fatally.

11.7.2 VEHICLE CONFIGURATION FAILURE

Without appropriate control authority throughout the flight, a mission failure can ensue. An event of this type occurred during the launch of Space Shuttle Orbiter Columbia. After almost 2 years of delays, Space Shuttle Columbia lifted off on January 16, 2003. At launch, some of the external tank's insulating foam broke away and struck the left wing as shown in Figure 11.10. One day after the launch, NASA employees reviewing launch footage noted that a piece of foam broke away from the external fuel tank and struck the wing. In the following days, many NASA engineers questioned the condition of the shuttle's thermal protection system. These engineers asked for high-resolution imaging from the Department of Defense satellites, as well as a spacewalk to

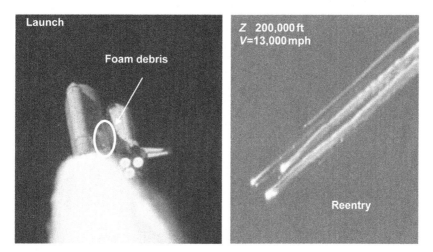

FIGURE 11.10

At the left foam debris from the external tank damaged the leading edge of Columbia's wing leading to breakup of the vehicle upon entry shown at the right (NASA).

determine the extent of the damage. NASA management turned down both requests in a controversial series of decisions. NASA managers claimed that because nothing could be done in terms of in-flight repair or a rescue mission, there was no reason to investigate the matter further. Compounding issues was the fact that pieces of foam had routinely broken off the external tank in previous launches with no catastrophic damage. However in this case the impact of the foam damaged the left wing's thermal protection tiles, ultimately resulting in an LOC and LOV on reentry on February 1, 2003. The shuttle broke up completely over the southern United States and the crew of seven was lost.

Stepaniak and Lane (2015) present an account of the aeromedical aspects of the Columbia accident and the investigation that followed. Among the conclusions reached are as follows:

> Crew survivability should be considered the primary mission success criterion and should be the main driver in vehicle design and mission architecture. Crew survivability should incorporate advanced technologies where feasible and should be simple, reliable, and attainable to address catastrophic failure modes. Lessons learned from related high-risk operations, as well as space mishaps, incidents, and close calls, can enhance crew survival by providing insight into failure modes, and improving procedures, design requirements, and occupant protection strategies.

A different configuration failure caused the crash of Soyuz-1. In this case, the drogue parachute of the landing system deployed improperly and was unable to pull the main decelerating parachute out of its container. Though the reserve parachute was released, it was prevented from filling with air by the still-attached drogue chute. The Soyuz-1 capsule hit the ground at 40 m/s (90 mph) killing the Soviet astronaut.

FIGURE 11.11

Containment failure: O-ring leak cause SRBM malfunction which destroyed Space Shuttle Orbiter Challenger (NASA).

11.7.3 ENERGETIC GAS AND DEBRIS CONTAINMENT FAILURE

Departure of gases and debris from planned and controlled paths may destroy vehicle components and lead to mission failure. An O-ring that was to contain combustion gases in one of the two solid rocket booster motors (SRBM) of the launch system for the Space Shuttle Orbiter Challenger permitted hot gases to compromise the external tank leading to an explosion and LOV. The initiation of this failure is shown in Figure 11.11. Another example of a containment failure occurred during the flight of Apollo 13. On April 13, 1970, Mission Control requested that the crew stir the oxygen tanks, a task required to prevent the oxygen "slush" from stratifying. The Teflon-insulated cables that powered the stirrer motor were damaged, causing a fire when electricity was passed through them. The fire heated the surrounding oxygen, increasing the pressure inside the tank above its nominal 1000 lb/in^2 (7 MPa) limit and causing the tank to explode. As is the case with most spacecraft failures, there were many contributing factors to the Apollo 13 accident. During assembly, the oxygen tank was dropped a short distance, perhaps several inches. This impact disturbed the internal plumbing of the tank, requiring the use of the tank's heater to purge it of oxygen. This was not considered to be a problem because the tank had a thermostat to help in controlling internal pressure. The thermostat was designed to operate on 28 volts, considerably less than the launch pad's 65 volts. The thermostat failed to shut off the heater, and it ran for days, heating the tank to over 430 °C. The tank's instrumentation only read to 38 °C, resulting in no indication of the tank's true internal temperature. At 430 °C, the insulation melted off the internal wiring, and when the tank was refilled with oxygen, it was a bomb waiting to go off. The situation after the explosion is illustrated in Figure 11.12.

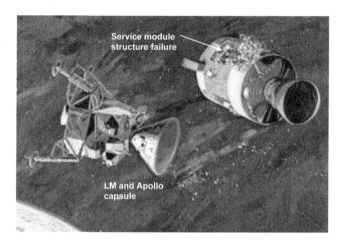

FIGURE 11.12

Containment failure: An oxygen tank in the service module overheated and blew up because its heater switches welded shut during excessive prelaunch electric currents (NASA).

11.7.4 **VEHICLE ENVIRONMENTAL SUPPORT FAILURE**

Without proper functioning of life support systems, the crew can be lost and mission failure will be certain. An example of an environmental support failure is the Apollo 1 fire in the command module on January 27, 1967. The crew was carrying out a standard test of the spacecraft's communication systems when a voltage spike was recorded. The cabin was designed to endure outward pressure when orbiting in the vacuum of space, so it was pressurized on the ground to about 9 kPa over ambient atmospheric pressure, a level similar to what the cabin would experience in space. The cabin atmosphere was pure oxygen because NASA staff was concerned about the management of nitrogen levels even though they knew that an oxygen-rich environment in the cabin elevated the risk of fire. An unknown source, possibly a spark, initiated a fatal fire killing three of the original seven astronauts in NASA's space program and almost terminating the space program. The fire was marked by a rapid rise in cabin temperature and pressure resulting in rupture of the cabin pressure vessel about 15 s after the verbal report of fire. The aftermath of the fire is shown in Figure 11.13.

Another example of an environmental support failure is that of Soyuz 11 which successfully docked with Salyut 1 space station on June 7, 1971, remained there for 22 days, and made a normal entry and landing on June 30, 1971. Although there were no signs of damage to the exterior of the capsule, the recovery team opened the capsule to find the three crew members dead. Apparently they asphyxiated because a ventilation valve failed in the open position when the entry capsule was separated from the orbital module by the firing of explosive bolts. The valve opened at an altitude of 168 km (104 mi), and the resulting pressure loss was fatal to the crew. The valve was located in an essentially inaccessible position between crew seats, but even so, later examination showed that manual closure of the valve required more time than was available during the emergency. The spacecraft was subsequently redesigned to accommodate only two astronauts so that the

FIGURE 11.13

Environmental support failure: The fire in the Apollo-1 capsule produced a rapid cabin temperature and pressure rise terminating in the rupture of the cabin pressure vessel (NASA).

increase in free volume would permit them to put on lightweight spacesuits for the stressful flight stages of launch and landing. There was a further upgrade of the Soyuz capsule, the Soyuz-T, which provided additional free volume so as to permit a crew of three once more.

11.7.5 EXTERNALLY CAUSED FAILURE

Events outside the vehicle and its command and control system, such as lightning, extreme wind shear, and space debris, can lead to mission failure. One instance of externally caused damage that could lead to failure is illustrated by the Space Shuttle Orbiter which routinely returned from orbit with its 5-cm-thick cabin windows scarred by small pits or craters caused by micrometeoroid and space debris strikes. It became a routine safety procedure to replace the windows each time rather than to try to polish out the damage.

Another instance of an externally caused contributor to failures is the recognition in the Challenger Commission report (NASA, 1986) that:

> At approximately 37 seconds, Challenger encountered the first of several high-altitude wind shear conditions, which lasted until about 64 seconds. The wind shear created forces on the vehicle with relatively large fluctuations. These were immediately sensed and countered by the guidance, navigation and control system.
>
> The steering system (thrust vector control) of the Solid Rocket Booster responded to all commands and wind shear effects. The wind shear caused the steering system to be more active than on any previous flight.

The bending experienced by the vehicle is likely to have exacerbated the sealing failure of the O-ring in the solid rocket booster and perhaps contributed to the LOV.

These functional failures are applicable to manned space missions in general and may be used in the construction of an MLD for specific cases. The latest evaluation of safety issues in NASA's aerospace programs is discussed by the NASA ASAP in ASAP (2014).

11.8 PRA OF THE SPACE SHUTTLE

The previous sections contained tools for the analysis of risk. Since the development of the space program, NASA has been interested in risk and reliability for obvious reasons. NASA has evaluated risk for every one of their manned missions, but the space shuttle risk assessment is of particular interest as NASA has had the time to refine their risk assessment tools and procedures. In fact, in the early development stages of the Apollo program it was estimated that it had less than 2% chance of success. As it turned out, 17 out of 18 Apollo missions were successful. A chronological view of NASA's risk predictions for the Space Shuttle is given in Table 11.6.

There are a few notes to make regarding Table 11.6. First of all, the program first started with an extremely low predicted risk, that only one in 7000 missions would result in LOV/LOC. As time progressed, and the Challenger tragedy unfolded, the risk increased dramatically, to one failure in only 70 launches. This shows that NASA was still learning and revising risk estimates. A PRA of the Space Transportation System presented by SAIC (1995) gave the distribution of LOV risk among the various elements of the system as shown in Figure 11.14 where it is seen that over half the risk lies in the propulsion elements. The study also points out that abort scenarios, like those shown in Figure 11.1, are also worthy of more serious study.

Table 11.6 NASA's Risk Predictions for the Space Shuttle

Year	$1 - R_{cs}$	R_{cs}	Notes
1981	0.0001429	0.9998571	1/7000, Prechallenger Explosion
1986	0.0128205	0.9871795	1/70, Postchallenger Explosion
1993	0.0111111	0.9888889	1/90
1997	0.0067568	0.9932432	1/148
2002	0.0040000	0.9960000	1/250
2003	0.0039370	0.9960630	1/254, Pre-Columbia TPS failure
2004	0.0081301	0.9918699	1/123, Discovery Launch
2007	0.0030769	0.9969231	1/325
2012	0.0020000	0.9980000	1/500, Future Goal

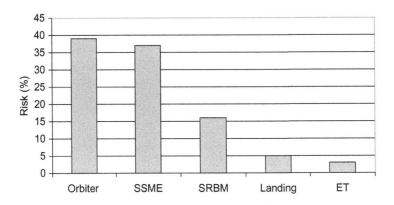

FIGURE 11.14

Distribution of LOV risk for the Space Transportation System.

11.9 CREW FLIGHT SAFETY

Efforts to ensure the safety of pilots in high-speed airplanes led to the development of rocket-boosted ejection seats for most high-speed aircraft and then to ejection capsules like those on the F-111 aircraft. The intent was to provide a means for safely removing the pilots from a compromised aircraft under all flight conditions, ranging from aircraft stationary on the ground to those flying at high altitude and speed. The same approach was used for space capsules by incorporating a rocket booster dedicated solely to carrying the crew capsule to a high enough altitude to ensure safe operation of the parachute recovery system and far enough downrange of the launch vehicle to preclude the possibility of sustaining damage from flying debris or an explosive fireball. A schematic diagram of such a launch escape system (LES) is shown in Figure 11.15. This type of LES was common to the Mercury, Gemini, and Apollo capsules and a similar configuration is intended for use on the Orion capsule. The main feature is the launch escape solid propellant rocket motor and the associated canted nozzles which can rapidly provide sufficient thrust to carry the capsule to an altitude of about 2000 m in 5 s. The smaller attitude control motor is intended to provide sufficient moment to pitch the LES over as it climbs so as to achieve the required downrange distance. The air data system, sometimes called the "q-ball," senses dynamic pressure (q) and attitude angle and provides it to the attitude motor control system. Once the appropriate altitude, down range distance, and attitude angle are achieved the tower jettison motor fires to separate the capsule from the LES. The standard capsule parachute recovery system described in Chapter 6 is then deployed to bring the capsule back to the ground safely. Under normal conditions of a safe launch, the LES is jettisoned some brief time after the first stage separates and the second stage motor fires.

The trajectory of the capsule may be computed using the equations developed in Chapter 7. We may consider a few simplifications to assist in basic sizing of the LES. From a safety point of view, the LES must transport the capsule to a safe position in a short time. A sound wave will travel about 1700 m in about 5 s and this would be appropriate baseline burn duration for our LES.

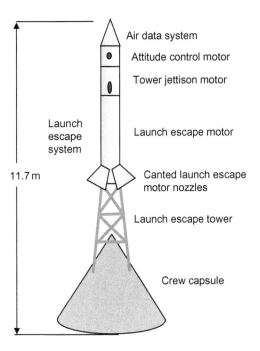

Air data system

Attitude control motor

Tower jettison motor

Launch escape motor

Launch
escape
system

Canted launch escape
motor nozzles

11.7 m

Launch escape tower

Crew capsule

FIGURE 11.15

Conventional launch escape system (LES) for a space capsule.

For a simple assessment assume a vertical launch ($\gamma = 90°$) with no drag so that the nondimensional acceleration is given by

$$\frac{1}{g_E}\frac{dV}{dt} = \frac{1}{g_E}\frac{d^2z}{dt^2} = \left[\left(\frac{F}{W}\right) - \sin\gamma\right] \tag{11.24}$$

Assuming $\sin\gamma \ll (F/W)$, $t/I_{sp} \ll 1$, and an initial velocity $V(0) = 0$ we may use Eqn (7.41) in the following form:

$$V \approx \left[\left(\frac{F_{vac}}{W}\right)_0 - 1\right] g_E t \tag{11.25}$$

Integrating this equation to find the altitude, assuming $z(0) = 0$, yields

$$z \approx \left[\left(\frac{F_{vac}}{W}\right)_0 - 1\right]\frac{1}{2}g_E t^2 \tag{11.26}$$

In order to have $z = 1700$ m at $t = 5$ s the thrust to weight ratio $(F_{vac}/W)_0$ of the LES must be about 14. From the result for burn-out time given in Eqn (7.49), we see that the propellant weight ratio is given by

$$\left(\frac{m_p}{m_0}\right) = \frac{t_{bo}}{I_{sp,vac}}\left(\frac{F_{vac}}{W}\right)_0 \tag{11.27}$$

For $t_{bo} = 5$ s, $(F_{vac}/W)_0 = 14$, and an assumed vacuum specific impulse $I_{sp,vac} = 270$ s, the propellant mass fraction $(m_p/m_0) = 0.259$. The mass of the capsule and the LES may be written as follows:

$$m_0 = m_{cap} + m_{LES,str} + m_p \approx m_{cap} + m_0 \left(\frac{m_p}{m_0}\right)\left(1 + \frac{m_{LES,str}}{m_p}\right) \tag{11.28}$$

Then the mass of the system is

$$m_0 = \frac{m_{cap}}{1 - \frac{m_p}{m_0}\left(1 + \frac{m_{LES,str}}{m_p}\right)} \tag{11.29}$$

As an example, the Apollo capsule mass is about 5300 kg so that, assuming the LES structural mass $m_{LES,str}$ is about 10% that of propellant mass m_p, yields $m_0 = 1.5 m_{cap} = 7414$ kg. Then the thrust of the LES motor, given by $F = 14(m_0 g_E)$, is about 14(72.7 kN) = 1017 kN (228.6 klb). This simple constant F analysis overestimates the actual Apollo LES motor which produced 650 kN (146.1 klb) of thrust. In practice the thrust is not constant but usually tailored to rapidly decrease to zero during the burn to maintain a reasonable acceleration for the crew as we shall subsequently show.

We may also make a simple assessment of an LES which provides flight path angle (γ) control that can pitch the LES over so as to achieve downrange as well as vertical distance from the site at which the escape emergency was initiated. Using Eqn (11.25), we make the following approximations:

$$\frac{dz}{dt} = V \sin \gamma \approx \left[\left(\frac{F_{vac}}{W}\right)_0 - 1\right] g_E t \sin \gamma \tag{11.30}$$

$$\frac{dx}{dt} = V \cos \gamma \approx \left[\left(\frac{F_{vac}}{W}\right)_0 - 1\right] g_E t \cos \gamma \tag{11.31}$$

Assuming the flight path angle (in degrees) is controlled by the attitude control motor according to the relation $\gamma = 90 - 12t$, we may integrate Eqns (11.30) and (11.31) to obtain the results for the trajectory shown in Figure 11.16. The solid line represents the trajectory during the burn and

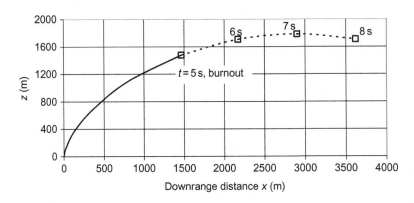

FIGURE 11.16

Approximate trajectories for an LES neglecting drag. Solid line describes the rocket firing phase while the dotted line describes the coast phase.

the dashed line represents the coasting flight of the capsule. Recall that drag is neglected in this approximate analysis. At some point after the burn-out of the LES rocket motor the tower jettison motor is ignited and the capsule coasts along on its own and the recovery parachutes may then be deployed. A photograph of a launch pad test of the Apollo LES is shown in Figure 11.17.

The acceleration of the LES as given by Eqn (11.24) is about $13g_E$ for the full 5 s of burn and this is quite high for humans as can be seen by referring back to Figure 6.2 in the discussion of atmospheric entry. In practice the thrust is modulated with time to account for mass loss so as to tailor an appropriate acceleration schedule while transporting the capsule to a safe distance. We may consider the decaying thrust case by using the previous equations in a series of quasistatic steps of constant thrust during 1-s intervals. For illustrative purposes, we use the variation of thrust shown in Figure 11.18. There is some small inaccuracy in the calculated property distributions because of the coarseness of the 1-s time steps. The constant thrust case has $F = 1017$ kN as described previously while the decaying thrust case has an initial thrust to weight ratio of $(F/W)_0 = 19$ and an initial thrust level of $F_0 = 1380$ kN. The total impulse is 3729 kN·s for the variable thrust case and 5085 kN·s for the constant thrust case. The trajectory performance of the two cases is shown in Figure 11.19.

FIGURE 11.17

An Apollo launch pad abort test on November 7, 1963 (NASA).

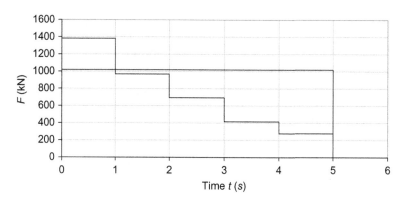

FIGURE 11.18

Constant and variable thrust histories for roughly equal total impulse.

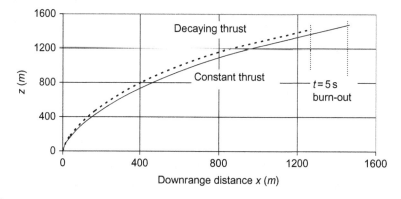

FIGURE 11.19

Trajectories with different thrust profiles.

The trajectories during powered flight are about equal with the burn-out points for the two cases pointed out in Figure 11.19. The acceleration histories of the two cases are illustrated in Figure 11.20 where it seen that the variable thrust case subjects the crew to higher acceleration for the first 2 s after which the acceleration rapidly drops. The constant thrust case shows increasing acceleration because the mass of the LES is dropping due to propellant consumption which is shown as a function of time in Figure 11.21. Note that the variable thrust case consumes about 22% less propellant than the constant thrust case making for a smaller mass penalty.

The acceleration levels are high in the cases used for illustration but we may consider the case of the Apollo launch system which has an initial thrust to weight ratio of approximately $(F/W)_0 = 12$. The trajectory achieved in this case, assuming constant thrust, is compared to the $(F/W)_0 = 14$ case treated previously in Figure 11.22, including 3 s of unpowered coasting. There is a difference in the performance but there is also a 5.6% savings in propellant with an acceleration history that is preferable for the crew, as can be seen in Figure 11.23. This of course can be

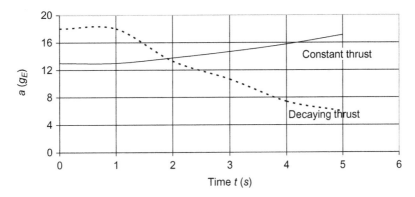

FIGURE 11.20

Acceleration histories for different thrust profiles in multiples of g_E.

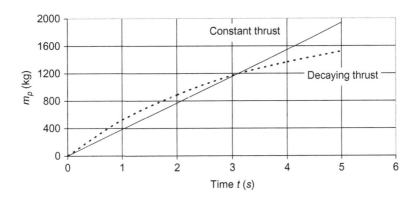

FIGURE 11.21

Propellant mass consumption histories for the two thrust profiles.

improved, as mentioned previously, by tailoring the thrust curve while keeping the total impulse approximately constant.

The velocity of the LES, as given by Eqn (11.25), neglects drag and reaches levels between $M = 1.6$ and $M = 2.2$ in the cases studied here. Drag effects will depress both the trajectory and the acceleration levels and should be considered in more refined developments of the LES design.

The launch abort system (LAS) for the Orion capsule shown in Figure 11.24 is about the same overall height as the Apollo LES but incorporates several improvements over the Apollo system. First, the launch rocket motor nozzles, which are canted 25° from the centerline and can produce a maximum total thrust of about 1750 kN, are seen to be located further forward on the Orion LAS than are those on the Apollo LES. This reduces the risk of impingement of nozzle plumes on the crew capsule. The difference in the nozzle positions arises because Orion's solid propellant in the rocket burns from the top of the casing down, rather than from the bottom up like Apollo's. The combustion gases in the Orion rocket motor rise vertically and then make a 155° turn to exit

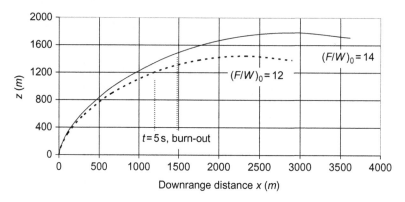

FIGURE 11.22

Comparison of the trajectories for two constant thrust cases showing burn-out at $t = 5$ s and three additional seconds of unpowered coasting.

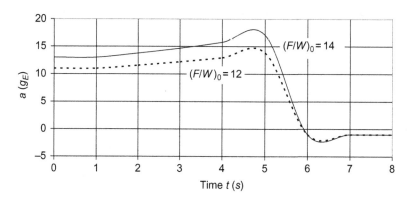

FIGURE 11.23

Comparison of the acceleration histories (in multiples of g_E) for two constant thrust cases with burn-out at $t = 5$ s and three additional seconds of unpowered coasting.

the nozzles. Second, Orion's attitude control motor has eight equally spaced ports around the periphery of the rocket as can be seen at the top of the LES in Figure 11.24. A pintle nozzle in that motor permits differential firing through those ports to more precisely direct the net thrust vector of the LES, unlike the single pitch-over port on the Apollo LES shown firing in Figure 11.25. Third, the Orion tower structure is housed within in a streamlined fairing which closes off the region between the rocket motor and the capsule reducing turbulence effects between them.

An alternate to the tower-based Orion LAS described above was developed by NASA as a risk mitigation exercise. It is called the Max Launch Abort System (MLAS) in honor of Max Faget, a NASA spacecraft design pioneer, and a schematic diagram illustrating the concept is shown in Figure 11.25. A version of the MLAS has been successfully tested although it is not expected to replace the LAS on Orion. The development of the concept is described by Gilbert (2015).

FIGURE 11.24

The Orion LAS showing attitude control jet ports at the top, canted main engine nozzles in the middle, and a fairing between the capsule and the launch escape rocket (NASA).

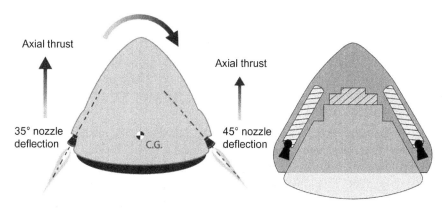

FIGURE 11.25

Schematic diagram of the MLAS showing powered fairing mounted over the Orion space capsule. Differential nozzle tilt provides both thrust and pitching moment (NASA).

11.10 HUMAN FACTORS IN RISK MANAGEMENT

The magnitude of the risk associated with an event is generally taken to be the product of the likelihood of occurrence of the event and the impact or consequences of the event. We may consider risk to be determined by the existence of hazardous conditions, the presence of possible initiating causes, and the magnitude of the possible consequences. It is impossible to eliminate risk entirely, if only because there are always some failure modes that are unknown. It is the responsibility of the design, development, and operational team to manage the risk so that it is maintained at reasonable levels. The broad categories of oversight may be described as follows:

- Engineering control: Hazard scenarios must be exhaustively defined and their likelihood and impact estimated to describe the risk. Efforts should be directed at defining failure modes and applying proven methods and criteria to "design out the risk." This approach, which is carried out prior to actually producing a vehicle, offers the greatest effectiveness in understanding and confining the risk.
- Administrative control: Consequences and acceptable risk level must be clearly defined. Control procedures and boundaries for safe operation must be equally well-defined to provide an effective risk mitigation policy. Implementation of these procedures must include means to verify actions, continue life cycle assessments, and to carry out periodic reviews.
- Personnel control: Dedicated efforts to provide methods for continual training of all project participants including channels for feedback identifying corrective actions aimed at reducing risk levels.

Often the greatest risk doesn't reside in the machinery, but rather in the humans operating them. For example, in the Space Shuttle Challenger disaster, it was well known that the SRB O-Rings were unlikely to seal properly below 12 °C, yet an administrative decision was made to launch even though the O-rings were at a temperature of around −13 °C. It is difficult to quantify the risk and reliability of a system, even more so when it is operated outside of its design conditions. Reason (1990) discusses four levels of failure: unsafe acts, preconditions for unsafe acts, unsafe supervision, and organizational influences. Unsafe acts are those directly related to a failure such as pilot error. Preconditions for unsafe acts include crew fatigue, poor communications practices, and insufficient training.

11.11 THE WEIBULL DISTRIBUTION

One of the random variables that have been found to be valuable in the modeling of the lifetime of components based on testing even relatively small samples is the Weibull random variable. Its probability density function is given by

$$f(t) = \frac{g}{t_c} \left(\frac{t}{t_c} \right)^{g-1} \exp\left[-\left(\frac{t}{t_c} \right)^g \right]$$ (11.32)

Here t represents time, t_c is a characteristic time for the event being modeled, and g is a shape factor that best captures the characteristics of the event being modeled and they are all positive quantities. The general shape of the Weibull distribution is shown in Figure 11.26 for several values of the parameters t_c and g.

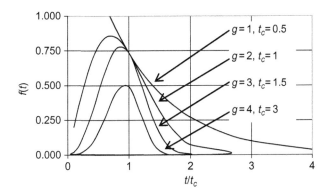

FIGURE 11.26

The Weibull distribution function as a function of the normalized time for various values of the parameters t_c and g.

The probability that the system being modeled fails before some time t is given by

$$\Pr(T \le t) = F_T(t) = \int_0^t f_T(\tau)dt \qquad (11.33)$$

Here T is the time to failure and $f_T(t)$ is the probability density function of the time to failure, while $F_T(t)$ is cumulative distribution function of the time to failure, T. Then the probability that the system will operate after the time t is given by

$$\Pr(T > t) = R(t) = 1 - F_T(t) \qquad (11.34)$$

The quantity $R(t)$ is called the reliability of the system being modeled. Then the mean, or expected, time to failure is given by the integral of the first moment of the failure probability density function as follows:

$$E\{t\} = \int_0^\infty tf_T(\tau)dt = \int_0^\infty t\frac{dF_T(t)}{dt}dt = -\int_0^\infty t\frac{dR(t)}{dt}dt \qquad (11.35)$$

Then, integration by parts yields

$$E\{t\} = \int_0^\infty R(t)dt \qquad (11.36)$$

For the Weibull distribution of Eqn (11.32), the expected time to failure and its variance are

$$E\{T\} = c\Gamma\left(1 - \frac{1}{g}\right)$$

$$\qquad (11.37)$$

$$\mathrm{Var}(t) = c^2\left[\Gamma\left(1 + \frac{2}{g}\right) - \Gamma^2\left(1 + \frac{1}{g}\right)\right]$$

The gamma function Γ is a standard mathematical function and is tabulated in handbooks and is generally available in computer mathematics libraries; it is defined as follows:

$$\Gamma(x) = \int_0^{\infty} y^{x-1} e^{-y} dy \tag{11.38}$$

The cumulative distribution of the Weibull probability density function is

$$F_T = 1 - \exp\left[-\left(\frac{T}{t_c}\right)^g\right] \tag{11.39}$$

Then the reliability of a system that follows a Weibull distribution is given by

$$R(t) = \exp\left[-\left(\frac{T}{t_c}\right)^g\right] \tag{11.40}$$

The failure rate at any time t is given by the ratio of the probability density function of the time to failure of the system to the reliability of the system and is given by

$$z(t) = \frac{f_T(t)}{R(t)} = \frac{g}{t_c}\left(\frac{t}{t_c}\right)^{g-1} \tag{11.41}$$

Failure rates for various shape factors and the same characteristic time are shown in Figure 11.27. Note that the failure rate for a shape factor $g = 1$ the failure rate is constant at $h(t) = 1/t_c$. A constant failure rate is typical of system behavior at times after the point of being "broken-in" but before the point of being "worn-out." For shape factors $g < 1$ the failure rate is inversely proportional to time so that failures are decreasing in number as time goes on. This is typical of early times, before the system is "broken-in," when the faulty components of the system are weeded out. Conversely, for shape factors $g > 1$ the failure rate increases with time, as you would expect as the system "wears out." For a discussion of the development, use, and illustrations of the Weibull distribution, see, for example, Abernethy (2015).

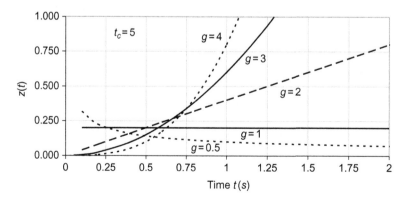

FIGURE 11.27

Failure rate $z(t)$ as a function for various shape factors and a constant value of characteristic time $t_c = 5$.

11.12 NOMENCLATURE

a	acceleration
E	expected time to failure
F	thrust or cumulative probability density function
FESD	functional event sequence diagram
f	probability density function
f_T	probability density function of the time to failure
g	Weibull distribution function shape factor
g_E	gravitational acceleration at the surface of the earth
I_{sp}	specific impulse
LH$_2$	liquid hydrogen
LAS	launch abort system
LES	launch escape system
LOC	loss of crew
LOV	loss of vehicle
LOX	liquid oxygen
M	Mach number
ME	main engines
MECO	main engines cut-off
MLD	master logic diagram
MLE	maximum likelihood estimate
$MTTF$	mean time to failure
m	mass
m_p	mass of propellant
N	number of test items
n	number of phases or number of components
OMS	orbital maneuvering system
OPBV	oxygen preburner valve
Pr	probability
PRA	probabilistic risk assessment
$Pr(X_i)$	probability of mission success in phase i
$Pr(x_i)$	probability of safe performance in phase i
Pr_{low}	lowest probability among systems
Pr_{same}	equal probability among systems
p	success parameter, Eqns (11.22) and (11.23)
R	reliability function
R_1	reliability of the unpowered part of mission, see Eqn (11.8)
R_2	reliability of the powered part of mission, see Eqn (11.8)
R_{ms}	probability of mission success
R_{cs}	probability of crew safety
r	number of successes, Eqns (11.22) and (11.23)
S/D	shut down
T	time to failure
t	time
t_c	characteristic time for an event
V	velocity

W	weight
z	failure rate function, Eqn (11.15) or altitude
X_i	mission success during phase i
x	downrange distance
x_i	mission is safe during phase i
Y_i	success of system i
y_i	safe performance of system i
λ	constant failure rate, see Eqn (11.18)
Γ	Gamma function
γ	flight path angle, measured positive up from the local horizon

11.12.1 SUBSCRIPTS

bo	conditions at burn-out of rocket motor
cap	relating to a space capsule
LES	relating to a launch escape system
str	relating to a structure
0	initial condition
vac	vacuum conditions

REFERENCES

Abernethy, R. (2015). *The new weibull handbook* (5th ed.). North Palm Beach, FL: Abernethy.

ASAP (2014). Annual Report for 2014, NASA Aerospace Safety Advisory Panel.

Gilbert, M. G. (2015). The max launch abort system—concept, flight test, and evolution. In T. Sgobba, & I. Rongier (Eds.), *Space safety is no accident* (pp. 343–354). New York, NY: Springer.

Isakowitz, S. J. (1991). *International reference guide to space launch systems* Reston, VA: American Institute of Aeronautics and Astronautics, 1991.

Musgrave, G. E., Larsen, A., & Sgobba, T. (2009). *Safety design for space systems* Burlington, MA: Elsevier.

NASA (1986). *Report of the presidential commission on the space shuttle challenger accident.* Washington, DC: NASA History Office.

Reason, J. (1990). *Human error.* New York, NY: Cambridge University Press.

SAIC (1989). *NASA quantifiable failure rate data base for space flight equipment.* Science Applications International Corporation Report No. SAICNY-89-10-43A and 43B, December.

Sforza, P. M., Shooman, M. L., & Pelaccio, D. G. (1993). A safety and reliability analysis for space nuclear thermal propulsion systems. *Acta Astronautica, 30,* 67–83.

Shooman, M. L. (1990). *Probabilistic reliability: An engineering approach* Melbourne, FL: Krieger.

SSH (2014). *NASA system safety handbook, Vol. 1. System safety framework and concepts for implementation.* NASA SP-2010-580.

Stepaniak, P. C., & Lane, H. W. (Eds.). (2015). *Loss of signal: Aeromedical lessons learned from the STS-107 Columbia space shuttle mishap.* NASA.

Welker, E. L., & Lipow, M. (1974). Estimating the exponential failure rate from data with no failure events. In *Proceedings of the annual reliability and maintainability symposium.* January, pp. 420–427.

ECONOMIC ASPECTS OF SPACE ACCESS

12.1 ELEMENTS OF SPACECRAFT COST

Cost evaluation is of fundamental importance to the space access enterprise. The scale of the expenditures involved, which are often problematical even for national governments, is particularly sensitive for the nascent commercial space business. Though national governments pursue space activities for a number of reasons, including geopolitical ones, aerospace companies look to earn money by developing and building spacecraft to sell to operators who, in turn, look to earn money by selling the use of the spacecraft to customers who desire access to space. The cost of a particular spacecraft is comprised of several basic elements:

- Development cost—The nonrecurring cost of developing and designing the spacecraft
- Production cost—The recurring cost of building each spacecraft
- Operating cost—The cost of using the spacecraft including ground operations
- Refurbishment cost—The cost of maintaining the spacecraft for continuing use. This is essentially a reusable spacecraft issue because expendable spacecraft are replaced, not refurbished
- Recovery cost—The cost of returning the crew to base. This is essentially an expendable spacecraft issue because reusable spacecraft are considered to return to base on their own
- Insurance cost—The cost of indemnification.

The various items will be discussed in the following sections in a manner that permits a rough order of magnitude (ROM) cost estimate for a spacecraft during the preliminary design stages. Manned spacecraft programs are considerably more expensive than unmanned programs and therefore experience with the Apollo and Space Shuttle programs is reviewed first to establish an overall appreciation for the magnitude of the costs involved.

12.2 COSTS OF THE APOLLO PROGRAM

A detailed breakdown of the Apollo budget appropriations is given by NASA (1978). The dollar values quoted for each budget year are given in the current dollars for that year and must be properly inflated for any comparisons to other times, such as the present. The cumulative cost over the life of the program, in 2015 dollars, is shown in Figure 12.2. The total cost of the program was about $132.1 billion in 2015 dollars. There were 11 Apollo flights so that the

Manned Spacecraft Design Principles. DOI: http://dx.doi.org/10.1016/B978-0-12-804425-4.00012-X

cost per flight, based on the overall project costs, was about $12 billion per flight. The maximum annual expenditure rate was about $18.75 billion (2015 dollars) per year over the period 1963–1969. The cost then leveled out over the 1969–1973 period when the program ended (Figure 12.1).

It is interesting to note that of the 26 budget categories listed in the Apollo program budgeting, 7 used up almost 87% of the total expenditure: spacecraft (command and service modules and the lunar module) 30.8%, engines (Saturn I, IB, and V stages) 39.9%, and flight operations (mission support and manned spaceflight operations) 16.4%. This distribution of costs is illustrated in Figure 12.2.

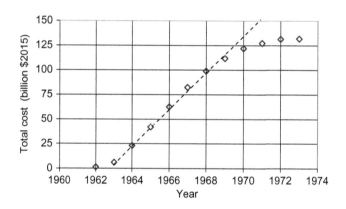

FIGURE 12.1

Cumulative costs of the Apollo program in billions of 2015 dollars are shown by year since inception in 1962. The dashed line represents an expenditure rate of about $18.75 billion (2015 dollars) per year.

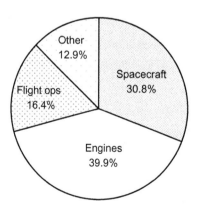

FIGURE 12.2

Distribution of total cost of the Apollo program in billions of $2015.

12.3 **COSTS OF THE SPACE SHUTTLE PROGRAM**

Space vehicle costs are broadly divided into research and development costs and operational costs. Generally the dividing event for a space access project is considered to be the first operational flight. An analysis of the Space Shuttle program by Pielke (1974) notes that the first flight of FY 1983 was also the first operational flight of the Space Shuttle program although it was actually the fifth flight overall. The total cost over the development period, 1972–1982, was placed at $32.5 billion, in 1992 dollars, or $53.03 billion in 2015 dollars.

A chart of the cumulative expenditures is shown in Figure 12.3. The maximum expenditure rate is about $8.33 billion (2015 dollars) per year, as shown by the dashed line and this occurred during the first half of the 40-year program. The maximum rate of expenditure slowed in the second half of the program to about half that value. The total cost through the retirement of the shuttle in 2011 is estimated to be around $228 billion in 2015 dollars. For a discussion of Shuttle costs see, for example, Pielke and Byerly (2011).

The cost per flight thus far, if figured solely on the basis of total cost, would be approximately $228 billion/135 flights, or about $1.69 billion (2015 dollars) per flight, while including the four flights prior to 1983 leads to approximately $228/139, or $1.64 billion per flight. The US General Accounting Office reported in GAO (1993) that NASA's budget request for FY 1993 showed an average cost per flight of $413.5 million with similar numbers (in constant dollars) projected out to FY 1997. As the GAO points out, this includes only those costs needed "to operate the space shuttle on a recurring and sustained basis for a given year divided by the number of flights planned for that year." Adjusting the average cost per flight to 2015 dollars yields a cost of about $696 million, or about 41% of the cost per flight based on total program cost. The point of the GAO report appeared to be clarification of what average cost per flight means in terms of overall program cost. This issue will be the subject of our economic analysis in a subsequent section. It is useful to note that GAO (1993) points out that the NASA quote of approximately $44.4 million per flight ($1993) for the marginal cost of a flight is about an order of magnitude less than the average cost of a flight

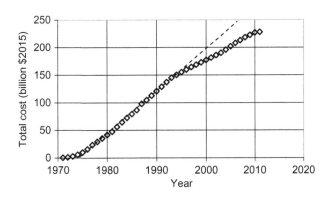

FIGURE 12.3

Cumulative costs of the Space Shuttle program in billions of 2015 dollars are shown by year since inception in 1971. The dashed line represents a maximum expenditure rate of about $8.33 billion per year.

because it "does not include any of the fixed costs that NASA says is necessary to maintain the capability to launch the shuttle eight or nine times a year." Thus it should be clear that one can get a wide variation in the cost per flight of the Space Shuttle depending upon the definitions used.

12.4 PRICE PER POUND TO ORBIT

The cost of placing a payload into orbit (LEO, MEO, and GTO, as well as highly elliptical orbits) is an important factor in the satellite launching business and has been the main subject of interest in the commercial space community. The all-up cost of launch vehicles is strongly dependent on payload size and satellite operators want a reliable launch vehicle that is just sufficient to place their payload in orbit with a reasonable factor of safety.

A metric for comparison that developed in the satellite launch business is the specific cost, which is the cost per pound to place a given payload into the desired orbit. However, launch prices are rarely made public, particularly since 2000. Futron (2002) developed and regularly updated a launch cost database which demonstrates trends in the specific cost of putting unmanned payloads into different kinds of orbits. The range of specific cost in 2002 dollars is shown in Figure 12.4 for Western and non-Western launchers. The cost also depends upon the mass of the payloads launched: small (less than 5000 lb or 2265 kg), medium (5,000−12,000 lb or 2265−5436 kg), intermediate (12,000−25,000 lb or 5,436−11,325 kg), and heavy (over 25,000 lb or 11,325 kg). In a study comparing launch costs from 1960 to 2010, Kendall and Portanova (2011) show that there has been little change in the specific cost (in constant dollars) since 2005 so that the data shown in Figure 12.4 is probably still reasonable if properly adjusted for inflation (see Section 12.6). Reducing mission costs is a major goal for any space program. A comprehensive review of methods for reducing such costs is given by Wertz and Larson (1996).

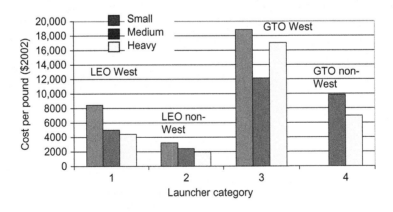

FIGURE 12.4

Specific cost of unmanned launches of various size payloads to different orbits shown for Western and non-Western launchers in $2002 (Futron, 2002). Costs depend on whether the launch vehicles are classified as small, medium, or heavy.

12.5 COMPONENTS OF LAUNCH COST

The objective of a space launch system is to successfully place a payload of a given mass into a specified orbit. The basic economic metric for such an enterprise is the cost per launch in which all the costs incurred in conceiving, developing, producing, and operating the vehicle are included. The cost per launch in constant dollars can be estimated according to the method of Wertz (2000) by starting with a breakdown of that cost into its constituent elements as follows:

$$C_l = C_d + C_p + C_o + C_r + C_f + C_i \tag{12.1}$$

The elements of the cost per launch on the right-hand side of Eqn (12.1) represent, respectively, the costs of design and development, of production, of flight operations, of recovery operations, of refurbishment, and of insurance. These cost centers will be addressed sequentially.

12.5.1 NONRECURRING COST OF DEVELOPMENT

The design and development process involves a one-time, or nonrecurring, cost that must be amortized over a number of years. That development cost may be divided by the number of launches over the life of the vehicle to provide the portion of the cost per launch assignable to designing and developing the vehicle. Assuming that there are a total of L launches within the N years over which the total nonrecurring development cost D is amortized at an annual interest rate i, the development cost per launch is given by

$$C_d = \frac{ND}{L} \frac{i}{[1 - (1+i)^{-N}]} = \frac{DA}{(L/N)} \tag{12.2}$$

Here DA is the annualized payment of the development cost and L/N is the average number of launches per year over the lifetime of the project. Equation (12.2) clearly illustrates the fundamental difficulty of cost estimation: all the terms in it seem to be unknown. From the outset of the cost estimation process, assumptions must be made about almost all the facets of the entire project. The first item to consider is the amortization period N which generally covers the lifetime of the project. In the case of the Apollo program the project lifetime was 10 years while for the Space Shuttle program it was about 40 years. Most cost estimating exercises that appear in the literature select $N = 15$ or 20 years as a typical value. However, the amortization period is a business decision closely tied to the availability and sources of funding and the complexity of the project.

The average number of launches per year L/N reflects the predicted demand of the market over the duration of the project. An assessment of actual space launches and forecast space launches for the period 2013−2022 is presented by Caceres (2015). The annual launches are aggregated by market, orbit, payload mass, and customer geographical region. For the particular case of heavy payloads (5500 kg or more) the number of all launches is forecast to vary between 16 and 44 per year with an average of 25 per year. However, the average number of launches per year for all manned spacecraft is shown in Figure 12.5 for each decade since the start of manned spaceflight.

It is clear that manned spacecraft launches are relatively rare, with an average of about 5 per year for the last 55 years. The experience of the Apollo and Shuttle programs shows that operational flights started about one half to one-third of the way through the project period, respectively. Several test flights were conducted in these programs but should be considered part of

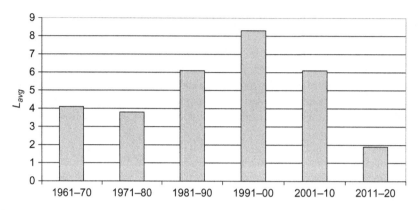

FIGURE 12.5

Average numbers of launches per year each decade for all manned spacecraft.

the development program and not part of actual flight operations. Therefore the total nonrecurring development cost, which may include some test flights, is amortized over a period longer than that during which operational flights are carried out. Thus the development cost per launch in Eqn (12.2) is based on D, the cost of development, which occurs during the early part of the program and L, the total number of operational launches, which occur in the later part of the program. For example, a nominal $N = 20$-year project may begin operational launches about 7 years into the program leaving the remaining 13 years with (using the average value quoted previously) 5 launches per year for a total of $L = 65$ launches yielding an average of $L/N = 65/20 = 3.25$ launches per year. Among the longer duration manned space projects the Space Shuttle program had $L/N = 135/40 = 3.375$ launches per year, the Apollo program had $L/N = 11/10 = 1.1$ launches per year, and Soyuz has, so far, $L/N = 122/48 = 2.54$ launches per year.

The interest rate i appearing in Eqn (12.2) depends on projected economic conditions over the lifetime of the project. The development of a spacecraft, particularly a manned spacecraft is considered risky so the interest rate is generally high, 15% or more. The effect of the interest rate enters through the amortization factor A in Eqn (12.2) and is shown in Figure 12.6.

It appears obvious that the design objective should be to minimize development cost while serving a market that requires a substantial number of launches annually. There seems to be little market demand for a large number of launches of manned spacecraft except, in the minds of some entrepreneurs, space tourism. Morring (2015) points out that "The diversity of the vehicle designs—capsules for Blue Origin, Boeing, Lockheed Martin and SpaceX; suborbital spaceplanes for Virgin Galactic and XCor Aerospace, and a lifting body for Sierra Nevada—is matched by that of the passengers lining up to fly on them." The attitude of many prospective space tourists suggests that space may be the latest theme park, one which can satisfy their interest in "being there." As a result there is a real business interest in manned spacecraft that is not necessarily matched in scope by government developers. The basic issue is the average launch frequency L/N which is the only important economic parameter dominated neither by technical nor financial issues. If the demand were great enough the other issues would likely be readily managed.

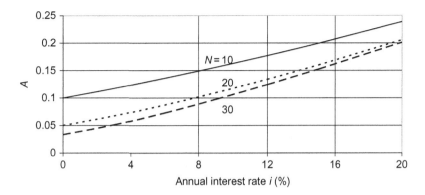

FIGURE 12.6

The amortization factor A in Eqn (12.2) is shown as a function of interest rate i for various project durations N.

Whitfield and Olds (1998) carried out Monte Carlo simulations which show that the volatility of the market plays an integral role in the viability of commercial advanced spaceflight vehicle programs further emphasizing the need for accurate market demand forecasts.

12.5.2 RECURRING PRODUCTION COST

Production of expendable vehicles may be considered to be carried out serially as launch vehicles are needed. Therefore producing the vehicle is a recurring cost and it may be estimated on the basis of a theoretical cost of the first unit CFU. The cost of the first unit is then discounted by the effects of a learning curve established in actually building units, so that as larger production runs are considered the unit cost declines. One approach involves a power law model for the total production cost P of n vehicles as follows:

$$P = (CFU)n^B \tag{12.3}$$

The parameter B, which discounts the cost of the production run as a result of lessons learned by manufacturing more and more units, depends upon the nature of the learning curve chosen and may be represented by

$$B = 1 - \frac{\ln\left(\frac{1}{S}\right)}{\ln 2} = 1 + \frac{\ln S}{\ln 2} \tag{12.4}$$

Here S is the learning curve fraction and depends upon n, the total number of units produced. Note that if $S = 1$, all that can be learned has been learned on the very first unit and no improvement is possible so that the total cost of production is simply the product of the cost of the first unit and the number of units produced. The total production cost may be written as follows:

$$P = (CFU)nf_l \tag{12.5}$$

Here f_l is a learning curve factor given by

$$f_l = n^{B-1} = n^{\ln S/\ln 2} \tag{12.6}$$

Table 12.1 Learning Curve Percentage

Number of Units Produced, n	Learning Curve Fraction, S
$1 < n < 10$	0.95
$10 < n < 50$	0.90
$n = 50$ or more	0.85

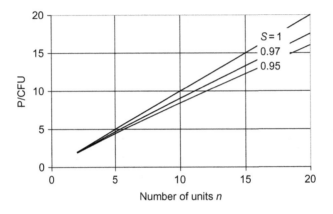

FIGURE 12.7

Effect of the size of the production run on the ratio of the total price of the first unit for various values of the learning curve fraction S.

Apgar et al. (1999) suggests graduating the values of S as shown in Table 12.1 rather than the continuous function as given by Eqn (12.6). The effect of the learning fraction on the ratio $P/CFU = nf_l$ is shown in Figure 12.7 where it is seen that small changes in S result in rather large changes in P/CFU.

The average production cost per launch of each vehicle for a production run of n vehicles, assuming all are ultimately launched so that $n = L$, is then simply given by

$$C_p = \frac{(CFU)nf_l}{L} \tag{12.7}$$

The cost of production of the nth vehicle CNU may be determined as follows:

$$CNU = (CFU)[nf_l(n) - (n-1)f_l(n-1)] = (CFU)n\left[f_l(n) - \left(\frac{n-1}{n}\right)f_l(n-1)\right] \tag{12.8}$$

The cost of the nth vehicle normalized by the cost of the first unit is shown in Figure 12.8 as a function of the size of the production run. For reasonable production runs, say, $n > 10$, Eqn (12.8) may be approximated by

$$CNU \approx (CFU)f_l = \frac{P}{n} \tag{12.9}$$

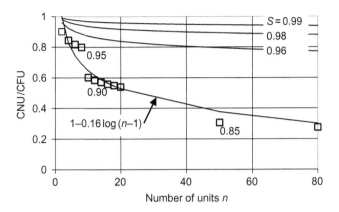

FIGURE 12.8

Cost of the nth unit normalized by the cost of the first unit as a function of the production run size according to the learning curve fraction S as given in Table 12.1 (data symbols) and Eqn (12.6). The analytic curve is added to illustrate the trend for Table 12.1 data.

Figure 12.8 shows that for larger production runs and fixed S the cost of production of the nth vehicle is approximately equal to the average cost of one unit. Examination of Figures 12.7 and 12.8 suggests that it is unlikely that the values of S in Table 12.1 are appropriate to advanced equipment like launch vehicles and engines because by the 30th unit the cost is about half the original cost. In the manufacture of such items, any new materials or techniques probably involve additional expense rather than less. It is more likely to expect that various details of the production processes will be improved resulting in some modest reductions in cost.

The insight afforded by Figure 12.7 is that unless relatively large numbers of vehicles are produced the cost of each will always be close to that of the "hand-made" first article ($S = 1$). In the previous section, we put forth a notional project that would likely have 65 or more launches over the life of the program. If the vehicle is expendable then 65 vehicles are needed. Using Eqn (12.3) or Figure 12.7 shows that the total cost is about $52CFU$ rather than the $65CFU$ it would cost with no learning curve. Because $n^B \sim (n-1)^B$ when n is large, Eqn (12.5) shows that the production cost per launch $C_p \sim (52/65)CFU$. The data points in Figure 12.8 do not follow a smooth curve because the learning percentage S in Table 12.1 is applied in a series of discrete steps. The cost of the nth vehicle as given by Eqn (12.7) is $CNU = 0.8CFU$ for $S = 0.96$. The logarithmic curve appearing in Figure 12.8 is included merely to illustrate the trend of the data points.

In the case of reusable vehicles, the cost model is somewhat more complex in that it is generally assumed that the vehicles would be produced essentially at the same time rather serially over time as in the case of expendable vehicles. This means that the production cost would be borne at the outset so it must be borrowed at some interest rate i, as is the case for the development cost. One may estimate the production cost using the previous model and use Eqn (12.3) with the same learning curve. Thus for reusable vehicles, the production cost per launch would be given by

$$C_{p,r} = (CFU)n^B \frac{i}{L[1-(1+i)^{-n}]} \tag{12.10}$$

The significant differences are that with a reusable vehicle the number of units produced will be substantially fewer that for expendable vehicles. Furthermore, though the cost of the expendable vehicles drops as more units are produced, the cost of producing the reusable vehicle is carried in a series of equal annual payments. Thus the cost of money and the rate of inflation have different importance for expendable and reusable vehicles.

In the discussion of number of launches in Section 12.5.1, the nominal example was assumed to have a total of 65 launches by conducting 5 launches per year for the last 13 years of the notional 20-year program. The Space Shuttle Orbiter was the only fully reusable part of the Space Transportation System (STS) and was designed for 100 flights and a 10-year operational lifespan. The STS fleet of Orbiters numbered 5 so that theoretically 500 launches could have been carried out, but the demand was not there. The major effort in reducing the cost of space access has been directed toward developing reusable launch systems because such systems deliver the best cost results, at least on paper and at a relatively high launch rate compared to historical data. It appears that even carefully conducted studies with reasonable assumptions indicate that in order to reap the apparent rewards of reusability there must be a demand for somewhere between 10 and 20 launches per year. Such a launch rate for manned spaceflight seems unlikely unless some commercial activity like space tourism, microgravity research and processing on the ISS, lunar or asteroid mining, etc. begins to offer a large return on investment.

12.5.3 COST OF FLIGHT OPERATIONS

One may use a learning curve model for estimating the cost of flight operations for each launch as well. Here however one must construct the cost of the first unit's flight operations $(CFU)_{ops}$. Considering the increased complexity of reusable vehicles it is likely that $C_{ops,e} < C_{ops,r}$. For (unmanned) expendable vehicles $C_{ops} = O(\$10^6)$, while for a reusable vehicle it might be an order of magnitude more. For example, in the extreme case of the manned Space Shuttle, flight operations cost are $O(\$10^8)$. For any type of vehicle, the flight operations can be reduced by doing additional work on this topic during development. However, this drives the development cost up and may not be considered a viable option by management. This is the same struggle found in designing out risk: it increases the development cost. In this sense, it becomes clear why all operations (like all risks) must be clearly defined.

12.5.4 COST OF REFURBISHMENT

In fully expendable vehicles, there is no refurbishment cost since the vehicle is lost at the end of the mission. However, reusable vehicles may have substantial costs associated with preparing the vehicle for the next launch: inspection, maintenance, replacement, recertification, retest, returning the vehicle to the launch pad, and the like. Typically the refurbishment cost per launch is estimated as a fraction of the vehicle production cost per launch, modified by a factor that expresses the wear and tear on of the vehicle as the vehicle ages. This might be represented as

$$C_f = F_1 F_2 C_p \tag{12.11}$$

Here F_1 is the refurbishment cost factor that may be 0.10–0.15, for example, while F_2 is an aging factor that might be 1.05–1.10.

12.5.5 **COST OF RECOVERY**

In the case of manned missions, there will always be a cost of recovery because upon the return of the astronauts to earth various efforts must be expended to retrieve them and bring them safely to base. With reusable vehicles, the craft can usually return to the desired landing spot under their own power and control and in this regard the cost of recovery may be minimal. On the other hand, if the landing occurs at a remote location (consider a shuttle landing at Edwards AFB in California rather than Cape Canaveral, FL) the vehicle must be recovered and return to the appropriate site. With expendable vehicles that carry capsules which land by parachute, there needs to be a recovery force of personnel and vehicles to recover the astronauts and the capsule. This can be very expensive in the case of water recovery operations.

12.5.6 **COST OF INSURANCE**

In manned missions, which up to now have been part of nationally or internationally sponsored programs, the governmental agency typically self-insures the missions. However, with the advent of space tourism this approach will change and the space insurance industry will have to branch out to consider underwriting these new commercial programs. In the case of commercial unmanned missions, the insurance rate has typically been considered as a percentage, say around 15%, of the cost of the launch. This launch insurance is separate from the on-orbit insurance applied to the operation of unmanned commercial missions.

12.6 **COST ESTIMATION RELATIONS**

The structure for estimating launch costs outlined in the previous section appears complete but there are a number of variables for which a price must be established. The most common approach to estimating the ROM cost of a launch system is the use of a cost estimation relation (CER) based on the mass of the component m_i and this is generally assumed to be of the form

$$\text{Cost} = a_i m_i^{b_i} \prod_{i=1}^{m} f_i \qquad (12.12)$$

This type of estimate for the cost of a hardware item enjoys a long and generally successful history in airplane development as discussed by Sforza (2014). The coefficients a_i and b_i in Eqn (12.12) are determined by fitting historical data for like items. As a consequence, the utility of estimates formed in this fashion depends greatly on the quality and applicability of the data in the historical record. In this sense the cost information for expendable launch vehicles is considerably more robust than for reusable launch vehicles because there is essentially no history of reusable launch vehicles. This is one reason to narrowly examine cost estimates for proposed reusable launch vehicles. To permit estimation of advanced concepts and technologies, Koelle (2013) and others use several multiplicative factors f_i to account for relatively subjective metrics like technical development status, technical quality factor, team experience factor, and various cost growth factors.

Because this cost estimating approach depends upon historical data the coefficients a_i and b_i are constantly evolving and it is often confusing to settle on some set with which to work. However, although CERs proposed by different researchers may look different, the differences are often misleading because a set of cost data with some scatter can be fitted with many different power laws which will all agree over the limited segment of the mass spectrum being fitted. Dryden and Large (1977) evaluated eight unmanned spacecraft cost estimation models against actual programs and found that mass was indeed the major driver of cost but the models were all quite different and gave different results. One of the interesting observations in their report is that researchers doing the cost estimating spent large amounts of time searching for the same technical and cost data, particularly the latter.

NASA had an online resource called the Spacecraft/Vehicle Level Cost Model which was accessible by students as well a higher level version called the NASA Air Force Cost Model (NAFCOM) that was available to contractors. NASA's Cost Analysis Division, a part of the Office of Evaluation, has discontinued these resources and has introduced a Cost Estimating Handbook (NAFCOM, 2014) which is designed to be an electronic resource for cost estimating issues. NASA is currently transitioning from NAFCOM to the Project Cost Estimating Capability (PCEC), which contains additional information and tools. PCEC incorporates NAFCOM models, as well as models developed by various NASA Centers and directorates. This consolidated model should provide a degree of consistency and traceability which will improve the accuracy of estimates of spacecraft development and production costs.

Koelle (2013) provides a continually updated cost estimation methodology with substantial detail that is a widely used tool for evaluation of the economics of new spacecraft. In order to eliminate consideration of the value of dollars in any given year, Koelle considers the cost as given by Eqn (12.12) to be in man-years. The dollar equivalent of man-years can be calculated for any year using an appropriate price deflator. Several different federal government deflators are listed in Table 12.2; beyond fiscal year (FY) 2014 the figures are government estimates. Snead (2006) presents a detailed application of this approach for the case of a fully reusable two-stage launch system and uses 1 man-year = 1m-y = $250,000 in 2005 dollars. Using, for example, the consumer price index (CPI) in Table 12.2 would set 1m-y to approximately $313,000 in 2015 dollars.

12.6.1 COST OF DEVELOPMENT

The nonrecurring development cost in man-years for each component is denoted by D_i. The set of coefficients for the CER shown in Table 12.3 was collected from several sources and may be used to estimate the development cost of each component. The mass of each component is in kilograms and the notes in Table 12.3 give further instructions related to the cost estimation calculation for each component.

To get an overall perspective on development costs, the CERs in Table 12.3 are plotted on a logarithmic scale in Figure 12.9 but for clarity each component curve is limited to its likely mass range. This figure shows that for a given mass the development costs for liquid rockets, space capsules, and launch stages are all rather close to one another while solid rocket booster costs are much lower, approximately one-fifth the cost of the others.

Liquid rocket motors and space capsules are all essentially limited to masses less than 10,000 kg (10 t) and for clarity their CER curves on Figure 12.9 are ended at that mass. The mass

Table 12.2 Various Federal Government Price Deflators

FY	GDP[a]	CPI[b]	DOD[c]	DOD[d]	DOD[e]
2005	83.28	79.98	80.99	83.6	81.15
2006	85.98	83.01	83.62	85.81	83.77
2007	88.31	84.84	85.92	87.75	86.00
2008	90.13	88.90	88.35	89.28	88.56
2009	91.19	88.21	88.87	90.47	89.77
2010	91.98	90.13	90.90	91.83	91.91
2011	93.78	92.84	92.99	93.5	93.75
2012	95.42	95.26	94.66	95.06	95.29
2013	96.86	96.78	96.45	96.57	96.91
2014	98.31	98.14	98.02	98.19	98.31
2015	100.00	100.00	100.00	100.00	100.00
2016	101.86	102.10	101.87	101.96	101.61
2017	103.88	104.35	103.93	104.00	103.36
2018	105.95	106.75	106.07	106.08	105.29
2019	108.08	109.20	108.28	108.20	107.37

[a]Gross Domestic Product deflator (OMB).
[b]Consumer Price Index CPI-W (OMB).
[c]DOD nonpay.
[d]DOD purchases, excluding pay, fuel, and medical factors.
[e]DOD Total.

Table 12.3 Coefficients for the Cost of Development (D)

Component	a_i	b_i	Notes
Launch stage	219	0.48	No engines
Liquid rocket	187	0.52	Turbopump-fed
Solid booster	10.4	0.60	Mass includes m_p
Space capsule	436	0.41	

range for liquid rocket motors used in Figure 12.9 easily encompasses motors like the 8600 kg (8.6 metric tons) Rocketdyne F-1A used on the Saturn V first stage. Similarly, space capsules are shown to range from 1 t (like the Mercury capsule) to 10 t which includes the 8.9 t Orion capsule. Because we are interested in manned spaceflight the mass of the launch stages is started off at 10 t but may be as large as 300 t. The mass of the first stage of the Saturn V is the Saturn S-IC whose dry mass is about 130,000 kg or 130 metric tons whereas the mass of the Atlas V first stage is about 35 metric tons. Solid rocket motor cost is based on the total mass including the propellant and this is also started off at 10 t but may be as large as 900 t (the 5-segment RSRM considered for the Ares launcher has a mass of about 860 t).

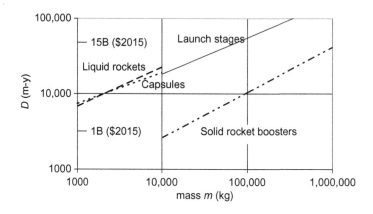

FIGURE 12.9

Variation of the development cost, in man-years, for various launch components as a function of component mass.

On the basis of CERs, it appears that development costs may be minimized by minimizing the mass of each of the components of the system to the point at which the integrated system will just accomplish the specified mission. However, the various subjective weighting factors f_i in Eqn (12.12) can influence the optimal cost at a mass greater than the minimum. For that reason, it is prudent to follow a consistent model like Koelle (2013) which includes specific suggestions for the f_i factors. To aid the economic perspective, the magnitude of the costs, in terms of $2015, is also indicated on Figure 12.9. It is seen that the costs range from roughly $1 to $15 billion in 2015 dollars.

12.6.2 PRODUCTION COST OF THE FIRST UNITS

The production cost in man-years for the first unit of each component is denoted by CFU_i. The set of coefficients for the first unit CER shown in Table 12.4 was collected from several sources and may be used to estimate the production cost of components. The mass of each component is in kilograms and the notes in Table 12.4 give further instructions related to the cost estimation calculation for each component.

To get an overall perspective on production costs of the first units of the various components, the CERs in Table 12.4 are plotted on a logarithmic scale in Figure 12.10 with each component limited to its likely mass range. Figure 12.10 shows that production costs appear to be somewhat more disparate than the development costs of Figure 12.9.

Two curves are given for liquid propellant rocket motors in Figure 12.10. One is for LH_2-fueled rocket motors and the other is for liquid rocket motors using other fuels such as kerosene-based RP-1 or hydrazine-based UDMH. The first unit LH_2-fueled rocket motors are seen to cost about 20–50% more than the rocket motors not using liquid hydrogen as a fuel. This difference represents the additional production requirements involved in constructing an engine with fuel turbopumps that are subjected to the extremely low temperatures of liquid hydrogen (20 K). We note

Table 12.4 Coefficients for the Cost of the First Unit (CFU)

Component	a_i	b_i	Notes
Launch stage	1.30	0.65	No engines, LH$_2$-fueled
Launch stage	1.02	0.63	No engines, not LH$_2$-fueled
Liquid rocket	5.31	0.45	LH$_2$-fueled
Liquid rocket	1.90	0.54	Not LH$_2$-fueled
Solid booster	2.30	0.40	Mass includes m_p
Space capsule	0.16	0.98	

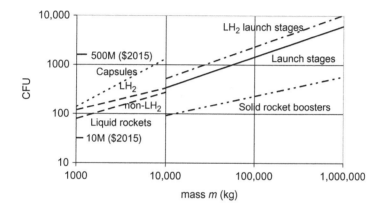

FIGURE 12.10

Variation of the production cost of the first unit, in man-years, for various launch components as a function of component mass.

again that the mass range shown in Figure 12.10 encompasses engine masses even larger than that of the 8600 kg (8.6 metric tons) F-1A built by Rocketdyne and used on the Saturn IC, the first stage of the Saturn V. The 5400 kg Russian-built NPO Energomash RD-180 motor that powers the Atlas V uses RP-1 kerosene-based fuel and LOX oxidizer as did the F-1A. On the other hand, the 3200 kg Aerojet Rocketdyne Space Shuttle Main Engine (SSME or RS-25) uses LH$_2$/LOX propellant while the heaviest LH$_2$/LOX engine is the 6600 kg Aerojet Rocketdyne RS-68.

Two curves for the production cost of the first unit are also given for launch stages in Figure 12.10. One is for stages carrying LH$_2$ fuel and the other is for stages carrying other fuels such as kerosene-based RP-1 or hydrazine-based UDMH. The low density of LH$_2$ fuel requires greater volume for the same mass so the stages carrying LH$_2$ fuel are larger than stages carrying other fuels and therefore may cost as much as 50−70% more.

The CER curves in Figure 12.10 suggest that avoiding LH$_2$-fueled launch systems, particularly in high mass first stages, can reduce production costs. Similarly, using solid rocket boosters in the first stage can reduce costs. The largest launchers actually do avoid the use of LH$_2$ fuel in the first stage. For example, the Saturn V and the Soviet Energia with lift-off masses of 2970 t and 2400 t, respectively, use RP-1 as a fuel in the first stage and LH$_2$/LOX in the upper stages while the Space Shuttle with a lift-off mass of 2030 t uses solid rocket boosters in the first stage and LH$_2$/LOX in the upper stage.

The production cost of the first unit of a capsule is essentially linear with mass at about $50,000 per kilogram. This is quite similar, though much larger, than the typical mature production cost of commercial aircraft which is about $2000 per kilogram (Sforza, 2014).

12.6.3 SPACE LAUNCH COST REDUCTION AND REUSABLE VEHICLES

The previous discussion of estimating the cost of manned space launch systems was restricted to conventional expendable launch systems. Detailed consideration of the cost savings to be realized with reusable systems as discussed by Snead (2006) is speculative enough so as to be beyond the scope of this book. Hertzfeld, Williamson, and Peter (2005) find two major barriers to reducing the cost of space launch systems: (i) "nobody yet has found and proved that a better and less expensive way of getting into orbit exists than the chemical propulsion multi-stage rockets now being used" and (ii) "overcoming the obvious high cost of new R&D programs for launch vehicles." Nevertheless, the idea of reducing the cost of space travel by means of a reusable space vehicle remains intuitively appealing and is the subject of continuing research. The high cost of new launch vehicle programs is addressed by Hogan and Villhard (2003) who discuss US government investment on reusable space vehicle development and particularly note eight specific projects carried out during the last half of the twentieth century as summarized in Table 12.5. The projects are essentially studies aimed at developing reusable space vehicles with the intent of reducing the cost of space access.

Table 12.5 Federal Investment in Reusable Vehicle Development (then-year dollars)

Project Designation	Investment	Start	End	Notes
X-20 Dynasoar	$3–$5B	1957	1963	Manned delta wing spaceplane canceled before prototype
Project START	$1B	1963	1975	Unmanned lifting body shapes tested in 80 flights
Space Shuttle	$11.4B	1972	2011	Manned double-delta spaceplane with 135 operational flights
X-30 National Aerospace Plane (NASP)	$3–$5B	1985	1993	Air-breathing SSTO spaceplane canceled before prototype was built
DC-XA Delta Clipper	$50M	1991	1996	Unmanned VTOL launch vehicle canceled after first flight failure
X-33 Advanced Technology Demonstrator	$1B	1996	2001	Lifting body orbital spaceplane canceled before prototype was 75% complete
X-34 Technology Testbed Demonstrator	$219M	1996	2001	Unmanned winged entry vehicle canceled prior to any flights
X-37 Advanced Technology Flight Demonstrator	$301M	1998	–	Unmanned winged vehicle with four operational flights Was on orbit up to 22 months

The cost estimating relations for expendable vehicles which were discussed previously usually provide a reasonably accurate view of costs. This is to be expected because they are based on substantial practical experience with real hardware. The validity of such cost estimating relations is problematical when applied to launch vehicle designs that deviate from those used in forming the cost database. In order to substantiate the coefficients used in cost estimating relations, one would need to have some data developed by building and operating the less conventional system. No reusable space launch system has ever been built and operated. However, there are three successful reusable spaceplanes which are launched by expendable systems but, upon return from space, operate like airplanes, flying back to and landing at the launch site for refurbishment and reuse. One is the 5000 kg Boeing X-37B, an operational unmanned spaceplane shown in Figure 12.11. The X-37B is 10 m long and has a wingspan of 4.5 m. It has been launched four times on an Atlas V into LEO, remained on-orbit for as long as 22 months, and then made successful automated entries and landings. A second reusable spaceplane is the unmanned Soviet Buran, shown in Figure 12.12, which closely resembles in size and shape the third reusable spaceplane, the Space Shuttle Orbiter.

The one component of the STS that is reusable is the manned Orbiter spaceplane. The interesting feature of the Orbiter is that it is essentially a conventional aluminum airplane with a TPS coating. A comparison of the component masses of the Orbiter with the US Air Force C-141 Starlifter transport was presented by NRC (1989) and an adaptation is presented in Table 12.6.

Table 12.6 shows that adding the 12,301 kg TPS of the Orbiter to the total mass of the C-141 yields a value of 109,757 kg, almost exactly equal to the mass of the Orbiter, illustrating that the Orbiter is a conventional transport with a TPS applied to its surface. The payload capabilities of the two vehicles are also quite similar with the Orbiter capable of carrying almost 90% of the payload attributed to the C-141.

FIGURE 12.11

An X-37B unmanned spaceplane sits on the runway after landing at Vandenberg AFB, California, at the close of its mission on December 3, 2010 (USAF).

FIGURE 12.12

The Soviet Space Shuttle Buran mounted on the back of a Soviet An-225 Mechta ferry aircraft at the 38th Paris International Air and Space Show in 1989. It made only one orbital flight before being canceled (USAF).

Table 12.6 Component Mass Comparison of the Space Shuttle Orbiter with the C-141 Starlifter (NRC, 1989)

Component	Orbiter m (kg)	C-141 m (kg)	Orbiter (m/m_0)	C-141 (m/m_0)	O/C-141
Wing	6667	15,561	0.061	0.160	0.382
Tail	1310	2676	0.012	0.027	0.436
Fuselage	18,818	15,639	0.172	0.160	1.072
Landing gear	5694	4954	0.052	0.051	1.024
Surface controls	1022	1654	0.009	0.017	0.550
Propulsion	15,414	11,421	0.141	0.117	1.203
APU	1785	237	0.016	0.002	6.716
Hydraulics	1039	1218	0.009	0.012	0.760
Electrical	1831	1215	0.017	0.012	1.342
Electronics	2294	1301	0.021	0.013	1.572
Furnishings	965	2273	0.009	0.023	0.378
Air conditioning	1653	1149	0.015	0.012	1.282
Thermal protection	12,301	0	0.112	0.000	–
Crew	567	1248	0.005	0.013	0.405
Fuel	10,807	5200	0.099	0.053	1.852
Payload	27,180	31,710	0.249	0.325	0.764
Total $= m_0$	109,345	97,456	1.000	1.000	1.000

Table 12.7 Size and Cost Comparison of the Space Shuttle Orbiter with the C-141 Starlifter

Item	Orbiter	C-141
Wingspan (m)	23.84	48.77
Length (m)	32.77	51.31
Height (m)	17.25	11.96
Cargo bay l (m)	18.29	31.78
Cargo bay w (m)	4.57	3.12
Cargo bay h (m)	4.57	2.74
Cargo bay volume (m^3)	300.19	272.35
Wing area (m^2)	249.91	299.89
Crew size	5–7	5–7
Units produced	5	285
Unit cost (million $1992)	1700	8.1
Operational period	1982–2011	1963–2006

One may carry the analogy a bit further by examining some other characteristics of the two vehicles as shown in Table 12.7. The cargo bay volume, wing area, and crew size are quite close and both vehicles had quite long operational periods. However, the great difference is in the number of units produced and the cost of each unit. The cost is expressed in $1992 because both cost quotes were given for almost exactly the same time frame and that was sufficient to illustrate the wide disparity in the cost of the two vehicles. Even if the total production cost (product of unit cost and number of units produced) was the quantity of comparison, the Orbiter's cost is almost four times that of the C-141.

It must be kept in mind that there is a great deal of leeway in making assumptions about costs for advanced systems that have no evidential support, as the array of canceled projects listed in Table 12.5 shows. As an example of successful development of a transformational technology, consider the development of the subsonic combustion jet engine. The progress from a detailed concept in the early 1930s to installation in an operational airplane in the early 1940s was carried out concurrently and independently by Great Britain and Germany. Furthermore, each nation started with a different fundamental concept for the mechanical design: centrifugal vs. axial flow turbomachinery. Within another two decades the first operational flights of a small number of military fighter aircraft had been supplanted by regular commercial transatlantic jet flights thereby changing the pattern of passenger travel forever. On the other hand, the supersonic combustion jet engine, seriously suggested around 1960 (Ferri, 1961), took about 50 years to get to an unmanned first test flight, the X-43A (hydrogen-fueled) in 2004 and the X-51 (JP-7 hydrocarbon-fueled) in 2010, and still is an intriguing but operationally elusive propulsion system.

12.7 **NOMENCLATURE**

A	amortization factor, see Eqn (12.2)
a_i	coefficient in the cost estimation relation, see Eqn (12.12)
B	learning curve parameter, see Eqn (12.4)
b_i	exponent in the cost estimation relation, see Eqn (12.12)
C_d	cost per launch for development
C_f	cost per launch for refurbishment
C_i	cost per launch for insurance
C_l	cost per launch
C_o	cost per launch for flight operations
C_p	cost per launch for production
$C_{p,r}$	cost per launch for production for reusable vehicles
C_r	cost per launch for recovery
CER	cost estimation relation, see Eqn (12.12)
CFU	cost of production of the first unit
CFN	cost of production of the nth unit
CPI	consumer price index
D	total nonrecurring development cost
DOD	US Department of Defense
F_1	cost factor, see Eqn (12.11)
F_2	aging factor, see Eqn (12.11)
f_i	factors in the cost estimation relation, Eqn (12.12)
f_l	learning curve factor, see Eqn (12.6)
GDP	gross domestic product
GTO	geosynchronous transfer orbit
i	interest rate
L	number of launches
LEO	low earth orbit
LH_2	liquid (cryogenic) hydrogen
LOX	liquid (cryogenic) oxygen
MEO	medium earth orbit
N	number of years to amortize the cost of the entire program
n	number of vehicles produced
OMB	US Office of Management and Budget
P	total production cost for n units
ROM	rough order of magnitude
RP-1	kerosene-based rocket propellant
TPS	thermal protection system

12.7.1 **SUBSCRIPTS**

avg	average
t	total

REFERENCES

Apgar, et al. (1999). In J. R. Wertz, & W. J. Larson (Eds.), *Space mission analysis and design* (3rd ed.). Torrance, CA/Dordrecht, the Netherlands: Microcosm Press/Kluwer Academic Publishers, 1999.

Caceres, M. (June, 2015). Expanding customer base for space payloads. *Aerospace America*, 22−24.

Dryden, J. A., & Large, J. P. (1977). A critique of spacecraft models. RAND Report R-2196-1- AF.

Ferri, A. (1961). *Possible directions of future research in high-speed air-breathing engines. Fourth AGARD colloquium—high mach number air-breathing engines.* New York, NY: Pergamon Press.

Futron (2002). *Space transportation costs: Trends in price per pound to orbit.* Bethesda, MD: Futron Corporation.

GAO (1993). *Space transportation: The content and uses of shuttle cost estimates.* U.S. General Accounting Office, GAO/NSIAD-93-115.

Hertzfeld, H. R., Williamson, R. A., & Peter, N. (September, 2005). *Launch vehicles: An economic perspective.* Washington, DC: Space Policy Institute, George Washington University.

Hogan, T., & Villhard, V. (October, 2003). *National space transportation policy: Issues for the future.* RAND Corporation, WR-105-OSTP.

Kendall, R., & Portanova, P. (2011). Aerospace corporation. *Crosslink Magazine, 11*(1).

Koelle, D. E. (2013). *Handbook of cost engineering and design of space transportation systems, Rev. 4, with Transcost Model 8.2.* TCS-TR-200.

Morring, F. (2015). Training day: Teaching regular folks to fly in space. *Aviation Week & Space Technology*, February 16−March 1, p. 26.

NAFCOM (2014). NASA cost estimating handbook version 4.0. <http://www.nasa.gov/offices/ooe/CAD/nasa-cost-estimating-handbook-ceh/#.VUvHcPlVhBd>.

NASA (1978). *The apollo spacecraft: A chronology.* NASA SP-4009, Vol. 1 (1969), Vol. 2 (1973), Vol. 3 (1973), and Vol. 4 (1978).

NRC (1989). *Hypersonic technology for military application.* Washington, DC: Air Force Studies Board, National Research Council, National Academies Press, Committee on Hypersonic Technology for Military Application.

Pielke, R. A. (February, 1974). Data on and methodology for calculating space shuttle programme costs. *Space Policy*, 78−80.

Pielke, R., & Byerly, R. (April, 7, 2011). Shuttle programme lifetime cost. *Nature, 472*(38). Available from http://dx.doi.org/10.1038/472038d.

Sforza, P. M. (2014). *Commercial airplane design principles.* Oxford, UK: Elsevier.

Snead, J. M. (2006). Cost estimates of near-term fully-reusable space access systems. AIAA 2006-7209, Space 2006 Conference, San Jose, CA.

Wertz, J. R. (2000). Economic Model of Reusable vs. Expendable Launch Vehicles. IAF Congress, Rio de Janeiro, Brazil, October 2−6, 2000.

Wertz, J. R., & Larson, W. J. (1996). *Reducing space mission cost.* Torrance, CA/Dordrecht, the Netherlands: Microcosm Press/Kluwer Academic Publishers.

Whitfield, J., & Olds, J. R. (1998). Economic uncertainty of weight and market parameters for advanced space launch vehicles. AIAA 98-5179. *1998 defense and civil space programs conference and exhibit.* Huntsville, AL, October 28−30.

Appendix A: Hypersonic Aerodynamics

A.1 ONE-DIMENSIONAL FLOW RELATIONS

Let us consider the high-speed flow of a gas under the simplest condition of steady ($\partial/\partial t = 0$) one-dimensional ($\vec{V} = u(x)\hat{i}$) flow. In this flow, we assume the gas is frictionless ($\mu = 0$), nonconducting ($k = 0$), nondiffusing ($D_{ij} = 0$), and nonreacting ($\dot{m}_i = 0$). Under these conditions, the equations describing the conservation of mass, momentum, and energy may be written as follows:

$$d(\rho u) = 0 \tag{A.1}$$

$$dp + \rho u\, du = 0 \tag{A.2}$$

$$dh + u\, du = 0 \tag{A.3}$$

We may readily integrate these equations to determine the constants of the motion to be the mass flow $\dot{m}$, the impulse function I, and the total enthalpy h_t, as given by

$$\dot{m} = \rho u \tag{A.4}$$

$$I = p + \rho u^2 \tag{A.5}$$

$$h_t = h + \frac{1}{2}u^2 \tag{A.6}$$

Furthermore, we note that the first and second laws of thermodynamics and the definition of enthalpy may be written as follows:

$$dq = de + pd\left(\frac{1}{\rho}\right) \tag{A.7}$$

$$T ds = dq \tag{A.8}$$

$$h = e + \frac{p}{\rho} \tag{A.9}$$

Manipulating these equations and noting that no heat can be transferred ($dq = 0$) yields

$$T ds = pd\left(\frac{1}{\rho}\right) + de = dh - \frac{dp}{\rho} \tag{A.10}$$

$$dh_t = T ds = 0 \tag{A.11}$$

The important result of Eqn (A.11) is that for the adiabatic, frictionless flows assumed here the total enthalpy and the entropy both remain constant.

Note that Eqns (A.1)−(A.11) involve six variables: u, p, ρ, h, s, and e for which there are five independent equations, i.e., Eqns (A.1)−(A.3), (A.7), and (A.10). To close the system of equations, we need a sixth equation and this role is filled by the equation of state, which may be generally written as

$$h = h(p, \rho) \tag{A.12}$$

A.1.1 **EFFECT OF CHEMICAL REACTIONS**

Even under these seemingly highly restrictive conditions, we may consider the effects of chemical reaction with neither excessive complication nor loss in generality. We will show that there are two important extremes of chemically reacting flows that fit into the current assumption of nondissipative flow, i.e., $ds = 0$. If the gas is considered a mixture of N different species, each denoted by the subscript i, then the enthalpy per unit mass of the flowing mixture depends on the enthalpy h_i and the mass fraction Y_i in the flowing stream and is given by

$$h = \sum_{i=1}^{N} Y_i h_i \tag{A.13}$$

$$Y_i = \frac{\rho_i}{\rho} \tag{A.14}$$

The differential of the enthalpy is

$$dh = \sum_{i=1}^{N} Y_i dh_i + \sum_{i=1}^{N} h_i dY_i \tag{A.15}$$

Using Eqn (A.15) in Eqn (A.10) leads to the following result:

$$Tds = dh - \frac{dp}{\rho} - \sum_{i=1}^{N} (h_i - Ts_i) dY_i \tag{A.16}$$

But combining Eqns (A.2) and (A.3) reveals that

$$dh - \frac{dp}{\rho} = 0$$

Therefore, Eqn (A.16) becomes

$$Tds = - \sum_{i=1}^{N} (h_i - Ts_i) dY_i \tag{A.17}$$

The quantity in parentheses on the right-hand side of Eqn (A.17) is the Gibbs free energy per unit mass

$$g_i = h_i - Ts_i \tag{A.18}$$

On a molar basis, the Gibbs free energy is often called the chemical potential

$$G_i = W_i g_i = H_i - TS_i \tag{A.19}$$

The capital letters here indicate that the properties indicated are per mole of constituent. There is no mass created in chemical reactions so the mass conservation equation (A.1) remains valid. However, the mass of each species present may change as a result of chemical reaction and the conservation of mass of each species may be written as

$$d(\rho_i u) = \dot{\rho}_i dx \tag{A.20}$$

Here the term $\dot{\rho}_i$ is the mass rate of production of species i per unit volume. Equation (A.20) simply states that the change in the mass flow of species i per unit area normal to the flow is equal to the mass rate of production of species i within a volume equal to the product of dx and the unit area. Using Eqns (A.1) and (A.14), it can be shown that Eqn (A.20) reduces to

$$dY_i = \frac{\dot{\rho}_i}{\dot{m}} dx \qquad (A.21)$$

Using Eqn (A.20) in the entropy relation of Eqn (A.17) yields

$$Tds = - \sum_{i=1}^{N} g_i dY_i = - \sum_{i=1}^{N} g_i \frac{\dot{\rho}_i}{\dot{m}} dx \qquad (A.22)$$

This may also be written in terms of the molar rate of production of species i per unit volume per unit time $\dot{n}_i$ as follows:

$$\rho u T \frac{ds}{dx} = - \sum_{i=1}^{N} G_i \dot{n}_i \qquad (A.23)$$

The entropy production given by Eqn (A.22) may be reformulated in a finite difference form to read

$$Tds \approx - \sum_{i=1}^{N} g_i \frac{1}{\rho} \frac{\Delta \rho_i}{\Delta t_c} \frac{\Delta x}{\underbrace{\frac{\Delta x}{\Delta t_r}}} = \frac{\Delta t_r}{\Delta t_c} \left[- \sum_{i=1}^{N} g_i \frac{\Delta \rho_i}{\rho} \right] \qquad (A.24)$$

The term $\Delta t_r / \Delta t_c$ is the ratio of the residence time of a fluid particle within the distance dx to the time for the production of species i to be complete. In the case of so-called "frozen" chemistry, reactions are so slow that $\Delta t_r / \Delta t_c \ll 1$ and the entropy change is essentially zero. In the case of equilibrium chemistry, production rates are infinitely fast such that equilibrium according to local conditions is always immediately satisfied, but the criterion for chemical equilibrium at a given temperature and pressure is that the Gibbs free energy, i.e., the chemical potential, be zero. Therefore the chemical potential term is zero and again we have

$$Tds = 0$$

A.1.2 ISENTROPIC FLOW

Thus for equilibrium or frozen flow of an inviscid, nonconducting, nondiffusing gas the entropy is constant. Assuming the gas to be perfect the equation of state may be written as

$$p = \rho R T = \frac{\rho R_u T}{W} \qquad (A.25)$$

The gas is comprised of a mixture of perfect gases that obey Dalton's law

$$p = \sum_{i=1}^{N} p_i \qquad (A.26)$$

We assume that all the species share the same temperature T so that the state equation (A.25) becomes

$$p = R_u T \sum_{i=1}^{N} \frac{\rho_i}{W_i} = \rho R_u T \sum_{i=1}^{N} \frac{Y_i}{W_i} \qquad (A.27)$$

Comparing Eqns (A.25) and (A.27) reveals that the mixture molecular weight is given by

$$W = \left[\sum_{i=1}^{N} \frac{Y_i}{W_i}\right]^{-1} \tag{A.28}$$

The enthalpy and energy differentials are given respectively by

$$dh = c_p dT \tag{A.29}$$

$$de = c_v dT \tag{A.30}$$

Substituting Eqns (A.25), (A.29), and (A.30) into Eqn (A.9) yields

$$c_p - c_v = R\left(1 - \frac{d \ln W}{d \ln T}\right) \tag{A.31}$$

Repeating this process with Eqn (A.10) yields

$$p\left(\frac{\rho}{W}\right)^{\frac{1}{R/c_p - 1}} = \text{constant} \tag{A.32}$$

Therefore we see that chemical reactions which change the molecular weight of the mixture do have an effect on the thermodynamic properties.

However if the molecular weight remains constant, then Eqns (A.31) and (A.32) reduce to the familiar forms

$$c_p - c_v = R = c_p\left(\frac{\gamma - 1}{\gamma}\right) \tag{A.33}$$

$$p\rho^{-\frac{c_p}{c_v}} = p\rho^{-\gamma} = \text{constant} \tag{A.34}$$

Here we have introduced the ratio of specific heats

$$\gamma = \frac{c_p}{c_v} \tag{A.35}$$

Introducing the isentropic relation of Eqn (A.34) into the momentum equation (A.2) and integrating results in the following relation, sometimes called the compressible Bernoulli equation:

$$\frac{\gamma}{\gamma - 1}\frac{p}{\rho} + \frac{u^2}{2} = \frac{\gamma}{\gamma - 1}\frac{p_t}{\rho_t} = \text{constant} \tag{A.36}$$

For $\gamma = \text{constant}$, using the isentropic relation of Eqn (A.34) in the definition of a^2, the square of the sound speed, yields the following result:

$$a^2 = \left(\frac{\partial p}{\partial \rho}\right)_s = \frac{\gamma p}{\rho} = \sqrt{\gamma RT} \tag{A.37}$$

Introducing $a^2 = \gamma p/\rho$ into the compressible Bernoulli equation (A.36) permits rewriting it as

$$\frac{a^2}{\gamma - 1} + \frac{u^2}{2} = \frac{a_t^2}{\gamma - 1} = \text{constant} \tag{A.38}$$

Now along a streamline, we may find $p, \rho, u,$ *and* h from the following four equations which represent the equations of mass, momentum, and energy conservation and the equation of state, respectively:

$$\dot{m} = \rho u$$

$$\frac{\gamma}{\gamma - 1} \frac{p}{\rho} + \frac{u^2}{2} = \frac{\gamma}{\gamma - 1} \frac{p_t}{\rho_t} \tag{A.39}$$

$$h + \frac{1}{2} u^2 = h_t \tag{A.40}$$

$$h = h(p, \rho)$$

The requirements here are that $\gamma = $ constant and that we know all the variables at a particular point along the streamline, e.g., at a stagnation point. The entropy may found from a state relation like

$$s = s(p, \rho)$$

A.1.3 ISENTROPIC FLOW EQUATIONS

For a perfect gas mixture with constant molecular weight and specific heats, the isentropic relation of Eqn (A.34) and the state equation (A.25) may be combined to yield

$$\frac{p}{p_t} = \left(\frac{\rho}{\rho_t}\right)^{\gamma} = \left(\frac{T}{T_t}\right)^{\frac{\gamma}{\gamma - 1}} \tag{A.41}$$

The enthalpy under these assumptions is simply $h = c_p T$ and the entropy may be found by integrating Eqn (A.10) to find

$$s = s^0 + c_p \ln\left(\frac{T}{T^0}\right) - R \ln\left(\frac{p}{p^0}\right) \tag{A.42}$$

In Eqn (A.42), the superscript 0 denotes standard conditions, e.g., $p^0 = 1$ atmosphere and $T^0 = 298.16$ K. Substituting the isentropic flow conditions of Eqn (A.41) into the entropy equation (A.42) will yield $s = s^0 = $ constant. Dividing Eqn (A.38) through by a^2 and rearranging, using Eqns (A.37) and (A.41), yields

$$1 + \frac{\gamma - 1}{2} M^2 = \left(\frac{p}{p_t}\right)^{-1} \left(\frac{\rho}{\rho_t}\right)$$

Here we have introduced the Mach number

$$M = \frac{u}{a}$$

Using $p\rho^{-\gamma} = $ constant in the rearranged equation provides the usual isentropic relations as follows:

$$\frac{p}{p_t} = \left(1 + \frac{\gamma - 1}{2} M^2\right)^{\frac{-\gamma}{\gamma - 1}} \tag{A.43}$$

$$\frac{\rho}{\rho_t} = \left(1 + \frac{\gamma - 1}{2} M^2\right)^{\frac{-1}{\gamma - 1}} \tag{A.44}$$

Meanwhile, from the energy equation (A.6), we obtain

$$T + \frac{u^2}{2c_p} = T + \gamma RT \frac{M^2}{2c_p} = T\left(1 + \frac{\gamma - 1}{2}M^2\right) = T_t$$

This last result gives us the general adiabatic relation for temperature

$$\frac{T}{T_t} = \left(1 + \frac{\gamma - 1}{2}M^2\right)^{-1} \tag{A.45}$$

Note that using the expression for T from Eqn (A.41) in Eqn (A.43) or (A.44) yields the same result as that shown in Eqn (A.45). That is because the energy and momentum equations are not linearly independent for isentropic flow. The result for T in Eqn (A.45), which is derived starting from the energy equation alone, is a more generally applicable result and holds as long as the flow is adiabatic, even if it is irreversible.

The fundamental definition of hypersonic flow is that $M \gg 1$. We shall see that when particular flow situations are considered, such as inviscid flow over a body or viscous flow near a surface, there are additional parameters which provide more specific definition of the case of hypersonic flow. We may determine the maximum velocity possible in an adiabatic flow by considering the integral of the energy equation (A.6) which requires that

$$u^2 = 2(h_t - h)$$

Then, if we allow the enthalpy, and therefore the temperature, go to zero all the internal energy of the gas goes into kinetic energy

$$\lim_{h \to 0} u^2 \to 2h_t$$

Because the stagnation enthalpy is fixed, there is a maximum velocity, as shown. But in this same process of permitting T to approach zero, the sound speed $a = \sqrt{\gamma RT}$ approaches zero and therefore the Mach number $M = u/a \to \infty$. Though we may permit T to approach zero, we may not do so for the density ρ because the mass cannot go to zero. Thus the perfect gas law requires that p and T must simultaneously approach zero.

A.2 NORMAL SHOCKS

Regions of isentropic flow may be separated by discontinuities within which the entropy jumps. Such jumps are called shock waves, and the one-dimensional conservation equations (A.1)–(A.3) and the integrals of the motion given by Eqns (A.4)–(A.6) are still applicable. Consider the discontinuity separating two regions of isentropic flow as depicted in Figure A.1.

Across a discontinuity, where subscript 1 denotes conditions upstream of the discontinuity and subscript 2 denotes conditions downstream, the integrals of the motion lead to the following relations:

$$\rho_1 u_1 = \rho_2 u_2$$
$$p_1 + \rho_1 u_1^2 = p_2 + \rho_2 u_2^2$$
$$h_1 + \frac{u_1^2}{2} = h_2 + \frac{u_2^2}{2}$$

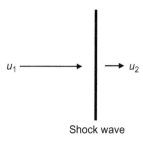

FIGURE A.1

A normal shock wave in one-dimensional flow separating two regions.

We denote the density ratio across the shock wave by

$$\frac{\rho_1}{\rho_2} = \varepsilon = \frac{u_2}{u_1}$$

Then the jump conditions across the shock wave may be written as

$$p_2 - p_1 = \rho_1 u_1^2 - \rho_2 u_2^2 = \dot{m} u_1 (1 - \varepsilon)$$

$$h_2 - h_1 = \frac{1}{2}(u_1^2 - u_2^2) = \frac{1}{2} u_1^2 (1 - \varepsilon^2)$$

$$u_2 - u_1 = -u_1(1 - \varepsilon)$$

Concentrating on the enthalpy jump we find that

$$h_2 - h_1 = \frac{1}{2} u_1 \cdot u_1 (1 - \varepsilon^2) = \frac{1}{2} \left[\frac{p_2 - p_1}{\dot{m}(1 - \varepsilon)} \right] u_1 (1 - \varepsilon)(1 + \varepsilon)$$

This may be rearranged as follows:

$$h_2 - h_1 = \frac{p_2 - p_1}{2\rho_1} \cdot (1 + \varepsilon) \tag{A.46}$$

Equation (A.46) is called the Hugoniot relation and it relates thermodynamic state properties *alone* across a normal shock—no velocities appear in the relation. Note that in the limit

$$\lim_{\varepsilon \to 1} h_2 - h_1 \to \frac{p_2 - p_1}{\rho_1}$$

Therefore, in the limit as ε approaches unity we have

$$dh = \frac{dp}{\rho}$$

The relationship above is exactly that given by Eqn (A.10) for the case of isentropic flow. Thus, this is the limiting case of a wave of infinitesimal strength, i.e., an isentropic Mach wave.

A.2.1 LIMITING SHOCK CONDITIONS IN HYPERSONIC FLOW

We do not know, at the present, of any other limiting value for ε, though we will find one subsequently. We may rewrite the Hugoniot equation (A.46) in the following form:

$$h_2 - h_1 = \frac{p_2 - p_1}{2}\left(\frac{1}{\rho_1} + \frac{1}{\rho_2}\right) \tag{A.47}$$

Solving Eqn (A.47) for ε leads to the following relation:

$$\varepsilon = \frac{(p_2 - p_1)/\rho_2}{2(h_2 - h_1) - (p_2 - p_1)/\rho_2}$$

Using $2h = h + h = h + e + p/\rho$ permits this equation to be rewritten as

$$\varepsilon = \frac{\dfrac{p_2}{\rho_2}\left(1 - \dfrac{p_1}{p_2}\right)}{(h_2 + e_2) - (h_1 + e_1) + \dfrac{p_1}{\rho_2}\left(1 - \dfrac{\rho_2}{\rho_1}\right)}$$

We may determine the effect of allowing $M_\infty \to \infty$, by permitting the enthalpy, internal energy, and pressure upstream of the shock to all approach zero for fixed non-zero upstream density, i.e.

$$h_1 = e_1 + \frac{p_1}{\rho_1} \to 0$$

This is proper since

$$e_1, p_1 \to 0$$

Note that this implies that $T_1 \to 0$. Meanwhile, since the mass cannot vanish, the density

$$\rho_1 = \frac{p_1}{RT_1} \to \frac{0}{0} \to \text{fixed}$$

Then, in the limit where $M \to \infty$, there is a corresponding limiting density ratio

$$\varepsilon_{\lim} = \frac{p_2/\rho_2}{h_2 + e_2} = \frac{p_2/\rho_2}{2h_2 - p_2/\rho_2} \tag{A.48}$$

This limiting density ratio depends only upon the state of the gas behind the shock. Substituting the sound speed from Eqn (A.37) and rearranging yields

$$\varepsilon_{\lim} = \frac{1}{2\dfrac{\gamma h_2}{a_2^2} - 1}$$

Here γ is the isentropic exponent in the relation $p\rho^{-\gamma} = $ constant and we may set $\gamma = c_p/c_v$ only if the gas is calorically perfect, i.e., c_p and c_v are constant. In such a case, we find

$$\varepsilon_{\lim} = \frac{1}{2\dfrac{h_2}{R_2 T_2} - 1}$$

Noting that the definition of the enthalpy is $h = e + p/\rho$, we can rewrite the limiting density ratio as

$$\varepsilon_{\lim} = \frac{1}{2\dfrac{e_2}{R_2 T_2} + 1}$$

The internal energy per unit mass e may have contributions from all the degrees of freedom of the molecules: translation, rotation, vibration, dissociation, and ionization. Thus the internal energy may be described as follows:

$$e = e_{trans} + e_{rot} + e_{rib} + e_{diss} + e_{ion}$$

This illustrates that activation of various degrees of freedom can affect the value of $\varepsilon_{\lim}$. Looking back at Eqn (A.48), we may rewrite it as follows:

$$\varepsilon_{\lim} = \frac{\left(\dfrac{p_2}{\rho_2}\right)}{h_2 + e_2} = \frac{\left(\dfrac{h_2}{e_2}\right) - 1}{\left(\dfrac{h_2}{e_2}\right) + 1} \tag{A.49}$$

No constraints have yet been placed on the properties of the gas but, noting that for the special case of constant specific heats $\gamma = c_p/c_v = h/e$, we arbitrarily define an effective ratio of specific heats as $\gamma_\varepsilon = h/e$ which transforms Eqn (A.49) into

$$\varepsilon_{\lim} = \frac{\gamma_\varepsilon - 1}{\gamma_\varepsilon + 1} \tag{A.50}$$

If the gas is calorically perfect $\gamma = \gamma_\varepsilon$ and the free stream conditions determine the value of the limiting value of the density ratio according to

$$\varepsilon_{\lim} = \frac{\gamma - 1}{\gamma + 1} \tag{A.51}$$

For a diatomic gas with a constant value of $\gamma = 7/5$, we have $\varepsilon_{\lim} = 1/6$ while for a monatomic gas with a constant $\gamma = 5/3$ we have $\varepsilon_{\lim} = 1/4$. Thus the shock wave can increase the density only so much and the pressure and temperature rise at the same rate to keep the limiting value of the density constant. This increasing temperature and pressure of the normal shock process can effect substantial changes to the internal degrees of freedom and thereby change γ_ε and $\varepsilon_{\lim}$. We may write the effective ratio of specific heats as

$$\gamma_\varepsilon = \frac{h}{e} = 1 + \frac{p}{\rho e} = 1 + \frac{RT}{e}$$

It must be kept in mind that the increasing temperature caused by the limiting shock process excites additional degrees of freedom in the molecules increasing the internal energy and generally resulting in reducing the value of γ_ε and bringing it closer to unity. In that case the magnitude of the limiting density ratio $\varepsilon_{\lim}$ given by Eqn (A.50) may be considerably reduced to values half, or even less, of those quoted for constant γ in Eqn (A.51), i.e., 1/6 for $\gamma = 7/5$ and 1/4 for $\gamma = 5/3$.

A.2.2 **THE NORMAL SHOCK RELATIONS**

The jump conditions across the normal shock and their limiting values in the case of $\gamma = $ constant are listed below. The development of these equations may be found in textbooks on gas dynamics such as Liepmann and Roshko (2002).

$$\frac{T_2}{T_1} = \left(\frac{a_2}{a_1}\right)^2 = \frac{[2\gamma M_1^2 - (\gamma - 1)][(\gamma - 1)M_1^2 + 2]}{(\gamma + 1)^2 M_1^2}$$

$$\lim_{M_1 \to \infty} \frac{T_2}{T_1} = \left(\frac{a_2}{a_1}\right)^2 \to 2\frac{\gamma(\gamma - 1)}{(\gamma + 1)^2} M_1^2 = \frac{2\gamma}{\gamma + 1}\varepsilon_{\lim} M_1^2$$

(A.52)

$$\frac{p_2}{p_1} = 1 + \frac{2\gamma}{\gamma + 1}(M_1^2 - 1)$$

$$\lim_{M_1 \to \infty} \frac{p_2}{p_1} \to \frac{2\gamma}{\gamma + 1} M_1^2 = \frac{2\gamma}{\gamma - 1}\varepsilon_{\lim} M_1^2$$

(A.53)

$$\frac{\rho_2}{\rho_1} = \frac{u_1}{u_2} = \frac{q_1}{q_2} = \frac{\gamma + 1}{(\gamma - 1) + 2/M_1^2}$$

$$\lim_{M_1 \to \infty} \frac{\rho_2}{\rho_1} = \frac{u_1}{u_2} = \frac{q_1}{q_2} \to \frac{\gamma + 1}{\gamma - 1} = \frac{1}{\varepsilon_{\lim}}$$

(A.54)

$$M_2 = u_2/a_2 = \left\{\frac{(\gamma - 1)M_1^2 + 2}{2\gamma M_1^2 - (\gamma - 1)}\right\}^{1/2}$$

$$\lim_{M_1 \to \infty} M_2 \to \sqrt{\frac{\gamma - 1}{2\gamma}} = \sqrt{\frac{\gamma + 1}{2\gamma}\varepsilon_{\lim}}$$

(A.55)

$$\frac{p_{t2}}{p_{t1}} = \left[1 + \frac{2\gamma}{\gamma + 1}(M_1^2 - 1)\right]^{-1/(\gamma - 1)} \left[\frac{(\gamma + 1)M_1^2}{(\gamma - 1)M_1^2 + 2}\right]^{\gamma/(\gamma - 1)}$$

$$\lim_{M_1 \to \infty} \frac{p_{t2}}{p_{t1}} = \left(\frac{2\gamma}{\gamma + 1}\varepsilon^\gamma M_1^2\right)^{-1/(\gamma - 1)}$$

(A.56)

$$C_p = \frac{p_2 - p_1}{q_1} = \frac{4}{\gamma + 1}\left(\frac{M_1^2 - 1}{M_1^2}\right)$$

$$\lim_{M_1 \to \infty} C_p \to \frac{4}{\gamma + 1} = \frac{4}{\gamma - 1}\varepsilon_{\lim}$$

(A.57)

Note that for air, with $\gamma = 7/5$ and $\varepsilon_{\ell im} = 1/6$, the limiting value for the temperature ratio across the shock is

$$\lim_{M_1 \to \infty} \frac{T_2}{T_1} = \frac{2(7/5)}{12/5}\frac{M_1^2}{6} = \frac{7}{36}M^2 \approx 0.2M_1^2$$

It was noted in Chapter 2 that the temperature in the middle atmosphere below 100 km altitude is $T_1 \sim 200$ K so that the temperature behind the shock would be $T_2 \approx 40M_1^2$. This suggests that at $M_1 = 10$ the temperature behind the shock would be 4000 K! However the real gas value for air is actually about 3000 K. Obviously real gas effects are important for hypersonic flows because the high temperatures arising from the shock compression at high Mach numbers cause variations in both the specific heats of the gas and the molecular weight.

It is also interesting to note that the limiting total pressure ratio across the shock varies with $M^{-2/(\gamma-1)}$ indicating that the total pressure loss going through a normal shock is great. The entropy change for a calorically perfect gas passing through a normal shock may be found by integrating Eqn (A.10). The entropy is typically written in terms of the change from the standard state as given in Eqn (A.42) may be put in the form

$$s - s^0 = c_v \ell n \left(\frac{p}{p^0} \left[\frac{\rho}{\rho^0} \right]^{-\gamma} \right)$$

Then across the normal shock the entropy rise is

$$s_2 - s_1 = c_v \ln \left(\frac{p_{t2}/p_{t1}}{(\rho_{t2}/\rho_{t1})^\gamma} \right)$$

But the shock wave is an adiabatic compression process in which the total temperature remains constant. Then the state equation requires that

$$\frac{p_t}{\rho_t} = RT_t = \text{constant}$$

Therefore the entropy rise reduces to

$$s_2 - s_1 = c_v \ln \left(\frac{p_{t2}}{p_{t1}} \right)^{1-\gamma}$$

This may also be written as

$$\frac{s_2 - s_1}{R} = -\ln \left(\frac{p_{t2}}{p_{t1}} \right) \tag{A.58}$$

Thus the entropy rise in the normal shock adiabatic compression process is directly related to the total pressure loss in the process.

A.3 STAGNATION PRESSURE ON A BODY IN HYPERSONIC FLOW

Consider the region in the vicinity of the stagnation point of a symmetric body in a hypersonic stream as shown in Figure A.2. The stagnation streamline is on the centerline of symmetry and the shock wave is locally normal to it. The streamlines on either side of the stagnation streamline must curve away from the stagnation point where the flow comes to rest and then accelerate while following the body contour. In the immediate vicinity of the point where the stagnation streamline crosses the shock the normal shock wave jump conditions of the previous section apply. On either side of this crossing the isentropic relations along a streamline apply, but each side has a different value for the entropy.

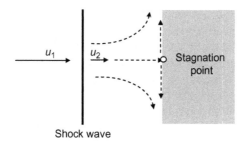

FIGURE A.2

Schematic diagram of flow in the stagnation region of a symmetrical body in a hypersonic stream.

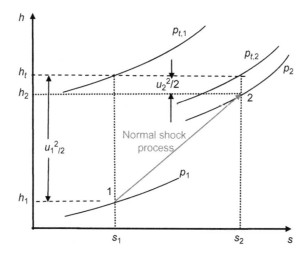

FIGURE A.3

The adiabatic normal shock process for the flow along the stagnation streamline is shown proceeding from state 1 in the free stream to state 2 behind the shock wave.

The flow process along the stagnation streamline may be readily visualized on an $h{-}s$ diagram, as illustrated in Figure A.3. Because the flow is adiabatic everywhere the stagnation enthalpy h_t is constant continuously along the entire streamline and is shown in Figure A.3 as a dashed line. The flow along the streamline meets the shock at point 1 with pressure, enthalpy, entropy, and velocity p_1, h_1, s_1, and u_1, respectively. Note that the sum $h_1 + u_1^2/2$ is the total, or stagnation, enthalpy h_t. Through the shock the properties jump to new values p_2, h_2, s_2, and u_2 and the total enthalpy $h_t = h_2 + u_2^2/2$. For a calorically perfect gas, the enthalpy may be replaced by temperature.

Along the stagnation streamline downstream of the shock, the flow is isentropic and with $\gamma = \gamma_2 = $ constant, Eqn (A.43) becomes

$$\frac{p_{t2}}{p} = \left[1 + \frac{\gamma_2 - 1}{2}M^2\right]^{\frac{\gamma_2}{\gamma_2 - 1}}$$

Here p_{t2} is the stagnation pressure immediately behind the normal shock, as given in Eqn (A.56), and it remains constant because the flow along the streamline is isentropic. The Mach number directly behind the shock $M_2 < 1$ and as the flow proceeds toward the stagnation point the Mach number continually decreases until it reaches $M = 0$ at the stagnation point. Obviously the pressure p continues to increase, from p_2 directly downstream of the shock as the flow continues to be isentropically compressed, until the stagnation point is reached where $p = p_{t2}$. Thus, between the free stream and the stagnation point there is an irreversible adiabatic compression process through a normal shock followed by an isentropic compression process along the streamline between the shock and the stagnation point.

A.4 OBLIQUE SHOCKS

Consider the normal shock wave of Figure A.1 to be moving parallel to itself with constant velocity $V_{1,t}$. To an observer on this shock wave the velocity field would appear as shown in Figure A.4.

The tangential component of velocity is, by definition, constant and because mass must be conserved we require the following conditions to be met

$$\rho_1 V_{1n} = \rho_2 V_{2n}$$
$$V_{1t} = V_{2t}$$

Using these conditions and applying the normal and tangential momentum conservation equations along with the energy equation to the flow across the shock yields

$$p_1 + \rho_1 V_{1n}^2 = p_2 + \rho_2 V_{2n}^2$$
$$\rho_1 V_{1n} V_{1t} = \rho_2 V_{2n} V_{2t}$$
$$h_1 + \frac{1}{2}(V_{1n}^2 + V_{1t}^2) = h_2 + \frac{1}{2}(V_{2n}^2 + V_{2t}^2)$$

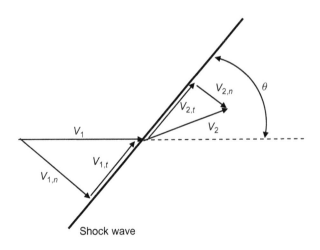

FIGURE A.4

Schematic diagram of an oblique shock wave.

The Hugoniot relation remains the same as before because it includes neither constants of the flow nor velocities. The relations for normal shocks given in Eqns (A.52)–(A.56) may be used for oblique shocks with M_1 replaced by $M_{1n} = M_1 \sin\theta$ and M_2 replaced by $M_{2n} = M_2 \sin(\theta - \delta)$. However, the ratios of static-to-stagnation pressure cannot be treated in this way, but instead are given by

$$p_2/p_{t,1} = \frac{2\gamma M_1^2 \sin^2\theta - (\gamma - 1)}{\gamma + 1} \left[\frac{2}{(\gamma - 1)M_1^2 + 2}\right]^{\frac{\gamma}{\gamma-1}} \tag{A.59}$$

$$p_2/p_{t,2} = \left\{\frac{2[2\gamma M_1^2 \sin^2\theta - (\gamma - 1)][(\gamma - 1)M_1^2 \sin^2\theta + 2]}{(\gamma + 1)^2 M_1^2 \sin^2\theta[(\gamma - 1)M_1^2 + 2]}\right\}^{\frac{\gamma}{\gamma-1}} \tag{A.60}$$

$$p_{t,2}/p_1 = \left[\frac{\gamma + 1}{2\gamma M_1^2 \sin^2\theta - (\gamma - 1)}\right]^{\frac{1}{\gamma-1}} \left\{\frac{(\gamma + 1)M_1^2 \sin^2\theta[(\gamma - 1)M_1^2 + 2]}{2[(\gamma - 1)M_1^2 \sin^2\theta + 2]}\right\}^{\frac{\gamma}{\gamma-1}} \tag{A.61}$$

Similarly, the pressure coefficient across an oblique shock must be written as follows:

$$C_p = \frac{p_1}{q_1}\left(\frac{p_2}{p_1} - 1\right) = \frac{2}{\gamma M_1^2}\left(\frac{2\gamma M_1^2 \sin^2\theta - (\gamma - 1)}{\gamma + 1} - 1\right) = \frac{4(M_1^2 \sin^2\theta - 1)}{(\gamma + 1)M_1^2} \tag{A.62}$$

Along a streamline behind the shock, the isentropic flow relations for pressure and density are those already given in Eqns (A.43) and (A.44), and the adiabatic relation for the temperature is that given by Eqn (A.45). It should be clear that the streamlines are all parallel behind the oblique shock so all flow properties are constant behind an oblique shock, as they are behind a normal shock, until some disturbance, like a solid body, influences the flow field. Recall that this was the case in Section A.3 where a body caused the flow to decelerate to a stagnation point and the streamlines to deviate from being uniformly parallel.

A.4.1 THE ENTROPY LAYER

For a calorically perfect gas, Eqn (A.58) shows that the entropy jump across a normal shock is given by

$$\frac{s_2 - s_1}{R} = -\ln\left(\frac{p_{t2}}{p_{t1}}\right)$$

The stagnation pressure ratio across an oblique shock may be obtained from that for a normal shock by replacing M_1 by $M_1 \sin\theta$ in Eqn (A.56) which yields

$$\frac{p_{t2}}{p_{t1}} = \left[\frac{(\gamma + 1)M_1^2 \sin^2\theta}{(\gamma - 1)M_1^2 \sin^2\theta + 2}\right]^{\frac{\gamma}{\gamma-1}} \left[\frac{\gamma + 1}{2\gamma M_1^2 \sin^2\theta - (\gamma - 1)}\right]^{\frac{1}{\gamma-1}} \tag{A.63}$$

In the limit $M_1 \to \infty$, with $M_1 \sin\theta > 0(1)$

$$\lim_{M_1 \to \infty} \frac{p_{t2}}{p_{t1}} = \varepsilon_{\lim}^{\frac{\gamma}{\gamma-1}}\left(\frac{2\gamma}{\gamma + 1}M_1^2 \sin^2\theta\right)^{\frac{-1}{\gamma-1}} \tag{A.64}$$

$$\frac{s_2 - s_1}{R} = \ln\left(\frac{2\gamma}{\gamma + 1}M_1^2 \sin^2\theta(\varepsilon_{\lim})^\gamma\right)^{\frac{1}{\gamma-1}} \tag{A.65}$$

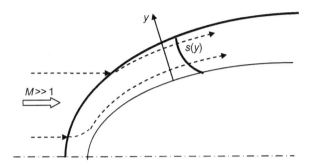

FIGURE A.5

Schematic diagram of hypersonic flow over a blunt body and the associated layer of high entropy close to the body.

The maximum entropy rise occurs for $\theta = 90°$ (normal shock) and decreases as θ decreases for a given M_1. A blunt body in hypersonic flight produces a curved shock like that shown in Figure A.5. The shock angle is normal to the flow ($\theta = 90°$) in the stagnation region and then decreases as the flow continues downstream. In hypersonic flow, the shock wave tends to follow the general shape of the body, as will be discussed in a subsequent section. Because the shock angle changes, the flow field normal to the body surface has associated with it an entropy gradient as illustrated in Figure A.5. The streamlines in Figure A.5 are depicted as dotted lines and the one that passes through the normal portion of the shock has higher entropy associated with it than does one that passes through a more oblique portion of the shock. Thus the shock wave generates an entropy layer, a region of hot gas with high entropy enveloping the body. The term entropy layer is used because this high entropy region has characteristics similar to that of a viscous boundary layer.

Crocco's theorem is derived from the conservation equations for steady, inviscid, nonconducting, nonreacting flow (Liepmann & Roshko, 2002). The theorem may be written as follows:

$$T\nabla s = \nabla h_t - \vec{V} \times (\nabla \times \vec{V}) \tag{A.66}$$

Thus an isentropic flow, with s and h_t constant, is irrotational. However the flow behind a curved shock is nonisentropic and therefore even for constant total enthalpy the flow field in the shock layer is rotational.

A.4.2 FLOW DEFLECTION AND THE SHOCK ANGLE

An explicit relationship between the shock angle θ and the flow deflection δ (Figure A.6) is given by Liepmann and Roshko (2002) as follows:

$$\tan \delta = \frac{2}{\tan \theta} \frac{M_1^2 \sin^2\theta - 1}{M_1^2(\gamma + 1 - 2\sin^2\theta) + 2} \tag{A.67}$$

Equation (A.67) shows that the flow deflection $\delta = 0$ for two shock wave angles: $\theta = 90°$ and $\theta = \sin^{-1}(1/M_1)$. When $\theta = 90°$ the flow is passing through a normal shock wave which obviously does not deflect the flow. However, when $\theta = \sin^{-1}(1/M_1)$ the flow is passing through a Mach wave which is generated by an infinitesimal disturbance. There is no convenient explicit relationship for the shock angle as a function of the deflection angle. For small shock angles, $\theta \ll 1$, in the hypersonic range where $M_1 \gg 1$, the relationship in Eqn (A.67) becomes

$$\lim_{M_1\theta \gg 1} \theta = \frac{\gamma + 1}{2}\delta \tag{A.68}$$

Note that for a uniformly valid limit to Eqn (A.67), we must have the product $M_1\theta \gg 1$ so that the case of a Mach wave is retained. Because $\gamma = O(1)$, Eqn (A.68) suggests the useful result that in hypersonic flow the shock wave lies very close to the body.

The pressure coefficient for flow across an oblique shock given in Eqn (A.62) reads as follows:

$$C_p = \frac{4}{\gamma + 1}\left(\frac{M_1^2 \sin^2\theta - 1}{M_1^2}\right)$$

In the hypersonic limit the pressure coefficient becomes

$$\lim_{M_1 \to \infty} C_p = \frac{4}{\gamma + 1}\left(\sin^2\theta - \frac{1}{M_1^2}\right) \to \frac{4}{\gamma + 1}\sin^2\theta = 2(1 - \varepsilon_{\lim})\sin^2\theta \tag{A.69}$$

If the limiting case of small deflections of the shock angle given in Eqn (A.68) is substituted into the above expression, we obtain

$$\lim_{M_1 \to \infty} C_p = (\gamma + 1)\delta^2 \tag{A.70}$$

We may rewrite Eqn (A.67) as follows:

$$M_1^2 \sin^2\theta - 1 = \tan\theta \tan\delta\left[M_1^2\left(\frac{\gamma + 1}{2} - \sin^2\theta\right) + 1\right]$$

Using the assumption that $\delta \ll 1$ and $\theta \ll 1$ while $M_1 \gg 1$ transforms this equation into

$$M_1^2\theta^2 - 1 = \delta\theta\left[M_1^2\left(\frac{\gamma + 1}{2} - \theta^2\right) + 1\right] = \frac{\gamma + 1}{2}M_1^2\delta\theta \tag{A.71}$$

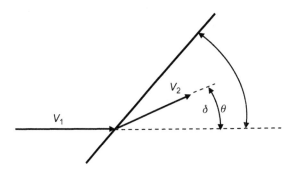

FIGURE A.6

Oblique shock wave showing shock wave angle θ and flow deflection angle δ.

This is a quadratic equation in θ^2 which has the solution

$$\frac{\theta}{\delta} = \frac{\gamma+1}{4} + \sqrt{\left(\frac{\gamma+1}{4}\right)^2 + \frac{1}{(M_1\delta)^2}} \tag{A.72}$$

Then using Eqns (A.71) and (A.72) in the hypersonic limit for C_p as given in Eqn (A.69) is

$$C_p = 2\delta^2 \left[\frac{\gamma+1}{4} + \sqrt{\left(\frac{\gamma+1}{4}\right)^2 + \frac{1}{(M_1\delta)^2}}\right] \tag{A.73}$$

The approximate result in Eqn (A.72) provides an explicit relationship for the shock angle in terms of the deflection angle for hypersonic flow. Equation (A.67) does not provide an explicit solution and considerable effort has been applied to achieving that end. An exact algebraic solution has been presented by Wolf (1993).

A.5 SMALL DISTURBANCE THEORY

Consider a slender pointed body in a hypersonic flow as illustrated in Figure A.7. In this case we assume that θ, δ, and $\theta - \delta$ are small enough such that $\sin\theta \approx \theta \approx \sin\delta \approx \delta$ and $\cos\theta \approx 1 \approx \cos\delta \approx 1$. In the hypersonic limit, Eqn (A.73) shows that

$$\frac{C_p}{\delta^2} = f(\gamma, M_1\delta) \tag{A.74}$$

This is in contradistinction to supersonic small disturbance theory where the Prandtl–Glauert rule is given by

$$\frac{C_{p,\text{sup}}}{\delta} = \frac{2}{\sqrt{M_1^2 - 1}} = f(M_1) \tag{A.75}$$

However if we permit $M_1 \rightarrow \infty$ in Eqn (A.75), the supersonic pressure coefficient may be written as

$$\lim_{M_1 \rightarrow \infty} \frac{C_{p,\text{sup}}}{\delta^2} = \frac{2}{\delta\sqrt{M_1^2 - 1}} \rightarrow \frac{2}{\delta M_1} = f(M_1\delta) \tag{A.76}$$

FIGURE A.7

A typical sharp-nosed slender body in hypersonic flow.

Thus the limiting value for the pressure coefficient for supersonic flow when $M_1 \to \infty$ is actually similar in form except for the appearance of γ in the hypersonic relation. Therefore, just as the quantity C_p/δ is a function of the Prandtl–Glauert similarity variable $\sqrt{M_1^2 - 1}$ for supersonic flow, the quantity C_p/δ^2 is a function of the hypersonic similarity variable $M_1\delta$ for hypersonic flow. Thus, flows over two different slender bodies at two different Mach numbers the value of C_p/δ^2 will be the same for both, provided $M_1\delta$ is also the same for both.

Referring to Eqn (A.55) for normal shocks, we may determine that the Mach number behind an oblique shock is given by

$$M_2 \sin(\theta - \delta) = \sqrt{\frac{(\gamma - 1)M_1^2 \sin^2\theta + 2}{2\gamma M_1^2 \sin^2\theta - (\gamma - 1)}} \tag{A.77}$$

For completeness, the solution for M_2 given by TR 1135 (1953) is included as follows:

$$M_2^2 = \frac{(\gamma + 1)^2 M_1^4 \sin^2\theta - 4(M_1^2 \sin^2\theta - 1)(\gamma M_1^2 \sin^2\theta + 1)}{[2\gamma M_1^2 \sin^2\theta - (\gamma - 1)][(\gamma - 1)M_1^2 \sin^2\theta + 2]} \tag{A.78}$$

However, evaluating Eqn (A.77) in the hypersonic limit and applying Eqn (A.68) yields

$$M_2 \approx \frac{1}{\theta - \delta}\sqrt{\frac{\gamma - 1}{2}} = \frac{1}{\theta}\frac{\gamma + 1}{\sqrt{2\gamma(\gamma - 1)}} = \frac{1}{\delta}\sqrt{\frac{2}{\gamma(\gamma - 1)}} \tag{A.79}$$

For small flow deflections $\theta \ll 1$ and $\delta \ll 1$, and therefore the flow can be hypersonic behind an oblique shock. Let u_2 denote the component of velocity behind the oblique shock in the free stream direction and v_2 that normal to it. TR 1135 (1953) presents the following relations for those velocity components normalized with respect to the free stream velocity:

$$\frac{u_2}{V_1} = 1 - \frac{2(M_1^2 \sin^2\theta - 1)}{(\gamma + 1)M_1^2}$$

$$\frac{v_2}{V_1} = \frac{2(M_1^2 \sin^2\theta - 1)}{(\gamma + 1)M_1^2}\cot\theta$$

In the hypersonic limit these components become

$$\frac{u_2}{V_1} \approx 1 - \frac{2}{\gamma + 1}\theta^2$$

$$\frac{v_2}{V_1} \approx \frac{2}{\gamma + 1}\theta$$

The change in velocity components across the oblique shock yields the perturbation velocity components

$$\frac{u'}{V_1} = \frac{u_2 - V_1}{V_1} \to -\frac{2}{\gamma + 1}\theta^2 = -\frac{\gamma + 1}{2}\delta^2$$

$$\frac{v'}{V_1} = \frac{v_2 - 0}{V_1} \to \frac{2}{\gamma + 1}\theta = \delta$$

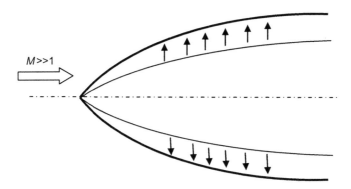

FIGURE A.8

Slender bodies in hypersonic flow act like a "piston" forcing transverse motions in the flow with little change in the streamwise direction.

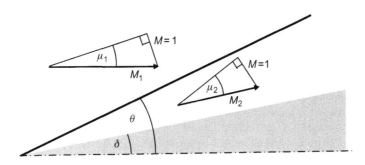

FIGURE A.9

Mach numbers and Mach lines in the small disturbance approximation.

Thus the x-direction perturbation is an order of magnitude smaller than the y-direction perturbation. The major effect is that of transverse rather than longitudinal motion as shown in Figure A.8. In small disturbance flow over slender bodies the large difference between the transverse and longitudinal motions gave rise to the so-called piston analogy. If the longitudinal velocity changes little from V_1, then the longitudinal coordinate x can be considered an effective time coordinate $t = x/V_1$ so that the problem is transformed from a steady flow to an unsteady flow in one less space coordinate. The body shape emulates the motion of a piston moving in the transverse direction generating a shock wave in front of it.

The fact that the shock wave angle and the body surface slope are very close in hypersonic flow extends to the Mach angle $\mu = \sin^{-1}(1/M)$, as can be seen in Figure A.9.

In the limit of $M_1 \gg 1$, Eqn (A.79) yields

$$\mu_2 \rightarrow \sqrt{\frac{\gamma(\gamma - 1)}{2}}\delta = \frac{\sqrt{2\gamma(\gamma - 1)}}{(\gamma + 1)}\theta$$

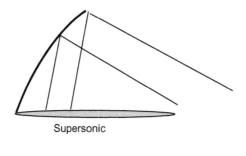

Supersonic

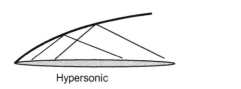

Hypersonic

FIGURE A.10

Characteristics network for supersonic and hypersonic flow over slender bodies illustrating the pronounced influence of reflected characteristics on the body surface in the case of hypersonic flow.

Therefore Mach waves will transmit information to shocks and can then reflect back with the results shown in Figure A.10. In supersonic flow, the Mach waves and the shock waves are characterized by relatively large shock and Mach angles. The Mach waves carry information about the body up to the shock, whereupon they reflect at similarly large angles. The net effect is that information from the shock is not likely to impinge on the body altering the flow there. On the other hand, in hypersonic flow the shock and Mach angles are very small and flow information from the shock can be communicated to the body.

A.6 PRANDTL–MEYER EXPANSION

We have discussed compression waves in hypersonic flow and now turn our attention to expansion waves. Consider the case of a simple flow deflected downward from the original free stream direction shown in Figure A.11. Let us start with a sonic flow in the free stream so that $M = 1$ and the corresponding Mach angle $\mu = \sin^{-1}(1/M) = \pi/2$. At some point, the horizontal wall suddenly turns down through an angle ν and a new Mach number M is reached.

At the vertex of the wall, an array of Mach waves are generated as the flow passes through a series of infinitesimal turns of magnitude $\Delta\nu$ and the Mach number continually changes until the flow passes through the last Mach line. The flow has $\mu = \pi/2$ associated with the uniform flow at $M = 1$. The first information the flow receives about the change in the wall slope may be considered to be carried by the first Mach wave indicated in the figure. We wish to determine the infinitesimal change in Mach number that is generated by the Mach wave.

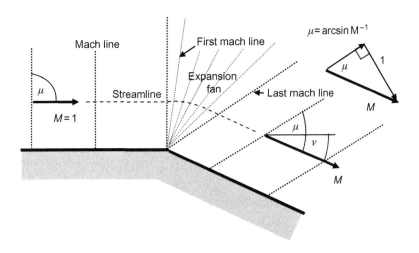

FIGURE A.11

Prandtl–Meyer expansion over a sharp corner.

Consider the flow passing through a wave of infinitesimal strength $\Delta p = dp$ that deflects the flow an infinitesimal amount $\Delta\nu = d\nu$. The process is isentropic and along a streamline the momentum conservation relation is simply

$$dp + \rho V dV = 0$$

This equation may be rearranged to yield

$$\frac{dp}{p} = -\gamma M^2 \left(\frac{dV}{V}\right) \tag{A.80}$$

To find the change in velocity across this weak disturbance, we may use the same geometry used in the consideration of oblique shocks and depicted in Figure A.12. Here now the flow deflects downward rather than upward and M_2 increases. From this diagram and the notion that $V_{1,t} = V_{2,t}$ and $V_2 = V_1 + dV$ we find

$$V_1 \cos(\mu_2 - d\nu) = V_2 \cos\mu_2 = (V_1 + dV)\cos\mu_2 \tag{A.81}$$

Invoking the usual relationship for the cosine of the difference of angles along with the requirement that $d\nu$ is vanishingly small leads to

$$\cos(\mu_2 - d\nu) = \cos\mu_2 \cos(d\nu) + \sin\mu_2 \sin(d\nu) \approx \cos\mu_2 + d\nu \sin\mu_2$$

Using this result in Eqn (A.81) yields

$$\frac{dV}{V_1} = d\nu \tan\mu_2 \tag{A.82}$$

The sign for $d\nu$ is taken as positive for a deflection that expands the flow and leads to an isentropic increase in velocity. If the deflection was upward, that is, compressive the sign of the right-hand side of Eqn (A.82) would be negative and the flow velocity would decrease, i.e., an isentropic

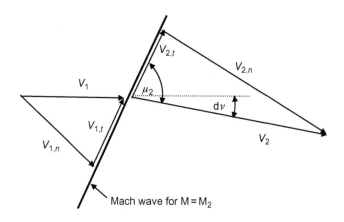

FIGURE A.12

Geometry of a Mach wave across which M changes from M_1 to M_2.

compression. The definition of the Mach number for a constant composition perfect gas may be written as

$$\frac{dM}{M} = \frac{dV}{V} - \frac{da}{a} = \frac{dV}{V} - \frac{\gamma - 1}{2\gamma}\frac{dp}{p} \tag{A.83}$$

Using Eqns (A.80) and (A.82) in Eqn (A.83) yields

$$\frac{dM}{M} = -\frac{1 + \frac{\gamma - 1}{2}M^2}{\sqrt{1 - M^2}}d\nu \tag{A.84}$$

The angle ν may be found by integrating Eqn (A.84) from $M = 1$, where we choose to set $\nu = 0$ at $M = 1$, to a general value M leading to the Prandtl−Meyer function defined as

$$\nu = \sqrt{\frac{\gamma + 1}{\gamma - 1}}\arctan\sqrt{\frac{\gamma - 1}{\gamma + 1}(M^2 - 1)} - \tan^{-1}\sqrt{M^2 - 1} \tag{A.85}$$

The Prandtl−Meyer function may be tabulated as a function of M for a given value of γ and is available, along with all the shock functions in TR 1135 (1953).

Note that in the hypersonic limit where $M \to \infty$ and $p \to 0$, there is a limiting value for ν given by

$$\nu_{max} = \left(\sqrt{\frac{\gamma + 1}{\gamma - 1}} - 1\right) \times 90° = 130.45°(\gamma = 1.4) \tag{A.86}$$

This equation shows that a specific gas may only be turned a finite amount before the Mach number becomes infinite. This maximum turning angle assumes practical significance for rocket exhausts in orbit where $p \sim 0$. Prandtl−Meyer angles and static-to-stagnation pressure ratios for $\gamma = 1.2$ and 1.4 and for Mach numbers up to 20 are given in Tables A.1 and A.2.

Table A.1 Prandtl–Meyer Angle and Static-to-Total Pressure Ratio for Two Values of γ and $1 < M < 2.5$

M	$\gamma = 1.4$ ν (deg)	$\gamma = 1.4$ p/p_t	$\gamma = 1.2$ ν (deg)	$\gamma = 1.2$ p/p_t
1.0	0	5.283E − 01	0	5.645E − 01
1.05	0.49	4.979E − 01	0.53	5.339E − 01
1.10	1.34	4.684E − 01	1.47	5.039E − 01
1.15	2.38	4.398E − 01	2.63	4.746E − 01
1.20	3.56	4.124E − 01	3.95	4.461E − 01
1.25	4.83	3.861E − 01	5.39	4.185E − 01
1.30	6.17	3.609E − 01	6.92	3.918E − 01
1.35	7.56	3.370E − 01	8.52	3.662E − 01
1.40	8.99	3.142E − 01	10.17	3.417E − 01
1.45	10.44	2.927E − 01	11.87	3.182E − 01
1.50	11.91	2.724E − 01	13.60	2.959E − 01
1.55	13.38	2.533E − 01	15.35	2.748E − 01
1.60	14.86	2.353E − 01	17.12	2.547E − 01
1.65	16.34	2.184E − 01	18.91	2.358E − 01
1.70	17.81	2.026E − 01	20.71	2.180E − 01
1.75	19.28	1.878E − 01	22.50	2.013E − 01
1.80	20.73	1.740E − 01	24.30	1.856E − 01
1.85	22.17	1.612E − 01	26.10	1.710E − 01
1.90	23.59	1.492E − 01	27.90	1.573E − 01
1.95	24.99	1.381E − 01	29.68	1.446E − 01
2.0	26.38	1.278E − 01	31.46	1.328E − 01
2.05	27.75	1.182E − 01	33.23	1.218E − 01
2.10	29.10	1.094E − 01	34.98	1.117E − 01
2.15	30.43	1.011E − 01	36.72	1.023E − 01
2.20	31.74	9.352E − 02	38.45	9.363E − 02
2.25	33.02	8.648E − 02	40.16	8.563E − 02
2.30	34.29	7.997E − 02	41.86	7.826E − 02
2.35	35.53	7.396E − 02	43.53	7.149E − 02
2.40	36.75	6.840E − 02	45.20	6.526E − 02
2.45	37.95	6.327E − 02	46.84	5.955E − 02
2.50	39.13	5.853E − 02	48.47	5.431E − 02

The static-to-stagnation pressure ratio becomes

$$\left(\frac{p}{p_t}\right)^{\frac{\gamma-1}{\gamma}} = \frac{1}{\gamma+1}\left\{1 + \cos\left[2\sqrt{\frac{\gamma-1}{\gamma+1}}\left(\nu + \arctan\sqrt{M^2-1}\right)\right]\right\} \tag{A.87}$$

Table A.2 Prandtl–Meyer Angle and Static-to-Total Pressure Ratio for Two Values of γ and $2.5 < M < 20$

M	$\gamma = 1.4$ ν deg	$\gamma = 1.4$ p/p_t	$\gamma = 1.2$ ν deg	$\gamma = 1.2$ p/p_t
2.5	39.13	5.853E − 02	48.47	5.431E − 02
2.6	41.42	5.012E − 02	51.66	4.512E − 02
2.8	45.75	3.685E − 02	57.82	3.102E − 02
3.0	49.76	2.722E − 02	63.66	2.126E − 02
3.2	53.48	2.023E − 02	69.19	1.455E − 02
3.4	56.91	1.512E − 02	74.42	9.957E − 03
3.6	60.10	1.138E − 02	79.36	6.826E − 03
3.8	63.05	8.629E − 03	84.01	4.692E − 03
4.0	65.79	6.586E − 03	88.41	3.237E − 03
4.2	68.34	5.062E − 03	92.55	2.243E − 03
4.4	70.71	3.918E − 03	96.47	1.561E − 03
4.6	72.93	3.053E − 03	100.16	1.092E − 03
4.8	74.99	2.394E − 03	103.65	7.687E − 04
5.0	76.93	1.890E − 03	106.95	5.440E − 04
5.2	78.74	1.501E − 03	110.07	3.872E − 04
5.4	80.44	1.200E − 03	113.03	2.773E − 04
5.6	82.04	9.643E − 04	115.83	1.998E − 04
5.8	83.54	7.794E − 04	118.48	1.448E − 04
6.0	84.96	6.334E − 04	121.00	1.055E − 04
6.2	86.30	5.173E − 04	123.40	7.741E − 05
6.4	87.57	4.247E − 04	125.68	5.710E − 05
6.6	88.77	3.503E − 04	127.84	4.236E − 05
6.8	89.90	2.902E − 04	129.91	3.160E − 05
8.0	95.63	1.024E − 04	140.47	6.090E − 06
10	102.32	2.356E − 05	153.10	5.645E − 07
12	106.89	6.922E − 06	161.88	7.497E − 08
14	110.19	2.428E − 06	168.29	1.309E − 08
16	112.69	9.731E − 07	173.18	2.823E − 09
18	114.64	4.327E − 07	177.02	7.203E − 10
20	116.20	2.091E − 07	180.11	2.105E − 10

An expansion process is isentropic under conditions of either equilibrium chemistry or frozen chemistry, i.e., in the absence of finite rate processes, friction, conduction, etc. Therefore the stagnation pressure is constant.

The case of a sudden expansion like that shown in Figure A.11 generates a fan of Mach waves called a Prandtl–Meyer expansion fan but the Prandtl–Meyer function is applicable to a more gentle continuous expansion like that shown in Figure A.13. Here the flow is known at station 1 whereupon it expands through a finite angle $\Delta\nu$ to reach station 2. Thus, for convex surfaces, the value

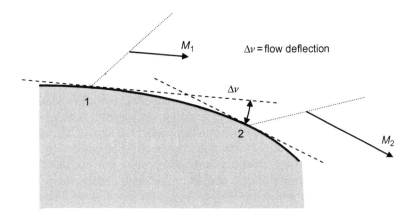

FIGURE A.13

Prandtl−Meyer flow over a smooth convex surface.

of v_2 is known if the flow deflection $\Delta\delta$ is known because $\Delta v = \Delta\delta$ and thus M_2 may be found, not directly, but generally by an iterative or interpolative process since it is not possible to rearrange the $v = f(\gamma, M)$ (Eqn A.85). The Prandtl−Meyer angle at station 2 is given by

$$v_2 = \Delta v + v_1$$

The value of M_2 may be found from the tabulated Prandtl−Meyer function. Knowing the Mach number at points 1 and 2 permits one to determine the pressure ratio directly by using the isentropic relation and the fact that the stagnation pressure is constant, $p_{t2} = p_{t1}$. The result is as follows:

$$\frac{p_2}{p_1} = \left(\frac{p_{t1}}{p_1}\right)\left(\frac{p_2}{p_{t2}}\right) = \left(\frac{1 + \dfrac{\gamma-1}{2}M_2^2}{1 + \dfrac{\gamma-1}{2}M_1^2}\right)^{\frac{-\gamma}{\gamma-1}}$$

An expansion of the pressure expression for small deflections $\Delta v = \delta \ll 1$ leads to

$$\frac{p_2 - p_1}{p_1} \approx \frac{-\gamma M_1^2}{\sqrt{M_1^2 - 1}}\Delta v = \frac{-\gamma M_1^2}{\sqrt{M_1^2 - 1}}\delta$$

The pressure coefficient

$$C_p = \frac{p_2 - p_1}{p_1}\frac{p_1}{q_1} = \frac{-2\delta}{\sqrt{M_1^2 - 1}}$$

This is the typical linearized supersonic result. As $M_1 \to \infty$

$$\frac{C_p}{\delta^2} \approx \frac{-2}{\sqrt{M_1^2\delta^2 - \delta^2}} \to \frac{-2}{M_1\delta}$$

And again the similarity parameter $M_1\delta$ appears and $C_p \sim \delta^2$ for small expansions as well as small compressions, as shown earlier for shocks. For small compressions ($\Delta v < 0$) the Prandtl−Meyer relations can again be used, since the flow remains close to isentropic.

A.7 CONICAL FLOW

We have been considering planar flows up to this point. A simple three-dimensional body is a right circular cone and one is shown in a supersonic flow in Figure A.14. The cone is assumed to generate a conical shock coaxial with the body.

The conical shock is straight and at any point on the shock the flow through it acts two-dimensionally so that the usual oblique shock equations apply locally. That means the flow properties are all constant immediately behind the shock surface. If the streamlines behind the shock were assumed to be parallel to the body surface, the distance between a streamline and the body surface constantly increases. Continuity could not be satisfied in this case because the flow properties immediately behind a straight shock must be constant but the flow area between a straight streamline and the conical body increases. Thus the streamlines bend toward the body, which itself is a streamline and the only one in the flow which is straight. Therefore, for a conical body with a conical shock the simplest presumption is that the flow is conical, i.e., properties are constant along rays. A typical conical ray and streamline are shown in Figure A.15.

The downstream surface of the shock is a ray and all properties on it are constant. Similarly, the cone surface is a ray and all properties are constant on it. It is the only ray which is also a streamline. As one proceeds from the shock each ray has a higher pressure than previous ray until one reaches the cone surface, which is the ray with the highest constant pressure. The cone acts as an isentropic compression surface and develops higher pressure on its surface than a wedge of equivalent shock angle.

Ferri (1949) covers the topic of conical flow in detail. Because of its fundamental nature it is always discussed in textbooks on gas dynamics (Anderson, 2003). Tables of data for conical flows appear in Sims (1964a and 1964b) and graphic data are presented in TR 1135 (1953). Rasmussen (1994) obtained an analytic solution for the hypersonic

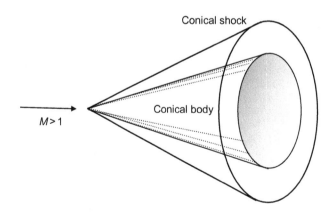

FIGURE A.14

Supersonic flow over a conical body at zero angle of attack.

small disturbance equations developed for conical flow by Van Dyke (1954) which can be expressed as

$$\frac{C_p}{\sin^2 \delta_c} = 1 + \frac{(\gamma + 1)K^2 + 2}{(\gamma - 1)K^2 + 2} \ln \left(\frac{\gamma + 1}{2} + \frac{1}{K^2} \right) \tag{A.88}$$

Here the parameter $K = M_\infty \sin \delta_c$ where δ_c denotes the semivertex angle of the cone. Equation (A.88) agrees very well with the exact results even up to $\delta_c = 30°$. Furthermore, results for air undergoing equilibrium chemical reactions show that p_c/p_∞ is an essentially unique function of $M_\infty \sin \delta_c$ and is practically equivalent to the calorically perfect result. Equilibrium chemistry in the conical shock layer is isentropic with a value for the entropy corresponding to that immediately behind the shock. If one knows p_c from $\gamma = 1.4$ results and the shock wave angle, then one can compute $s_2 = s_c$ and use it with p_c to find ρ_c and T_c. For large values of K, the cone pressure is given by

$$\frac{p_c}{p_\infty} \approx \frac{\gamma}{2} K^2 \left[1 + \left(\frac{\gamma + 1}{\gamma - 1} \right) \ln \left(\frac{\gamma + 1}{2} \right) \right] \tag{A.89}$$

For a gas with $\gamma = 1.4$, the ratio of cone surface pressure to free stream pressure is approximated by

$$\frac{p_c}{p_\infty} \approx 1.467 \, K^2$$

The shock wave angle for a cone of semivertex angle δ_c was found by Rasmussen (1994) to be, for small angles, given by

$$\theta = \delta_c \sqrt{\frac{\gamma + 1}{2} + \frac{1}{K^2}} \tag{A.90}$$

In order to find a reasonable value of shock angle θ to use to in calculating the entropy behind the shock it is usually sufficient to use the approximation of Eqn (A.90) to arrive at

$$M_1^2 \sin^2 \theta \approx M_1^2 \sin^2 \left(\delta_c \sqrt{\frac{\gamma + 1}{2} + \frac{1}{M_1^2 \sin^2 \delta_c}} \right)$$

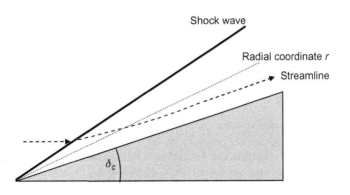

FIGURE A.15

Streamline curvature in flow over a cone of semivertex angle δ_c.

Hayes and Probstein (1959) present the following results for flow over a cone in the constant density approximation:

$$\frac{1}{2}C_{p_c} = \frac{p_c - p_\infty}{q_\infty} = \frac{\sin^2 \delta_c}{\left(1 - \frac{1}{4}\varepsilon\right)\cos^2(\theta - \delta_c)}$$

$$\frac{p_c}{p_\infty} = 1 + \frac{\gamma M_\infty^2 \sin^2 \delta_c}{\left(1 - \frac{1}{4}\varepsilon\right)\cos^2(\theta - \delta_c)}$$

$$1.043 > \frac{1}{1 - \frac{1}{4}\varepsilon} > 1.025 \text{ for } 1.4 \geq \gamma \geq 1.1$$

We may therefore assume

$$\frac{p_c}{p_\infty} = 1 + \frac{\gamma M_\infty^2 \sin^2 \delta_c}{\left(1 - \frac{1}{4}\varepsilon\right)} \tag{A.91}$$

The shock angle is given by the equation

$$\sin \delta_c = \left(1 - \frac{1}{2}\varepsilon\right)\sin \theta \cos(\theta - \delta_c)$$

or

$$\theta - \delta_c = \frac{1}{2}\varepsilon \tan \theta$$

For a given shock angle θ, one may calculate ε using the oblique shock approach. Then

$$\delta_c = \theta - \frac{1}{2}\varepsilon \tan \theta$$

One may then calculate

$$\frac{p_c}{p_\infty} = 1 + \frac{\gamma_\infty M_\infty^2 \sin^2 \delta_c}{\left(1 - \frac{1}{4}\varepsilon\right)\cos^2(\theta - \delta_c)} \tag{A.92}$$

Then using p_c and $s_2 = s_c = $ constant one may find T_c and ρ_c.

A.8 NEWTONIAN FLOW

This method is based upon the notion that as $M \to \infty$ and $\gamma \to 1$, the shock layer is coincident with the body surface. Momentum considerations can then be used to deduce a relationship between the pressure on the surface and the flow inclination. Of course we have seen that shock layers in hypersonic flow can be rather thin, but they are not coincident with the body. Indeed it is really correct to deal with Newtonian flow only when one is dealing with shock inclination, where we have found that

$$\lim_{\varepsilon \to 0} C_p = 2(1 - \varepsilon)\sin^2\theta \to 2 \sin^2\theta \tag{A.93}$$

and

$$\lim_{\theta \to \delta, \varepsilon \to 0} C_p = \frac{2 \sin^2\delta}{(1-\varepsilon)\cos^2(\theta-\delta)} \to 2\sin^2\delta \tag{A.94}$$

In a general sense, at a point on the surface of a shock wave this becomes

$$C_p = 2\frac{(\vec{V}_\infty \cdot \hat{n})^2}{\vec{V}_\infty \cdot \vec{V}_\infty} \cdot (1-\varepsilon) \tag{A.95}$$

The general geometry of an elemental portion of the shock wave is shown in Figure A.16. Here $\hat{n}$ is the outward normal to the shock surface, i.e., pointing toward the upstream side. When the shock is normal, this relation suggests that, in the limit as $\varepsilon \to 0$, the shock and the body become coincident, $p_2 = p_{t2}$, and therefore

$$\lim_{\varepsilon \to 0} C_{p,max} = \frac{p_2 - p_1}{q_1} = \frac{p_{t2}-p_1}{q_1} = 2 \tag{A.96}$$

However, behind the shock in a finite shock layer the stagnation pressure p_{t2} is given by

$$\frac{\gamma}{\gamma-1}\frac{p_2}{\rho_2} + \frac{1}{2}u_2^2 = \frac{\gamma}{\gamma-1}\frac{p_{t2}}{\rho_{t2}}$$

The subscript 2 denotes conditions immediately downstream of the shock. The pressure sensed by the body is the stagnation pressure downstream of the shock, and therefore the pressure coefficient on the body at the stagnation point is

$$C_p = \frac{p_{t2}-p_1}{q_1} = \frac{(p_2-p_1)+(p_{t2}-p_2)}{q_1}$$

In Section A.2, we showed that the difference in the static pressure across the shock is

$$\frac{p_2 - p_1}{q_1} = 2(1-\varepsilon) \tag{A.97}$$

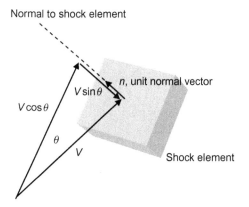

FIGURE A.16

Geometry of an elemental portion of the shock wave.

The difference between the stagnation pressure and the static pressure downstream of the shock is

$$\frac{p_{t2} - p_2}{q_1} = \left(\frac{p_{t2}}{p_2} - 1\right)\frac{p_2}{q_1} = \left(\frac{p_{t2}}{p_2} - 1\right)\frac{p_2}{q_2} \cdot \frac{q_2}{q_1}$$

$$\frac{p_{t2} - p_2}{q_1} = \left[\left(1 + \frac{\gamma - 1}{2}M_2^2\right)^{\frac{\gamma}{\gamma-1}} - 1\right]\frac{2}{\gamma M_2^2}\frac{\rho_2}{\rho_1}\frac{V_2^2}{V_1^2}$$

Since the one-dimensional continuity equation requires $\rho_1 V_1 = \rho_2 V_2$, this last form of the equation may be written as follows:

$$\frac{p_{t2} - p_2}{q_1} = \left[1 + \frac{\gamma}{2}M_2^2 + \cdots - 1\right]\frac{2}{\gamma M_2^2}\varepsilon \approx \varepsilon \qquad (A.98)$$

Therefore the maximum pressure coefficient on the body is given by

$$C_{p_{\max}} = \frac{p_2 - p_1}{q_1} + \frac{p_{t2} - p_2}{q_1} = 2(1 - \varepsilon) + \varepsilon = 2 - \varepsilon = 2 - \varepsilon \qquad (A.99)$$

Thus, for real flows past blunt bodies the value of $C_{p,\max} < 2$. The same result is obtained for both two-dimensional and axisymmetric flows. Furthermore, the same first terms appear in a constant density analysis.

Lees (1956) introduced the idea that one should modify the Newtonian flow rule of Eqn (A.95) in the case of blunt bodies by using the actual value of $C_{p_{\max}}$, i.e.,

$$C_p = (2 - \varepsilon)\frac{(\vec{V}_\infty \cdot \hat{n})^2}{\vec{V}_\infty \cdot \vec{V}_\infty} \qquad (A.100)$$

This is found to give quite useful results. Of course, as the body inclination δ becomes parallel to the free stream direction $C_p \to 0$, and the results are no longer accurate. Accuracy falls off when the flow deflection $\delta > 35°$, as indicated in Figure A.17.

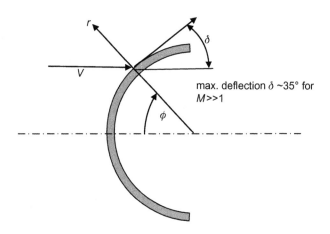

FIGURE A.17

Flow over a cylindrical or spherical nose. The Newtonian approximation is good up to a flow deflection θ of about 30°.

For pointed bodies, as depicted in Figure A.18, a similar generalization is made, i.e.,

$$C_p = \frac{C_{p_N}}{\sin^2\delta} \frac{(\vec{V}_\infty \cdot \hat{n})^2}{\vec{V}_\infty \cdot \vec{V}_\infty} \tag{A.101}$$

For wedges of half-angle δ or cones of half-angle δ_c, we have the following nose pressure coefficients:

$$C_{p,N,w} = \frac{2\sin^2\delta}{(1-\varepsilon)\cos^2(\theta-\delta)}$$

$$C_{p,N,c} = \frac{2\sin^2\delta_c}{\left(1-\dfrac{\varepsilon}{4}\right)\cos^2(\theta-\delta_c)} \tag{A.102}$$

Here the correct $C_{p,\max}$ values are greater than 2. This approach gives useful results until the body becomes nearly parallel to V_∞. For two-dimensional bodies the $C_{p,N,w}$ for the wedge would be used while for axisymmetric bodies $C_{p,N,c}$ for the cone would be used. These methods work reasonably well and lend themselves to application to bodies of quite arbitrary shape. Of course, we can apply the Prandtl−Meyer expansion relations at an appropriate point on the body to continue the calculation beyond that region where the Newtonian approximation fails as shown in Figure A.19.

The Prandtl−Meyer expansion may be used for two-dimensional bodies in general, while for axisymmetric (or three-dimensional) bodies it should be used only where the transverse curvature of the body is small. That is, one should expect that streamline curvature due to the shape of the body will influence the pressure development to an extent beyond that described by the two-dimensional Prandtl−Meyer expansion alone.

At high Mach numbers, the shock expansion method for pointed bodies works very well with the matching point for the Prandtl−Meyer expansion fixed anywhere downstream of the shock. If the body still has substantial positive slope at its aft end, it usually is sufficient to use the Newtonian method alone.

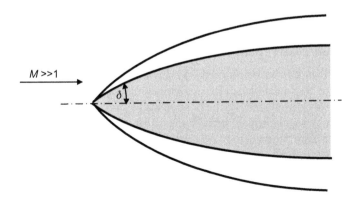

FIGURE A.18

Pointed slender body in hypersonic flow showing nose angle δ.

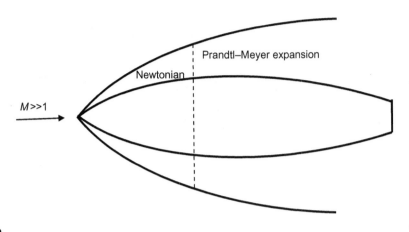

FIGURE A.19

Configuration of pointed slender body in hypersonic flow indicating regions where the Newtonian and Prandtl−Meyer approaches apply.

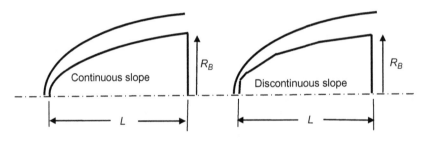

FIGURE A.20

Schematic diagram illustrating similar bodies with continuous and discontinuous slopes both producing shocks with continuous slope.

A.9 INFLUENCE OF BODY SHAPE

Since the Newtonian method is based primarily on shock shape and this is relatively insensitive to details of the body shape one must be careful in applying the method. One may design a body with discontinuities but the shock produced will be continuous in slope, as suggested by Figure A.20.

In addition, there may be centrifugal effects caused by the longitudinal curvature of the body which alter the pressure distribution across the shock layer as illustrated in Figure A.21. Streamline curvature introduces a centrifugal pressure change given by

$$\Delta p \approx +\rho V^2 \frac{\Delta r}{R_c}$$

The symbol R denotes the radius of curvature of the streamline. Bodies with rapid changes in curvature do not always lend themselves to Newtonian analyses; consider the example shown in Figure A.22.

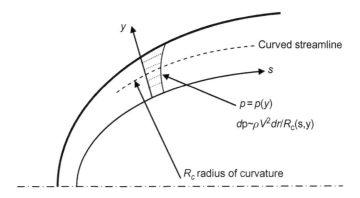

FIGURE A.21

Schematic diagram illustrating the effect of centrifugal acceleration.

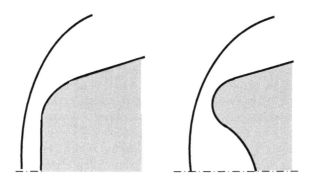

FIGURE A.22

Illustration of how different nose shapes can lead to very similar shock waves.

The difference in the detailed shape of the nose has little effect on shock shape and application of the Newtonian approximation based on body slope in these cases will not be as successful as in those where the body shape changes more gradually.

A.10 EFFECTS OF ANGLE OF ATTACK

We have been considering symmetric bodies to carry out the theoretical developments and such flows are of importance for symmetrical flight applications, i.e., those which do not require lift. However, there are many practical situations in which the generation of a lift force normal to the flight path is essential. Such lift forces are generally produced by rotating a symmetric body to some angle of attack.

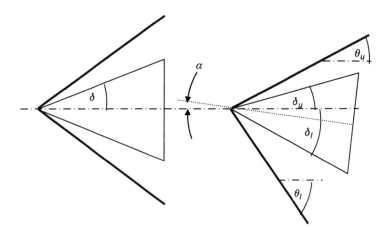

FIGURE A.23

Symmetrical wedge at angles of attack $\alpha = 0$ on the left and $\alpha > 0$ on the right.

A.10.1 WEDGE AT ANGLE OF ATTACK

The simplest case to consider is that of a wedge of half-angle δ where both it and the angle of attack α are small enough to ensure that the shock wave remains attached to the vertex of the wedge as shown in Figure A.23. The two surfaces of the wedge cannot communicate with each other so each one acts independently at its individual effective wedge angle: $\delta_u = \delta - \alpha$ for the upper surface and $\delta_l = \alpha + \delta$ for the lower surface. Associated with each effective wedge angle is the corresponding shock angle θ_u or θ_l. Any angle of attack α which permits an attached shock can be calculated. For example, shock detachment occurs at an wedge angle of $\delta = 45.4°$ when $\gamma = 1.4$. The exact two-dimensional shock result is

$$C_{p_N} = \frac{2 \sin^2 \delta}{(1 - \varepsilon)\cos^2(\theta - \delta)} \tag{A.103}$$

A.10.2 SMOOTH BLUNT BODIES AT ANGLE OF ATTACK

A common hypersonic shape is that of a smooth, slender, symmetrical, blunt-nosed body like that shown in Figure A.24. In two-dimensional form, the body might be considered a wing and the axisymmetric counterpart a fuselage. An approximate stagnation point must be found to proceed. When using the Newtonian approximation, one would choose the point where $\vec{V}_1 \cdot \hat{n}_{body} = -1$.

The maximum pressure coefficient in the hypersonic limit will occur at the assumed stagnation point with the exact normal shock value

$$C_{p_{max}} = 2\left(1 - \frac{\varepsilon}{2}\right) \tag{A.104}$$

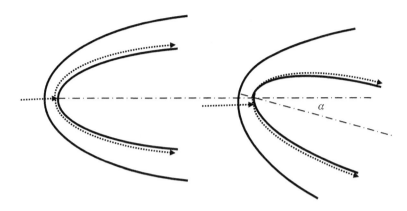

FIGURE A.24

Symmetrical blunt-nosed slender body at angles of attack $\alpha = 0$ on the left and $\alpha > 0$ on the right. A sketch of the stagnation streamline is shown as a dotted line in both cases.

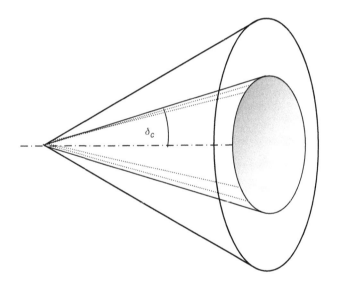

FIGURE A.25

A sharp cone of semivertex angle δ_c at zero angle of attack in hypersonic flow. The shock wave is attached at the vertex of the cone.

A.10.3 SHARP CONES AT ZERO ANGLE OF ATTACK

An approximation for the pressure coefficient for sharp cones at zero angle of attack in hypersonic flow, like that in Figure A.25, is given by Rasmussen (1994) as follows:

$$C_{p_N} = \sin^2\delta_c\left[1 + \frac{(\gamma+1)K^2+2}{(\gamma-1)K^2+2}\ln\left(\frac{\gamma+1}{2} + \frac{1}{K^2}\right)\right] \qquad (A.105)$$

This result is for flows in which the shock is attached and gives good agreement with experiments. For a gas with $\gamma = 1.4$ shock detachment occurs when the cone semivertex angle $\delta_c = 58°$. The value of C_p is not greatly affected by either $K = M_1 \sin \delta_c$ or γ.

Using similarity rules, Linell and Bailey (1956) developed an approximate equation for the pressure coefficient on a cone with an attached shock in hypersonic flow as follows:

$$C_{pN} = 4 \sin^2 \delta_c \frac{2.5 + 8\beta \sin \delta_c}{1 + 16\beta \sin \delta_c} \tag{A.106}$$

Here the quantity $\beta = \sqrt{M_1^2 - 1}$ is Prandtl–Glauert factor. The Linell–Bailey equation is applicable to both the supersonic and hypersonic range. It does not depend on γ, which, as we have mentioned before, does not influence pressure much. The constant density theory for flow over a sharp cone yields a solution that doesn't depend directly on the value of K provided $K \gg 1$. For this equation to provide reasonably accurate results, the value of $K \geq 4$.

$$C_{pN} = 2 \sin^2 \delta_c \frac{2 \sin^2 \delta_c}{\left(1 - \dfrac{\varepsilon}{4}\right) \cos^2(\theta - \delta_c)} \tag{A.107}$$

A.10.4 SHARP CONES AT SMALL ANGLES OF ATTACK

Cones at angle of attack generate three-dimensional flow fields. The general configuration of such a flow field is illustrated in Figure A.26. An approximate technique for calculating the pressure coefficient on the surface of the cone is based on the idea that in hypersonic flow the shock wave lies close to the body. The assumption then is that the pressure coefficient on a ray of the cone corresponding to a specific azimuthal angle ϕ is given by

$$C_{pN} = C_{pN,eff} - (C_{pN,eff} - C_{pN,eff}^*)f(M)$$

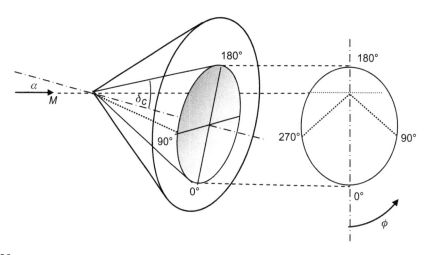

FIGURE A.26

Flow field over a sharp cone at a small angle of attack α. The azimuthal angles ϕ are shown on the left in a quartering view and on the right in a view looking upstream from behind the base of the cone.

Here $C_{pN,eff}$ is the Newtonian pressure coefficient for a cone whose semivertex angle δ_c is the same as that of the ray of the cone in question. Thus $C_{pN,eff}$ is a function of the azimuthal angle ϕ. The pressure coefficient $C^*_{pN,eff}$ is defined as the pressure coefficient $C_{pN,eff}$ at $\phi = 90°$. An effective cone angle at any azimuthal station is assumed to be

$$\delta_{c,eff} = \sin^{-1}(\sin \delta_c \cos \alpha + \cos \delta \sin \alpha \cos \phi)$$

$$\frac{M_c}{M_1} = \cos \delta_{c,eff} \left(1 - \frac{\sin \delta_{c,eff}}{M_1}\right)^{1/2} \left([1 + e^{-(1+1.52K_{eff})}]\right)\left[1 + \left(\frac{K_{eff}}{2}\right)^2\right]^{-1/2} ; K_{eff} < 1$$

$$\frac{M_c}{M_1} = \cos \delta_{c,eff} \left(1 - \frac{\sin \delta_{c,eff}}{M_1}\right)^{1/2} \left(1 + 0.35K_{eff}^{3/2}\right)^{-1/2} ; K_{eff} \geq 1$$

Here $K_{eff} = M_1 \sin \delta_{c,eff}$ and the Mach number function $f(M) = (M_c^{-3/2})_{\phi = \pi/2}$. This set of equations permits one to determine C_{pN} all around a conical tip from which one may continue with a Newtonian or Newtonian plus Prandtl−Meyer Solution.

A.10.5 **UNUSUAL SHAPES**

The use of the approximations presented previously must be applied with caution. Some degree of physical insight should be brought to bear on the process of considering methods to be used. The fundamental reasoning in the Newtonian approximation is that the shock wave lies close to the body. Two examples of body shapes for hypersonic flight that aren't well-suited for the Newtonian approximation are shown in Figure A.27.

The first case, a sharp cone with large semivertex angle, will not have an attached conical shock. Instead the shock detaches from the cone forming the smooth curved shock more typical of

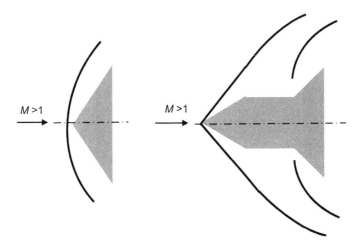

FIGURE A.27

Some body shapes that are not well-suited to the Newtonian approximation because the shocks they produce in practice do not lie close to the surface.

a blunt body, like a sphere, for example. The second case is a cone—cylinder—flare configuration which does have a conical shock attached to the vertex of the conical forebody. However, the flare of the afterbody acts much like a cone with a large semivertex angle that produces a strong curved shock standing out in front of the flare.

A.11 **NOMENCLATURE**

a	sound speed
C_p	pressure coefficient
c_p	specific heat at constant pressure
c_v	specific heat at constant volume
e	internal energy
h	enthalpy
G_i	Gibbs free energy per mole
g_i	Gibbs free energy per unit mass
I	total impulse
K	$M_\infty \sin \delta_c$
M	Mach number
$\dot{m}$	mass flux
$\hat{n}$	unit normal vector
p	pressure
R	gas constant
q	heat
R_u	universal gas constant
R_c	radius of curvature
R_B	radius of base
r	radial coordinate
s	entropy
T	temperature
t_c	characteristic chemical reaction time
t_r	characteristic flow residence time
u	velocity component in the x-direction
v	velocity component in the y-direction
V	velocity
V_n	velocity component normal to a shock
V_t	velocity component parallel to a shock
W	molecular weight
W_0	molecular weight prior to any chemical reaction
x	space coordinate in the free stream direction
Y_i	mass fraction of species i, see Eqn (A.14)
y	space coordinate normal to the free stream direction or normal to a body surface
Z	compressibility function W_0/W
α	angle of attack
β	Prandtl—Glauert factor $\sqrt{M^2 - 1}$
δ	flow deflection angle
ε	ratio of density upstream of a shock to that downstream of a shock
ϕ	azimuthal angle
γ	ratio of specific heats

γ_ε effective ratio of specific heats h/e
μ Mach wave angle
ν Prandtl−Meyer angle
ρ density
θ shock wave angle

A.11.1 SUBSCRIPTS

c conditions at the surface of a cone
eff   effective
lim   limiting case
i denoting the chemical species i
sup   supersonic flow conditions
N conditions at the nose of a body
t total, or stagnation, conditions
1   conditions upstream of a shock
2   conditions downstream of a shock
∞ free stream conditions

A.11.2 SUPERSCRIPTS

0   standard conditions: 1 atm pressure and 298.16 K temperature
$()'$ perturbation quantity

REFERENCES

Anderson, J. D. (2003). *Modern compressible flow: With historical perspective* (3rd ed.). New York, NY: McGraw-Hill.

Blick, E. F., & High, M. D. (1965). Cone pressure distribution at large and small angles of attack. *AIAA Journal*, *2*(11), 2054−2055.

Ferri, A. (1949). *Elements of aerodynamics of supersonic flows.* New York, NY: Macmillan.

Hayes, W. D., & Probstein, R. F. (1959). *Hypersonic flow theory.* New York, NY: Academic Press.

Lees, L. (1956). Laminar heat transfer over blunt-nosed bodies at hypersonic flight speeds. *Jet Propulsion*, *26*, pp. 259−264 and 274.

Liepmann, H. W., & Roshko, A. (2002). *Elements of gasdynamics.* New York, NY: Dover.

Linell, R. D., & Bailey, J. Z. (1956). Similarity rule estimation methods for cones and parabolic noses. *Journal of the Aeronautical Sciences*, *23*(8), 796−797.

Rasmussen, M. (1994). *Hypersonic flow.* New York, NY: John Wiley & Sons.

Sims, J. (1964a). *Tables for supersonic flow around right circular cones at zero angle of attack.* NASA SP-3004.

Sims, J. (1964b). *Tables for supersonic flow around right circular cones at small angle of attack.* NASA SP-3007.

TR 1135 (1953). *Equations, tables, and charts for compressible flow.* NACA Report 1135.

Van Dyke, M. (1954). *A study of hypersonic small-disturbance theory.* NACA Report 1194.

Wolf, T. (1993). Comment on approximate formula of oblique shock wave angle. *AIAA Journal*, *31*(7), 1363.

Appendix B: Spaceplane Coordinates

Dimensional data for several spaceplanes have been collected from a number of sources and using the information available, such as drawings and photographs, coordinates defining their geometry have been estimated. This data is presented here for general use although no claim is made as to the accuracy of the dimensions. The spaceplanes considered include the following:

Vehicle	Tables	Figures	Notes
Rockwell International Space Shuttle Orbiter (USA)	B.1–B.3	B.1–B.3	135 flights between 1986 and 2011
USAF/NASA X-24C (USA)	B.4–B.7	B.4–B.7	Wind tunnel tests
North American X-15 (USA)	B.8–B.10	B.8–B.11	199 flights between 1959 and 1968
Soviet spaceplane Bor-4 (USSR)	B.11	B.12–B.17	4 orbital flights between 1982 and 1984 (unmanned)
Northrop HL-10 lifting body (USA)	B.12–B.13	B.18–B.21	35 flights between 1966 and 1975 ($M < 1.6$)
CNES/ESA Hermes (European Union)	B.14–B.15	B.22–B.26	Wind tunnel tests
Institute of Space and Astronautical Sciences HIMES (Japan)	B.16–B.18	B.27–B.29	Wind tunnel tests

The dimensions presented are in inches unless otherwise noted. The x, y, and z coordinates represent the longitudinal, spanwise, and vertical directions as shown in Figure 8.3 and as may be deduced from the figures accompanying the tables. Where possible, photographs of test models or full-scale vehicles are shown.

Several of these vehicles have been assessed aerodynamically using the panel methods described in Chapter 8 and some results are shown in Table B.19 and Figure B.30. The panels used in the calculations are those defined by the geometrical information contained in the data presented here.

Similarities in configuration between the successful X-15 research aircraft and the proposed X-24C spaceplane are discussed and illustrated in Figures B.31 and B.32. In addition, inertia properties for these two vehicles and the HL-10 are discussed and illustrated in Figure B.33.

B.1 SPACE SHUTTLE ORBITER

The contours of the Space Shuttle Orbiter at 10 stations normal to the longitudinal axis are given in Tables B.1 and B.2. Coordinates for planform and elevation views are given in Table B.3. From these tables one may construct the views of the Orbiter shown in Figures B.1–B.3. A photograph of the Orbiter appears in Figure 6.75 and a drawing of the Orbiter is shown in Figure 8.70.

Table B.1 Contour Coordinates for the Space Shuttle Orbiter Fuselage Stations 1−5

Fuselage Station	1	1	2	2	3	3	4	4	5	5
x (in)	6.06	6.06	29.1	29.1	93.1	93.1	197.1	197.1	249	249
Ray	y (in)	z (in)	y (in)	z (in)	y (in)	z (in)	y (in)	z (in)	y (in)	z (in)
1	0	−17	0	−30.7	0	−48.5	0	−58.2	0	−60.6
2	6.8	−15.6	12.1	−30	21.4	−48.5	31.1	−58.2	33.5	−60.6
3	12	−12	26	−22	42.8	−41	62.2	−55	67	−60.6
4	15.6	−6.8	33	−12	60	−20	93.3	−35	100.6	−50.9
5	17	0	36	0	64.2	0	93.3	0	100.6	0
6	16.8	3	36	7	64.2	6.1	93.3	55.2	100.6	55.7
7	16	6	34	13	62	24	91	70	96	70
8	14.6	8.6	31	18.5	57	35	86	83	90	85
9	13	11	28	23	51	45	77	95	83	100
10	11	13	23	28	43	52	66	103	73	115
11	8.6	14.6	17.5	32	34	60	52	109	60	135
12	6	16	12.5	34	23	66	37	111.5	40	160
13	3	16.8	6.5	36	11	69.5	15	111.5	20	162.4
14	0	17	0	36.4	0	70.3	0	111.5	0	162.4

Table B.2 Contour Coordinates for the Space Shuttle Orbiter Fuselage Stations 6−10

Fuselage Station	6	6	7	7	8	8	9	9	10	10
x (in)	349	349	1071	1071	1100	1100	1150	1150	1292	1292
Ray	y (in)	z (in)	y (in)	z (in)	y (in)	z (in)	y (in)	z (in)	y (in)	z (in)
1	0	−63	0	−77.6	0	−77.6	0	−75	0	−75
2	35.3	−63	35.3	−77.6	35.3	−77.6	35.3	−75	35.3	−75
3	70.6	−63	70.6	−77.6	70.6	−77.6	70.6	−75	70.6	−75
4	106.1	−63	106.1	−77.6	106.1	−77.6	106.1	−75	106.1	−75
5	106.1	0	106.1	0	106.1	0	106.1	0	106.1	0
6	106.1	56	106.1	56	106.1	56	106.1	56	106.1	56
7	106.1	75.1	106.1	75.1	106.1	75.1	106.1	75.1	106.1	75.1
8	106.1	120.1	106.1	120.1	106.1	120.1	135	109	135	109
9	100	135.1	100	135.1	110	143	135	155	135	155
10	80	146.1	80	146.1	100	170	115	185	115	185
11	60	155.1	60	155.1	70	190	77	202	77	202
12	40	159.1	40	159.1	37	187	33	195	33	195
13	15	162.4	15	162.4	15	162.4	15	162.4	15	162.4
14	0	162.4	0	162.4	0	162.4	0	162.4	0	162.4

Table B.3 Planform and Elevation Coordinates for the Space Shuttle Orbiter

Fuselage Planform and Elevation	x	y	z_{upper}	z_{lower}	Wing Planform	x	y	z
	0	0	0	0		285	106	−19
	6	17	17	−17		733	181	−33
	29	36	36	−31		797	208	−35
	93	64	70	−49		1063	467	−43
	197	93	112	−58		1198	467	−48
	249	101	162	−61		1268	106	−49
	349	106	162.	−63				
	1071	106	162	−78	Vertical Tail Planform	x	y	z
	1100	106	162	−78		1040	0	162
	1150	106	162	−75		1359	0	471
	1292	106	162	−70		1467	0	471
	1292	0	162	−63		1337	0	250
Body Flap	1292	106						
	1377	106						
	1377	0						
	1292	0						
OMS Pod	1071	120.1						
	1071	162.4	120.1					
	1100	190	75.1					
	1150	202	75.1					
	1270	202	75.1					
	1270	202	75.1					
	1270	75.1						

Wing Elevation (Inboard)	x	z_{upper}	z_{lower}	%c	Wing Elevation (Outboard)	x	z_{upper}	z_{lower}	%c
	285	−19	−19	0		797	−35	−35	0
	334	6	−35	5		820	−23	−43	5
NACA 64-409	383	16	−41	10	NACA 64(1)-012	842	−18	−55	10
	482	28	−48	20		887	−14	−62	20
	580	34	−52	30		978	−13	−69	40
	733	31	−54	45.5		1023	−15	−63	50
	797	26	−53	52		1113	−23	−61	70
	1063	−12	−43	79		1180	−41	−53	85
	1198	−35	−52	92		1248	−50	−49	100
	1268	−49	−49	100					

Vertical Tail Elevation	x/c	x (Root)	y/c	y (Root)	x (Tip)	y (Tip)
	0	1040	0	0	1358.8	0
	0.05	1053.23	0.0281	7.43526	1364.255	3.06571
	0.1	1066.46	0.03871	10.24267	1369.71	4.223261

(Continued)

Table B.3 (*Continued*)						
Vertical Tail Elevation	*x/c*	*x* (Root)	*y/c*	*y* (Root)	*x* (Tip)	*y* (Tip)
	0.15	1079.69	0.0462	12.22452	1375.165	5.04042
	0.2	1092.92	0.05173	13.68776	1380.62	5.643743
	0.3	1119.38	0.05844	15.46322	1391.53	6.375804
	0.4	1145.84	0.05981	15.82573	1402.44	6.525271
	0.5	1172.3	0.0548	14.50008	1413.35	5.97868
	0.6	1198.76	0.04548	12.03401	1424.26	4.961868
	0.7	1225.22	0.0335	8.8641	1435.17	3.65485
	0.8	1251.68	0.0209	5.53014	1446.08	2.28019
	0.9	1278.14	0.00786	2.079756	1456.99	0.857526
	1	1304.6	0	0	1467.9	0

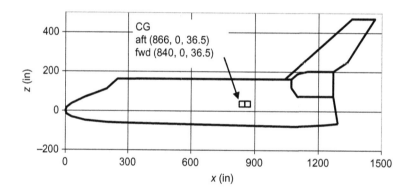

FIGURE B.1

Elevation view of the Space Shuttle Orbiter showing center of gravity range.

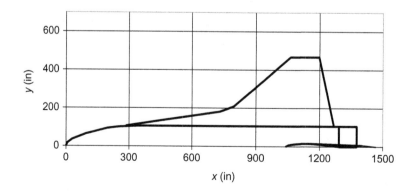

FIGURE B.2

Planform view of the Space Shuttle Orbiter.

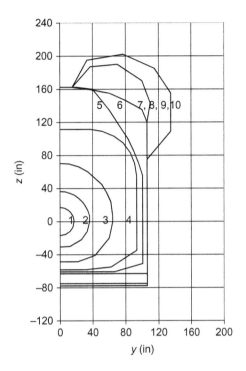

FIGURE B.3

Contours of the Space Shuttle Orbiter showing the fuselage stations. Any differences between contours 7 and 8 are indistinguishable on this scale and the same is true for contours 9 and 10.

B.2 USAF/NASA X-24C

The contours of the X-24C at 9 stations normal to the longitudinal axis are given in Tables B.4 and B.5. Coordinates for planform and elevation views are given in Tables B.6 and B.7. From these tables one may construct the views of the X-24C shown in Figures B.4–B.6. The X-24C was never built as a flight vehicle but a wind tunnel model is shown in Figure B.7. The X-24C design was an outgrowth of lifting body flight vehicles tested by NASA, several of which appear in Figures 6.67 and 6.73.

Table B.4 Contour Coordinates for the X-24C Fuselage Stations 1–4								
Fuselage Station	**1**	**1**	**2**	**2**	**3**	**3**	**4**	**4**
x (in)	41.4	41.4	120	120	144.8	144.8	177.9	177.9
	y (in)	**z (in)**	**y (in)**	**z (in)**	**y (in)**	**z (in)**	**y (in)**	**z (in)**
1	0	−8.3	0	−12.4	0	−13.7	0	−15.4
2	2	−8.1	5	−12.4	5.5	−13.7	6.5	−15.4

(Continued)

Table B.4 (*Continued*)

	y (in)	z (in)	y (in)	z (in)	y (in)	z (in)	y (in)	z (in)
3	5	−7.5	15.5	−12.4	18	−13.7	20.5	−15.4
4	7.5	−5	25	−12.4	31	−13.7	38	−15.4
5	9	−3	27.9	−9	34.1	−10.5	41.4	−12
6	10.3	0	27.9	0	34.1	0	41.4	0
7	10.3	5	27	8	33	8.5	40	9.5
8	8.5	8	23	18	29	19	36	20
9	5	11	13	27	21	29	29	31
10	1	12.3	2	29	12	34	21	37
11	0.75	12.4	1.5	29	10	38	17	43.5
12	0.5	12.4	1	29	8	42	12	48
13	0.25	12.4	0.5	29	4	44.5	6	49.5
14	0	12.4	0	29	0	45.5	0	49.7

Table B.5 Contour Coordinates for the X-24C Fuselage Stations 5–9

Fuselage Station	5	5	6	6	7	7	8	8	9	9
x (in)	215.2	215.2	331	331	384.8	384.8	480	480	579.3	579.3
	y (in)	z (in)	y (in)	z (in)	y (in)	z (in)	y (in)	z (in)	y (in)	z (in)
1	0	−17.3	0	−23.3	0	−26.1	0	−31	0	−4.1
2	7.5	−17.3	10.5	−23.3	12	−26.1	14.5	−31	14.5	−4.1
3	24	−17.3	34	−23.3	39	−26.1	46	−31	46	−4.1
4	46	−17.3	69	−23.3	86.9	−26.1	86.9	−31	86.9	−4.1
5	49.7	−14	74.5	−19	86.9	−22	86.9	−28	86.9	−2
6	49.7	0	74.5	0	86.9	0	86.9	0	86.9	0
7	48	10	73	12	82	14	82	14	82	14
8	45	22	67	26	76	27	76	27	76	27
9	39	33	59	37.5	71	40	71	40	71	40
10	32	42	48	47	66	54	66	54	66	54
11	26	48	37	52	46	54	46	54	46	54
12	18	51	24	54	54	54	24	54	24	54
13	10	53	11	54	11	54	11	54	11	54
14	0	54	0	54	0	54	0	54	0	54

Table B.6 Planform Coordinates for the X-24C

Fuselage Planform	x (in)	y (in)	Wing Planform	x (in)	y (in)	z (in)
	0	0		385	86.9	−25
	41.4	10.3		530	149	19
	120	27.9		633	149	13
	144.8	34.1		627.5	86.9	−39
	177.9	41.4		385	86.9	−25
	215.2	49.7				
	331	74.5				
	384.8	86.9				
	480	86.9				
	579.3	86.9	Tip Chord	103		
	579.3	0	Root Chord	243		

Table B.7 Fin and Tail Coordinates for the X-24C

Off C.L. Fin Planform	x	y	z	Vertical Tail Planform	x	y	z
	507	62	54		478	0	54
	542	89	84		594	11.5	54
	585	89	84		594	3.2	108
	569	62	54		548	0	108
	507	62	54		594	3.2	108
		66	54		548	0	108
		92	84		478	0	54
		89	84				

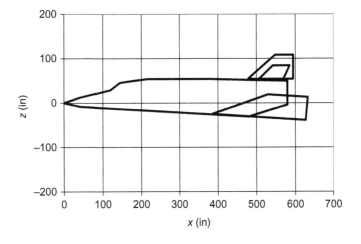

FIGURE B.4

Elevation view of the X-24C spaceplane.

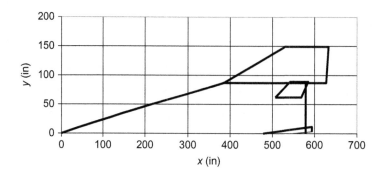

FIGURE B.5

Planform view of the X-24C spaceplane.

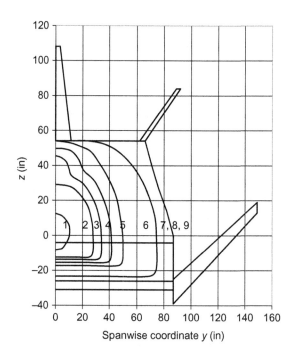

FIGURE B.6

Contours of the X-24C showing the fuselage stations. Any differences between contours 7, 8, and 9 are indistinguishable on this scale.

FIGURE B.7

Model of X-24C in an US Air Force wind tunnel.

B.3 NORTH AMERICAN X-15

The contours of the X-15 at 8 stations normal to the longitudinal axis are given in Table B.8. Coordinates for planform and elevation views are given in Tables B.9 and B.10. From these tables one may construct the views of the X-15 shown in Figures B.8–B.10. A photograph of the X-15 appears in Figure B.11.

Table B.8 Contour Coordinates for the X-15 Spaceplane								
Fuselage Station	**1**	**1**	**2**	**2**	**3**	**3**	**4**	**4**
x (in)	88.6	88.6	114.6	114.6	160.6	160.6	204.8	204.8
	y (in)	z (in)	y (in)	z (in)	y (in)	z (in)	y (in)	z (in)
1	0.0	-19.6	0.0	-23.3	0.0	-25.8	0.0	-28.0
2	4.2	-19.3	3.8	-22.7	3.5	-25.4	3.1	-27.8
3	8.2	-17.7	8.0	-21.3	8.5	-24.3	8.0	-27.0
4	12.0	-15.6	12.6	-19.0	13.4	-22.0	12.5	-25.0
5	15.0	-12.4	16.5	-16.0	17.9	-18.6	16.6	-22.3
6	17.4	-9.0	19.6	-12.3	21.0	-14.5	20.4	-19.0
7	19.0	-5.0	21.7	-7.6	24.0	-10.0	23.5	-15.3

(Continued)

Table B.8 (*Continued*)

	y (in)	z (in)	y (in)	z (in)	y (in)	z (in)	y (in)	z (in)
8	19.6	−1.0	23.0	−1.1	25.8	−1.2	25.8	−11.0
9	19.6	−0.5	23.0	−0.6	25.8	−0.7	31.3	−7.6
10	19.6	0.0	23.0	0.0	25.8	0.0	33.5	0.0
11	19.6	0.5	23.0	0.6	25.8	0.7	31.3	7.6
12	19.6	1.0	23.0	1.1	25.8	1.2	25.8	11.0
13	19.0	5.0	21.7	7.6	24.0	10.0	23.5	15.3
14	16.0	11.5	18.6	13.6	20.8	15.3	20.3	19.3
15	12.0	15.5	14.7	17.7	17.0	19.3	16.0	23.0
16	7.0	18.3	10.5	20.4	14.3	21.5	9.0	26.5
17	1.5	19.6	10.0	20.6	13.7	21.8	1.5	28.0
18	1.0	19.6	7.6	29.0	8.8	29.4	1.0	28.0
19	0.5	19.6	4.0	31.0	4.0	31.0	0.5	28.0
20	0.0	19.6	0.0	31.0	0.0	31.0	0.0	28.0
Fuselage Station	**5**	**5**	**6**	**6**	**7**	**7**	**8**	**8**
x (in)	323.9	323.9	456.8	456.8	595	595	609	609
	y (in)	z (in)	y (in)	z (in)	y (in)	z (in)	y (in)	z (in)
1	0.0	−28.0	0.0	−28.0	0.0	−25.0	0.0	−24.0
2	3.1	−27.8	3.1	−27.8	4.9	−24.6	5.0	−23.3
3	8.0	−27.0	8.0	−27.0	9.5	−23.3	10.0	−21.6
4	12.5	−25.0	12.0	−25.0	14.4	−20.5	14.4	−18.9
5	16.6	−22.3	16.6	−22.3	18.3	−17.2	18.3	−15.2
6	20.4	−19.0	20.4	−19.0	21.4	−13.2	21.0	−11.3
7	23.5	−15.3	23.5	−15.3	23.8	−8.0	23.0	−6.3
8	23.8	−14.9	23.8	−14.8	25.0	−1.3	24.0	−1.2
9	44.0	−3.5	44.0	−3.5	25.0	−0.7	24.0	−0.6
10	44.0	0.0	44.0	0.0	25.2	0.0	24.0	0.0
11	44.0	3.5	44.0	3.5	25.0	0.7	24.0	0.6
12	23.8	14.8	23.8	14.8	25.0	1.3	24.0	1.2
13	23.5	15.3	23.5	15.3	23.8	8.0	23.0	6.3
14	20.3	19.3	20.3	19.3	20.3	15.0	19.5	14.0
15	16.0	23.0	16.0	23.0	15.0	20.0	14.5	18.9
16	9.0	26.5	9.0	26.5	8.7	23.6	8.3	22.4
17	1.5	28.0	1.5	28.0	1.5	25.0	1.4	24.0
18	1.0	28.0	1.0	28.0	1.0	25.0	1.0	24.0
19	0.5	28.0	0.5	28.0	0.5	25.0	0.5	24.0
20	0.0	28.0	0.0	28.0	0.0	25.0	0.0	24.0

Table B.9 Planform Coordinates for the X-15

Fuselage Planform Coordinates	x	y	Wing Planform Coordinates	x	y	z
	0	0		288.7	0	0
	88.6	19.6		323.9	44	0
	114.6	23		395.8	134.16	0
	160.6	25.8		431.6	134.16	0
	204.8	33.5		455.8	44	0
	323.9	44		467.7	0	0
	456.8	44		455.8	44	0
	581.3	44		323.9	44	0
	595	25.2				
	609	24				
	609.05	0				

Horizontal Tail Planform	x (in)	y (in)	z (in)	Vertical Tail Planform	x	y	z
	506.5	44	−5.6		472	0	28
	575.7	108.5	−22.9		595	10.8	83
	601	108.5	−22.9		595	0	83
	581.3	44	−5.6		504.3	0	28
	506.5	0	−5.6		595	7.9	83
	450	0	0		595	0	83
	570	0	0		458	0	0

Table B.10 Planform Coordinates and Frontal Area for the X-15

Ventral Fin Planform	x (in)	y (in)	z (in)	Frontal Area as a Function of x	x (in)	A (ft^2)
	470.6	0	−28		0	0.0
	595	10.9	−28		86	4.2
	595	0	−28		114.6	6.2
	499	0	−74		160.6	7.6
	595	8.4	−74		204.8	9.1
	595	0	−74		323.9	10.5
	458	0	0		456.8	10.5
					581.3	10.5
					595	7.0
					609	6.2
					609	

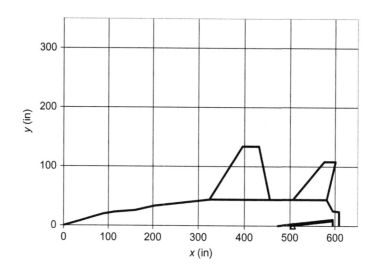

FIGURE B.8

Plan view of the X-15 spaceplane.

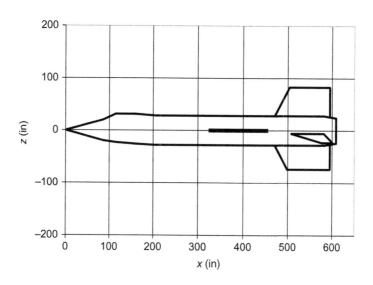

FIGURE B.9

Elevation view of the X-15 spaceplane.

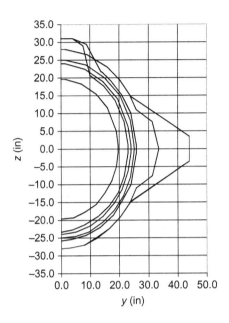

FIGURE B.10

Contours of the X-15 spaceplane. The fuselage stations are very close so are not identified.

FIGURE B.11

Photograph of the X-15 landing at the NASA Dryden Flight Test Center (NASA).

B.4 SOVIET SPACEPLANE BOR-4

The contours of the Bor-4 at 8 stations normal to the longitudinal axis are given in Table B.11. Coordinates for planform and elevation views are also provided in Table B.11. From this table one may construct the views of the Bor-4 shown in Figures B.12–B.14. Some lift and drag characteristics of the Bor-4 calculated by the methods of Chapter 8 are shown in Figures B.15 and B.16. A photograph of the Bor-4 is shown in Figure B.17.

The Soviet Bor-4 spaceplane is assumed to have the following characteristics: $S = 55.2 \text{ ft}^2$, $c_{mac} = 87$ in, and $x_0 = 109$ in. These are taken from, or based on, photographs and drawings from various sources.

Table B.11 Contour Coordinates for Soviet Spaceplane

Fuselage Station	1	1	2	2	3	3	4	4
x (in)	1.6	1.6	7.9	7.9	29.5	29.5	35.8	35.8
Ray	y (in)	z (in)	y (in)	z (in)	y (in)	z (in)	y (in)	z (in)
1	0	−3.2	0	8.7	0	−15	0	−15
2	2.3	−3.1	5.9	−8.6	12.5	−15	17	−15
3	5	−2.4	9.2	−7.7	16.4	−13.5	19	−13.8
4	6.9	−1	12	−5	19	−10.6	21.2	−11.1
5	7.1	0	13	−2	20.4	−7.5	22	−8.8
6	6.9	1	12.5	0	20	−4	21.7	−4.5
7	6.9	1.9	11	1.9	15	1.5	19.2	1
8	5.5	2.4	9	3.5	12	4.7	15.6	6
9	4	3	7	4.8	9.8	6.5	13.2	8.1
10	2	3.1	3.8	5.8	6.3	9.2	10.5	10.1
11	1.1	3.2	2.1	6	4.7	10.2	5.7	17
12	0	3.2	0	6	0	10.2	0	18.2
Fuselage Station	5	5	6	6	7	7	8	8
x (in)	42.1	42.1	77.2	77.2	102.8	102.8	156.3	156.3
Ray	y (in)	z (in)	y (in)	z (in)	y (in)	z (in)	y (in)	z (in)
1	0	−15	0	−15	0	−15	0	−15
2	19	−15	28	−15	33	−15	33	−15
3	21	−14	31	−14	36	−14	36	−14
4	23.2	−11.2	33.1	−11.2	37.9	−11.2	37.9	−11.2
5	24	−8.8	33.9	−8.8	38.6	−8.8	38.6	−8.8
6	23.7	−5	33.9	−5	38.6	−5	38.6	−5
7	22.4	0.7	31	0.7	36.2	0.7	38.6	4.5
8	19	6.7	25	8	32.5	4.5	32.5	4.5
9	15.7	10.2	16.4	12.8	21	9	21	4.5

Table B.11 (*Continued*)

Ray	y (in)	z (in)	y (in)	z (in)	y (in)	z (in)	y (in)	z (in)
10	13	15	11.6	12.7	12	9.6	12	4.5
11	7.1	19.5	7	18	7	15	7	7.8
12	0	20.5	0	18.5	0	16.5	0	8.5

Fins	x (in)	y (in)	z (in)	Vertical Tail	x (in)	y (in)	z (in)
x,z plane	102.8	38.9	−6.5	x,z plane	120	0	13.3
	156.3	58.3	29.1		156.3	0	24.3
x,y plane	165.8	58.3	29.1		164.3	0	24.3
	156.3	38.9	−6.5		151.5	0	10
	102.8		−6.5	y,z plane	151.5	0	10
y,z plane		38.9	−6.5		151.5	2	10
		58.3	29.1		156.3	2	24.3
		56.3	29.1		156.3	0	24.3
		38.9	4.5	x,y plane	120	0	
		38.9	4.5		120	2	
		56.3	29.1		164.3	2	
		58.3	29.1		164.3	0	

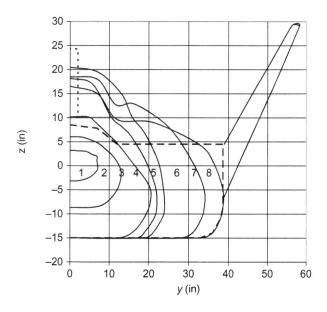

FIGURE B.12

Contours of the Soviet spaceplane showing the central and outboard fins.

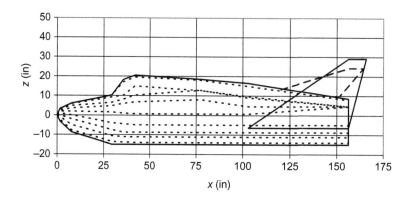

FIGURE B.13

Elevation view of the Soviet spaceplane showing portside rays.

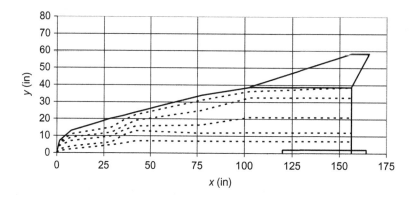

FIGURE B.14

Plan view of Soviet Bor-4 spaceplane showing upper surface rays.

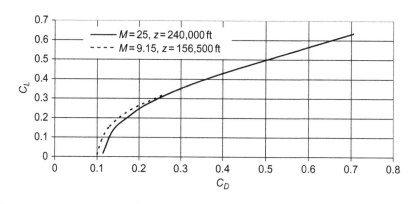

FIGURE B.15

Drag polar for the Soviet Bor-4 spaceplane as calculated by the methods of Chapter 8.

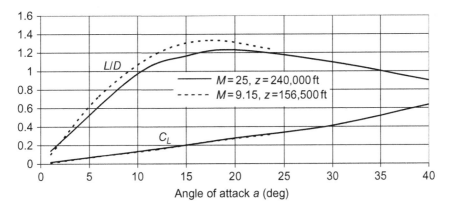

FIGURE B.16

L/D and C_L as a function of α for the Soviet Bor-4 spaceplane as calculated by the methods of Chapter 8.

FIGURE B.17

Recovery at sea of an unmanned Soviet Bor-4 spaceplane photographed by Australian P-3 Naval Reconnaissance Aircraft (NASA).

B.5 NORTHROP HL-10 LIFTING BODY

The contours of the HL-10 at 8 stations normal to the longitudinal axis are given in Table B.12. Coordinates for planform and elevation views are given in Table B.13. From these tables one may construct the views of the HL-10 shown in Figures B.18–B.20. A photograph of the HL-10 flown by NASA appears in Figure B.21.

Table B.12 Contour Coordinates for the HL-10 Spaceplane

Fuselage Station	1	1	2	2	3	3	4	4
x (in)	5.5	5.5	11.7	11.7	31.2	31.2	62.3	62.3
Ray	y	z	y	z	y	z	y	z
1	0	−11	0	−15	0	−23.5	0	−29.6
2	2	−10.5	3	−13.5	6	−23.5	9	−29.6
3	5	−9.5	6.5	−12	11	−22	17	−29.6
4	7.5	−7.5	9	−10	14	−19	23	−26
5	9	−5	11	−7	16	−15	26	−23
6	9.2	−3	12	−5	17.1	−10	26.5	−16
7	9.4	0	12.5	0	17.1	0	26.5	0
8	9.2	2.5	12	3.5	16.5	4.5	26	6
9	9	5	11	7	15	8.5	22	11
10	7.5	7.5	9.5	9.5	13	13	17	16
11	5	9.5	7	12	10	16	12	19
12	2	10.5	3	13.5	4	18	6	20
13	0	11	0	14	0	18.7	0	20.3
Fuselage Station	5	5	6	6	7	7	8	8
x (in)	98.2	98.2	154.3	154.3	202.5	202.	250.9	250.9
Ray	y	z	y	z	y	z	y	z
1	0	−34.3	0	−27.3	0	−15.6	0	−3.1
2	15	−34.3	16	−27.3	17	−15.6	20	−3.1
3	26	−34.3	30	−27.3	36	−15.6	40	−3.1
4	34	−33	40	−27.3	49	−15.6	58	−3.1
5	36.6	−27	48	−27.3	58	−15.6	67	−3.1
6	36.6	−18	53	−22	63.9	−10	70	−1.7
7	36.6	0	53	0	63.9	0	72	0
8	35.5	7.5	52	11	63.9	11	74	11
9	30	14	50	16	51	14	53	11
10	22	19	22	19	27	14	31	11
11	14	20	14	19	14	14	14	11
12	6	20.3	6	20.3	6	20.3	6	20.3
13	0	20.3	0	20.3	0	20.3	0	20.3

Table B.13	Planform and Elevation Coordinates for HL-10						
Fins	x	y	z	**Vertical Tail**	x	y	z
	202.5	63.9	11		199.1	0	20.3
	229	80	36.6		227.5	0	78
	260.9	90.6	50.4		250.5	0	78
	250.9	74	11		250.9	0	20.3
					199.1	3.7	20.3
y,z plane	202.5	61.9	12		227.5	2.35	78
	229	78	36.6		250.5	2	78
	260.9	88.6	50.4		250.9	0	78
	260.9	90.6	50.4		199.1	0	
	202.5	63.9			250.9	3.7	
					250.9	0	
					227.5	0	
					250.9	2.35	
					250.9	0	

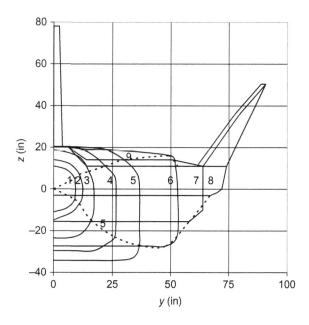

FIGURE B.18

Contours of the HL-10 spaceplane (1–8), the central and outboard fins and two of the longitudinal rays (5 and 9).

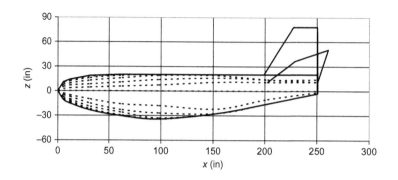

FIGURE B.19

Elevation view of the HL-10 spaceplane showing portside rays.

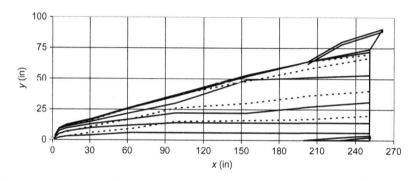

FIGURE B.20

Plan view of HL-10 spaceplane showing upper surface and lower surface (dashed lines) rays.

FIGURE B.21

HL-10 lifting body on the ramp at NASA Dryden Research Center 1966 (NASA).

B.6 HERMES SPACEPLANE

The contours of the Hermes spaceplane at 8 stations normal to the longitudinal axis are given in Table B.14. Coordinates for planform and elevation views are given in Table B.15. From these tables one may construct the views of the Hermes spaceplane shown in Figures B.22–B.24. Nominal flight conditions planned for the Hermes spaceplane are shown in Figure B.25 and the calculated L/D determined by the methods of Chapter 8 is shown in Figure B.26. The Hermes spaceplane was never built as a flight vehicle and photographs of a wind tunnel model were unavailable.

Table B.14 Contour Coordinates for the Hermes Spaceplane

Fuselage Station	1	1	2	2	3	3	4	4
x	7.7	7.7	26.4	26.4	135.6	135.6	160.1	160.1
Ray	y	z	y	z	y	z	y	z
1	0	15.1	0	28.2	0	62.4	0	92.4
2	2	15	4	28	13	62	13	91
3	5	14.3	8	27	18	60	25	87
4	7.5	13.7	13	26	30	56	37.5	70
5	11	12.1	19	23	40	50	47	58
6	13.5	10.4	22	21	46.5	43	53	49
7	16	8	26	17	53	34	58	38
8	18	4.5	32	9	60	17	64	18
9	18.8	0	33.9	0	61.2	0	65.1	0
10	18	−2.5	33	−3	60	−7	63	−8
11	17	−4.5	32	−5	57	−12	60	−14
12	16	−6	30	−8	54	−15	56	−17
13	15	−7	27.5	−10	50	−18	52	−20
14	12.5	−8.5	25	−12	44	−22	46	−22
15	10	−9.7	21	−14	38	−23	40	−23
16	5	−11	14	−16	21	−24	20	−24
17	0	−11.3	0	−17	0	−24	0	−24
Fuselage Station	5	5	6	6	7	7	8	8
x	177	177	190	190	234	234	678	678
Ray	y	z	y	z	y	z	y	z
1	0	103.6	0	105.6	0	113	0	113
2	15	101	15	103	18	112.5	18	112.5
3	27	91.5	29	96	35	110	35	110
4	40	73	43	79	55.5	96	55.5	96
5	48	59	51.5	64	64	81	64	81

(Continued)

Table B.14 (*Continued*)

Ray	y	z	y	z	y	z	y	z
6	54	50	57	53	65.9	60	65.9	60
7	59	39	61	40	65.9	43	65.9	43
8	65	18.5	65	18	65.9	18	65.9	18
9	65.9	0	65.9	0	65.9	0	65.9	0
10	65	−9	65	−10	65.9	−10	65.9	−10
11	61.5	−14	63	−15	65	−16	65.9	−16
12	57	−17.5	60	−20	61.5	−21	65.9	−24.5
13	53	−20.5	56	−22	57.5	−24	60	−25.2
14	47	−22.5	50	−24	50	−25	50	−25.2
15	40	−23.5	40	−25	40	−25.2	40	−25.2
16	20	−24.5	20	−25.2	22	−25.2	22	−25.2
17	0	−25.2	0	−25.2	0	−25.2	0	−25.2

Table B.15 Planform and Elevation Coordinates for the Hermes Spaceplane

	x	y	z		y	z		
Inboard Wing	233.5	65.9	0		65.9	20		
	516	152.5	0		150	6.5		
	661	152.5	0		155	−6.5		
	678	65.9	0		65.9	−20		
	678	0	0					
					150	6.5		
Outboard Wing	516	152.5	0		198	75		
	602.6	201.4	75		201	75		
	644	201.4	75		155	−6.6		
	661	152.5	0					

	x/c (%)	x	z/c (%)	z_{up}	z_{low}	x	z_{up}	z_{low}
Wing Elevation and Root & Tip Airfoil Contours	0.0	233.5	0.00	−5.30	−5.30	516.0	0.00	0.00
	0.5	235.7	0.98	−2.04	−8.56	516.7	1.06	−1.06
	0.8	236.8	1.18	−1.37	−9.23	517.1	1.28	−1.28
	1.3	239.1	1.49	−0.34	−10.26	517.8	1.62	−1.62
	2.5	244.6	2.04	1.48	−12.08	519.6	2.21	−2.21
	5.0	255.7	2.81	4.06	−14.66	523.3	3.06	−3.06
	7.5	266.8	3.39	6.00	−16.60	526.9	3.69	−3.69
	10.0	277.9	3.87	7.59	−18.19	530.5	4.21	−4.21
	15.0	300.1	4.62	10.08	−20.68	537.8	5.02	−5.02
	20.0	322.3	5.17	11.93	−22.53	545.0	5.63	−5.63
	25.0	344.5	5.58	13.27	−23.87	552.3	6.06	−6.06
	30.0	366.7	5.84	14.16	−24.76	559.5	6.36	−6.36
	35.0	388.9	5.98	14.61	−25.21	566.8	6.50	−6.50
	40.0	411.1	5.98	14.62	−25.22	574.0	6.50	−6.50
	100.0	677.5	5.98	14.62	−25.22	661.0	6.50	−6.50

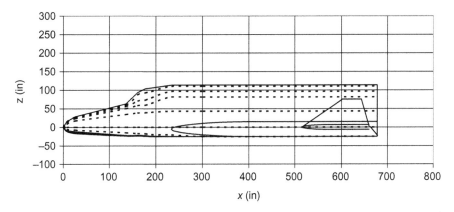

FIGURE B.22

Elevation view of the Hermes spaceplane showing portside rays.

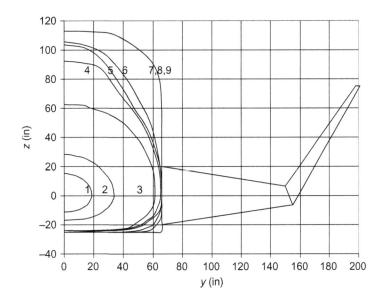

FIGURE B.23

Contours of the Hermes spaceplane showing the outboard fins. Contours 7, 8, and 9 are almost the same.

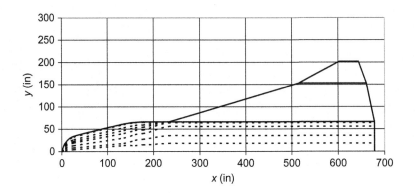

FIGURE B.24

Planform view of the Hermes spaceplane showing several surface rays.

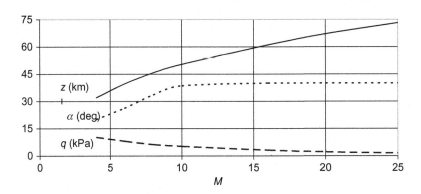

FIGURE B.25

Nominal flight conditions for the Hermes spaceplane.

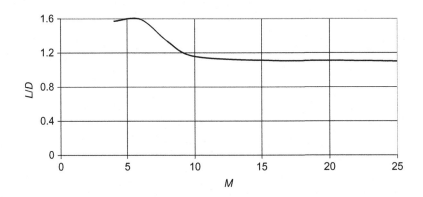

FIGURE B.26

Lift to drag ratio for the Hermes spaceplane as determined for the flight conditions in Figure B.9.

B.7 INSTITUTE OF SPACE AND ASTRONAUTICAL SCIENCES HIMES SPACEPLANE (JAPAN)

The contours of the HIMES spaceplane at 8 stations normal to the longitudinal axis are given in Tables B.16 and B.17. Coordinates for planform and elevation views are given in Table B.18. From these tables one may construct the views of the HIMES spaceplane shown in Figures B.27–B.29. The HIMES spaceplane was never built as a flight vehicle and photographs of a wind tunnel model were unavailable.

Table B.16 Contour Coordinates for HIMES Spaceplane Stations 1−4

Fuselage Station	1	1	2	2	3	3	4	4
x	5.5	5.5	21.9	21.9	46.6	46.6	192	192
Ray	y	z	y	z	y	z	y	z
1	0	10.95	0	21.9	0	30.1	0	58.9
2	1.7	10.7	3	21.7	4.5	30.1	9	58.5
3	3.7	10.5	7	21	10.5	29.5	20	57.5
4	5.5	10.2	10.5	20	15	28	30	54.5
5	7	9.8	14	18	20	26	39	49.5
6	9	9	17	16	24	22.5	46	43
7	10.5	7.6	19	13.5	27.5	19	51	35.5
8	12	6	21	10.5	29.5	14	55	26.5
9	13.2	3	22.5	5.5	31.5	8	58	14.5
10	13.7	0	23.3	0	32.9	0	58.9	0
11	13.5	−1.2	22.8	−2.5	32.5	−3	57.5	−5.5
12	13.3	−2.1	22.5	−3.5	32	−5	55	−9
13	13	−2.8	22	−4.5	31.5	−6.5	53.5	−11
14	12.7	−3.4	21.8	−5.5	31	−8	51	−13.5
15	12.5	−3.8	21.5	−6.5	30	−9	49	−15
16	12.2	−4.5	21	−7	29.5	−10.5	46	−16.5
17	11.8	−4.8	20.5	−8.5	28	−12	43	−17.5
18	10.5	−6	19	−10	26	−14	35.5	−19.5
19	9	−7	15.5	−13	21	−17.5	25	−21
20	5.2	−8.1	10	−15	13	−20	14	−21.5
21	0	−8.2	0	−16.4	0	−20.5	0	−21.9

Table B.17 Contour Coordinates for HIMES Spaceplane Stations 5−8

Fuselage Station	5	5	6	6	7	7	8	8
x	318	318	371	371	424	424	476	476
Ray	y	z	y	z	y	z	y	z
1	0	83.6	0	83.6	0	83.6	0	83.6
2	13	83	13	83	13	83	13	83

(Continued)

Table B.17 (*Continued*)

Ray	y	z	y	z	y	z	y	z
3	28	80	28	80	28	80	28	80
4	41.5	75	41.5	75	41.5	75	41.5	75
5	53.5	67	53.5	67	53.5	67	53.5	67
6	63	58.5	63	58.5	63	58.5	63	58.5
7	70	48.5	70	48.5	70	48.5	70	48.5
8	75.5	36	75.5	36	75.5	36	75.5	36
9	78.5	19	78.5	19	78.5	19	78.5	19
10	79.5	0	79.5	0	79.5	0	79.5	0
11	79	−7.5	79.5	−8	79.5	−8	79.5	−8
12	78	−12.5	79	−13	79.5	−13	79.5	−13.5
13	76.5	−16	78.5	−17.5	78	−17.5	79.5	−19
14	74	−19	76	−20.5	76	−21	79.5	−26
15	71	−21.5	73	−23.5	73	−23.5	75	−26
16	67	−23.5	70	−25	69	−24.5	71	−26
17	62	−25	62.5	−26	62.5	−26	62.5	−26
18	47	−26	47	−26	47	−26	47	−26
19	30.5	−26	30.5	−26	30.5	−26	30.5	−26
20	17	−26	17	−26	17	−26	17	−26
21	0	−26	0	−26	0	−26	0	−26

Table B.18 Planform and Elevation Coordinates for HIMES Spaceplane

Wing Planform	x	y			Fins	x	y	z
	337	79.5				411	54.8	68.5
	444	184				471	84.9	151
	476	184				515	84.9	151
	476	79.5				475	54.8	68.5
	476	0				411	54.8	68.5

Wing Elevation	x/c (%)	x	z/c (%)	z_{upper}	z_{lower}	x	z_{upper}	z_{lower}
	0	337.0	0.00	−17.80	−17.80	444.0	−17.80	−17.80
	1	337.7	0.98	−16.46	−19.14	444.2	−17.51	−18.09
	1	338.0	1.18	−16.18	−19.42	444.2	−17.45	−18.15
	1	338.7	1.49	−15.76	−19.84	444.4	−17.35	−18.25
	3	340.4	2.04	−15.01	−20.59	444.8	−17.19	−18.41
	5	343.9	2.81	−13.95	−21.65	445.5	−16.96	−18.64
	8	347.3	3.39	−13.15	−22.45	446.3	−16.78	−18.82
	10	350.7	3.87	−12.50	−23.10	447.0	−16.64	−18.96
	15	357.6	4.62	−11.47	−24.13	448.5	−16.41	−19.19
	20	364.4	5.17	−10.71	−24.89	450.0	−16.25	−19.35
	25	371.3	5.58	−10.16	−25.44	451.5	−16.13	−19.47
	30	378.1	5.84	−9.79	−25.81	453.0	−16.05	−19.55
	35	385.0	5.98	−9.61	−25.99	454.5	−16.01	−19.59

Table B.18 (*Continued*)

Wing Elevation	x/c (%)	x	z/c (%)	z_{upper}	z_{lower}	x	z_{upper}	z_{lower}
	40	391.8	5.98	−9.61	−25.99	456.0	−16.01	−19.59
	100	474.0	5.98	−9.61	−25.99	474.0	−16.01	−19.59
Wing Front		y	z		**Fin Front**	y	z	
		79.5	−26			54.8	68.5	
		184	−19.59			84.9	150.6	
		184	−16			80.8	150.6	
		79.5	−9.6			48.7	68.5	

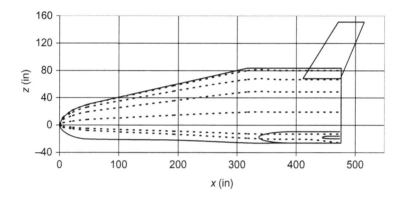

FIGURE B.27

Elevation view of HIMES spaceplane showing portside rays.

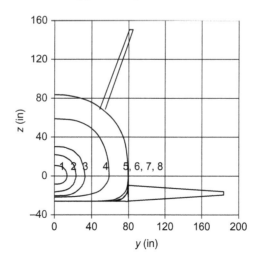

FIGURE B.28

Contours of Japan's HIMES spaceplane and its canted fins. Note that contours 5−8 are almost identical.

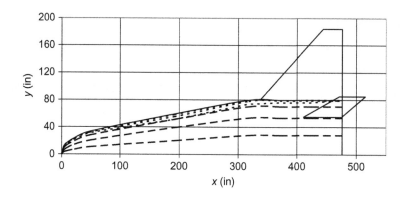

Plan view of HIMES spaceplane showing upper surface rays.

B.8 ESTIMATED LIFT DRAG AND MOMENT DATA FOR SEVERAL SPACEPLANES

Four of the spaceplanes have been assessed aerodynamically using the simplified panel method described in Chapter 8. A typical trajectory is used to specify the variation of Mach number with altitude and representative angles of attack are used. The lift to drag ratio variation given in Table B.19 is illustrated in Figure B.30 wherein it is clear that at high M and the attendant high angle of attack all vehicles give $L/D \sim 1$ with some variation showing for $M < 10$.

Table B.19 Lift Drag and Moment Data for Several Spaceplanes

Vehicle	M	z (kft)	z (km)	α (degrees)	C_L	C_D	C_m	L/D
Orbiter	25	240	73.15	40.0	0.77	0.70	0.039	1.10
	20	220	67.06	40.0	0.76	0.69	0.039	1.10
	16	200	60.96	40.0	0.75	0.68	0.040	1.10
	10	165	50.29	38.5	0.70	0.61	0.044	1.15
	8	150	45.72	33.5	0.59	0.44	0.049	1.34
	6	130	39.62	26.2	0.42	0.25	0.048	1.66
	4	102	31.09	20.0	0.40	0.19	0.047	2.10
BOR-4	25	240	73.15	40.0	0.66	0.68	0.221	0.97
	20	220	67.06	40.0	0.66	0.68	0.221	0.97
	16	200	60.96	40.0	0.66	0.68	0.221	0.97
	10	165	50.29	38.5	0.64	0.63	0.212	1.01
	8	150	45.72	33.5	0.55	0.47	0.184	1.15
	6	130	39.62	26.2	0.39	0.29	0.141	1.34
	4	102	31.09	20.0	0.26	0.18	0.103	1.40

Vehicle	M	z (kft)	z (km)	α (degrees)	C_L	C_D	C_m	L/D
Hermes	25	240	73.15	40.0	0.53	0.48	0.102	1.11
	20	220	67.06	40.0	0.53	0.48	0.102	1.11
	16	200	60.96	40.0	0.53	0.47	0.102	1.11
	10	165	50.29	38.5	0.50	0.43	0.098	1.16
	8	150	45.72	33.5	0.41	0.31	0.081	1.35
	6	130	39.62	26.2	0.28	0.17	0.056	1.65
	4	102	31.09	20.0	0.16	0.09	0.034	1.80
HIMES	25	240	73.15	40.0	0.92	0.87	− 0.016	1.06
	20	220	67.06	40.0	0.92	0.87	− 0.015	1.06
	16	200	60.96	40.0	0.92	0.87	− 0.015	1.06
	10	165	50.29	38.5	0.88	0.80	− 0.039	1.11
	8	150	45.72	33.5	0.75	0.58	0.022	1.30
	6	130	39.62	26.2	0.54	0.33	0.047	1.63
	4	102	31.09	20.0	0.35	0.18	− 0.055	1.92

Table B.19 (*Continued*)

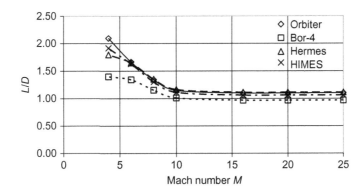

FIGURE B.30

L/D as a function of Mach number for several vehicles.

B.9 SIMILARITIES IN HYPERSONIC SPACEPLANES

Similarities in configuration between the successful X-15 flight research aircraft and the proposed X-24C spaceplane are illustrated in Figures B.31 and B.32. The vehicles are similar in size and weight suggesting that the X-24C would be a good candidate for a flight vehicle. However the X-24C was never built but was the subject of intensive wind tunnel testing and CFD analysis. The geometric similarities should lead to similar inertia characteristics.

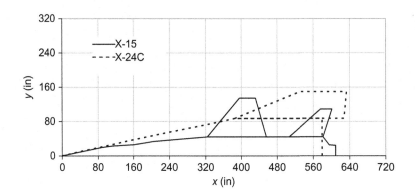

FIGURE B.31

Planform view of X-15 (solid lines) and X-24C (dotted lines).

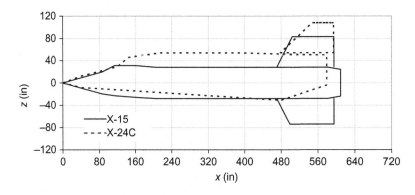

FIGURE B.32

Elevation view of X-15 (solid lines) and X-24C (dotted lines).

The nondimensional radius of gyration about the x, y, and z axes given by

$$\frac{r_x}{b} = \sqrt{\frac{I_x}{mb^2}}$$

$$\frac{r_y}{l} = \sqrt{\frac{I_y}{ml^2}}$$

$$\frac{r_z}{l} = \sqrt{\frac{I_z}{ml^2}}$$

Values for the radii of gyration of the X-15, X-24C, and HL-10 are given for different weights in Table B.20. The results are also illustrated in Figure B.33.

Note that the nondimensional radius of gyration in pitch r_y/l (i.e., around the y-axis) and in yaw r_z/l (i.e., around the z-axis) are very close in magnitude for all three vehicles. Similarly,

Table B.20 Radii of Gyration for Several Vehicles						
	W (lbs)	**b (ft)**	**l (ft)**	**r_x/b**	**r_y/l**	**r_z/l**
X-15	15,560	22.4	50.8	0.123	0.253	0.256
X-24C	16,137	24.8	48.3	0.122	0.253	0.260
HL-10	6466	13.6	21.2	0.191	0.266	0.286
X-15	32,000	22.4	50.8	0.101	0.207	0.209
X-24C	55,400	24.8	48.3	0.0885	0.209	0.215

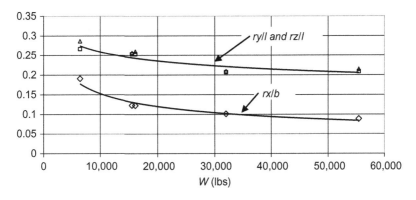

FIGURE B.33

Nondimensional radii of gyration for X-15, X-24C, and HL-10. Curves added only to illustrate trends.

nondimensional radius of gyration in roll r_x/b (i.e., around the x-axis) is much the same for all three vehicles. This suggests that the dynamic response characteristics of the three vehicles are very much alike. The Boeing 747 is at least an order of magnitude heavier than the vehicles considered here, yet its nondimensional radii of gyration are given approximately by $r_x/b = 0.149$, $r_y/l = 0.18$, and $r_z/l = 0.22$. Only in roll is the Boeing 747 response likely to be much different than that of the spaceplanes. It is clear that handling qualities have been considered in the design of these vehicles.

B.10 NOMENCLATURE

b	wingspan
CG	center of gravity
C_D	drag coefficient
C_L	lift coefficient
C_m	moment coefficient about the center of gravity
D	drag
c	local wing chord
I	moment of inertia

L   lift
l   length
M   Mach number
r   radius of gyration $(I/m)^{1/2}$
W   weight
x   longitudinal coordinate of fuselage centerline
y   spanwise coordinate from fuselage centerline
z   vertical coordinate from fuselage centerline or altitude

B.10.1 **SUBSCRIPTS**

x   referred to the *x*-axis
y   referred to the *y*-axis
z   referred to the *z*-axis

Index

Note: Page numbers followed by "*f*" and "*t*" refer to figures and tables, respectively.

Printed in the United States
By Bookmasters